ISBN 978-1-330-32417-2
PIBN 10027358

1 MONTH OF
FREE
READING

at
www.ForgottenBooks.com

By purchasing this book you are eligible for one month membership to ForgottenBooks.com, giving you unlimited access to our entire collection of over 1,000,000 titles via our web site and mobile apps.

To claim your free month visit: www.forgottenbooks.com/free27358

A

COMPENDIUM OF CHEMISTRY

INCLUDING

GENERAL, INORGANIC, AND ORGANIC CHEMISTRY

BY

Dr. CARL ARNOLD

*Professor of Chemistry in the Royal Veterinary School
of Hanover*

*AUTHORIZED TRANSLATION FROM THE ELEVENTH ENLARGED
AND REVISED GERMAN EDITION*

BY

JOHN A. MANDEL, Sc.D.

*Professor of Chemistry, Physics, and Physiological Chemistry in the
University and Bellevue Hospital Medical College*

FIRST EDITION

SECOND THOUSAND

NEW YORK
JOHN WILEY & SONS, Inc.
London: CHAPMAN & HALL, Limited
1914

PRESS OF
BRAUNWORTH & CO.
BOOKBINDERS AND PRINTERS
BROOKLYN, N. Y.

PREFACE.

THE successive editions of Prof. Carl Arnold's "Repetitorium der Chemie" have served during the past twenty years as most useful condensed statements of chemical knowledge. The work has had a large circulation in Germany among students and professional chemists. The numerous editions, each of which has been larger than its predecessors, attest both the popularity of the work and the care with which the author has kept in touch with the advances in chemical science. The eleventh and last edition contains concise but clear statements of the most important theories and facts, especially in the recently developed domain of physical chemistry, as well as a classified review of the most important inorganic and organic compounds, including statements of the constitution and derivation of these substances. A very complete index forms a valuable feature of the work and renders its contents readily accessible.

The feeling that this work should be within the reach of American chemists and students has prompted me to undertake this translation.

As I have not been able to find an English equivalent of the German word "Repetitorium," I have been obliged to make use of the word "Compendium," which seems to express the nature of the work.

J. A. M.

May, 1904.

v

CONTENTS.

PART FIRST. GENERAL CHEMISTRY.

PART SECOND. INORGANIC CHEMISTRY.

PART THIRD. ORGANIC CHEMISTRY.

PART FIRST.

GENERAL CHEMISTRY.

INTRODUCTION.

The aim of the natural sciences is the investigation of the objects and processes of nature. They are divided into the special sciences, each of which is devoted to a certain kingdom of nature or a portion thereof, and the general sciences, which are confined to no special kingdom of nature. The general sciences are divided into physics and chemistry.

The earlier division of the natural sciences into descriptive (including natural history, botany, zoology, mineralogy, and astronomy) and exact (including natural philosophy, chemistry, physics, and biology) is no longer in accordance with the conditions, since chemistry, on the one hand, as it must also take into consideration the external characteristics of chemical substances, is at once a descriptive science, while, on the other hand, botany, etc., since they must investigate not only the external reality but also the chemical and physical changes transpiring within, are exact as well as descriptive sciences.

Matter, material, or substance may be defined as anything which can be weighed without reference to its configuration.

Body is the name given to anything having a definite form. For example, iron, glass, and marble are forms of matter, while a knife, a drinking-glass, and a marble statue are bodies.

Chemistry is the science of matter, its properties and its changes; its foundation is the law of the conservation of matter (p. 10), which states that no loss of matter can take place in any chemical change. All phenomena which accompany an alteration of matter belong to the domain of chemistry.

For example, sulphur and iron mixed together give an apparently homogeneous, gray powder, in which, however, the separate particles of iron and sulphur can be detected with the aid of a microscope and can be separated from each other by a magnet. If, however, the mixture be heated, a black mass is formed in which no lack of uniformity can be noticed with the microscope, from which neither iron nor sulphur can be separated by any mechanical treatment, and which no longer possesses the properties of either iron or sulphur. The mixture obtained in the first case is called a mechanical mixture, since by a purely mechanical process it is possible to separate it into its constituents, while that obtained in the second case is called a chemical compound, since it is possible to separate it into its components only by a chemical process and not by any mechanical treatment.

Physics is the science of the alteration of the state of a body; its foundation is the law of the conservation of force (e.g. energy): "No loss of force (e.g. energy) occurs in any physical process, but only the transformation of one form of energy into another form." All phenomena in which the matter undergoes no alteration in composition belong to the domain of physics.

For example, a rod of glass or a piece of sulphur after being rubbed with a cloth attracts light bodies, a steel rod after rubbing with a magnet attracts objects of iron, and ice is transformed by warming into water and then into steam, but neither the glass, the sulphur, the steel, nor the water undergoes any change in composition.

The law of the conservation of energy constitutes the first principle of the mechanical theory of heat. The second principle of this theory applies particularly to the transformation of heat into mechanical energy: "Only when heat passes from a warmer to a cooler body can it be transformed into mechanical work, and indeed this is true for only a certain portion of the heat which passes. That portion of the heat energy which can be converted into work is called the available energy."

Divisions of Chemistry.

A distinction is made between pure chemistry, which is engaged with chemistry solely for its own sake, and applied chemistry, which considers chemistry in relation to certain special interests; and according to the purpose which it serves it is denoted as medical, pharmaceutical, technical, agricultural, physiological, etc., chemistry.

Analysis and Synthesis.—Analytical chemistry or analysis serves for the development of pure and applied chemistry and is employed for the purpose of resolving the more complex substances into simpler substances or elements   Analytical chemistry is divided into qualitative analysis, by which only the nature of the components is learned, and quantitative analysis, which determines the quantities by weight of the various components present. A further service is rendered by synthetical chemistry or synthesis, by which the more complex substances are built up from the simpler substances or elements. By synthesis it is possible to test the results obtained by analysis, and by the preparation from simpler substances of more complicated compounds, occurring in nature or artificially produced, to arrive at the mode of formation of the latter, and thereby to discover a method for preparing natural products.

Pure chemistry is divided into:

1. *General or theoretical chemistry;* this treats of the laws governing chemical reactions, as well as the theoretical points of view which have found recognition in chemistry. It may be divided into:

a. *Stoichiometry* ($\sigma\tau o\iota\chi\epsilon\hat{\iota}o\nu$, a first principle, $\mu\epsilon\tau\rho\epsilon\hat{\iota}\nu$, a measure); this treats of the relation between the properties of existing substances and their composition, their atomic and molecular weight (p. 13), and their constitution (p. 29).

The designation stoichiometry, i.e., the art of calculating chemical elements, originated from the fact that this part of general chemistry had its beginning in the consideration of volume and weight relations of chemical processes.

b. *Chemical affinity;* this treats of the laws governing the mutual chemical attractions of the elements under different external influences, as well as the rules governing the formation and transformation of chemical compounds. The chief purpose of the doctrine of affinity is to give an explanation of the causes acting in the changes which matter may be made to undergo.

Chemical processes are always accompanied by physical changes, namely, by the production or absorption of energy in the form of heat, electricity, or light, the quantity standing always in direct relation to the nature of the reacting substances, so that it is possible to follow the course of a chemical reaction by physical methods of measurement.

That branch of general chemistry which employs the resources of physics is called physical chemistry. Frequently, however, the whole of general chemistry is referred to as physical chemistry.

2. *Special or systematic chemistry;* this treats of the knowledge pertaining to separate chemical substances and classifies them according to general systems. It may be divided into

a. *Inorganic chemistry;* this treats of the elements (p. 5) and their compounds with the exception of those of carbon.

b. *Organic chemistry;* this treats of the compounds of carbon. Such great numbers of these are known, and their relationships to one another are so complex, that for practical reasons they are separated from the first class and treated after the other compounds. Only the compounds of carbon with oxygen, sulphur, and the heavy metals, because of their peculiar behavior, etc., are discussed under inorganic chemistry.

Organic chemistry was formerly restricted to the combustible compounds of the vegetable and animal organisms, which were believed to owe their origin solely to vital processes. However, after the synthesis of urea (a product of animal life) by Wöhler in 1828, the conclusion was reached that similar laws applied to the products of living and inanimate nature, and that the assumption of a special vital force was without foundation. Since then a great number of organic products of plant and animal origin have been artificially prepared and also a great number of compounds closely related to these, so that now a classification into organic and inorganic chemistry according to the older views would be quite impossible.

Many solid organic substances, e.g., wood, muscle, leaves, etc., are organized, i.e., they have a characteristic structure, the simplest form of which is the cell. The cell is the result of the vital process and cannot be artificially produced, while it will ultimately be possible to prepare the chemical constituents, in common with all other organic substances.

I. STOICHIOMETRY.

Simple and Compound Substances.

All existing substances can be divided into compound and simple substances.

1. *Complex substances* or *compounds* can be separated into two or more substances differing from one another and from the original substance, namely, into two or more elements.

2. *Simple substances* or *elements* are those substances which, up to the present, have resisted all efforts to resolve them into more simple constituents. When elements enter into chemical combination their properties disappear entirely or partly and compounds with new properties appear, the actual substance of the elements themselves remaining unaltered, which is shown by the fact that the elements can be again separated by chemical means, and numerical quantities of the compound result as the sum of the numerical quantities of the elements which have formed them (additive property, p. 30). In spite of the enormous number of compounds occurring on the earth, the number of the elements from which they are built up is not great and the compounds essential to mankind are formed from only about one-half of the known elements. At present 78 elements (table, p. 25) are known, and in addition to these 8–10 substances are designated as elements which in all probability are mixtures of unknown elements whose separation will be possible by the use of improved methods.

That this small number of elements can form all known compounds will be understood by a consideration of the law of multiple proportions (P. 11). Only about a third of the elements occur free in nature. Certain elements are universally distributed; for example, oxygen is contained in the air, in water, and in the solid crust of the earth in such quantity that it constitutes nearly one-half of the weight of the globe, while other elements, like cerium, lanthanum, etc., appear only in certain p   e and in very small quantities. The elements are very unequally distributed on the earth; up to the present 7 have been found in the air and 30 in the water of the ocean, while all are more or less

5

widely scattered throughout the earth's crust. The bulk of the earth's crust (the crystalline rocks) consists on the average of only the following 8 elements in 100 parts, while the other elements are present only as tenths or hundredths of a per cent.:

Oxygen, 47.3   Aluminium, 8.2   Calcium,   3.7   Sodium,   2.8
Silicon, 27.9   Iron,   4.8   Magnesium, 2.8   Potassium, 2.5

As shown by spectrum analysis (p. 46), the elements of which the earth is formed constitute also the other heavenly bodies. Without question, however, in addition to the positively known elements there are others of whose existence we are at present entirely ignorant.

According to geological hypothesis the centre of the earth is a mass of molten substances, small quantities of which escape to the surface during volcanic eruptions. Since the radius of the earth is 4000 miles, while the solid crust is only about 25 miles in thickness, it is possible that elements exist in the interior which, owing to their high specific gravity, have not yet reached the surface; this appears the more probable since the specific gravity of the globe itself is 5.6, while that of the crust is only 2.5. Moreover, the periodic system of the elements (p. 54) anticipates the existence of about 18 elements as yet undiscovered.

Chemical Processes.

The processes involving a change in substance are called chemical processes or transformations, or, in general, chemical reactions.

In the restricted sense the name chemical reaction is applied to those chemical processes which serve for the identification of given substances, and the substances employed for the purpose of such tests are called reagents.

Every element and every compound can take part in a chemical reaction, but the ability to do this varies within wide limits. While many compounds offer extreme resistance to an alteration in their composition, other compounds are stable only under very special conditions. The following classification of chemical reactions is based upon the final products:

1. If two or more different elements or substances unite to form a single new substance, this is called a *union* or a *combination*, e.g., $A + B = AB$; $AB + CD = ABCD$.

When an element unites with a compound already in existence, this is called *addition*, e.g., $A + BC = ABC$.

2. If one element replaces another in a compound and sets it free, the process is called *substitution*, e.g., $A + BC = AB + C$.

3. If a compound is broken up into simpler compounds or into elements, the process is known as *decomposition*, e.g., $AB = A + B$; $ABC = AB + C$.

If, when the causes which promote the decomposition are removed, the decomposition products recombine to form the original substance, the decomposition is called *dissociation*.

4. When compounds mutually exchange certain of their constituents, the process is known as *multiple decomposition* (double, treble, etc., decomposition), e.g., $AB+CD = AC+BD$.

The chemical processes known as *condensation, polymerization, allotropic* and *isomeric transformation* will be discussed later.

Of course several chemical processes can take place simultaneously, as a result of which the entire process becomes very complicated, and the formation of a compound is almost always dependent on a previous decomposition of the reacting substances, and a decomposition with a resulting combination of the decomposition products.

Chemical reactions which lead to the formation of compounds more complex than the original substances are called synthetic processes, and those which lead to simpler compounds are called analytical processes.

Causes of Chemical Change.

The chief cause of the combination of reacting substances is a force known as affinity, by which is understood a force of attraction acting between the separate elements, tending to bring them together and to hold them combined in the resulting compound. The affinity between various elements is of different magnitude, and indeed with one and the same element varies greatly under different conditions.

Chemical affinity is not identical with chemical energy, since the former denotes the force which causes the chemical work, while the latter denotes the work which is performed in chemical reactions by or against the force of affinity (and other forces which are also involved). The relation of chemical affinity to chemical energy is similar to that of the weight of a body to the work which the body could perform through the medium of its weight; this work depending both on the weight of the body and on the height through which it falls.

The force of affinity differs from other forces, as, for example, the force of gravitation and magnetism, in that it does not act at a distance, for which reason substances which are to combine must be brought into intimate contact, and accordingly gases and liquids react with one another more readily than solids, and the latter act on one another more readily when powdered or amorphous than when compact or crystalline.

It is usually necessary to start or to continuously support chemical affinity by heat, electricity, light, or mechanical agitation.

Frequently a number of these forces act at the same time, heat being almost always present. The action of these forces, which are indeed only modes of motion, consists in imparting motion to the chemical substance (to the atoms, see p. 13), which not only-excites this to enter into combination, but under some circumstances can be so violent as to cause through the motion a decomposition of the compound which is formed at first.

Affinity can, however, be accelerated or modified by certain other causes, namely, through the relative proportions of the reacting substances in solution, by the degree of electrolytic dissociation, by the solubility and volatility of the products formed, and also by the circumstance as to whether the substances are already extant or whether they enter into reaction at the moment of their production. Affinity is also modified by the presence of so-called catalytic agents, certain substances which, without undergoing any change themselves, influence the course of chemical reactions.

The chief causes which operate to promote the decomposition of complex substances into simpler ones, namely, those which tend to reduce or counteract the force of affinity, are likewise heat, electricity, light, and mechanical shock, since these forces can so increase the motion of the particles of chemical substances that the compounds break to pieces.

1. *Heat.* In general an increase of temperature up to a certain limit increases the combining power (affinity) of substances; on the other hand heating beyond a certain degree causes the decomposition of compound substances. For example, if mercury is heated in the air, it extracts oxygen from it and is converted into the red oxide of mercury; if the latter is heated to glowing, however, it decomposes again into mercury and oxygen (see further under Thermochemistry).

2. *Light* can cause both combination and decomposition (see further under Photochemistry).

3. *Electricity.* The action of the electric spark, through the increase in temperature which accompanies it, can cause either combination or decomposition. The electric current produces apparent decomposition; in reality its effect is to separate from one another the products of decomposition already set free by other forces (see further under Electrochemistry).

4. *Mechanical shock* can promote the combination of many substances, but it can also produce decomposition by destroying the structure of many compounds, often with an explosion. In many cases, however, the cause of the chemical change is due directly to the heat produced by the mechanical shock, or to the more intimate contact due to the jar and pressure.

5. *Electrolytic Dissociation* (see under Theory of the Ions).

6. The relative quantities of the reacting substances present often exert a pronounced influence on the course of the reaction. For example,

chlorine sets bromine free from potassium bromide and forms potassium chloride, while a proportionately large quantity of bromine sets chlorine free from potassium chloride and forms potassium bromide. A solution containing a small quantity of an active substance usually acts more feebly than one containing a large quantity of the active substance, while in many cases the weaker solution has no perceptible action whatever. For example, copper is dissolved only by concentrated sulphuric acid (see further under Chemical Mechanics).

7. *The influence of the solubility and volatility* of the resulting substances is always very strong and characteristic. When brought together in solution only those substances combine which can form insoluble or difficultly soluble compounds. If, for example, acetic acid is added to an aqueous solution of potassium carbonate, there is formed potassium acetate, and carbon dioxide is set free; but if carbon dioxide is passed into an alcoholic solution of potassium acetate, potassium carbonate and free acetic acid are formed, since potassium carbonate is insoluble in alcohol.

Non-volatile or slightly volatile substances at higher temperatures displace more volatile substances having stronger ability to react. For example, when silicic acid is fused with a salt of sulphuric acid, the weaker silicic acid displaces the stronger sulphuric acid, since the former is not volatile, while in aqueous solutions salts of silicic acid are decomposed by sulphuric acid, silicic acid being set free.

8. *Catalytic Agents.* The presence of small quantities of certain substances promotes or retards the course of certain chemical reactions without any apparent change taking place in the substances themselves. Such substances are known respectively as positive or negative catalytic agents or contact substances, and their action is known as catalysis or catalytic action. For example, phosphorus or carbon do not burn when heated in absolutely dry oxygen, and many metals when placed in contact with it enter into no combination, but these substances combine as soon as even a trace of water is present. Cane-sugar is split into two kinds of sugar by the action of dilute acids, the acids themselves undergoing no alteration. The probable action of many positive catalytic agents is that they first combine with one substance and then give this up to the other; i.e., the action depends upon the formation of unstable intermediate products (see further under Hydrogen and Chloride of Lime). The action of most positive catalytic agents can be compared to the action of fresh oil on a clock which previously ran with much friction and very slowly because of old viscous oil on the pinions.

9. *The Nascent State* (status nascens). Substances show an increased affinity at the moment that they are set free from their compounds; for example, oxygen does not bleach vegetable colors, but when set free from water by chlorine does so instantly (see p. 17).

Weight and Volume Relations in Chemical Reactions.

The combination of elements follows certain laws which are called stoichiometrical laws. The study of the laws governing the combination of the elements and their application to chemical calculations is called stoichiometry in the restricted sense (p. 3). -

1. Law of the Conservation of Matter.

The total weight of matter resulting from a combination or decomposition is always equal to the sum of the weights of all the substances taking part in the reaction.

Superficial observation appears to contradict this law, since, for example, when a candle is burned, matter seems to disappear. However, more careful investigation shows that when a candle burns, gaseous products are formed which are not directly perceptible (carbon dioxide and water vapor), and that the weight of these products corresponds exactly with the weight of the candle burned plus the weight of the oxygen of the air consumed in the combustion.

2. The Law of Constant Proportions.

The elements do not combine with one another to form compounds in indefinite, but in absolutely fixed and unalterable, relative proportions by weight.

Every definite compound therefore contains the elements of which it is composed in absolutely fixed, unalterable proportions by weight; for example, water, irrespective of its source, always consists of 8 parts of oxygen and 1.01 parts of hydrogen, common salt always of 23 parts of sodium and 35.4 parts of chlorine.

If the proportions by weight in which one element combines with others is known, then the proportions by weight in which the other elements combine with one another is known also. *The constancy of the proportions by weight sharply distinguish chemical compounds from mixtures.*

The constant combining proportions have been investigated for all compounds; the values obtained from analysis or synthesis were first calculated on the basis of 100 parts of the compound, but the attempt was soon made to find a simpler expression for these relations and to bring them into conformity with the combining relations of all other elements.

This can be accomplished, for example, when oxygen, which combines with almost all other elements, is taken as the point of departure. In order to allow a comparison with the combining weights now in use, namely, the atomic weights, the figures given below are calculated on the basis of how much of the different elements combines with 8 parts of oxygen, rather than with 1 part of oxygen.

Thus 8 parts of oxygen combine with 1.01 parts of hydrogen to form water, with 35.4 parts of chlorine to form chlorine monoxide, with 16 parts of sulphur, with 4.68 parts of nitrogen to form nitrous anhydride, with 3 parts of carbon to form carbon dioxide, with 27.5 parts of manganese to form manganous oxide, with 23 parts of sodium to form sodium oxide, with 31.8 parts of copper to form cupric oxide, with 100 parts of mercury to form mercuric oxide, and with 28 parts of iron to form ferrous oxide.

Consequently 1.01 parts of hydrogen will combine with 35.4 parts of chlorine, 16 parts of sulphur, 4.68 parts of nitrogen, and 3 parts of carbon, and further 35.4 parts of chlorine or 16 parts of sulphur will combine with 27.5 parts of manganese, 23 parts of sodium, 31.8 parts of copper, and 100 parts of mercury, etc.

The proportions by weight in which the elements enter into combination are called *combining weights* and since they have the same chemical values as the proportions by weight in which they replace or displace one another in chemical decompositions, they are also called *equivalent weights* (p. 12).

For example, if a rod of copper is placed in a solution of mercuric chloride (a compound of 100 parts of mercury with 35.4 parts of chlorine), copper passes into solution until all of the mercury has separated out. For every 100 parts of mercury which separate it will always be found that 31.8 parts of copper enter into combination with 35.4 parts of chlorine.

3. Law of Multiple Proportions.

Many elements can combine with one another in more than one proportion by weight, that is, they can form more than one compound, a circumstance which appears to contradict the law of constant proportions. If, however, these proportions by weight are examined more closely, it is found that they are always a whole multiple of the lowest quantity by weight of the given element which enters into combination.

For example, nitrogen forms five compounds with oxygen:

14 (3×4.68) pts. N. with 8   pts. O. gives nitrous oxide.
14 (3×4.68) " " " 16 (2×8) " " " nitric oxide.
14 (3×4.68) " " " 24 (3×8) " " " nitrous anhydride.
14 (3×4.68) " " " 32 (4×8) " " " nitrogen dioxide.
14 (3×4.68) " " " 40 (5×8) " " " nitric anhydride.

Manganese and oxygen also form five compounds:

27.5 pts. Mn. with 8   pts. O. form manganous oxide.
27.5 " " " 16 (2×8) " " " manganese peroxide.
55.0* " " " 24 (3×8) " " " manganic oxide.
82.5† " " " 32 (4×8) " " " manganous-manganic oxide.
55.0‡ " " " 56 (7×8) " " " permanganic anhydride.

*2×27.5.   †3×27.5.   ‡2×27.5.

If, in a chemical process, the relative weights of the reacting elements present do not correspond to the laws of constant or multiple proportions, then the excess of the particular element remains uncombined.

For example, if 100 parts of copper are heated with 100 parts of sulphur every 31.8 parts of copper combine with 16 parts of sulphur, and accordingly 100 parts of copper with 50.3 parts of sulphur, then $100 - 50.3 = 49.7$ parts of sulphur will remain uncombined, since 31.8 copper : 16 sulphur = 100 copper : x sulphur ($x = 50.3$).

The ability of the elements to combine with one another in several proportions by weight, as well as the tendency to combination of the smallest parts (the atoms, p. 13) of the elements, explains the enormous number of compounds.

4. Law of Simple Proportions by Volume.

The combination of elements or compounds existing in the form of gases takes place according to certain simple relations by volume: *The measured volume of a gaseous compound produced by the combination of two or more gaseous substances is equal either to the sum of the volumes of its components or else is smaller than this in a ratio expressible by whole numbers* (Gay-Lussac's law of volumes). If the volume of that gas which enters into the reaction in the proportion of one volume be taken as unity, then the volume of gas resulting from the combination of elementary gases will occupy the space of two volumes, irrespective of the variation of the sum of the volumes of the reacting gases.

1 vol. (1.01 parts) of hydrogen +1 vol. (35.4 parts) of chlorine give 2 vols. (36.41 parts) of hydrogen chloride.

2 vols. (2.02 parts) of hydrogen +1 vol. (16.0 parts) of oxygen give 2 vols. (18.02 parts) of water vapor.

3 vols. (3.03 parts) of hydrogen +1 vol. (14.04 parts) of nitrogen give 2 vols. (17.07 parts) of ammonia gas.

4 vols. (4.04 parts) of hydrogen +1 vol. (12.0 parts) of carbon give 2 vols. (16.04 parts) of marsh-gas.

Since the gases, like all other substances, can combine only in definite proportions by weight, therefore the weights of the volumes of the combining gases must stand to one another in the same ratio as the combining weights of the elements which constitute the gases.

The combining or equivalent weights calculated on the basis of oxygen =8, which were given above as examples, were formerly used as the foundation of all chemical calculations, but have been abandoned, since under certain circumstances different combining weights were obtained for one and the same element so that the choice became optional and uncertainty resulted. For example, manganous oxide consists of 27.5 parts of manganese and 8 parts of oxygen, manganic oxide of 18.33 parts of manganese and 8 parts of oxygen (p. 11); further, nitrous anhydride of 4.68 parts of nitrogen and 8 parts of oxygen, nitrogen dioxide of 3.51 parts of nitrogen and 8 parts of oxygen (p. 11); further, methane of 3 parts of carbon and 1.01 parts of hydrogen, ethane of 4 parts of carbon and 1.01 parts of hydrogen. This variation in composition is explained by the law of multiple proportions (since 27.5 manganese=3×9.17 and 18.33 manganese=2×9.17; 4.68 nitrogen=4×1.17 and 3.51 nitrogen

$=3\times1.17$); but with respect to the choice of the combining weights manganese can be taken either as 27.5 or 18.33, nitrogen as 4.68 or 3.5, carbon as 3 or 4, etc.

These difficulties were overcome by the atomic theory (likewise through the adoption of the atomic weights).

Theory of Atoms and Molecules.

The natural sciences first investigate the separate manifestations of nature, then attempt to discover the natural laws which are the basis of these phenomena, and finally try to find the causes (natural forces) which are disclosed by the laws and phenomena. Since, however, the actual nature of things is beyond the scope of human intellect, it is necessary to assume certain suppositions or hypotheses, on the basis of which we are able to explain the separate phenomena and the laws which produce them.

If an hypothesis is applicable to the greater number of the observed phenomena, it becomes a theory.

The laws of constant and multiple proportions are facts, but the subsequent assumption of the existence of atoms and molecules is, on the contrary, only a theory which, however, has much apparent truth, since without it not only a great number of chemical but also physical phenomena would be entirely incomprehensible.

1. Atoms.

In explanation of the fact that of every element only a definite quantity by weight or a whole multiple of this quantity can take part in the formation of a chemical compound, it is assumed that the elements consist of very small particles which are mechanically or chemically not further divisible. These particles are called atoms (α, privative and $\tau\acute{\epsilon}\mu\nu\omega$, cut). The atoms of the same element are absolutely alike, and also of equal weight and size. The atoms of the various elements differ from one another in weight and size. There are as many different kinds of atoms as there are elements.

It is evident that if the weight of an atom of hydrogen is 1.01 and the weight of an atom of chlorine is 35.4, these two elements can combine with one another only in these proportions by weight; and further, that if the weight of an atom of oxygen is 8, in a series of compounds formed from nitrogen and oxygen the quantities by weight of the latter will increase by increments of 8 or by multiples of this number. This likewise depends upon the indivisibility of the atoms (Dalton's atomic theory).

From the foregoing considerations it is impossible to determine whether the equivalent weights or combining weights which have been given also represent the atomic weights. For example, if water consisted of 1 atom of hydrogen with 1 atom of oxygen, then the atomic weight of hydrogen would be equal to 1.01 if the atomic weight of oxygen used as the basis is taken as 8. However, water can be formed from 2 atoms of hydrogen and 1 of oxygen, in which case the atomic weight of hydrogen would be equal to one-half of its equivalent weight, namely, 0.505. In order to settle questions of this nature it is necessary to know the further facts underlying the determination of atomic weights, through which such uncertainties can be avoided.

2. Molecules.

The chemical compounds are formed by the combination of two or more dissimilar elements, and if we imagine any chemical compound whatsoever broken up by mechanical force (cutting, pounding, etc.) into not further divisible, smallest particles, then these smallest particles will always consist of a group of atoms, which can be divided further only by chemical and not by mechanical action. These particles are called molecules (*molecula*, small mass).

For example, a molecule of sodium chloride is a particle of sodium chloride which is not further divisible by mechanical forces: if, however, it is acted upon by chemical forces, then it is further split up into a sodium particle and a chlorine particle, that is, its molecules can be decomposed into an atom of sodium and an atom of chlorine.

In the case of the undecomposable substances, the elements, the assumption might seem justified that a molecule is likewise an atom, since no further similar elementary components can be separated by the action of chemical forces from the smallest, not further mechanically divisible, elementary particles; nevertheless, for the reasons first discussed on p. 16, *d*, it is necessary to distinguish between the concept *molecule* and *atom*.

Determination of the Molecular Weight.

Since it is possible to determine for the different elements the smallest relative quantities by weight in which they enter into combination, it must also be possible to determine the relative weights of the different molecules, since the weights of the molecules must likewise stand in some relation to one another.

From the quantitative chemical analysis of a compound it is possible to calculate only the relation in which the separate atoms are present, but the number of the latter which are present in one molecule of the compound must be determined by special methods.

The most useful methods of determination are of physical nature, since these exclude question as to the size of the molecules; but in many cases the chemical investigation also gives a sufficient indication for decision (see Part III, Determination of the Molecular Formula).

In general, substances have the same molecular weight whether in a gaseous, liquid, or solid condition, so that it does not generally seem objectionable to use the substances in all three states of aggregation for determining their molecular weights.

The numerically expressible properties of substances, i.e., the specific gravity, the refraction, dispersion, and rotation of the plane of polarized light, and the heat of fusion, are calculated on the basis of the molecular weight of substances whenever it is desired to express the relation between the chemical composition and the properties of substances. Concerning the further significance of the molecular weight in chemistry see p. 26 and 42, *b*, and Isomerism, Part III.

The absolute values of the atomic and molecular weights are without practical significance for chemistry, but can be approximately determined from various physical phenomena.

1. Molecular Weight of Volatile Substances.

This, in the case of all substances which can be converted undecomposed into the gaseous form, can be deduced from their gas densities (p. 43) on the basis of the following considerations:

a. According to Boyle's law,* *the volume of all gases at a constant temperature is inversely proportional to the pressure.*

It follows as a consequence that the pressure (the tension) which a gas exerts on the surrounding walls on compression at a constant temperature must increase in the same measure that the volume is decreased or proportional to the concentration of the gas. If the volume is decreased to one-half, the concentration of the gas being accordingly doubled, the pressure is also doubled.

b. Gay-Lussac's law † states that, *if the pressure remains constant, the volume of all gases on warming increases in the same proportion for every* 1°. The coefficient of expansion is $\frac{1}{273}$ or 0.003665, that

---

* Also called Mariotte's law.
† Also called Charles' law.

is to say, for every increase in temperature of 1° C. the volume of the gas expands $\frac{1}{273}$ of its volume at 0° C.

Consequently the pressure that a gas on warming at constant volume exerts on the surrounding walls for every 1° increase in temperature must increase $\frac{1}{273}$ of its pressure at 0° centigrade.

Both laws are explainable by

c. Avogadro's hypothesis: *Equal volumes of all gases at the same pressure and temperature contain an equal number of molecules.* Therefore by a comparison of the weights of equal volumes of different gases at the same pressure and temperature the relative weights of the molecules can be determined.

Concerning the apparent deviation of many gases from Avogadro's hypothesis see "Thermochemistry and Dissociation."

d. The smallest particles of the free elements consist generally, like the compounds, of molecules and not of free atoms.

By Avogadro's hypothesis this can be demonstrated as follows:

$$\boxed{\frac{100}{H_2}} + \boxed{\frac{100}{Cl_2}} = \boxed{\frac{100}{HCl}\,\frac{100}{HCl}}$$

1 vol. + 1 vol.   2 volumes.

$$\boxed{\frac{100}{H_2}\,\frac{100}{H_2}} + \boxed{\frac{100}{O_2}} = \boxed{\frac{100}{H_2O}\,\frac{100}{H_2O}}$$

2 vols.   + 1 vol.   2 volumes.

$$\boxed{\frac{100}{H_2}\,\frac{100}{H_2}\,\frac{100}{H_2}} + \boxed{\frac{100}{N_2}} = \boxed{\frac{100}{NH_3}\,\frac{100}{NH_3}}$$

3 vols.   + 1 vol.   2 volumes.

If a certain volume of hydrogen contains 100 molecules of hydrogen, then an equal volume of chlorine contains the same number of chlorine molecules. By the combination of this 1 volume of hydrogen with the 1 volume of chlorine, 2 volumes of hydrogen chloride (p. 12) are obtained, which consequently must contain 200 molecules of hydrogen chloride. But 200 molecules of hydrogen chloride must contain 200 atoms of hydrogen and 200 atoms of chlorine; therefore in the formation of hydrogen chloride each molecule of hydrogen and chlorine has split into 2 parts, that is to say, each of the molecules of hydrogen and chlorine consists of 2 atoms. Similarly 1 volume of oxygen with hydrogen furnishes 2 volumes of water vapor, 1 volume of nitrogen with hydrogen 2 volumes of ammonia-gas, so that here also each molecule of oxygen or nitrogen has separated into two parts.

There is much evidence to support the theory that the molecules of many elements consist of a number of atoms (exceptions, . 21), such as the existence of allotropic modifications of the elements (see Ozone), certain chemically characteristic reactions (see Hydrogen Peroxide), as well as the energetic action of the elements at the moment that they are released from their compounds (p. 9), which can be explained in the following manner: In the free state the atoms have already combined to form molecules and their affinity is already partly satisfied; therefore before an atom of the free element can enter into a compound, the force must first be overcome by which it is held in the molecule by the other atoms. However, at the moment that an element is released from one of its compounds its atoms are quite free and can then act with their entirely unweakened affinity much more energetically upon other molecules present.

From the foregoing considerations it is evident that

Molecular weight is the smallest relative quantity by weight of an element or a compound which appears in the free condition.

Atomic weight is the smallest relative quantity by weight of an element which is to be found in the molecular weights of any of its compounds.

The atomic weight of an element holds good only so long as no compound of it is known which contains in the molecule a still smaller quantity of the element than that previously found.

In order to ascertain the relative weights of atoms and molecules it is before all things necessary to select one substance as a starting-point of the comparison, as unit, and to compare all other substances with respect to their atomic and molecular weights with this one substance taken as the standard.

The unit which has been chosen is 1 *volume of oxygen* = 1 *atom of oxygen* = 16 *parts by weight of oxygen.*

It might seem to be the simplest plan to refer all the quantities by weight of the elements entering into compounds to that element which enters into compounds in the smallest quantities by weight, namely, to hydrogen, and as a matter of fact hydrogen was for a long time taken as the unit of the atomic weights. But since hydrogen combines with only a few other elements to form compounds which are suitable for analysis, the relation of its atomic weight to the atomic weights of the other elements must usually first be determined through the medium of oxygen, so that also formerly oxygen really served as the basis. This makes no difference so long as the proportion $H:O = 1:16$ is held to be correct, but it has been demonstrated that it is extremely difficult to accurately determine this relation since hydrogen is the lightest element and therefore the accurate determination of its weight is greatly influenced by errors of observation and by the small quantities of im_ purities, which are in practice very difficult to remove.   In addition to this the proportion by weight in which hydrogen combines with oxygen is not yet established with complete certainty, but as a result of many experiments it can be assumed that H:O=1:15.88 or 1.088:16 is ap_ proximately correct.   If now O=16 is changed to O=15.88, then the numbers expressing the atomic weights of all the other elements must be altered, while by retaining O=16 the only change necessary is to make H=1.008 (or in round numbers 1.01).   If, as a result of further refinements in the methods of experiment, the ratio between hydrogen and oxygen is again demonstrated to be incorrect, the only change neces_ sary will be that of the atomic weight of hydrogen and not that of all the other elements.

If 1 atom of oxygen occupies 1 volume, then 1 molecule of oxygen, since it consists of 2 atoms (p. 16), must occupy 2 volumes, and since an equal number of molecules are contained in equal volumes of all gases, the molecule of every gaseous substance must occupy the same space as 2 atoms (= 32 parts by weight = 2 volumes) of oxygen.

The molecular weight is therefore that number which expresses how many times lighter or heavier a molecule of an element or a compound is than one molecule (= 32 parts by weight) of oxygen.   The molecular weight of a substance which can be converted into vapor without decomposition is therefore found by determining its specific gravity (gas density p. 43), that is, by determining how much a given volume of its gas weighs when the weight of an equal volume of oxygen is taken as 32 units.

1 litre of oxygen weighs 1.429 grams and 1 litre of hydrogen chloride weighs 1.678 grams; therefore the molecular weight of the latter is ob- tained from the proportion 1.429:32::1.678:x ($x=36.5$).   The molecular weight of a gas can also be determined by multiplying the weight of 1 litre by 22.4, since $32 \div 1.429 = 22.4$.   The correctness of the molecular weight as determined by the preceding method is demonstrated further:

a. By Gay-Lussac's law of volumes (p. 12), which states that the volume of gas which results from the combination of elementary gases occupies the same volume as 2 volumes=2 atoms of oxygen:

1 vol.=1 atom of oxygen with 2 vols. of hydrogen gives 2 vols. of water vapor;

1 vol.=1 atom of oxygen with 2 vols. of nitrogen gives 2 vols. of nitrous oxide;

2 vols.=2 atoms of oxygen with 1 vol. of nitrogen gives 2 vols. of nitrogen dioxide;

2 vols.=2 atoms of oxygen with 1 vol. of sulphur gives 2 vols. of sul- phur dioxide.

b. The smallest molecular weight which can be taken for hydrogen chloride is 36.41, since, according to chemical analysis, this quantity contains 1.01 parts of hydrogen, namely, the smallest quantity by weight of hydrogen which enters into combination, corresponding to one atom of hydrogen. Likewise the smallest molecular weight which can be taken for water is 18.02, since in this quantity there are contained 16 parts of oxygen, namely, the smallest quantity by weight (corresponding to one atom) of oxygen which enters into combination. Moreover, 36.41 parts of hydrogen chloride and 18.02 parts of water vapor occupy the same volume as 2 volumes, equal to 32 parts, of oxygen.

2. Molecular Weight of Soluble Substances.

The molecular weight of all substances which dissolve without decomposition can be determined from their behavior in very dilute solutions. In such solutions these substances act as if they were present as gases in the volume which is occupied by the solution, so that the laws governing gases apply also to dilute solutions if the osmotic pressure (a property of dissolved substances, p. 47, *b*) is considered instead of the gas pressure.

Every substance which is soluble without decomposition shows, in very dilute solutions, an osmotic pressure which corresponds to the gas pressure which it would exert if it we e contained as gas at the same temperature in a volume equal to that of its solvent (van't Hoff's theory).

a. The osmotic pressure at constant temperature is inversely proportional to the volume of liquid in which a given quantity of the dissolved substance is contained; the osmotic pressure must therefore be proportional to the concentration of the solution (analogy to Boyle's law, p. 15).

b. On warming the solution, the increase in the osmotic pressure is the same for every 1°, namely, $\frac{1}{273}$ of the osmotic pressure at 0° C. (analogy to Gay-Lussac's law, p. 15).

c. Equal volumes of solutions which, at the same temperature, show equal osmotic pressures, contain an equal number of molecules of the dissolved substance (analogy to Avogadro's law, p. 16).

The osmotic pressure can be determined only with difficulty, but other properties of dilute solutions are known whose magnitudes stand in a close relation to the osmotic pressure (p. 47, *b*) and are proportional to it, namely, the lowering of the vapor pressure, the depression of the freezing-point, and the elevation of the boiling-point, of which the two latter especially can be readily determined.

Dilute solutions of different substances, which contain an equal number of molecules of the dissolved substance in equal quantities of the same solvent (equimolecular solutions), have the same osmotic pressure (are isotonic: ἴσος, *equal,* τόνος, *tension*), the same depression of the freezing-point, the same lowering of the vapor pressure, and correspondingly the same elevation of the boiling-point. The molecular weight of a substance which dissolves without decomposition is therefore determined by dissolving a small, accurately weighed quantity of the substance in a known quantity of a liquid, and determining the osmotic pressure or the vapor pressure or, what is simpler, the freezing- or boiling-point. The value thus obtained is compared with that which is obtained when a substance of known molecular weight in a corresponding gram-quantity is dissolved in the same quantity of the same liquid.

For example, a quantity of a substance of known molecular weight, corresponding to its molecular weight when dissolved in 1000 grams of benzene (if soluble therein), will lower the freezing-point of the solvent by 4.9°. Therefore if the addition of a known quantity x of a substance of unknown molecular weight to 1000 grams of benzene lowers the freezing-point 9.8°, the molecular weight of the unknown substance must be equal to one-half of x, since $9.8 = 2 \times 4.9$.

If the osmotic pressure, etc., are determined in aqueous solutions, the solutions of acids, bases, and salts show greater osmotic pressures, etc., than would be expected from the theory, and therefore for the determination of the molecular weight of these substances other solvents than water (e.g., acetic acid, benzene, pyridine, etc.) must be employed. Concerning the cause of these exceptions see Theory of the Ions.

Properties, like the osmotic pressure, the expansion and pressure of gases, which can assume similar values for chemically comparable quantities of unlike substances, are called colligative properties (*colligare,* to bind); these, in contradistinction to the additive properties (P. 30), are influenced only by the number of the molecules, not by their nature or constitution.

3. Molecular Weight of Substances which are Volatilized with Difficulty.

The molecular weights of the metals which come under this class are determined similarly to the molecular weights of dissolved substances. If small equimolecular quantities of other metals are melted together with larger and equal quantities of certain metals (bismuth, lead, cadmium, tin), the resulting products (alloys) correspond to dilute solutions and show a corresponding equal lowering of the melting-point. Therefore by determining the melting-point of an alloy prepared from a metal of unknown molecular magnitude, its molecular weight can be calculated in the manner described in 2.

DETERMINATION OF THE ATOMIC WEIGHT.

1. Determination from the Molecular Weight.

Since the atomic weight is the smallest quantity of an element that is present in the molecular weight of any of its compounds, it is only necessary to determine, by the analysis of all compounds of the given element, in which of these compounds the smallest quantity by weight of the element is present in a quantity of the compound corresponding to its molecular weight.

For example, the atomic weight of carbon was determined in this manner: the molecular weight of a compound of carbon with hydrogen, which is called marsh-gas, is 16.04; in 16.04 parts of this gas there are contained 12 parts of carbon and 4.04 parts of hydrogen. The atomic weight of carbon could be equal to 12, 6, 3, etc., that is, in one molecule of marsh-gas there could be 1, 2, 4, etc., atoms of carbon combined with the 4 atoms of hydrogen, but since of all the compounds of carbon which have been examined not a single one of these contains less than 12 parts of carbon in the molecule, the number 12 must be taken as the minimum or atomic weight of carbon.

From the assumption that a molecule of every substance in a gaseous condition occupies the same volume as 2 volumes = 2 atoms of oxygen, it might be concluded that 1 atom of any elementary gas occupies the same volume as 1 volume = 1 atom of oxygen, so that the gas densities of the elements would also be their atomic weights, and that by a comparison of the weights of equal volumes of the given gas and of oxygen the relative weights of the atoms could be determined. With certain elements, however, the weights thus found do not correspond to the atomic weights; for example, different atomic weights are found for sulphur and iodine at different temperatures (see Dissociation); moreover, the molecular weight of phosphorus as determined from its gas density is 124, that of arsenic 300, and accordingly their atomic weights must be 62 and 150 respectively. But the minimum weights which have been found by chemical analysis in any of their compounds (i.e., in molecular quantities of phosphine and arsine) correspond to only 31 for phosphorus and 75 for arsenic. When in a gaseous condition their molecules therefore consist of 4 atoms.

In the case of metals the minimum weight which is found in the molecular weights of the corresponding compounds is of the same magnitude as the molecular weight (p. 20, 3) of the given metal. Their molecules therefore consist of only 1 atom.

The monatomicity of the molecules of the metals can be demonstrated by Avogadro's law, as well as the diatomic character of the molecules of the non-metals; for example, 2 volumes of mercury vapor and 1 volume of chlorine gas give 2 volumes of mercurous chloride vapor; 2 volumes of mercury vapor and 2 volumes of chlorine give 2 volumes of mercuric chloride vapor; if the 2 volumes of mercury vapor contain 200 molecules, then the 2 volumes of resulting mercurous or mercuric chloride vapor must likewise contain 200 molecules (p. 16, *d*), and therefore in the chemical combination of the mercury no splitting up of mercury molecules can have taken place.

The compounds of all metals in a gaseous condition occupy the same space as the vapors of the metals which are contained in them, for since the molecules of the vaporized metals are not further divisible, when they enter into chemical combination the number of the molecules and consequently the space which they occupy cannot be increased.

2. Determination from the Specific Heat.

This method depends upon a relation which exists between the atomic weight and the specific heat of every solid element, and also serves to check the atomic weights found by other methods.

Specific heat is the number of heat units (see Thermochemistry) which it is necessary to add to 1 gram or·1 kilogram of a substance in order to raise its temperature by 1° C. In the case of solids this quantity of heat can be readily determined.

In order to warm equal quantities of two substances to the same temperature two different quantities of heat are necessary. Thus, for example, in order to warm a given quantity of water to a certain temperature, a quantity of heat is required which is thirty-one times as great as that which would be required in order to produce the same increase in temperature in a quantity of platinum of equal weight. The specific heat of platinum is therefore only $\frac{1}{31}$ of that of water, or expressed as a decimal fraction 0.032. The specific heat of one and the same substance varies according to its state of aggregation.

If the specific heat of the elements in the solid state is calculated on the basis of the atomic weight, instead of on the basis of equal quantities, it is found that quantities of the elements taken in proportion to their atomic weights require equal quantities of heat in order to be raised to the same temperature, or that all elements have the same atomic heat (law of Dulong and Petit).

Beryllium, germanium, and the solid non-metals boron, carbon, phosphorus, sulphur, and silicon have the proper atomic heat only at very high temperatures.

The atomic heat is obtained by multiplying the specific heat (H) by the atomic weight (A).

Since 1 gram of platinum requires 0.032 heat-units to raise its temperature 1°, then the weight of platinum corresponding to its atomic weight (i.e., 194.8 grams) will require 194.8 times this quantity of heat, namely $194.8 \times 0.032 = 6.2$.

Since the average atomic heat is 6.4 $(A \times H = 6.4)$, the atomic weight (A) of an element can be found by dividing the quantity 6.4 by the specific heat of the element; for example, the specific heat of lead is 0.031, therefore its atomic weight is $6.4 \div 0.031 = 206.9$.

The smallest quantities by weight of copper and silver which combine with 1 atom of oxygen are 63.6 of the former and 216 of the latter. Do these quantities represent one or more atoms? Since the atomic heat of all elements is equal, therefore the atomic heat of these elements must correspond with the atomic heat of those elements whose atomic weights have been determined by other methods. For example:

	Specific Heat.	Atomic Weight.	Atomic Heat.
Lead..........	$0.0310 \times 206.9 =$		6.4
Copper........	$0.0951 \times 63.6 =$		6.0
Silver.........	$0.0570 \times 215.8 =$		12.2

If these atomic heats of copper and silver are compared with those of other metals, it will be found that copper has nearly the average atomic heat 6.4, i.e., that its atomic weight is correctly determined, while the atomic heat of silver is twice as great. If the atomic weight of silver is taken as half the value, namely 107.9, then the value obtained for the atomic heat is 6.1.

The atomic heat of the elements which cannot be examined in the solid state can be calculated for this condition, since the elements in their solid compounds possess the same atomic heat as in the solid, free condition. The molecular heat corresponds to the sum of the atomic heats of the elements which constitute the molecule (law of Neumann-Kopp). For example:

	Sp. Heat.	Mol. Wt.	
Silver chloride (1 atom silver, 1 atom chlorine)...	0.089	$\times 143.3 =$	2×6.3
Mercuric iodide (1 atom mercury, 2 atoms iodine)	$0.0423 \times$	$453.9 =$	3×6.3

3. Determination from Isomorphism.

Many substances which have an analogous chemical composition i.e., a similar number and arrangement of the atoms in the molecule, when the number of atoms of which they are constituted is equal, possess the property of appearing in the same crystalline form (Isomorphism). There are a large number of isomorphic compounds known in which different elements have replaced one another in proportions corresponding to their atomic weights. Therefore it is assumed that that quantity by weight shall be considered as the sought atomic weight of an element which can replace a quantity by weight of another element corresponding to the known atomic weight of the latter in an isomorphic compound without the crystalline form of the compound undergoing an alteration.

For example, in potassium-aluminium sulphate the aluminium (27.1 parts) can be replaced by chromium (52.1) parts without altering its crystalline form, and therefore the stated quantities by weight stand to one another in the same ratio as their atomic weights. Isomorphism (p. 35) can be of use for proving the accuracy of atomic weights found by other methods, but is of secondary importance as a separate method for the determination of atomic weights, since substances which are not of analogous composition exhibit isomorphism, etc.

4. Determination from the Periodic System.

If a determined atomic weight does not fit into the periodic system (p. 54), then the assumption is safe that the atomic weight has been incorrectly determined and a redetermination is in order.

Symbols, Formulas, Equations.

1. Chemical Symbols.

For the sake of simplicity in chemistry the elements are denoted only by the initial letters of their names, but in cases where the names of several elements begin with the same letter, a second letter from their name is also added. These symbols have also a quantitative significance since they denote not only the element but also its atomic weight.

2. Chemical Formulas.

The compounds formed by the combination of elements are denoted by placing together the symbols of the elements of which they are constituted, and a combination of symbols of this character is called a chemical formula; for example, HCl represents hydrogen chloride. If several atoms of one element are contained in the molecule of the compound, this fact is denoted by a small number written as a subscript of the given symbol; for example, H_2O = water, NH_3 = ammonia. A large figure placed in front of a formula or a small figure written after a formula enclosed in brackets refers to the for-

ELEMENTS WITH ATOMIC WEIGHTS.

Name.	Symbol.	Atomic Weight.	Name.	Symbol.	Atomic Weight.
Aluminium............	Al	27.1	Neon...............	Ne	20.0
Antimony (Stibium)...	Sb	120.2	Nickel.............	Ni	53.7
Argon...............	A	39.9	Niobium............	Nb	94.0
Arsenic..............	As	75.0	Nitrogen............	N	14.04
Barium..............	Ba	137.4	Osmium.............	Os	191.0
Beryllium...........	Be	9.1	Oxygen.............	O	16.0
Bismuth.............	Bi	208.5	Palladium..........	Pd	106.5
Boron...............	B	11.0	Phosphorus.........	P	31.0
Bromine.............	Br	79.96	Platinum...........	Pt	194.8
Cadmium............	Cd	112.4	Potassium (Kalium)..	K	39.15
Cæsium..............	Cs	132.9	Praseodymium.......	Pr	140.5
Calcium.............	Ca	40.1	Radium.............	Ra	225.0
Carbon..............	C	12.0	Rhodium............	Rh	103.0
Cerium..............	Ce	140.25	Rubidium...........	Rb	85.4
Chlorine.............	Cl	35.45	Ruthenium..........	Ru	101.7
Chromium...........	Cr	52.1	Samarium...........	Sa	150.0
Cobalt..............	Co	59.0	Scandium...........	Sc	44.1
Copper (Cuprum)......	Cu	63.6	Selenium...........	Se	79.2
Erbium..............	Er	166.0	Silicon.............	Si	28.4
Fluorine.............	F	19.0	Silver (Argentum).....	Ag	107.93
Gadolinium..........	Gd	156.0	Sodium (Natrium)....	Na	23.05
Gallium.............	Ga	70.0	Strontium..........	Sr	87.6
Germanium..........	Ge	72.5	Sulphur............	S	32.06
Gold (Aurum)........	Au	197.2	Tantalum..........	Ta	183.0
Helium.............	He	4.0	Tellurium..........	Te	127.6
Hydrogen...........	H	1.01	Terbium............	Tb	160.0
Indium..............	In	114.0	Thallium...........	Tl	204.1
Iodine..............	I	126.85	Thorium...........	Th	232.5
Iridium.............	Ir	193.0	Thulium...........	Tu	171.0
Iron (Ferrum)........	Fe	55.9	Tin (Stannum).......	Sn	119.0
Krypton.............	Kr	81.8	Titanium...........	Ti	48.1
Lanthanum..........	La	138.9	Tungsten..........	W	184.0
Lead (Plumbum)......	Pb	206.9	Uranium...........	U	238.5
Lithium.............	Li	7.03	Vanadium..........	V	51.2
Magnesium..........	Mg	24.36	Xenon.............	X	128.0
Manganese..........	Mn	55.0	Ytterbium.........	Yb	173.0
Mercury(Hydrargyrum)	Hg	200.0	Yttrium...........	Y	89.0
Molybdenum.........	Mo	96.0	Zinc..............	Zn	65.4
Neodymium..........	Nd	143.6	Zirconium.........	Zr	90.6

mula taken as a whole or to that portion enclosed in the brackets only; for example, $3H_2SO_4$ or $(H_2SO_4)_3$ denotes 3 molecules of sulphuric acid; $Fe_2(SO_4)_3$ is the formula for an iron salt of sulphuric acid and indicates that this substance consists of 2 atoms of iron and 3 atomic complexes SO_4.

The formulas used express the molecular magnitudes of the

given compounds, as well as the number of atoms in the molecules, for which reason they are also called atomic molecular formulas. They give in addition to the qualitative composition also the relations by weight and volume in which the elements are present in the molecules of the compounds.

If the formula of a substance is given, it is possible to calculate the weight of a liter of the substance in a gaseous condition, as well as its specific gravity with respect to any other gas taken as unity (p. 42, b), and from these quantities to further calculate its gram-volume and molecular volume, if only the molecular weight (=32) and the weight of a liter (=1.4291) of oxygen are known. It must, however, be borne in mind that all considerations involving the volumes of gases must be based on the volume occupied by gases under a pressure of 760 mm. and a temperature of 0°. Since all specific gravities of gases are now referred to the molecular weight of oxygen=32 as unit (p. 43), their specific gravities are therefore directly expressed by their molecular weights. The formula NH_3, for example, consequently expresses the following concerning one molecule of ammonia:

1. That it consists of nitrogen and hydrogen.

2. That it consists of 1 atom=14.04 parts of nitrogen and 3 atoms =3×1.01 parts of hydrogen and therefore has the molecular weight 17.07.

3. That it consists of 1 volume of nitrogen and 3 volumes of hydrogen.

4. That it has the specific gravity 17.07 with respect to oxygen taken as unit; further, that its specific gravity with respect to hydrogen=1 is 8.53, and with respect to air=1 is 0.59 (see p. 44).

5. That 1 liter of it weighs 0.76 gram, that its molecular volume is 22.4 liters, and that its gram-volume is 1.31 liters.

For the basis of these calculations see pp. 43, 44.

In addition to the empirical formulas mentioned above, in which only the symbols of the elements constituting the molecules are expressed, there are also the so-called rational formulas which indicate the grouping of the atoms in the molecules (p. 29).

3. Chemical Equations.

This term is applied to those expressions which are used to represent chemical reactions by the symbols and formulas of the reacting substances. In these equations the reacting substances are placed on the left (of the sign of equality) and the resulting substances are placed on the right. The sum of the atoms of each element on both sides must of course be equal, since a chemical equation is at the same time an expression of the principle of the conservation of matter (p. 10).

For example, iron (Fe) reacts with mercuric sulphide (HgS) to form iron sulphide and mercury. This process can be briefly expressed as follows:

$$HgS + Fe = FeS + Hg,$$

which expresses that

$$
\begin{matrix}
\text{Mercuric} & Hg = 200.3 \\
\text{sulphide} & S = 32
\end{matrix} \Bigg\} 232.3
\qquad
\begin{matrix}
Fe = 56 \\
S = 32
\end{matrix} \Bigg\} 88
\begin{matrix}
\text{iron} \\
\text{sulphide}
\end{matrix}
$$

$$
\begin{matrix}
\text{and} & + \\
\text{iron} & Fe = 56
\end{matrix}
\qquad \text{gives}
\begin{matrix}
+ \\
Hg = 200.3
\end{matrix}
\begin{matrix}
\text{and} \\
\text{mercury.}
\end{matrix}
$$

From this equation it is therefore known that for every 232 parts of mercuric sulphide which are used 200 parts of mercury will be obtained, and that for every 232 parts of mercuric sulphide 56 parts of iron must be taken, which are afterwards obtained as 88 parts of iron sulphide. For these reasons the study of chemical proportions is of importance not only for the theoretical development of chemistry, but for applied chemistry as well.

For the sake of simplicity, chemical equations are mostly written in terms of the atomic quantities; but since free elements enter into reactions in molecular quantities, it is more in accord with the facts to employ only the latter in equations; for example $H_2 + Cl_2 = 2HCl$ instead of $H + Cl = HCl$. Equations are often only the expression of an ideal reaction, since many reactions do not proceed quantitatively, Secondary reactions can take place along with the main reaction expressed by the equation, thus producing other products, which, owing to the small quantities in which they occur, can be neglected (see, for example, the preparation of ethyl alcohol). It is further an important fact that reactions proceed only to a certain final state of equilibrium (see Chemical Mechanics).

Theory of Valence.

1. Valence.

Although the atomic theory affords an explanation of the law of constant proportions, when applied to the law of multiple proportions it merely serves to show that the atoms can combine in several relations.

Nothing, however, is stated by the atomic theory as to how many compounds the same elements can form with one another, or as to why certain elements combine more readily in certain proportions than in others.

These matters are explained by the theory of valence, or atomicity, which attributes to every atom the power of combining with only a perfectly definite number of atoms of every other element.

The power which the atoms of any element have to combine with the atoms of every other element is called the valence or atomicity of the element, and it is measured by the number of hydrogen atoms with which one atom of the given element can combine, or which one atom of the given element can replace in a compound. The valence of hydrogen is thus taken as unity.

If an element does not combine with hydrogen, then its valence is determined from the number of other atoms, equivalent to an atom of hydrogen, with which it can combine.

1 atom of chlorine combines with 1 atom of hydrogen to form HCl,
1 " " oxygen " " 2 atoms " " " " H_2O,
1 " " nitrogen " " 3 " " " " " H_3N,
1 " " carbon " " 4 " " " " " H_4C.

Accordingly the valence of chlorine = 1, of oxygen = 2, of nitrogen = 3, of carbon = 4; and chlorine is therefore said to be uni- or monovalent, oxygen bi- or divalent, nitrogen trivalent, and carbon quadri- or tetravalent. It is also often customary to say that chlorine possesses one and oxygen two bonds or units of affinity, because it used to be believed that chemical affinity (p. 7) was connected with the valency.

The valence is denoted by Roman numerals or horizontal dashes placed near the symbols (p. 29).

The valence is not, like the atomic weight, an inherent property of the element, but is dependent upon the properties of the elements which react upon one another. Every element exhibits, however, a maximum valence. For example,

$$\overset{\text{II}}{CO} \quad \overset{\text{III}}{PCl_3} \quad \overset{\text{III}}{P_2O_3} \quad \overset{\text{II}}{SCl_2} \quad - \quad \overset{\text{I}}{Cl_2O} \quad -$$

$$\overset{\text{IV}}{CO_2} \quad \overset{\text{V}}{PCl_5} \quad \overset{\text{V}}{P_2O_5} \quad \overset{\text{IV}}{SCl_4} \quad \overset{\text{IV}}{SO_2} \quad \overset{\text{III}}{Cl_2O_3} \quad -$$

$$\overset{\text{VI}}{SO_3} \quad \overset{\text{V}}{\cdot Cl_2O_5} \quad \overset{\text{VII}}{Cl_2O_7}$$

Those compounds in which an element appears with a valence less than its maximum valence are called unsaturated compounds, e.g., CO, PCl_3, etc.

Owing to the variable valence of the elements, the theory of valence does not afford an insight into all the compounds which elements can form with one another, but nevertheless a knowledge of the maximum value is an important assistance in approximately determining the number of possible compounds.

2. Atomic Linking.

As a result of the assumption of the valence for the elementary atoms, it follows that the separate atoms in a molecule of a chemical compound will not be held together by a force exerted in common by all the other atoms present, but that the force of attraction will act only from atom to atom, i.e., that each atom will be attached to its neighbors only.

Monovalent atoms can be compared to spheres which have only one ring or hook and can therefore be attached to only one other atomic sphere; if this second atomic sphere is likewise a monovalent atom, then no more atoms can be attached, and there results a saturated molecule, the chain consisting of atomic spheres being a closed

one; for example,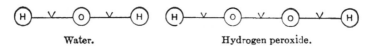

If the second atom has a valence greater than one, then only one of its valences will be used in combining with the monovalent atom, and the remainder can hold additional atoms and thus lengthen the chain of atomic spheres. For example,

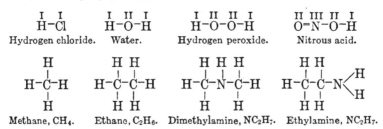

Water.   Hydrogen peroxide.

The attachment of two valencies is ordinarily denoted not by hooks, but only by a dash, thus:

$$\overset{\text{I}\ \ \text{I}}{\text{H--Cl}} \qquad \overset{\text{I}\ \ \text{II}\ \ \text{I}}{\text{H--O--H}} \qquad \overset{\text{I}\ \ \text{II}\ \ \text{II}\ \ \text{I}}{\text{H--O--O--H}} \qquad \overset{\text{II}\ \ \text{III}\ \ \text{II}\ \ \text{I}}{\text{O}={\text{N--O--H}}}$$

Hydrogen chloride.   Water.   Hydrogen peroxide.   Nitrous acid.

$$\begin{array}{c} \text{H} \\ | \\ \text{H--C--H} \\ | \\ \text{H} \end{array} \qquad \begin{array}{c} \text{H}\ \ \text{H} \\ |\ \ | \\ \text{H--C--C--H} \\ |\ \ | \\ \text{H}\ \ \text{H} \end{array} \qquad \begin{array}{c} \text{H}\ \ \text{H}\ \ \text{H} \\ |\ \ |\ \ | \\ \text{H--C--N--C--H} \\ |\ \ \ \ \ \ | \\ \text{H}\ \ \ \ \ \ \text{H} \end{array} \qquad \begin{array}{c} \text{H}\ \ \text{H} \\ |\ \ | \\ \text{H--C--C--N} \end{array}$$

Methane, CH_4.   Ethane, C_2H_6.   Dimethylamine, NC_2H_7.   Ethylamine, NC_2H_7.

The bonding of the atoms in molecules according to their valency is called the chemical constitution or structure of the substances. The basis of chemical structure is founded on the view that every valence of an atom is attached to a valence of some other atom. The formulas constructed on this principle, which give an idea as to the arrangement of the atoms in the molecules, are called *structural* or *constitutional formulas*.

Structural formulas, in contradistinction to empirical formulas, are called rational formulas. They are particularly indispensable in the case of the carbon compounds, since there are considerable numbers of these which, although of similar qualitative and quantitative composition, have quite different properties. This is known as isomerism.

For example, the empirical formula NC_2H_7 stands for two such compounds, so that from this it cannot be determined which of the compounds is denoted, while the rational formulas mentioned above make it instantly possible to distinguish between them. The study of chemical transformations, the different methods of preparation, and the isomerism of organic compounds all lead to the assumption of a definite arrangement of the atoms in the molecules (see Part III, Isomerism and Investigation of the Constitution).

The method of writing these formulas suggests the arrangement of the atoms in a single plane; but since in reality the separate molecules must have definite volume, the separate atoms must be considered as distributed through all three dimensions of space (concerning the arrangement of the atoms in space see Part III, Stereochemistry).

The unvarying arrangement of the atoms in the molecule as a result of their linking does not necessitate that the atoms are fixed immovably with respect to one another in the molecule, but it can be assumed that notwithstanding this they move about a point of equilibrium.

The constitution of salts containing water of crystallization and of other similar double salts (see Magnesium Sulphate) cannot be explained by the assumption of an additional valency belonging to the atoms, for which reason it is supposed that in these compounds several molecules are present and that these exert a mutual force of attraction on one another. Such compounds are therefore called molecular compounds. Example, $MgSO_4,K_2SO_4 \cdot 6H_2O$.

Properties of substances which depend upon the nature and number of the atoms in the molecule are called *additive properties;* for example, the molecular weight is an additive property since it is equal to the sum of the weights of the atoms which form the molecule.

Properties of substances which depend not only on the nature and number of the atoms in the molecule, but also on the manner in which the atoms are attached to one another, are called *constitutive properties;* for example, the isomerism and the optical activity of organic compounds are constitutive properties. Colligative properties, see p. 20.

From the foregoing considerations it is evident that chemistry can be defined as that science which treats of the structure of the molecule and the investigation of the arrangement of the atoms within the molecule, while physics treats of the molecule as such and its properties. *Chemistry is the study of equilibrium and the motion of the atoms within the molecule; physics, the study of equilibrium and the motion of the molecule itself.*

3. Equivalence.

Many chemical processes proceed in such a manner that one element enters in the place of another in the molecule of a compound. This is called substitution (p. 6). The quantities of the elements thus involved depend upon their valence, which in the case of polyvalent elements denotes only a fraction of their atomic weight. But since the atoms are indivisible, the substitution can only proceed under such conditions that one atom of a bivalent element replaces two atoms of a univalent element, one atom of a trivalent replaces three of a univalent, or one atom of a bivalent plus one atom of a univalent. In other words, the sums of the atoms replacing one another must be of equal value (equivalent). For example,

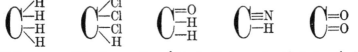

| Methane. | Chloroform. | Formaldehyde. | Hydrocyanic acid. | Carbon dioxide. |

The quantities by weight of the elements which show equal values (equivalence) are called *equivalent* or *substitution weights*, and are the fractions of the atomic weights corresponding to the valence of the given atoms. Therefore in the case of monovalent elements the equivalent weights are equal to the atomic weights, in the case of polyvalent elements they are fractions of the atomic weights.

The equivalent weight of an element is equal to its atomic weight divided by its valence (with respect to hydrogen taken as the unit).

If chlorine is allowed to act on hydriodic or hydrobromic acid, the iodine or bromine is set free, and the chlorine combines with the hydrogen, viz., $HI + Cl = HCl + I$; $HBr + Cl = HCl + Br$.

One atom of chlorine (35.4 parts) replaces 1 atom of bromine (80 parts) or 1 atom of iodine (126.8 parts); and bromine also sets free the iodine in hydriodic acid: $HI + Br = HBr + I$. The atoms of Cl, Br, I, and H have therefore the same value.

Oxygen also replaces iodine in hydriodic acid and sets it free: $2HI + O = H_2O + 2I$. It is evident from this equation that oxygen has not the same chemical value as chlorine and iodine, and that, as has already been observed, oxygen is bivalent, while chlorine and iodine are univalent. Only 1 atom ($= 16$ parts) of oxygen has entered in place of 2 atoms ($= 2 \times 126.8$ parts) of iodine; therefore 2 atoms of iodine have the value of 1 atom of oxygen, or, since each atom of iodine $= 126.8$ parts, the equivalent quantity of oxygen is $\frac{1}{2}$ atom $= 8$ parts of oxygen.

One atom of nitrogen $= 14.04$ parts is equivalent to 3 atoms (3.03 parts) of hydrogen, and therefore 1 atom (1.01 parts) of hydrogen is equivalent to $\frac{1}{3}$ atom ($= 4.68$ parts) of nitrogen.

The combining weights employed on p. 10 in explaining the laws of constant and multiple proportions are identical with the equivalent weights discussed above, and were formerly used in chemistry instead of the atomic weights. They are still of significance in electrochemistry (under which see Faraday's Law).

Properties of Molecular Aggregations.

Those masses of matter which are capable of perception are produced by aggregations of molecules. These aggregations can be of different kinds, which determines the state of aggregation of the substance, i.e., its appearance in the solid, liquid, or gaseous condition.

The alteration of the state of aggregation of a substance is produced, not by the alteration of the state of the molecules themselves, but through the nature of the motion and the alteration of the amount of space between them. It is assumed that a certain amount of space exists between the separate molecules, and that in the case of gases this intermediate space is so large that the molecules themselves are incomparably small in comparison with the space which separates them. But since the molecular weights of many substances in the same states of aggregation undergo an alteration with a change of temperature (see Dissociation), it is not impossible that at times the alteration of the state of aggregation is accompanied by a change in the size of the molecule. Usually, however, the molecular weights determined for a substance in different states of aggregation are found to correspond with one another (p. 15). The mutual coherence of the molecules in solid and liquid substances is produced by a mechanical force of attraction called cohesion.

The state of aggregation is dependent on the temperature and pressure. By increasing the temperature most solid substances can be liquefied, and all liquids (within certain limits of pressure) can be converted into gases; by lowering the temperature all gases can be liquefied and all liquids converted into solids. Many solid substances cannot alter their state of aggregation by an increase in temperature without undergoing a chemical decomposition. By a lowering of the pressure liquids can pass over into gases, and, conversely, by an increase in the pressure (within certain limits of temperature) gases can pass over into liquids.

The molecules, as well as the atoms which compose them, are considered as being in a state of constant vibration. The molecules of solid substances vibrate about a point of equilibrium, and on warming the vibrations become quicker and of greater amplitude (expansion of warmed substances), until finally the cohesion is so far overcome that it can no longer retain the molecules in a position of equilibrium and the substance becomes liquid (melts). In overcoming the cohesion a certain quantity of heat is used up (heat of fusion), so that the melting-point remains constant until all is melted. If the warming is continued, then both on the surface (evaporation) as well as within the liquid the motion of the molecules will be so increased that the cohesion will be more and more overcome and the molecules will fly out into the space above the liquid. Finally the motion of the molecules will become so violent that they will entirely overcome the pressure of the liquid and that of the air pressing upon this and the liquid will begin to boil, whereupon the temperature (boiling-point) will remain constant until all is vaporized, since now all the added heat (the heat of vaporization) is used up in the work of severing the last bond of the force of cohesion.

If heat acts still further on completely gasified substances, the vibrations of the atoms in the molecules can become so violent that the molecules split up into atoms or molecules of simpler constitution (dissociation, see Thermochemistry).

1. Solids.

These have a characteristic appearance; within them the molecules move about a fixed point of equilibrium, vibrating or rotating.

All solid substances occur organized (p. 4), crystallized, or amorphous; many are known in both the latter forms, but possess in both cases different physical properties.

Very many substances, when they pass from the liquid, dissolved, or gaseous state into the solid condition, assume a regular form bounded by planes; i.e., they appear as crystals.

Every crystallizable substance has a perfectly definite crystalline form which can facilitate its identification. If it separates from a solution, the smallest crystal has the same form as the largest, and as they grow in the solution the forms of the crystals undergo no alteration. It is very seldom, however, that the crystal form is regularly developed, so that it is very often possible that in one and the same crystalline substance the shape and development of the separate planes can be different. This is never the case with the position of the planes with respect to the axes. The angles which the planes of every crystallized substance make with one another are unalterable (law of the constancy of corresponding interfacial angles). In addition to their regularly bounded forms, crystals possess a regular internal structure and (with the exception of those of the regular system) exhibit in different directions within them different physical properties with respect to cohesion, hardness, elasticity, conductivity of heat, and the transmission of light. Frequently, especially on rapid cooling, the crystals are disturbed in their development so that they grow through one another and a crystalline body is obtained. The crystalline character of a body is determined by breaking it, after which it can be determined from the surface of the fracture whether the texture is laminated, radiated, or granular.

If a substance shows no evidence of crystalline structure and breaks with a conchoidal fracture, it is called amorphous. Amorphous substances have the same physical properties in all directions within them, and can be compared to strongly supercooled liquids (p. 35), since many substances can be obtained in an amorphous state if, while in the fused condition, they are cooled rapidly. The amorphous state is almost always unstable, and amorphous substances after standing for some time pass over spontaneously into the crystalline condition.

a. Crystallography.

The study of crystal forms is called crystallography. Many thousands of crystal forms are known, but they can all be referred to six classes or systems. This can be accomplished by comparing the crystals according to their directions of development, called axes, i.e., we imagine a series of lines (axes) passing through the middle points of the crystals, so placed that the crystal faces lie symmetrically about these axes. From the number of axes, their length and inclination, all crystals can be classified into six systems.

The classification of all crystal forms of the six systems can be accomplished also by the assumption of symmetry planes instead of axes. A symmetry plane denotes a plane which can be imagined as so dividing a crystal into two halves that the one half stands in the same relation to the other half as the one half stands to its image as seen reflected in a mirror. The differences between the six crystal systems are therefore the following:

1. Regular or isometric system: 3 axes of equal length, all at right angles to one another.—9 planes of symmetry.

2. Quadratic or tetragonal system: 3 axes, 2 of equal length, the third (principal axis) longer or shorter, all at right angles to one another. —5 planes of symmetry.

3. Orthorhombic or trimetric system: 3 axes of different lengths, all at right angles to one another.—3 planes of symmetry.

4. Monoclinic, monosymmetric, or clinorhombic system: 3 axes of

different lengths, 2 of these at an oblique angle with one another but at right angles to the third.—1 plane of symmetry.

5. Triclinic, asymmetric, or clinorhomboidal system: 3 axes of different lengths, all forming oblique angles with one another.—No planes of symmetry.

6. Hexagonal system: 4 axes, 3 of these of equal length and cutting one another at angles of 60°, and a fourth axis longer or shorter and perpendicular to the plane of the other three.—7 planes of symmetry.

A compound which can appear in crystal forms belonging to more than one crystal system is called polymorphic, namely, dimorphic, trimorphic, etc., according to whether the forms belong to 2, 3, etc., crystal systems. In these cases the different crystal forms always exhibit a difference in their physical properties (calcium carbonate as aragonite is orthorhombic, as calcite hexagonal). Elements also can exhibit polymorphism (sulphur appears both orthorhombic and monoclinic, although this is usually denoted as allotropism (which see).

Various compounds can often have the same crystal form and are then called isomorphic, e.g., the carbonates of calcium, magnesium, zinc, manganese, and iron. These can form mixed crystals (p. 50) and often have a similar constitution, a fact which is made use of in the determination of the atomic weights (p. 23).

b. *Fusion and Vaporization.*

Solid substances which can be melted are called fusible, and when they can be vaporized they are called volatile. Every solid substance, like every liquid, at any given temperature has a definite vapor pressure, which, however, at ordinary temperatures is generally very small. It is a fact, however, that many substances are volatile even at ordinary temperatures. The melting-point is that temperature at which a substance becomes liquid, and has a fixed value for every pure crystalline substance that melts without decomposition. It is therefore a characteristic of substances and is important for determining their purity. Many substances do not melt without undergoing decomposition. Amorphous substances have no definite melting-point.

The melting-point corresponds with the freezing- or solidifying-point of the melted substance. Many substances can be cooled under favorable conditions below their solidifying-point without becoming solid. For example, water, if protected from every mechanical disturbance, can be cooled below 0° without freezing; on being shaken it instantly changes into ice and the temperature rises to the melting-point (p. 36).

Substances which under certain conditions remain melted at temperatures which are lower than their normal points of solidification are said to be supercooled or superfused (analogy to supersaturated solutions, p. 52).

By increasing the external pressure the melting-points of most substances which melt with an increase of volume is raised. In the case of ice and certain other substances which undergo a reduction in volume on melting, an increase in the external pressure causes a lowering of the melting-point.

By a reduction of the external pressure all fusible substances can pass directly over into the gaseous condition, as soon as the pressure has become less than the vapor pressure of the substance at its melting-point. For·example, ice melts at 0°, and at this temperature has a vapor pressure of 4.6 mm. If the external pressure is reduced to less than 4.6 mm., the ice passes into the form of vapor before it reaches its melt‐ing-point.

Many solid substances (mercurous chloride, ammonium chloride, metallic arsenic, and many carbon compounds, on being heated at normal atmospheric pressure, vaporize without melting, since their vapor pressures at their melting-points are higher than the atmospheric pressure. If, therefore, the external pressure is increased until it equals the vapor pressure, then these substances can be melted.

The pressure below which many solid substances no longer melt, but pass directly into the form of vapor, is called their sublimation pressure.

The transformation of a solid substance into vapor and the condensation of the vapor into the solid state is called sublimation; it serves to separate volatile from non-volatile substances. The ordinarily slight variations in the atmospheric pressure can be neglected in melting-point determinations.

After a substance has begun to melt, the quantity of heat which is added until all is melted (the heat of fusion, p. 33) does not produce any increase in the temperature. In like manner a fused substance on solidification yields the same quantity of heat which it took up on melting without undergoing any reduction in temperature (the heat of solidification).

The heat of fusion or solidification is the number of heat-units (see Thermochemistry) which are required in order to transform 1 gram or 1 kilogram of a solid substance into liquid at the same temperature. If the heat of fusion is multiplied by the molecular weight, the molecular heat of fusion is obtained.

In regard to the alteration of the melting-point in mixtures of solid substances see p. 50. For the relation of the lowering of the melting-point to the molecular weight see p. 20, and for the relation of the melting-point to the constitution of organic compounds see Part III.

c. Specific Heat.

of solid substances and its relation to the atomic and molecular weight. See p. 22.

d. Specific Gravity.

By dividing the atomic weight of an element by its specific gravity we obtain the atomic volume, i.e., the volume in cubic centimeters occupied by a quantity of the substance equal to its atomic weight expressed in grams (gram-atom). For example, lithium, atomic weight 7.01, specific gravity 0.59, therefore atomic volume is 11.9; that is to say, 7.01 grams of lithium occupy a volume of 11.9 cubic centimeters (0.59 gram Li:7.01 gram Li::1 c.c.:x). For the different modifications of an element the atomic volume is different.

In regard to the relation of the atomic volume to the properties ·of the atoms see p. 56.

The molecular volume of a compound is the volume in cubic centimeters occupied by a quantity of the compound equal to its molecular weight taken in grams (gram-molecule). It is obtained by dividing the gram-molecule of the compound by the specific gravity of the compound.

For the relation of the molecular volume to the constitution of organic compounds see Part III. The atomic and molecular volumes do not correspond to the actual volumes occupied by the atoms and molecules, since in determining the specific gravity not only the volumes of the atoms and molecules are measured, but also the space between them. Atomic and molecular volumes are accordingly the volumes which contain equal numbers of atoms and molecules.

The gram-volume, or specific volume, of a solid substance is the volume occupied by 1 gram of the substance, and is determined by dividing 1 by the specific gravity of the given substance. For example, the specific gravity of lithium is 0.59, i.e., 1 c.c. of lithium weighs 0.59 gram, therefore 1 gram of lithium will occupy a volume of 1.69 c.c. (0.59 gram Li:1.0 gram Li:: 1 c.c.:x c.c.).

e. Optical Properties.

These will be treated under liquids.

·2. Liquids.

These have no fixed form, but assume that of the vessel which contains them. Their molecules have no longer a definite position of equilibrium, but have a progressive motion in addition to one of vibration and rotation, so that each molecule is from time to time surrounded by different neighboring molecules.

a. Boiling and Evaporation.

On being cooled to a definite temperature (freezing-point) all liquids become solid; on being heated to a definite temperature (boiling-point) liquids are vaporized. The latter is true only for such liquids as vaporize without decomposition, but in the case of all liquids there is a definite formation of vapor even at temperatures below the boiling-point.

If the formation of vapor takes place only at the surface of a liquid, this is called evaporation; but if it proceeds also within the liquid, it is called boiling. The boiling-point is lower the lower the external pressure, and an increase in the external pressure causes an elevation of the b iling-point.

Every liquid boils when the tension of its vapor equals the external pressure. The boiling-point is always taken as that temperature at which the tension of the vapor is equal to an external pressure of 760 mm. of mercury. Many substances which decompose when boiled under atmospheric pressure can be boiled under diminished pressure without undergoing any decomposition. The boiling-point is constant for every pure substance and is therefore a valuable means for the identification, as well as for the determination of the purity, of a substance.

Under certain suitable conditions the external pressure can be made much lower than the vapor tension of the liquid without causing the latter to boil (superheated liquids, analogous with supercooled substances, p. 35).

When a pure liquid begins to boil the temperature remains constant as long as the boiling continues. The heat of vaporization is the number of heat-units which are required to transform 1 gram or 1 kilogram of liquid into vapor having the same temperature, or is the number of heat-units set free when 1 gram or 1 kilogram of vapor is converted into liquid having the same temperature (see Water Vapor).

For the relation of the boiling-point to the constitution of organic compounds see Part III.

b. Specific Gravity.

The calculation of the specific, atomic, and molecular volumes from the specific gravity is carried out as in the case of solids (p. 36).

For the relation of the molecular volume to the constitution of organic compounds see Part III.

c. Spectrum.

The property which many colored solutions have of absorbing a part of the light split up by a prism. For the application of this property to the determination of their chemical nature see p. 45.

d. Refraction of Light.

Every transparent substance possesses a definite power of refracting light, and the quotient of the sine of the angle of refraction (r) into the sine of the angle of incidence (i), called the coefficient or index of refraction (n), is a constant for every given substance. $\dfrac{\text{Sine } i}{\text{Sine r}} = n.$ This is of great assistance in identifying a substance and for determining its purity. It is, however, dependent on the density (specific gravity, d) of the given substance (also on the pressure and temperature), as well as on its state of aggregation. By the use of the formula $\dfrac{n^2-1}{n^2+2} \cdot \dfrac{\text{l as}}{d} = R$ a factor R is obtained which for liquids and gases is dependent only on their chemical constitution. This is known as the specific refractive power or specific refractive constant.

The product of the specific refractive power of an element with its atomic weight is called its atomic refraction.

The product of the specific refractive power of a compound with its molecular weight is called its molecular refraction.

From the specific refractive power of a liquid mixture (i.e., a solution), when the specific refractive power and the constitution of one of the substances is known, that of the other can be calculated. On the other hand, in a liquid mixture of several substances of known refractive power the proportions by weight in which the different substances are present can be calculated from the specific refractive constant of the mixture.

For the relation of the specific refractive power to the constitution of organic compounds see Part III.

e. Rotation of the Plane of Polarization.

Many organic and inorganic compounds rotate the plane of polarized light to the right or left and are therefore called dextrorotatory and lævorotatory respectively, or, in general, optically active or circularly polarizing. A small number of these compounds show this behavior only in the crystalline state, and the rotation ceases when they are dissolved or pass over into the liquid condition. Very few organic compounds belong to this class. A much larger number of compounds, of which all are organic, exhi it optical activity when in the liquid condition, some indeed not only when existing as liquids, but also when in the form of gases.

In the case of those substances which possess this property only in the crystalline state, their behavior must be attributed to the arrangement of the molecules in the crystal. With those which show this behavior in the fused, dissolved, or gaseous condition it cannot be assumed that the effect is due to any arrangement of the molecules, since these are free to move about in all directions, but it must be assumed that it is caused by some particular arrangement of the atoms in the molecules. Now it has been shown that all those organic compounds which are optically active when fused, dissolved, or gaseous contain at least one asymmetric carbon atom, namely, a carbon atom which is attached to four dissimilar atoms or groups of atoms; for example,

$$\text{lactic acid} \quad \begin{array}{c} H_3C \\ H \end{array} \!\!\! >\!\!C\!\!<\!\!\! \begin{array}{c} OH \\ COOH \end{array}, \quad \text{malic acid} \quad \begin{array}{c} HO \\ H \end{array} \!\!\! >\!\!C\!\!<\!\!\! \begin{array}{c} COOH \\ CH_2COOH. \end{array}$$

The optical activity remains through all reactions which do not alter the asymmetry, but if, as shown in the above formulas, only one OH group is replaced by an H atom, the optical activity vanishes with the asymmetry.

On the other hand, however, the presence of an asymmetric carbon atom does not always produce optical activity, since every optically active compound is also known in an optically inactive modification which can be split up into two optically active modifications. In the case of compounds containing two asymmetric carbon atoms it is moreover possible to obtain still another inactive modification which cannot be further resolved into two active forms (see Stereoisomerism).

The inactive resolvable modification can be either a mechanical mixture of equal numbers of molecules of the dextro- and lævorotatory variety, which then, with the exception of the optical activity, retains all of the characteristics of its components, or it is a true compound of the dextro- and lævorotatory modifications which differs from the two active forms in its melting-point, boiling-point, etc., and is called a racemic modification (since it was first observed in the case of the tartaric acids). See further under Isomerism, Part III.

Compounds with asymmetric carbon atoms can be obtained only by separating them from natural products or by fermentation processes, or by preparing them from other optically active substances. From inactive compounds by synthetical processes inactive modifications are always obtained, and these must first be split up in order to become optically active. The separation is carried out with the assistance of certain bacteria which attack one modification but leave the other, or by the preparation of compounds with certain optically active substances like strychnine, brucine, morphine, and cinchonine, when on evaporation the compounds of one modification crystallize out first (see further under Aspartic Acid, Malic Acid, Lactic Acid, Tartaric Acid, Asparagine, Artificial Sugars, etc.).

On the other hand many optically active compounds can be converted into inactive ones if they are heated alone or with water, or when certain substances are added to their solutions. Moreover, many dextrorotatory substances can be transformed into lævorotatory, and *vice versa*, by heating them with water, pyridine, or quinoline to 141–170°.

The magnitude of the rotation of a compound (the angle of rotation) has a constant value for every substance and is therefore of valuable assistance for the identification of a substance and for determining its purity. It depends, however, on the temperature, on the thickness, of the layer through which the light passes, and on the wave length of the light itself.

The specific rotatory power (α) is the rotation of the plane of polarization produced by a layer of liquid 1 decimeter in thickness which contains 1 gram of the substance in 1 cubic centimeter at a temperature of 20°. The light used is the yellow sodium light having a wave length corresponding to the D line of the solar spectrum. This is denoted thus: $(\alpha)_D$. If the length of the tube containing the solution is l, if α is the angle of rotation measured for a layer 1 decimeter in thickness, and if d is the specific gravity of the liquid, then for liquids $(\alpha)_D = \dfrac{\alpha}{l \cdot d}$, and for solutions containing p grams of substance in 100 grams of solution $(\alpha)_D = \dfrac{\alpha \cdot 100}{l \cdot p \cdot d}$ By the use of the latter formula it is possible to calculate how much of an optically active substance is contained in 100 grams of a solution, if all the other quantities are known, since $p = \dfrac{\alpha \cdot 100}{l \cdot d \cdot (\alpha)_D}$ (quantitative analysis by polarization).

3. Gases.

Matter in the gaseous state is capable of completely filling any given space. In gases the molecules do not exert any mutual attraction on one another, but move about in straight lines until they encounter one another or are stopped by some other resistance, whereupon they again proceed in a straight line in some other direction. They thus distribute themselves about in all directions unless they are prevented by some impenetrable barrier. When such an obstacle is encountered, the walls of the containing vessel, for example, they exert a pressure upon this which increases with the number of rebounding molecules as well as with their mass and velocity. This is known as the expansive force or tension of the gas.

The theory which explains the behavior of gases and their pressure on the walls of the containing vessel on the assumption of a movement of the molecules is called the kinetic theory of gases ($\kappa\iota\nu\eta\sigma\iota\varsigma$, *motion*). Gases which can be liquefied by cooling to the ordinary temperature of the air are called vapors (see below). The name vapor is often applied in common usage to colorless aeriform bodies which are visible and which contain innumerable little drops of liquid carried along in the process of evaporation. In scientific usage this meaning of the word is excluded. The collecting of gases is carried out, according to their solubility, over water, salt solution, or mercury (p. 51), and with specifically heavy gases by the direct displacement of air. Because of the evaporation of the water, gases evolved from aqueous solutions or collected over water are always moist; in order to dry them they are conducted through tubes filled with substances which absorb water vapor (alkali hydroxides, calcium oxide, calcium chloride, sulphuric acid, phosphorus pentoxide, etc.).

a. Liquefaction.

All gases can be liquefied by cooling, although the necessary temperature is often more than a hundred degrees below the zero-point. All gases can also be liquefied by increasing the pressure, but only when at the same time they are cooled to a certain definite temperature, which for every gas is different. That temperature above which a gas, even under the greatest pressure, can no longer be liquefied is called its *critical temperature*. The pressure exerted by a gas at its critical temperature is called its critical pressure. In order to liquefy a gas the process must therefore be carried out at a temperature somewhat below its critical temperature. At the critical temperature and the critical pressure the volume (the critical volume) of a gas is the same as that of an exactly equal weight of the liquefied gas. Through the critical temperature it is possible to sharply define gas and vapor. Vapor is a gas at any temperature below its critical temperature, i.e., vapors can be liquefied by increasing their pressure, while gases can be liquefied only by an increase in pressure accompanied by a simultaneous decrease in temperature. It is, however, under certain conditions possible for a vapor to remain unaltered at pressures under which it would ordinarily liquefy (supercooled vapors, analogy to supercooled liquids, p. 35). The critical temperature (Γ) and the critical pressure (P) of a number of gases are given below:

	T.	P.		T.	P.
Hydrogen........	−220°	15 atm.	Chlorine........	+146°	94 atm.
Oxygen..........	−118	50 "	Carbon monoxide	−140	35 "
Nitrogen.........	−146	36 "	Carbon dioxide...	+ 31	77 "
Air..............	−140	39 "	Ammonia	+130	115 "
Nitrous oxide.....	+ 39	175 "	Hydrogen chloride	+ 53	86 "
Nitric oxide	− 93	71 "	Hydrogen sulphide	+100	92 "
Argon............	−121	51 "	Sulphur dioxide ..	+155	79 "

If, by removing the pressure to which it is exposed, a liquefied gas is allowed to rapidly evaporate, the gas evolved will remove from its surroundings a quantity of heat equal to that given out by it on liquefaction (heat of vaporization). As a result of this the liquefied gas still remaining is so strongly cooled that it generally solidifies.

The liquefaction of the difficultly liquefiable gases is carried out on a large scale by the use of Linde's regenerative apparatus, which employs the strong cooling resulting from the expansion of compressed gases.

The apparatus consists of a pump which draws the gas from the generating apparatus and then compresses it to the required pressure. From the pump the compressed and therefore heated gas passes through an iron pipe surrounded by cold water. The continuation of this pipe is a spiral tube several hundred meters in length surrounded by a larger iron pipe. The compressed gas flows through the inner tube to a vessel at the end, where it is allowed to expand suddenly, and then cooled by expansion it flows back again through the outer pipe to the compressor. By this process the gas entering through the inner tube is cooled, the cycle beginning anew at the pump, until finally after repeated cooling and expansion in the apparatus the temperature falls to the critical temperature and liquefaction commences.

Many gases are supplied to the trade strongly compressed or liquefied in wrought-iron cylinders, which under these conditions are attacked very little or not at all by the gases within them.

b. Volume Relations.

The volumes of all gases at any given temperature vary inversely with the pressures to which they are subjected, and with unvarying pressure undergo an expansion of $\frac{1}{273} = 0.003665$ of their volume at 0° for every increase in temperature of 1° and a corresponding contraction for every decrease of 1° in temperature. For the explanation of this behavior by Avogadro's hypothesis see p. 16.

Gases follow the laws of pressure and warming only when in a dilute condition; strongly compressed or cooled gases show more or less marked deviation from the laws, according to whether by pressure or cooling they can be more or less readily converted into liquids; that is, they show these deviations when near to their transformation into the liquid state.

These variations lead to the conclusion that as the molecules come closer together their own volumes become of greater moment and the attraction of one for the other (cohesion) can be no longer neglected. As a result of this mutual attraction the pressure necessary to produce a given compression is lessened, and this reduction of the pressure is found to be directly proportional to the square of the density, and inversely proportional to the square of the volume (Van der Waals' theory).

In the ordinary chemical operations gases are not weighed but, since it is more convenient, their volume is measured, and from this their weight is calculated. Since the volume of a gas is dependent on the atmospheric pressure and temperature (p. 15, 1), as well as on the water vapor which it carries (see Water), every measured gas volume (V) must be reduced to the normal volume (V_0) which the dried gas would occupy at 0° and a barometric pressure of 760 mm. This normal gas volume is often merely a mathematical assumption, since many substances can no longer exist as gases at 0°. The reduction is carried out by means of the following formula: $V_0 = \dfrac{V(B-W)}{760(1 + 0.003665T)}$, in which 0.003665 is the coefficient of expansion of gases (p. 15, 1), T the observed temperature, and B the observed pressure (height of barometer) in millimeters. Since the removal of water vapor is inconvenient, the gas to be measured is brought in contact with water in order that it may become completely saturated with water vapor, and then from the observed pressure in millimeters (B) a quantity is subtracted equal in millimeters to the tension of aqueous vapor at the temperature of observation.

The most important quantities, with relation to the volume of a gas, can be calculated from the knowledge of its molecular weight and the weight of a liter of oxygen ($= 1.429$ grams).

The absolute weight of a liter of any gas is found by multiplying its molecular weight by 0.04466, since

Mol. wt. of oxygen 32	:	Mol. wt. of the given gas M	::	Wt. of liter of oxygen 1.429	:	Wt. of liter of given gas x

$x = \dfrac{M \cdot 1.429}{32}$, therefore $x = 0.04466$. For example, the weight of a liter of ammonia gas, NH_3, is $17.07 \times 0.04466 = 0.76$ gram; a liter of hydrogen gas, H_2, is $2.02 \times 0.04466 = 0.09$ gram; a liter of air (calculated molecular weight $= 28.95$, p. 44) is $28.95 \times 0.04466 = 1.293$ grams.

The calculated weights per liter generally differ somewhat from those determined by direct weighing, since the gases do not exactly follow the laws on which the calculation is based.

The gram-volume or specific volume of gases, namely, the volume in cubic centimeters occupied by 1 gram of the given gas, is found by dividing 1000 by the weight of 1 liter of the gas. For example, the gram-volume of ammonia $NH_3 = 1316$ c.c., since 0.76 g. $NH_3 : 1.0$ g. $:: 1000$ c.c.$:x$ $(x = 1316)$.

In order to obtain the molecular volume or mol.-volume of a gas, i.e., the volume in cubic centimeters occupied by one gram-molecule, the molecular weight is multiplied by the gram-volume, or the molecular wight is divided by the weight of 1 liter. For example, the molecular volume of ammonia $NH_3 = 22400$ c.c., since 1 g. $NH_3 : 17.07$ g. $NH_3 :: 1316$ c.c.$:x$, or 0.76 g. $NH_3 : 17.07$ g. $NH_3 :: 1000$ c.c.$:x$ $(x = 22400$ c.c.$)$.

The molecular volume of all gases is 22400 c.c. $= 22.4$ liters, which is a necessary consequence of Avogadro's hypothesis.

A simple calculation of the quantities which are related to the gas volume is possible by the use of the constant 22.4 in combination with the molecular weight. For example, the weight of a liter of ammonia is given by the proportion 22400 c.c. : 1000 c.c. :: 17.07 g. $NH_3 : x$, the gram-volume of ammonia by the proportion 17.07 g. $NH_3 : 1$ g. $NH_3 :: 22400 : x$.

c. Specific Gravity (Density).

The specific gravity or density of a gas is a number which expresses how many times heavier or lighter a gas is than an equal volume of a gas taken as unity at the same pressure and temperature. The comparison between equal volumes of different gases is always made at 760 mm. pressure and 0° temperature (p 42).

As unit for the determination of the specific gravity of gases the weight of an equal volume of air was first employed, but since air is a mixture of different gases, the unit next chosen was hydrogen. Since, however, the latter gas can be purified only with difficulty, and since, moreover, it is the lightest of all gases and therefore difficult to weigh (P. 17), oxygen has now been taken as the unit of specific gravity, just as it is used as the basis of atomic and molecular weights. In order to express the intimate relation between the specific gravity and the molecular weight of gases, the unit chosen is not oxygen = 1, but oxygen = 32 (= the molecular weight of oxygen). This offers the advantage that the specific gravity and the molecular weight are then identical, since the molecular weight is likewise based on the assumption of oxygen = 32 as unit (p. 18).

The specific gravity of a gas with respect to another gas taken as unit is found by dividing its molecular weight by that of the unit gas. For example, it is found that the specific gravity of ammonia with respect to hydrogen as unit $\frac{17.07}{2.02} = 8.53$, since

Mol. wt. of hydrogen 2.02	:	Mol. wt. of ammonia 17.07	::	Sp. gr. of hydrogen 1	:	Sp. gr. of ammonia x ($x=8.53$).

It is often important to know the specific gravity of a gas with respect to air. The gas density of air corresponds to the molecular weight 28.95, since

Wt. of 1 liter of oxygen 1.429	:	Wt. of 1 liter of air 1.293	::	Mol. wt. of oxygen 32	:	Mol. wt. of air x ($x=23.95$).

The specific gravity of a gas with respect to air=1 is therefore found by dividing the molecular weight of the given gas by 28.95; for example, in the case of ammonia $\frac{17.07}{28.95} = 0.59$.

But the specific gravity of a gas can also be found by dividing the weight of 1 liter of the gas by the weight of 1 liter of the gas taken as unit; therefore oxygen is $\frac{1.429}{0.09} = 15.9$ times and air $\frac{1.293}{0.09} = 14.4$ times heavier than hydrogen; and further, oxygen is $\frac{1.429}{1.293} = 1.105$ times and ammonia $\frac{0.76}{1.293} = 0.59$ times heavier than air. The specific gravity of a gas with respect to water is one one-thousandth of the weight of a liter of the gas, namely, for oxygen $\frac{1.429}{1000}$, and for ammonia $\frac{0.76}{1000}$, since

Wt. of 1 liter of water 1000	:	Wt. of 1 liter of ammonia 0.76	::	Sp. gr. of water 1	:	Sp. gr. of ammonia x.

d. Spectra and Spectrum Analysis.

α. Emission Spectra.

Glowing solids and liquids (e.g., calcium light, Welsbach light, the light of ordinary illuminants which contain glowing carbon, incandescent platinum, etc.) send out rays of light of all refractivities, and therefore when their light is dispersed and split up by a prism a continuous spectrum, which is not dependent on the nature of the material, is obtained.

Incandescent gases and vapors emit but few light rays and these of definite refractivity, and give therefore, when their light is split up by a prism, a partial spectrum (emission spectrum), consisting of bright, vertical lines or bands, which depends on the chemical nature of the given substance and therefore serves for its identification (spectrum analysis). Each of these lines or bands corresponds to rays of light of definite wave length, and is therefore differently refracted from the other rays of light. For this reason the lines in the spectrum appear in the same place and no one line can cover another, so that any one vaporized substance can be identified in the presence of any number of others.

Every element has a characteristic spectrum, and likewise every compound if it is stable at the temperature at which the incandescent gas is formed. Unstable compounds give the spectra of their decom-position products. Compounds usually show a banded spectrum, free elements one consisting of lines.

The apparatus which serves for observing the spectrum and for measur-ing the position of the lines and bands is called a spectroscope, and by its use quantities of substances can be detected which are far too small to be weighed on a balance.

The observation of new lines, not previously noticed, can lead to the detection of new elements, as was the case with cæsium, rubidium, thallium, indium, gallium, germanium, etc.

In order to convert substances into the gaseous state for the purposes of spectrum analysis, they are brought into a hot, non-luminous flame. In the case of compounds of the metals, the spectrum obtained is usually that of the free metal, since the compounds are decomposed by the flame. When the temperature of the flame is insufficient to vaporize the sub-stance, then in the case of metals electric sparks are passed between thin rods composed of the metal. With compounds of the metals the electric sparks are passed between a platinum point and the compound. Gases are brought to incandescence by enclosing them in a highly rare-fied state in glass tubes through which the current from an induction coil is conducted. The light emitted by a flame colored with sodium or one of its compounds gives a spectrum consisting of a single yellow line, potassium and its compounds give a spectrum consisting of a red and a violet line, etc.

β. Absorption Spectra.

If light giving a continuous spectrum is passed through colored trans-parent solids or liquids (for example, colored glass plates or colored liquids) and if a spectrum is formed from this light with the help of a prism, then only that part of the spectrum appears which has the same color as the substance penetrated, while all other parts of the spectrum are absorbed by the passage of the light through the colored substance.

Many colored solids, liquids, and gases, when white light is passed through them, do not absorb definite colors, but only a part of them, and therefore give a discontinuous spectrum (absorption spectrum) con-taining dark lines or bands. The position of the absorption bands thus obtained is an invariable characteristic of each of the given substances and therefore serves to identify a given substance and distinguish it from others.

Reversal of the Spectrum. If the spectrum of white light which has been passed through a flame containing sodium vapor is examined, a dark line will be noticed in a position exactly corresponding to that of the yellow sodium line; if the white light has passed through potassium vapor, then two dark lines appear in the positions of the red and violet potassium lines; and in a similar manner the reversed spectra of all the elements can he obtained. The cause of this reversal of the spectra is given by Kirchhoff's law: A gas or vaporized substance absorbs all light rays having the same periods of vibration as those which the substance itself sends out, while it is transparent to all other light rays, so that under the given conditions the absorption spectrum of a substance corresponds to its emission spectrum. The dark lines which appear in the spectra of the self-luminous heavenly bodies are explained by Kirchhoff's law, if it be assumed that these bodies consist of an incandescent liquid or solid nucleus surrounded by glowing gases or vapors. By comparing the absorption spectra of the heavenly bodies with the emission spectra of the elements it is possible to determine what elements exist in the atmospheres of the different heavenly bodies. If the direct light sent out from the sun is cut off, as is the case in a total eclipse, then the only spectrum obtained is that of the glowing gases surrounding the sun, which then consists of bright lines corresponding to the different elements constituting the solar atmosphere.

e. *Refraction and Rotation of Light.*

This is described under liquids.

4. Physical Mixtures.

Physical mixture is the name applied to a physically and chemically homogeneous complex of different substances which cannot be separated into its components by purely mechanical means. Physical mixtures, in contrast to chemical compounds as well as mechanical mixtures, have an uncertain composition and are produced by a uniform mixing together of the molecules of the different substances constituting the mixture. The separation of their constituents is therefore more or less difficult, in contrast to the mechanical mixtures, whose constituents can readily be separated.

Physical mixtures, without reference to their actual state of aggregation, are considered as solutions, and are accordingly distinguished as gaseous, liquid, and solid solutions. They occupy an intermediate position between mechanical mixtures and chemical compounds.

a. *Diffusion and Osmosis.*

When two liquids, solutions, or gases are brought into contact a gradual, spontaneous mixing takes place until the composition of both is the same. This phenomenon is known as diffusion.

While the diffusion of liquids is determined by their nature, all gases when brought into contact can mix in every proportion without relation to the specific gravity of the gases. If, for example, a vessel containing carbon dioxide is covered by one containing hydrogen, although the carbon dioxide is twenty-two times heavier than the hydrogen it will work upward through the hydrogen and the latter will work downward through it, until the mixture of the two becomes perfectly uniform and homogeneous. The rate of diffusion of a gas is inversely proportional to the square root of its density, so that, for example, the rate of diffusion of oxygen, which is sixteen times as dense as hydrogen, is only one-fourth of that of hydrogen.

If a heavier liquid is covered with a lighter one with which it is miscible, or if a solution is covered with the solvent, then the lighter liquid will pass into the heavier and the dissolved substance will pass into the lighter solvent, until the liquid, throughout every part, has the same composition. The dissolved substance therefore acts with respect to the pure solvent exactly as a gas behaves with respect to empty space or to another gas, namely, both strive to fill the space offered to them (gases the surrounding walls, solutions the surrounding liquid).

Similarly, if two miscible solutions are placed in direct contact with one another, the substance in one solution diffuses into the solution of the other substance until the dissolved substances are uniformly distributed through all parts of the liquid. The rate of diffusion of dissolved substances is dependent on the nature of the substance and the nature of the solvent, as well as on the concentration and temperature of the latter.

In the case of diffusible liquids, as well as in the case of all gases, the diffusion is not prevented if the substances are separated by a porous partition.

The diffusion of gases is independent of the nature of the separating diaphragm; the diffusion of liquids and solutions through a porous partition is called osmosis (ὠσμός, pushing) and is dependent on the nature of the partition and of the liquid. Those substances which can undergo osmosis in solution, among which are included many acids and all crystallizable substances, are called crystalloids in contrast to those substances which apparently cannot undergo osmosis, the so-called colloids or gelatinoids. The substances included in the latter class all belong to the amorphous substances (p. 34) and dissolve in water only to form the so-called colloidal solutions (p. 53); for example, glue (colla, gelatine), gums, albumins, dextrins, the hydroxides of the heavy metals, as well as the aqueous solutions of certain heavy metals (p. 53). The osmosis of the colloids, owing to their high molecular weights, takes place so slowly that it does not come into practical consideration.

Osmosis can serve for the separation of aqueous solutions of the crystalloids from the colloids; this process is called *dialysis* and is carried out with an apparatus known as a dialyser.

This consists of a vessel open at top and bottom, having a bottom formed by a permeable membrane (an animal bladder or parchment-paper), and dipping with its lower part in a large vessel filled with water. If an aqueous solution of a colloid and a crystalloid is placed in the vessel with the membrane, the crystalloid diffuses into the outer water and the outer water diffuses into the inner vessel until the water within and without is of the same strength with respect to the crystalloid. If the dialyser is now brought into a vessel containing fresh water, the process begins anew, and after repeated renewal of the external water the dialyser finally contains an aqueous solution of the colloid only.

b. Osmotic Pressure.

Just as the molecules of a gas have a tendency to disperse and thereby exert a pressure on the walls of the containing vessel, so the molecules of a dissolved substance behave with respect to their solvent.

In order to determine the pressure of dissolved substances on their solvents, it is necessary to employ a dialyser with a special membrane which is permeable to the solvent but not to the dissolved substance.

Such a membrane is called a semipermeable membrane, and consists of a porous earthenware cylinder containing in its pores precipitated copper ferrocyanide. Other semipermeable membranes are also found in the vegetable and animal world.

If a solution is placed in a dialyser having a semipermeable membrane, and the whole is placed in a vessel containing the pure solvent, then the molecules of the dissolved substance, in their futile efforts to diffuse into the solvent, will exert on the membrane a pressure which is called the osmotic pressure. If the semipermeable membrane is considered as a piston which can be moved in a vessel, then the piston will be lifted by the osmotic pressure and the pure solvent will penetrate into the solution until both liquids have the same composition. If, however, the piston is so weighted that it cannot be moved, then this counterbalancing of the piston is equivalent to the osmotic pressure exerted by the molecules of the dissolved substance.

In very dilute solutions the molecules of the dissolved substances are so widely separated from one another that only those properties which depend upon the number of the molecules are exhibited. Therefore dilute solutions completely correspond with gases in their general behavior. The osmotic pressure of dilute solutions follows the same laws as the pressure of gases.

As shown on p. 43, 1 gram-molecule of every gas at 0° and 1 atmosphere pressure occupies a volume of 22.4 liters. If the gas is now compressed to a volume of 1 liter, it would exert a pressure of 22.4 atmospheres, and exactly the same osmotic pressure is produced when 1 gram-molecule of any substance is contained in 1 liter of solution. If the weight of a substance and its osmotic pressure are known, then the quantity by weight which dissolved in 1 liter would give an osmotic pressure of 22.4 atmospheres can be readily calculated. On account of various difficulties the osmotic pressure is generally determined from other properties of solutions which stand in close relation to the osmotic pressure, namely, from the vapor pressure, boiling-point, and freezing-point of solutions. This is of great importance, not only for the simplification of molecular weight determinations, but also in medical chemistry for determining the osmotic pressure of the different animal fluids, which cannot be carried out directly. If a solution evaporates, only the solvent is given off, and if a solution freezes, then, at first, only the solvent separates, while as a result the dissolved substance in the liquid remaining is compressed into a smaller volume. Just as the compression of gases increases their pressure, so the concentrating of the solution increases here the osmotic pressure, whereby the vaporization and freezing of the solvent becomes more difficult, and the vapor pressure and freezing-point are correspondingly lowered.

Since the boiling-point is the temperature at which the vapor pressure of a liquid is in equilibrium with the atmospheric pressure, therefore when the vapor pressure of a liquid is lowered by dissolving some substance in it, it requires the addition of a greater quantity of heat, namely, a higher temperature, in order that its vapor pressure may again equal the atmospheric pressure.

From this follows the correspondence between equimolecular solutions with respect to osmotic pressure, vapor pressure, boiling-point, and freezing-point as mentioned on p. 19.

Just as certain gases show exceptions to the gas laws (see Dissociation), so there are a series of compounds, the so-called electrolytes, which give a higher osmotic pressure than corresponds to their molecular weights (see Electrolytic Dissociation).

c. Gaseous Solutions.

In contrast to other substances, gases can mix with one another, i.e., dissolve in one another, in all proportions. In gas mixtures every gas behaves with respect to its own particular properties just as if it alone were present and retains its individual characteristics, namely, its expansive force, refraction of light, solubility in liquids, specific heat, etc. For example, the pressure exerted by a quantity of gas occupying a given volume is equal to the sum of the separate pressures which the different gases would exert if they alone occupied the given volume (Dalton's law *). The volume of a gas mixture is equal to the sum of the volumes of its constituents.

Liquids and solids dissolve in gases only when they can evaporate, in which case the evaporation, following Dalton's law, takes place just as if an air-free space were present.

d. Solid Solutions.

Solutions of solid substances in other solid substances are represented by the alloy, which are homogeneous (mostly solid) mixtures of metals in indefinite proportions by weight. They are formed by melting the metals together or by mixing the fused metals. In the case of most of the metals, the mixing can take place only within certain limited proportions by weight.

Under solid solutions are also included the isomorphic mixtures (mixed crystals), namely, the individual crystals which separate out from mixtures of solutions of isomorphic salts and to whose formation the given salts can contribute in indefinitely varying proportions (p. 23).

Examples of solutions of liquids in solids are furnished by the zeolites, certain hydrated silicate minerals from which the water can be removed in indefinite quantities or be replaced by other liquids without producing an alteration in their external appearance.

Solutions of gases in solids can be formed by many porous substances, such as charcoal, or certain finely pulverized metals, as platinum, palladium, and gold, since these substances can condense gases in their pores. Because of its resemblance to absorption (the condensation of gases in liquids) this phenomenon is known as absorption. The absorption phenomena also include the general property of all solid substances to condense gases on their surface, as well as the property of many porous substances (i.e., substances with large surfaces), for example charcoal, to remove dissolved substances from their solutions.

When solid mixtures are melted either the whole mixture or only one constituent becomes liquid. The melting-point (solidification-point) is lower than that of the most difficultly fusible, often indeed lower than that of the most readily fusible, constituent, since the melting-

---

* Also called Henry's law.

point of a substance is lowered (p. 48) when another substance is dissolved in it. The lowering of the freezing-point is a distinctive feature of the alloys.

If a solid mixture is repeatedly melted and then allowed to solidify, it often happens that one of the constituents partially separates, but finally, after the removal of the separated material, a mixture of constant composition is obtained. This mixture always has a lower melting-point than the original mixture, and on further fusion and solidification this melting-point remains constant and the mixture does not further alter in composition. Such a mixture is called a *eutectic* mixture or cryohydrate, and the temperature at which it commences to separate is called a *eutectic point*.

e. Liquid Solutions.

These are the solutions in a restricted sense, and are formed when liquids combine with either gases or other liquids or solids to form a new, entirely homogeneous liquid having new properties.

The solubility is different and usually limited for different soluble substances in the same solvent and for the same soluble substance in different solvents. It is dependent on the temperature and, especially in the case of gases, on the pressure. Indeed the same chemical compound can have a different solubility according to its water of crystallization. The greater number of inorganic compounds are insoluble in alcohol, while carbon compounds are mostly soluble in alcohol and ether, and their solubility decreases with an increasing content of carbon.

Solutions closely resemble chemical compounds; chemical compounds of the dissolved substance with the solvent often separate from them; for example, many anhydrous salts separate from their aqueous solutions with water of crystallization.

A more or less complete separation of the constituents of liquid solutions can be effected by distillation.

When a liquid is converted into vapor and the vapor again condensed by cooling the process is called *distillation*. In this process dissolved gases are given off entirely or partially (p. 53) and solid substances are left behind.

Rectification is the repeated distillation of a liquid for the purpose of completely freeing it from admixtures.

Dry distillation is the decomposition of non-volatile organic compounds by heat out of contact with the air, as a result of which solid, liquid, and gaseous decomposition products are obtained.

Fractional distillation, see p. 53.

A solution which contains so much of the dissolved substance that, under the given conditions, the solvent can take up no further quantity is called a *saturated solution*.

α. Solutions of Gases in Liquids.

The solution of gases in liquids is also called absorption, and since the gas is liquefied (p. 41) the process of absorption is always accompanied by the evolution of heat. The most difficultly liquefiable gases are therefore the least soluble. Every gas dissolves in a liquid already saturated with another gas exactly as if the other gas were not present. The quantities of dissolved gases are measured according to their volumes at 0° and 1 atmosphere pressure. The power of liquids to dissolve gases always decreases with an increase in temperature, but also on the solidi-fication of the solvent the dissolved gases are set free. Liquids which already contain dissolved liquids or solids are able to dissolve gases less than pure liquids (collecting of gases soluble in water over hot water or salt solution).

The quantities by weight of gases which dissolve are directly propor-tional to the pressures of the gases during the solution process (Law of Henry-Dalton, not holding in the case of very readily soluble gases, since here chemical combination usually occurs), so that 1 volume of a liquid dissolves at 2 atmospheres pressure double the quantity by weight (but the same volume, see Boyle's law, p. 15) of gas that would be dis-solved at 1 atmosphere pressure. If the pressure is reduced, then a corresponding quantity of the gas is set free (foaming of champagne, soda-water, etc.). A solution of ammonia gas in water will give off the former until its pressure in the space above the liquid has become sufficiently great. If sulphuric acid is now introduced into this space, it will combine with all the ammonia gas present, so that fresh quantities of ammonia will be evolved by the solution until all the ammonia which it contains will have been removed.

Notwithstanding the fact that the total pressure which is exerted by a mixture of gases is equal to the sum of its partial pressures—namely, the pressures which each of the gases alone exert—from a gaseous mix-ture only so much of each gas will be absorbed as corresponds to its partial pressure. If, therefore, another gas is conducted through or over a solution of a gas, all of the dissolved gas which has separated will be removed, as a result of which its partial pressure will be constantly re-duced and the gas at first dissolved in the liquid will finally be entirely removed from it. If the solution of a gas is boiled, the vapor of the sol-vent sweeps off the quantities of the dissolved gas which have separated until the liquid no longer contains any of the gas dissolved in it.

β. Solutions of Solids in Liquids.

Such solutions are usually formed with the absorption of heat, since the molecules of the dissolved substance are driven into the solvent with a certain pressure, the solution pressure, and accordingly energy in the form of heat is withdrawn from the surroundings. The case is analogous to the evaporation of a liquid or a solid where the molecules enter into the space offered to them with a certain pressure, the vapor pressure. Just as a liquid evaporates until the pressure of the vapor formed is equal to the vapor tension of the liquid, so a solid substance passes into solution in a liquid until the osmotic pressure of the solution comes into equilibrium with the solution pressure of the solid substance. If, how-ever, a salt combines with a part of the solvent to form a chemical com-pound, then heat will be evolved, since the quantity of heat which is set free by the combination is greater than that required for liquefaction. For example, anhydrous calcium chloride dissolves in water with the evolution of heat, since it combines chemically with a part of the water. On the other hand calcium chloride containing water of crystallization dissolves in water with cooling.

The absorption of heat is stronger the more rapidly the solution takes place; and further in the case of aqueous solutions when, instead of being dissolved in water, the substance is mixed with ice or snow, since these are then liquefied by the effort made by the soluble substance to pass into solution, and their heat of fusion (p. 33) produces a further cooling. The action of freezing mixtures depends on this principle; for example. in preparing a saturated aqueous solution of ammonium chloride the temperature sinks about 13°, of calcium chloride about 23°, of ammonium sulphocyanide about 31°; on mixing 1 part of common salt with 2 parts of snow the temperature sinks to −20°, on mixing 2 parts of calcium chloride and 1 part of snow to −42°.

If heat is absorbed on dissolving a solid, then generally the addition of heat increases the solubility of the substance. In the case of many substances, however, the increase in the solubility is only very slight (e.g., in the case of common salt), in the case of many the solubility actually decreases (e.g., gypsum, calcium hydroxide), and in some cases the solubility increases up to a certain temperature and then decreases (Glauber's salt).

Hot saturated solutions of salts, when they are allowed to cool without being disturbed and are protected from dust, do not separate out any of the dissolved substance, although after cooling they contain much more of the dissolved substance than is possible under ordinary conditions. Such solutions are called *supersaturated solutions* (analogy to superfused solids, p. 35).

If into such a solution there is introduced even an infinitely small quantity of the salt which is contained in the solution or of some other salt which is isomorphic with this, then the excess of the dissolved salt separates until a saturated solution is formed (see Sodium Sulphate). The separation of the excess of the dissolved salt can also be caused by dust falling into the solution or by stirring the solution with a glass rod, but this only happens when the objects introduced into the solution are contaminated by traces of the salt which is contained in the solution.

Saturated solutions are in stable equilibrium, supersaturated solutions are in unstable equilibrium. Mother-liquor is the name given to a salt solution from which by evaporation or cooling a part of the dissolved salt has been separated, while a part of the salt, and particularly all the more readily soluble salts, still remains dissolved in it.

The freezing-point of liquids is lowered when solid substances are dissolved in them, and the boiling-point is raised. The laws governing this, which can serve for the determination of the molecular weights and their relation to the osmotic pressure of dilute solutions, have been already mentioned on p. 48.

These laws apply only approximately in the case of concentrated solutions, just as the gas laws no longer accurately apply to gases when they approach their points of liquefaction.

γ. *Solutions of Liquids in other Liquids.*

These solutions usually take place with the evolution of heat. The limit of the mutual solubilities of two liquids in the case of many liquids is greater with increasing temperature, in the case of many others is less. Many liquids mix in all proportions; for example, alcohol with water. The alteration of the composition of mixed liquids by distillation (p. 50) always proceeds in such a manner that the vapor pressure decreases, namely, the boiling-point rises.

Two liquids insoluble in one another arrange themselves, according to their specific gravities, one above the other; for example, water and benzene. If a mixture of volatile substances of this character is distilled, the mixture which passes over has a constant composition so long as both constituents are present in the distilling vessel. The boiling-point of the mixture is lower than that of either of the constituents.

Two liquids, of which each is only slightly soluble in the other, likewise give a mixture containing two layers each of which consists chiefly of one of the liquids. On distillation the same relations pertain as in the case of non-miscible liquids; the boiling-point of the mixture can be lower than the boiling-point of the most readily volatile liquid; it can also be higher, but not higher than the boiling-point of the most difficultly volatile constituent.

Two liquids which are miscible in all proportions show on distillation the following complex relations:

The boiling-point of the mixture can be lower than that of the more readily volatile constituent; on distillation a mixture having a constant composition and a low and constant boiling-point passes over, so that the residue finally consists of that constituent which was present in excess.

The boiling-point of the mixture can also be higher than that of either of the constituents; on distillation at first that constituent which is present in excess passes over until the remainder has reached such a composition that its boiling-point rises, after which a constant mixture of constant higher boiling-point distils over. Such constant boiling mixtures occur in the case of the aqueous solutions of volatile acids (see hydrochloric, sulphuric, and nitric acids). If, for example, a solution of nitric acid be heated, then in the case of a dilute solution water passes over, in the case of a concentrated solution nitric acid passes over, until a temperature of 121° is reached, after which a constant mixture of 68 parts nitric acid and 32 parts water distils over. Such mixtures were formerly considered to be chemical compounds, but their content of water does not correspond to the law of multiple proportions and their composition alters with the pressure.

The boiling-point of the mixtures can be higher than that of the more readily volatile, but lower than that of the more difficultly volatile, constituent, so that the boiling-points of all possible mixtures therefore lie between those of the pure constituents. Therefore in the process of distillation, since the boiling-point always rises, that portion passing over first will contain more of the lower boiling constituent, that part going over last will contain more of the higher boiling constituent, so that by submitting the first portion to repeated distillation the separate constituents can be separated more and more from one another (fractional distillation).

δ. Colloidal Solutions.

Of the colloids, i.e., substances apparently incapable of undergoing osmosis, only a few are directly soluble in water. On the other hand, however, it is possible to dissolve the remainder of them in water in the presence of salts, acids, etc., and by dialysis to convert these solutions into pure aqueous solutions, the so-called *colloidal solutions*. Gold, silver, mercury, and various other metals can form colloidal solutions when an electric arc is formed under water between electrodes of the given metal, which causes the metal to be finely pulverized and dissolved. All colloidal solutions have on the one hand the properties of true solutions, on the other hand the properties of substances very finely divided in water. Many colloidal solutions can gelatinize or coagulate on the addition of foreign substances, namely, of salts, and with many this process can take place spontaneously. Others, for example gelatine, agar-agar, etc., solidify only below certain definite temperatures and above this temperature become again liquid. The gelatinized colloidal solutions are called geles or hydrogeles; they swell up strongly in water by absorbing it, and many dissolve on the addition of much water (e.g., glue). They can dissolve crystalloids, but not colloids, and permit their diffusion (use of gelatine for bacteria culture, photographic dry plates, etc.). The liquid colloidal solutions are also called soles; certain soles of the heavy metals, for example platinum and gold, exhibit the properties of enzymes and are therefore also called inorganic enzymes. Like organic enzymes, the action of these can be weakened by certain poisons.

RELATIONS BETWEEN ATOMIC WEIGHT AND PROPERTIES OF THE ELEMENTS.

If the elements are compared with one another with respect to their properties and their compounds, it is evident that they can be divided into groups or families, whose members show great similarity to one another. Other separate members of these groups show relations to other groups and thus form connecting links between the groups. These relations of the elements are most clearly expressed when they are arranged according to the magnitude of their atomic weights, when it is found that the succeeding elements exhibit apparently irregularly increasing properties, but that, after the passage of a given period, the chemical and physical behavior of

		Group I.	Group II.	Group III.	Group IV.
Hydrogen Compounds		Unknown	MH	MH$_2$	MH$_3$
Highest Oxygen Compound		Unknown	M$_2$O	MO	M$_2$O$_3$
Period.	**Series.**				
I.	1.	Helium 4	Lithium 7	Beryllium 9˙	Boron 11
II.	2.	Neon 20	Sodium 23	Magnesium 24	Aluminium 27
III.	3.	Argon 40	Potassium 39	Calcium 40	Scandium 44
	4.	—	Copper 64	Zinc 65	Gallium 70
IV.	5.	Krypton 82	Rubidium 85	Strontium88	Yttrium 89
	6.	—	Silver 108	Cadmium 112	Indium 114
	7.	Xenon 128	Cæsium 133	Barium 137	Lanthanum, etc., 138.
V.	8.	—	—	—	Ytterbium 173
	9.	—	Gold 197	Mercury 200	Thallium 204
	10.	—			—

the following elements suggests or indeed repeats that of the elements which precede them. *The properties of the elements are periodic functions of their atomic weights* (Periodic law). If the elements are so arranged according to the magnitude of their atomic weights that those similar elements which recur after certain periods stand one below the other, then these vertical rows form the groups or families, while the horizontal rows contain the periods, that is, the elements whose atomic weights lie between the successive members of one family. The arrangement of the elements according to this system is called the periodic or natural system of the elements and is shown in the table below. In this the elements fall into five periods consisting of ten rows, and in eight groups, of which the second to the eighth each forms a side group as indicated by a horizontal displacement to the right of certain of the elements included in the vertical rows. In this way those elements of a group which show the greatest similarity to one another are brought together. The periods are distinguished as large and small periods according to the number of elements which are contained in them. The elements of every group (vertical row) show differences between the atomic weights which closely correspond to those between the atomic weights of the other groups, and the same is the case in the periods (horizontal rows).

The periodic system makes it possible to divide the elements comprehensively, to check the accuracy of the atomic weight determinations, and to predict the existence and properties of elements which are still undiscovered.

Group V.	Group VI.	Group VII.	Group VIII.
MH_4	MH_3	MH_2	MH
MO_2	M_2O_5	MO_3	M_2O_7
Carbon 12	Nitrogen 14	Oxygen 16	Fluorine 19
Silicium 28	Phosphorus 31	Sulphur 32	Chlorine 35
Titanium 48	Vanadium 51	Chromium 52	Manganese, etc., 55
Germanium 72	Arsenic 75	Selenium 79	Bromine 80
Zirconium 91	Niobium 94	Molybdenum 96	Ruthenium,etc..102
Tin 118	Antimony 120	Tellurium 128	Iodine 127
—	—	—	—
—	Tantalum 183	Tungsten 184	Osmium, etc., 191
Lead 207	Bismuth 208	—	—
Thorium 232	—	Uranium 239	—

The four elements with "etc." affixed each represent a subgroup of elements in the system having almost equal atomic weights and showing great similarity to one another. For example:

Manganese, etc. =Manganese 55, Iron 56, Cobalt 59, Nickel 58.7.
Ruthenium, etc. =Ruthenium 102, Rhodium 103, Palladium 106.
Osmium, etc. =Osmium 191, Iridium 193, Platinum 195.

The elements standing below one another in these subgroups form many similarly constituted compounds.

Lanthanum, etc. =Lanthanum 138, Cerium 140, Praseodymium 141, Neodymium 144.

Samarium 150, Gadolinium 156, Erbium 166, and Thulium 171 cannot be satisfactorily inserted in the system.

Among the physical properties of the elements which show a direct relation to the atomic weight are included the specific gravity and atomic volume (p. 36), the extensibility, fusibility, and volatility, the specific refractive power, the specific heat, the conductivity for heat and electricity, and the electrochemical character. All of these properties show a maximum or minimum in the middle of the periods (horizontal rows). For example:

Third Period.	K	Ca	V	Cr	Mn	Fe	Co	Ni
Sp. gr.	0.87	1.6	5.5	6.8	7.2	7.9	8.5	8.8
Atomic vol.	45.40	25.2	9.3	7.7	6.9	7.1	6.9	6.6
Melting-point.	62°	760°	3000°	3000°	1900°	1800°	1800°	1600°

Third Period.	Cu	Zn	Ga	Ge	As	Se	Br
Sp. gr.	8.8	7.1	5.9	5.5	5.6	4.5	2.9
Atomic vol.	7.2	9.0	11.6	13.1	13.2	17.5	26.9
Melting-point.	1100°	420°	30°	900°	500°	217°	−8°

In the case of the elements of many groups (vertical rows) the specific gravity, the fusibility, etc., increases with the atomic weight. For example:

Second Group.	Li	Na	K	Rb	Cs
Sp. gr.	0.59	0.97	0.87	1.5	1.8
Melting-point.	180°	96°	62°	38.5°	26°
Boiling-point.	900°	740°	670°	500°	270°

Third Group.	Be	Mg	Zn	Cd	Hg
Sp. gr.	1.8	1.7	7.1	8.6	13.5
Melting-point.	1000°	700°	420°	320°	−39°
Boiling-point.	1600°	1300°	950°	750°	357°

In the successive periods (horizontal rows) the elements continually assume a more metallic character; the first period contains 2 metals, the second 3, the third and fourth periods 11, the fifth only metals.

In the case of the chemical properties the dependence of these on the atomic weight is shown in the following.

The groups (vertical rows) contain the elements which show the greatest chemical similarity. The first group contains the chemically indifferent elements of unknown valence, He, Ne, A, Kr, X; the two following groups contain the elements which form the strong bases Li, Na, K, Rb, Cs, and Be, Mg, Ca, Sr, Ba; this basic character falls off in the middle groups and gradually passes over into an acid-forming character.

The valence of the elements increases up to the fourth group and then decreases proportionately.

In the case of the oxygen compounds a steady increase in valence is shown; the increase in oxygen corresponds to one-half atom from member to member. For example:

I	II	III	IV	V	VI	VII
Na_2O	MgO	Al_2O_3	SiO_2	P_2O_5	SO_3	Cl_2O_7
$\left(NaO_{\frac{1}{2}}\right)$	$\left(MgO_1\right)$	$\left(AlO_{1\frac{1}{2}}\right)$	$\left(SiO_2\right)$	$\left(PO_{2\frac{1}{2}}\right)$	$\left(SO_3\right)$	$\left(ClO_{3\frac{1}{2}}\right)$

In the case of the hydroxides the number of the hydroxide groups at first increases and then falls off, while a number of oxygen atoms and the valence steadily increase. For example:

$$Na(OH)_1.Mg(OH)_2.Al(OH)_3.Si(OH)_4.PO(OH)_3.SO_2(OH)_2.ClO_3(OH)_1.$$

The periodic system assumes the existence of elements as yet unknown, as is shown by the gaps in the table. Since the properties of these elements are influenced by their position in the system, these can be predicted with comparative accuracy; in fact such gaps have already been filled by the discovery of germanium, gallium, and scandium. Their properties have been found to completely agree with those which were predicted for them.

The periodic system serves to check the atomic weights. Since the properties of an element bear a close relation to its atomic weight, so the properties of an element, as well as its atomic weight, can be made use of in arranging the element in the periodic system. If now an element, according to its determined atomic weight, would occupy a position in the system into which it would not fit according to its properties, then it can generally be assumed that there is an error in the determination, and a redetermination of the atomic weight should be undertaken (p. 25).

The fact that the properties of the elements are functions of their atomic weights has led to the conjecture that the elements are built up from one or more fundamental substances and are therefore at present indivisible ($\overset{\prime}{\alpha}\tau o\mu o\varsigma$) only because there is no means at hand for splitting them up into smaller particles.

On the other hand it is asserted that the properties of the elements are not only influenced by their atomic weight, but also by their molecular weight and the energy relations in the molecule. Thus, for example, through the appearance of the allotropic modifications of the elements (see Ozone) a molecular substance of quite different properties can be formed by the combination of a different number of similar atoms. The present form of the periodic system cannot be final, since it exhibits many exceptions; it also offers no explanation of the fact that many elements form several series of compounds which differ from one another much more than they differ from series of compounds formed by entirely different elements occupying positions in other periods.

II. AFFINITY.

CHEMICAL AFFINITY.

The study of affinity includes the study of chemical mechanics and the transformation of chemical energy into mechanical energy (volume, surface, and motion energy), or into non-mechanical energy (heat, electric, and radiant energy).

In chemical reactions the production or absorption of heat, electricity, or light takes place, and these different sorts of energy are produced from all or a part of the chemical energy. Conversely, heat, electricity, and light are often the cause of chemical reactions, and during their progress an equivalent transformation of the given energy into chemical energy takes place.

Chemical energy is the most complex of all forms of energy and the least understood, and the only method by which it can be more closely investigated is by transforming it into other forms of energy. It is simplest to transform it into heat.

The knowledge of the actual nature of chemical affinity is at present as unfathomable as that of gravity, but great progress has been made in the study of its action, as well as its dependence on mass, temperature, and pressure. The consideration of the different forms of chemical reactions leads to the conclusion that the assumption of a force of affinity considered as a force of attraction is not only of little advantage in explaining chemical reactions, but is often directly contrary to experience.

For example, when water (H_2O) acts on glowing iron (Fe), ferro-ferric oxide (Fe_3O_4) and hydrogen (H) are formed, from which it must be assumed that the chemical affinity of oxygen is greater for iron than for hydrogen: $3Fe + 4HOH = Fe_3O_4 + 4H_2$. If, however, hydrogen is allowed to act on glowing ferro-ferric oxide, iron and water are produced, from which it must be assumed that the chemical affinity of oxygen is greater for hydrogen than for iron.

Since these assumptions cannot both be right, the theory which has led to these assumptions, namely, the theory of chemical affinity, must be wrong.

58

The idea of atoms having attractive forces has therefore been more and more abandoned, and it has become usual to consider the atoms, as well as the molecules built up from them, as *actively moving* masses, whose relations to one another are determined by the form and magnitude of the motion (kinetic nature of affinity).

CHEMICAL MECHANICS.

Chemical mechanics treats of the rate or velocity of chemical reactions (chemical kinetics or dynamics) and the equilibrium relations which are established after the progress of chemical reactions (chemical statics).

Mechanics teaches that every system of bodies strives to approach a state in which the amount of energy which can be transformed into work (the available energy) is as small as possible. Chemical reactions therefore often take place spontaneously on bringing the given elements together; for example, arsenic and chlorine combine immediately to form arsenic trichloride.

The starting of a chemical reaction often requires the action of some external influence, such as heat, electricity, or light. This involves a loosening of the chemical energy, since a part of the molecules must first be split up into atoms, or the union of the atoms in the molecule must first be severed before the chemical reaction becomes possible. If this has been started in this manner, then it proceeds of itself. In those chemical reactions which proceed with the absorption of heat, on the other hand, energy must be continually supplied, since otherwise the reaction stops.

The phenomena of chemical kinetics and also chemical statics are governed by the law of mass action (law of Guldberg and Waage), as well as by the nature of the reacting substances and the temperature. This law states *that the chemical action of a substance taking part in a chemical reaction is proportional to its active mass* (likewise the concentration of the mass, i.e., the quantity contained in unit volume; for example, the number of gram-molecules of the reacting substances contained in a liter).

The chemical transformation of many substances does not therefore proceed completely to new substances, when the existing conditions are such that the newly formed substances can likewise act on one another. For example, if the two compounds *AB* and *CD* only react to form the compounds *AD* and *BC*, then in the normal course of the reaction not only will these be formed, but the original substances *AB* and *CD* will be 'present in the system or structure (i.e., in the number of substances which take part in the reaction), since *A* has affinity for *B* as well as for *D*. Since these affinities act simultaneously, the system comes into equilibrium when certain quantities of the four possible compounds have been formed.

The law of mass action shows the signification of the concentration of substances on the progress of chemical reactions, lays the foundation for the investigation of chemical statics and dynamics, and is of wide significance, since with the help of this law, with proper consideration of the nature of the substances involved, the temperature and pressure, it is possible to deduce mathematically the relations between the quantities of the reacting substances and their action and to reach important conclusions in respect to the velocity of reactions and in respect to chemical equilibrium.

1. Chemical Statics.

Chemical statics treats of the equilibrium relations which ensue after a chemical reaction has proceeded for a certain period. Since, according to the law of mass action, the tendency with which a substance strives to undergo transformation increases with its concentration, therefore in a chemical reaction, as a result of the decrease in the original substances, their tendency to transformation becomes constantly weaker, and on the other hand, as a result of the increase in the products, their tendency to produce retransformation becomes constantly greater; finally, in this way the mutual action of the substances will cease, since now the products, in the quantities in which they have formed, will have a tendency to again reproduce the original substances. The chemical reaction therefore comes to a standstill before it is completed, that is, the system reaches a state of equilibrium.

Every such state of equilibrium is really not to be considered as a static but as a dynamic system, since the transformation of materials does not cease in the state of equilibrium, but the tendencies in both directions compensate one another, so that as a result no further alteration in the given chemical system can be detected.

The state of equilibrium can therefore be considered as that state in which the velocities of reaction on both sides of the given system have become equal, and from this it follows that in all reactions which lead to a state of equilibrium this will be reached irrespective of whether the reaction starts with the original substances or with the final products.

Those reactions which lead to the same final states, whether they start with the original substances or with the products, are called reversible reactions. They are denoted by substituting the symbol$\rightleftharpoons$ for the symbol $=$ in the equations representing the reaction.

If, for example, an alcohol is mixed with an acid, an ester and water are formed, but a part of the alcohol and the acid always remains unaltered (provided that neither the acid nor the alcohol are present in very great excess, p. 62) no matter how long the action is permitted to continue. The quantities of ester and water which will be formed are dependent on the original quantities of alcohol and acid taken, and approach a certain limit which is different for every different mixture. The same limit is reached when an ester and water are brought together in a proportion which is equivalent to that of the alcohol and acid. For example:

$$CH_3COOH + \quad C_2H_5OH \quad \rightleftharpoons CH_3COOC_2H_5 + HOH$$
Acetic acid + Ethyl alcohol$\rightleftharpoons$ Acetic ester + Water.

If finely divided, glowing iron (Fe) and water vapor (H_2O) are allowed to act on one another in an enclosed space, an equilibrium system results which always contains iron and water vapor in addition to ferroferric oxide and hydrogen: $3Fe + 4H_2O \rightleftharpoons Fe_3O_4 + 4H_2$ The same system of equilibrium is obtained when glowing ferro-ferric oxide and hydrogen are heated in a closed vessel.

However, a chemical reaction does not only proceed to a state of equilibrium, but is complete, if through the mutual action of the reacting substances there are formed insoluble or volatile substances, since in such cases the given products are removed from the sphere of action, so that the same reaction which causes the formation of the first quantity of insoluble or volatile substance can again take place until the reaction is complete.

When sodium bicarbonate ($NaHCO_3$) is heated, it is converted completely into sodium carbonate (Na_2CO_3) if the carbon dioxide (CO_2) which is formed can escape: $2NaHCO_3 = Na_2CO_3 + CO_2 + H_2O$. However, if the carbon dioxide is prevented from escaping a state of equilibrium is reached, since sodium bicarbonate will be again partially reformed:

$$2NaHCO_3 \rightleftharpoons Na_2CO_3 + CO_2 + H_2O.$$

If water vapor is allowed to act on glowing iron in an open tube so that the hydrogen formed can escape, then no state of equilibrium is reached, but the reaction is complete: $3Fe + 4H_2O = Fe_3O_4 + 4H_2$. The case is the same when hydrogen is allowed to act on glowing ferro-ferric oxide: $Fe_3O_4 + 4H_2 = 3Fe + 4HOH$, since in this case the resulting water vapor can escape (above). The quantities of solids present in a system exert no influence on the final state of equilibrium, since the alteration in concentration of solids under ordinary conditions is too insignificant; therefore for the equilibrium between iron, water, ferro-ferric oxide, and hydrogen it is immaterial how much of the two solid substances iron and ferro-ferric oxide and in what proportions they are present.

When no insoluble or volatile substances are formed it is also possible for the reaction to be complete or almost complete, if a great excess of one of the reacting substances is present.

If an acid and an alcohol are mixed there are formed water and an ester: for example, $CH_3COOH + C_2H_5OH \rightleftarrows CH_3COOC_2H_5 + HOH$. This reaction at ordinary temperatures reaches a state of equilibrium very slowly, but if the substances are heated in a closed vessel at 100° then the state of equilibrium is attained after several hours. The proportion of the quantities of the four reacting substances present at any interval until the final state is reached is such that the original substances are constantly decreasing, the products constantly increasing. If the quantity (i.e., the concentration) of one of the reacting substances is increased, then the reaction can be carried as desired in one or the other direction beyond the state of equilibrium. For example, by the action of much acid on a little alcohol, or of much alcohol on a little acid, the formation of ester is almost complete, and on the other hand by the action of much water on a little ester the latter is almost completely split up into alcohol and acid.

Most chemical reactions involve a final state of equilibrium; complete retransformations are only occasionally encountered; indeed in the latter, in most cases, the reactions come to a stop before the limit of possible transformation is reached, although the quantities of materials which remain finally unaltered are so small that they escape direct notice.

Concerning complete, non-reversible reactions, see endothermic compounds, p. 69.

At constant pressure (and also at constant volume), on an increase in temperature a shifting of the equilibrium takes place with an increase of that substance of the system whose formation takes place with the absorption of heat. At constant temperature an increase in the pressure produces a shifting of the equilibrium to that side where the reaction proceeds with a decrease in volume (principle of variable equilibrium).

In the system $CaCO_3$ (solid) $\rightleftarrows CaO$ (solid) $+ CO_2$ (gas), the pressure remaining constant, an increase in temperature causes an increased splitting up of the calcium carbonate ($CaCO_3$) into calcium oxide (CaO) and carbon dioxide (CO_2). On the other hand, an increase in the pressure at constant temperature promotes the formation of calcium carbonate until the original pressure is again established. Salts which are in contact with their saturated solutions are dissolved in greater quantity by an increase in pressure if the total volume of the salt and the quantity of water necessary to dissolve it is greater than the volume of the resulting solution.

Equilibrium of the first order, namely, that in which only one substance (constituent) is present, from the physical standpoint is that between the different states of aggregation of a substance, e.g., between water and water vapor, between ice and water and water vapor; from the chemical standpoint it is that which occurs when complex molecules split up into simple molecules of similar composition, e.g., $N_2O_2 \rightleftharpoons 2NO$.

Equilibrium of the second, third, etc., order, namely, when two, three, etc., substances (constituents) are present, from the physical standpoint is the equilibrium between two, three, etc., substances which can form a physical mixture (p. 46) with one another, e.g., the mixture of two, three, etc., gases or liquids with one another, the solution of one, two, etc., solid substances in a liquid; from the chemical standpoint the equilibrium which occurs from the action of two, three, etc., substances on one another.

As substances or components of a chemical equilibrium not all atoms or atomic complexes can be considered as the same, but only the smallest number of those of which all the bodies taking part in the equilibrium are composed. Therefore the system $N_2O_2 \rightleftharpoons 2NO_2$ has only one component, the system $CaCO_3 \rightleftharpoons CaO + CO_2$ only two components. As a general rule for any given system as many components are to be taken as the number of members of the system minus one.

Homogeneous equilibrium is that in which all the components form a physically and chemically homogeneous system, e.g., as gases, as mixed liquids, as solutions, etc. This form of equilibrium exists in the gaseous chemical systems $H_2O \rightleftharpoons H_2 + O$ and $N_2O_4 \rightleftharpoons 2NO_2$, or in the liquid chemical system $C_2H_5OH + CH_3COOH \rightleftharpoons CH_3COOC_2H_5 + HOH$ (p. 61).

The existence of solid solutions (p. 49) and the power possessed by solids to diffuse into other solids (p. 46) allow the assumption that in homogeneously solidifying mixtures (e.g., in alloys) a condition of mutual chemical action and final equilibrium can exist, although in such cases the chemical action proceeds so slowly that it has been impossible to measure it except in certain individual cases.

Heterogeneous or non-homogeneous equilibrium is that in which the components are present in different states of aggregation and thus form a heterogeneous system. It is represented by the chemical system $CaCO_3$ (solid) $\rightleftharpoons CaO$ (solid) $+ CO_2$ (gas), and further by the physical systems KNO_3 (solid) $\rightleftharpoons KNO_3$ (dissolved), H_2O (liquid) $\rightleftharpoons H_2O$ (vapor), H_2O (solid) $\rightleftharpoons H_2O$ (liquid).

Condensed equilibrium is that form of heterogeneous equilibrium in which no gaseous components are present, or if present can be neglected.

The chemically and physically homogeneous parts of which a heterogeneous system is built up are called *phases;* they can be mechanically separated from one another and the separate phases can be physical mixtures as well as chemical substances.

For example, the system Water + Water vapor consists of two phases, the system Ice + Water + Water vapor of three phases, the system $CaCO_3$ (solid)$\rightleftarrows CaO$ (solid) + CO_2 (gas) of three phases, the system 3Fe (solid) +4HOH (vapor)$\rightleftarrows Fe_3O_4$ (solid) + $4H_2$ (gas) of three phases, since the water vapor and the hydrogen gas, like all gases, constitute only one phase, since they represent in all parts chemically and physically homogeneous mixtures.

The degree of freedom stands in a definite relation to the phases of a heterogeneous system. By degree of freedom is understood the conditions (i.e., pressure, temperature, and volume relations), which can be optionally chosen without thereby altering the equilibrium of the system. For example, the system Water + Water vapor, consisting of two phases, has one freedom, namely, either the pressure or the temperature can be chosen as desired without altering the system. If the pressure is fixed then the system can exist only at a perfectly definite temperature, if the temperature is fixed then the system can exist only at a perfectly definite pressure. The system Ice + Water + Water vapor, consisting of three phases, has no freedom and can therefore exist only at a perfectly definite pressure (4.6 mm.) and a definite temperature (+0.007°).

Also every elementary substance has two freedoms for every state of aggregation, i.e., of its three variable quantities, pressure, temperature, and volume, two can be altered as desired without changing the state of aggregation.

Every system of equilibrium consisting of one component has with two phases one freedom, with three phases no freedom; for every further component the number of freedoms increases by one; therefore the system $CaCO_3 \rightleftarrows CaO + CO_2$, consisting of two constituents (P. 62) and three phases (above), has one freedom. Hence follows the phase rule of Gibbs: *Equilibrium exists when the sum of the phases* (P) *and the degrees of freedom* (F) *of a system is equal to two more than the components* (B) (see p. 63), that is, when P+F=B+2.

With the assistance of the phase rule, the number of degrees of freedom on the one hand, and the number of phases on the other, which for the given components of a system must be present in a state of equilibrium, can be readily calculated beforehand, since F = B + 2 − P and P = B + 2 − F. From the latter it follows: If a heterogeneous system is in a state of complete equilibrium then one phase more must be present than the number of the components, if more phases are present then equilibrium of the separate phases is only possible at a definite temperature, pressure, and concentration. If less phases are present, then the equilibrium is incomplete.

All systems of equilibrium with equal degrees of freedom show complete agreement in their behavior if the factors (i.e., pressure, temperature, or volume), which influence the equilibrium, are allowed to continuously alter. Since the degree of freedom can be calculated with the help of the phase rule, this rule also serves for dividing systems of equilibrium according to the number of their degrees of freedom into non-, uni-, di-, etc., variant systems. The system of equilibrium consisting of Water vapor + Water + Ice, which can coexist only at 0.007° and 4.6 mm. pressure, is a nonvariant system; this system has one component and three phases, therefore $F = 1 + 2 - 3 = 0$. Neither the pressure nor temperature can be altered without destroying the system.

A monovariant system is represented by $CaCO_3$ (solid)$\rightleftarrows$CaO (solid) + CO_2 (gas), since it has 2 components and 3 phases, namely, $= 2 + 2 - 3 = 1$; if the temperature is varied, then for every given temperature, the pressure under which the system stands is a perfectly definite pressure, or if the pressure is varied, then a definite temperature corresponds to every given pressure.

The number of phases present in a heterogeneous system of equilibrium can be determined with the assistance of the phase rule, which is of importance in those cases in which the nature of the phases cannot be determined by direct observation. For example, metallic palladium absorbs large quantities of hydrogen and it is a difficult matter to tell whether a chemical compound, Pd_2H, or a solid solution is formed. In the case of chemical combination, owing to partial decomposition, there would exist a system of equilibrium, viz., Pd_2H (solid)$\rightleftarrows$Pd$_2$ (solid) + H (gas), comprising 2 components and 3 phases. According to this the degree of freedom would be $2 + 2 - 3 = 1$, which is contrary to fact. Therefore a solid solution and not a chemical compound must be formed in this process.

2. Chemical Kinetics.

Chemical kinetics treats of the rate of progress of chemical reactions (the velocity of reaction), namely, the ratio of the quantities (calculated in gram-molecules) of substance transformed to the time required for the process.

The velocity of reaction depends on the nature of the substances, the temperature of the reacting mixture, and the quantities of the reacting substances (namely, their concentration, p. 59), but the pressure (except in the presence of gases) does not come into consideration.

All chemical reactions have the same general characteristic with respect to the velocity of reaction in that they begin with the greatest velocity and that this becomes constantly smaller.

Temperature increases the velocity of reaction.

The heating of many organic substances which are to be combined is carried out in closed vessels in order that the substances may be retained in a liquid condition even at such high temperatures by the resulting pressure and in order that the velocity of reaction may be sufficiently great.

The influence of temperature cannot be explained simply on the assumption that an increase in this causes the molecules of the reacting substances to encounter each other more frequently, since this molecular movement in gases, and apparently also in liquids, at the average temperature increases only about $\frac{1}{6}$ per cent. for every degree of heat, while the actual increase of the velocity of reaction for every degree is 10–12 per cent. An explanation of this phenomenon has not yet been discovered.

The velocity of reaction differs greatly. It is very great in the case of the formation of salts in aqueous solutions (since this is a reaction between ions, p. 83); in the case of the formation and decomposition of organic compounds it is so much slower that it can be readily followed quantitatively. This is also the case in chemical reactions in gases when the temperature employed is not too high. On the other hand, for example, hydrogen and oxygen act so slowly on one another at ordinary temperature that a measurement of the velocity of reaction is impossible. In such cases it is possible, by raising or lowering the temperature, so to alter the conditions of experiment that the velocity is brought within the range of measurement.

In many cases the presence of certain substances has an accelerating or retarding action on the velocity of reaction, although these substances apparently do not take part in the chemical change. This action is called *catalytic action* and the substances are called *catalytic agents* (p. 9). There are special catalytic agents which are only active in the case of certain reactions, and general catalytic agents of which the acids are the most conspicuous examples. Practically all slowly proceeding reactions are accelerated by the presence of small quantities of acids, provided that the acid does not combine with one of the reacting substances. Catalytic agents play an important part not only in technical chemistry but also in biology (see Ferments).

Up to the present the velocity of reaction has been investigated chiefly for chemical reactions in homogeneous systems (p. 63), while in heterogeneous systems (p. 63), where it is dependent on the size and character of the contact-surfaces, in addition to the temperature, and on other conditions (e.g., the diffusion), it still offers great difficulties.

If a chemical reaction proceeds not only to a state of equilibrium, but as almost or quite complete, then, in general, after a time which is ten times as great as that required for the reaction to become half completed, the still unaltered portion of the system will have sunken to a quantity which is no longer measurable. In such reactions the only factor which comes into consideration is the velocity of reaction of the original substances, which in every moment is proportional to the product of the concentration (the active mass) of the still undecomposed original substances and a constant coefficient of velocity, which, however, has a different value for mono-, di-, tri-, etc., molecular reactions (i.e., for reactions which can be considered as proceeding from one, two, three, etc., molecules).

With the help of the coefficient of velocity it can be determined whether a reaction is monomolecular or dimolecular or trimolecular, etc. For example, is the assumption correct that the decomposition of arseniuretted hydrogen is a tetramolecular reaction, since the separated arsenic in a solid state contains at least four atoms in the molecule (p. 21): $4AsH_3 = As_4 + 6H_2$; the determination of the velocity coefficient gives a result which corresponds to a monomolecular reaction. The reaction must therefore proceed as follows: $AsH_3 = As + H_3$, so that each molecule first splits into its atoms and then two hydrogen atoms combine to form a hydrogen molecule and the unknown number of arsenic atoms combine to form an arsenic molecule of the solid arsenic.

If a chemical reaction proceeds only to a state of equilibrium, then in this state the velocity of reaction must have become equally great on both sides of the system, namely, in this state the original substances of the reaction must combine just as fast as the products react on each other to form the original materials. For example, if one gram-molecule of alcohol and acetic acid are mixed, then a state of equilibrium results just as soon as two-thirds of the total quantity of ethyl acetate, which would be formed if the reaction was complete, has been produced. In this case neither the velocity of the transformation of the original materials nor the simultaneous transformation of the products can be measured, but this process can be considered as if the opposed reactions proceeded independently of one another (principle of coexistence); the difference of these two velocities can be measured, since it must correspond to the velocity with which the reaction approaches the state of equilibrium. Since the chemical change observed in every moment of time is equal to the change in one direction diminished by the change in the other direction, therefore by determining this difference the difference in the velocities of reaction can be arrived at. For example, the quantity of ethyl acetate formed in every moment of time from acetic acid and ethyl alcohol can be measured, and likewise the quantity of ethyl alcohol and acetic acid formed from ethyl acetate and water. From the difference of the quantities formed in both cases the product of the velocity of reaction can be calculated.

THERMOCHEMISTRY

is the study of the relation between chemical energy and heat.

1. General.

Of all transformations of chemical energy into other forms of energy, the transformation into heat proceeds the most readily. The production or absorption of heat (and often at the same time of other forms of energy) accompanies all chemical reactions.

· The changes in energy consisting in the absorption and production of heat supply important explanations of the course of a reaction and the manner of formation, stability, and nature of the resulting compounds. However, on the combination of substances which show the greatest affinities (p. 6) for one another it is by no means true that the greatest quantities of heat are set free, but the quantity of heat set free depends also on the change in the state of the substances which take part in the given chemical process. Therefore the quantity of heat produced or absorbed in a chemical process cannot, as was formerly assumed, serve as a measure of the affinity of the given substances

According to whether heat is produced or absorbed in a chemical reaction the reaction is called *exothermic* or *endothermic*. Similarly a substance is called exothermic or endothermic according to whether in its formation heat is set free or absorbed. Exothermic changes are positive heat reactions, endothermic are negative heat reactions. The heat is measured in heat-units or calories, a large calorie (= Cal.) being the quantity of heat required to warm 1 kilogram of water 1°, a small calorie (= cal.) the quantity of heat required to warm 1 gram of water 1°.

In order to allow the simplest comparison the heat is calculated, not on the basis of the gram-unit, but on the basis of the gram-quantity corresponding to the number of atoms which combine to form a molecule. The heat values are mostly based on the state in which the substances exist at 18°

The heat is distinguished as the heat of solution, the heat of dilution, the heat of formation, the heat of decomposition, the heat of neutralization, and the heat of combustion; further, as the heat of hydration (i.e., the heat which results on the combination of a substance with a definite number of water molecules) and as the heat of dissociation (i.e., the heat which results from the splitting up of a dissolved substance into its ions). The heat calculated on the basis of the gram-molecule of the given substance is called the molecular heat of solution, heat of formation, etc.

The mixing-calorimeter is used in measuring the heat in liquids or solutions. It consists of a vessel of metal or glass in which the reacting substances are mixed, their temperature having been accurately measured beforehand. The quantity of heat is calculated from the alteration in temperature which is produced in the quantity of water present, this having been carefully measured.

The combustion-calorimeter serves for measuring the heat produced in combustions. It consists of a tightly closed steel sphere (calorimetric bomb) lined on the interior with platinum or enamel. Within this the substance under investigation is placed and the bomb is filled with oxygen under a pressure of about 25 atmospheres. The substance is then ignited by a wire brought to incandescence by an electric current and is burned. The quantity of heat produced is calculated from the elevation in the temperature of the water with which the steel sphere is surrounded.

Exothermic compounds contain less energy than their components and are therefore, at ordinary temperatures, more stable than these. Their formation, having been once started, proceeds without the further addition of heat (energy) from without (p. 59) with greater or less velocity, depending on the quantity of heat developed. The violence of the formation can in some case be such as to cause an explosion (for example, the formation of water from $H_2 + O$, of HCl from $H + Cl$). The decomposition of exothermic compounds requires the continuous addition of heat (or other form of energy) and is therefore an endothermic reaction; it proceeds slowly, never with an explosion, and is limited by the opposing tendency of the components to again enter into combination, since these components, because of the absorption of heat, are more energetic than the original compound and can combine to produce this (see Dissociation).

Endothermic compounds contain more energy than their components and can therefore more or less readily split up into them. Their formation proceeds only slowly with the continuous addition of heat (or other form of energy) from without. As a result of the elevation in temperature the combination often takes place only partially, owing to the tendency of the resulting compound to break down again into its components. The decomposition of endothermic compounds does not require the addition of heat (or other form of energy), but needs only an external stimulus in order to spontaneously proceed both rapidly and completely (often so rapidly that an explosion results, p. 73) with the evolution of heat. The decomposition is therefore an exothermic reaction, and the decomposition products formed cannot of themselves again combine when the temperature is lowered, since, because of the evolution of heat, they are poorer in energy than the original compound. This explains the existence of many non-reversible reactions.

Every compound has a certain heat of formation which is equal to its heat of decomposition (Law of Lavoisier and Laplace). If 2.02 grams of hydrogen and 16 grams of oxygen combine to form water, a quantity of heat equal to 68 Cal. is set free: $2H + O = H_2O + 68$ Cal., while 126.8 grams of iodine and 1.01 grams of hydrogen combine with an absorption of heat equal to 6 Cal.: $I + H = HI - 6$ Cal. If one molecule of water ($= 18$ grams) is decomposed into its elements, then the quantity of heat evolved, equal to 68 Cal., must be again added to it, which is then contained as chemical energy in the elements (molecules) which are set free: $H_2O = H_2 + O - 68$ Cal., while in the decomposition of 1 molecule of hydrogen iodide ($= 127.8$ grams) an evolution of heat equal to 6 Cal. takes place: $HI = H + I + 6$ Cal.

Equations which include the energy set free or absorbed, measured in the form of heat, are called thermochemical equations.

The evolution of heat which accompanies a chemical process is always the same whether the process takes place in one step or whether it passes through a number of intermediate processes (Law of Hess).

For example, on dissolving 39 grams of potassium in 36.4 grams of dissolved hydrochloric acid a quantity of heat equal to 61.8 Cal. is evolved, and the final result is the same whether the process takes place in one step: $K + HCl = KCl + H + 61.8$, or whether it takes place in two reactions: $K + H_2O = KOH + H + 41.8$ Cal. and $KOH + HCl = KCl + H_2O + 13.7$ Cal. The law is very important, since it makes it possible to calculate the heat in many cases where it cannot be determined directly; for example, the heat of formation of carbon monoxide (CO) $= 26.3$ Cal. is deduced from the combustion of carbon to carbon dioxide: $C + 2O = CO_2 + 94.3$ Cal., and from the combustion of carbon monoxide to carbon dioxide: $CO + O = CO_2 + 68$ Cal.; therefore 94.3 Cal. $- 68$ Cal. $= 26.3$ Cal.

In many chemical reactions there is a tendency to produce those substances whose formation is accompanied by the evolution of the greatest quantity of heat (principle of maximum work). For example, in a system consisting of potassium+chlorine+bromine potassium chloride (KCl) and not potassium bromide (KBr) is formed, since $K + Br = KBr + 90$ Cal.; $K + Cl = KCl + 106$ Cal.

Accordingly, a chemical reaction generally takes place more readily if the resulting products have an opportunity to enter into a second reaction, since in this case a greater quantity of heat is produced. For example, chlorine decomposes water only very slowly: $H_2O + 2Cl = 2HCl + O + 10$ Cal.; but if the oxygen can immediately exert a further chemical action (as is the case, for example, in the presence of sulphur dioxide (SO_2), then a rapid decomposition of the water takes place $2H_2O + 2Cl + SO_2 = H_2SO_4 + 2HCl + 74$ Cal.

The principle of maximum work has not the universal application which was at first attributed to it; it applies to such reactions as proceed at ordinary conditions of temperature and pressure and to such substances as are stable under increase in temperature. It does not apply at high temperatures; indeed under such conditions the contrary is generally true and endothermic compounds are formed, as is illustrated by the many dissociation phenomena (see below).

2. Transformation of Heat into Chemical Energy.

This occurs on the formation of endothermic compounds and on the decomposition of exothermic compounds by the addition of heat. The latter is a dissociation phenomenon.

Dissociation is the name applied to the splitting up of the molecules of certain substances into simpler constituents which takes place under certain favorable conditions. Dissociation exists only so long as the dissociating influences are acting; when they cease the dissociation products again combine to form the original substances.

A distinction is made between the thermal dissociation produced by heating, which for short is called dissociation, and that which appears on the solution of certain substances, the electrolytes, which is known as electrolytic and hydrolytic dissociation (p. 77).

Thermal dissociation does not appear suddenly in the whole quantity of the given substance, but begins gradually, and steadily increases with increasing temperature. With decreasing temperature the recombination of the dissociation products proceeds similarly.

For example, hydrogen and oxygen begin to combine to form water vapor (H_2O) at 200°, but at 1200° a splitting up of the water into the two elements begins. At 2500° half of the water is dissociated and finally with increasing temperature all of the water is split up and the two elements exist side by side, as at 200°. If the temperature is now lowered the formation of water again takes place: $H_2O \rightleftarrows 2H + O$, and at 1200° the elements have again completely combined with one another.

Crystallized, anhydrous sulphuric acid (H_2SO_4) even at 40° begins to split up into sulphur trioxide (SO_3) and water (H_2O); this dissociation is complete at 416°; on cooling the combination of the dissociation products gradually takes place. •

The dissociation phenomena constitute a class of reversible reactions (p. 61), and at constant pressure (see below) for every given temperature the degree of dissociation of a substance is perfectly definite, so that a definite state of equilibrium exists between the undecomposed substance present and its decomposition products.

If a closed vessel filled with hydrogen iodide (HI) is heated to 518°, then 21 per cent. of the gas dissociates into H + I. If quantities of H + I corresponding to their atomic weights are heated in a closed vessel to 515°, then 21 per cent. of the gaseous mixture remains uncombined, while the remainder is transformed into HI gas: $2HI \rightleftarrows 2H + 2I$.

If the dissociation takes place with an increase in volume, which is usually the case, then it is dependent on the pressure as well as on the temperature. If the pressure is increased the dissociation decreases, if the pressure is lowered then the dissociation increases. For every given pressure (the dissociation pressure or dissociation tension) the degree of dissociation of a substance at constant temperature has a definite value, so that a state of equilibrium exists also at the given pressure.

Hydrogen iodide when sufficiently heated dissociates without increase in volume: $2HI = 2H + 2I$ (2 molecules (4 volumes) = 2 volumes + 2 volumes), so that on heating this substance to 518° (see above) the dissociation will be only 21 per cent. whether the pressure be increased or lowered. When barium dioxide (BaO_2) is heated to 700° at reduced pressure it completely dissociates into barium oxide and oxygen, $BaO_2 = BaO + O$, which causes an increase in volume. Under sufficiently high pressure, however, at the same temperature barium oxide combines completely with oxygen to form barium dioxide.

The abnormal (i.e., apparently contradictory to Avogadro's law) gas densities which many substances show by exhibiting in a gaseous state a smaller specific gravity (= vapor density, p. 43) than corresponds to their molecular weights, as determined by chemical methods, are due to dissociation phenomena. Such substances, on being converted into gases, suffer a more or less complete dissociation into simpler molecules or even into atoms, as a result of which at a constant pressure the volume is increased (in proportion to Gay-Lussac's law, p. 15) and the specific gravity correspondingly diminished, or if the volume remains constant the pressure on the surrounding walls is increased. For example, the vapor density of iodine at 600° = 127, corresponding to the molecular weight $I_2 = 254$; above 600° this value decreases steadily and at 1500° has only half the value and then remains constant. This is explained by the gradual splitting up of the iodine molecule I_2 into the free atoms $I + I$.

The vapor density of sulphur at 190–300° is 128, at 500° it is 96, but at 1000° it becomes constant and then has the value 64. Since the atomic weight of sulphur is 32 and the molecular weight at 190–300° is 256, the molecule at this temperature contains 8 atoms. At 500° the molecular weight is 192 and the molecule contains 6 atoms. At 1000° the molecular weight is 64 and the molecule therefore contains 2 atoms.

The increase in the number of atoms in the molecules of certain elements when the temperature is lowered makes it appear probable that many elements in the liquid or solid state contain more atoms in the molecule than when in the gaseous condition.

Through the alteration in the volume and the specific gravity (vapor density) of a gaseous substance it is possible to calculate the degree of dissociation.

For example, the specific gravity of vaporized ammonium chloride (NH_4Cl) does not correspond to its molecular weight = 53.4 (p. 43), but is only equal to 26 7; therefore the complete dissociation of the molecule NH_4Cl ($=2$ vols.) into one molecule of NH_3 ($=2$ vols.) and one mole- cule HCl ($=2$ vols.) has taken place, and a doubling of the volume and a corresponding reduction in the specific gravity to the half of that which would agree to the molecule NH_4Cl is produced.

In order to explain the dissociation it is assumed that this is caused not only by the molecular motion (p. 59), but also by the motion of the atoms within the molecule; on warming both motions are increased and finally the atomic vibrations are so violent that the atoms are sep- arated from one another and the molecules finally break up into atoms. Moreover at any given temperature, as a result of the irregularity of the encounters of the molecules, all of the molecules will not have the same velocity; those which move the most rapidly will be the warmer, those which move the slowest will be the cooler, and therefore the dis- sociation is gradual and increases with the temperature, since the disso- ciation only occurs among the more highly heated molecules, the number of which increases with the temperature.

Concerning the acceleration of chemical processes by heat see Chemical Kinetics, p. 65.

Concerning the shifting of the equilibrium by heat see Chemical Statics, p. 62.

3. Transformation of Chemical Energy into Heat.

This takes place on the formation of exothermic compounds and on the decomposition of endothermic compounds; frequently during these processes a portion of the resulting heat-energy is con- verted into radiant energy (p. 91).

The evolution of heat on the formation of exothermic compounds and on the decomposition of endothermic compounds can sometimes take place with great velocity and with the development of great pressure (as a result of the expansion by heat of the gases present before or after the reaction). This is called an explosion.

The mechanical action of an explosion is strongest when it is pro- duced by the decomposition of a solid or liquid substance, since in such cases the resulting gases occupy a much greater volume with respect to the exploding substance than when this is already gaseous before the explosion (see Water, Gunpowder, Glycerine Nitrates, and Cellulose Nitrates).

If the chemical reaction is once started at any one point the formation of exothermic compounds and the decomposition of endo- thermic compounds proceeds without the addition of any further heat (or other form of energy) if the heat resulting from the reaction is sufficient to raise the temperature of the neighboring parts to the temperature of ignition. If, however, the heat at first produced is removed by radiation or conduction more rapidly than it can be transmitted to the neighboring parts (see Combustion), then the reaction apparently ceases, since the velocity of reaction assumes a minimum value (p. 66).

In general exothermic compounds are stable on heating, exposure to shock, etc., and this stability increases the greater the heat of formation. Water vapor and hydrogen chloride gas, whose heats of formation are $+68$ Cal. and $+22$ Cal., respectively, are completely dissociated only at very high temperatures and not at all affected by pressure, shock, etc. Endothermic compounds, on the contrary, are mostly unstable. Nitrogen chloride, whose heat of formation is -38 Cal., decomposes into chlorine and nitrogen under the slightest disturbing influence. Many endothermic compounds, however, are in most cases stable. For example, acetylene gas, although its heat of formation is -58 Cal., can nevertheless be subjected to many manipulations without undergoing decomposition. It has, however, been observed in the case of this substance that it is unstable when acted upon simultaneously by a sudden high pressure and a high temperature.

ELECTROCHEMISTRY

is the study of the relation between electrical energy and chemical energy.

1. Transformation of Electrical Energy into Chemical Energy.

The electric arc and spark discharge can cause chemical combination or decomposition, but judging from all appearances this is due only to the heat evolved or from the mechanical disturbance produced.

The electric current, on the contrary, causes only chemical decomposition (and spatial separation, p. 77), which occurs when it is allowed to pass through a conductor of the second order. This phenomenon is called *electrolysis* ($\lambda \acute{v} \epsilon \iota \nu$, to loosen) and it can be very diversified.

Conductors of the first order are those substances which conduct electricity without undergoing decomposition; e.g., metals, peroxides, carbon, selenium.

Conductors of the second order or *electrolytes* are those substances which conduct electricity only by undergoing a simultaneous chemical alteration. To this class belong acids, bases, and salts when they are fused or dissolved in water.

Non-conductors or *non-electrolytes* are substances which permit no passage, or only a very slight passage, of electricity. To non-conductors belong the aqueous solutions of organic compounds with the exception of the typical organic acids, bases, and salts, as well as the solutions of almost all substances in the greater number of organic solvents.

Electrodes (ὁδός, path) is the name given to those conductors of the first order through which the electricity enters and leaves in the electrolysis of the conductors of the second order. The electrode through which the positive electricity enters the electrolyte is called the *anode* (ἀνά, above). The electrode through which the negative electricity enters the electrolyte is called the *cathode* (κατά, below).

Ions.—As a result of the electrical tension (=difference of tension, electrical potential, difference of potential, electromotive force) which is produced by the electric current at the electrodes, certain components of the electrolyte, the ions (ἰών, to migrate), move with an electric charge to the electrodes, where they then undergo a chemical alteration. The negative electric components which move to the anode are called *anions*, the positive electric components which go to the cathode are called *cations*.

The chemical alteration of the ions at the electrodes is of the following character: Either the elementary ions combine with one another to form molecules, or the elementary ions and the complex ions, which can no longer exist free as fractions of molecules, enter into chemical reaction with the water which serves as solvent, with molecules which are still decomposed, or with the metal of the electrodes.

Such so-called secondary processes were the cause of the erroneous idea that water to which sulphuric acid had been added was decomposed on electrolysis, because in this case hydrogen is evolved at the cathode and oxygen at the anode. Pure water does not conduct the electric current at all, but conducts only when it contains salts, bases, or acids; these are then decomposed and their ions exert a decomposing action on the water. For example, sulphuric acid, H_2SO_4, decomposes on electrolysis into $H_2 + SO_4$, but the anion SO_4 cannot exist free at the anode and immediately forms sulphuric acid again by abstracting from the water the necessary quantity of hydrogen: $SO_4 + HOH = H_2SO_4 + O$. Potassium sulphate, K_2SO_4, dissolved in water splits up on electrolysis into the cation K, which immediately decomposes water at the cathode, $K_2 + 2HOH = 2KOH + H_2$, and into the anion SO_4, which likewise decomposes water at the anode $SO_4 + HOH = H_2SO_4 + O$. The oxygen and hydrogen separated from water on electrolysis therefore originate only indirectly from the water.

In equal periods of time a current of equal strength separates the ions from solutions of electrolytes in quantities by weight which stand in the same ratio to one another as their equivalent weights, i.e., the atomic weight divided by their valence (Faraday's law).

It should be noted that the equivalent weight of many elements is variable according to the valence which they have in the given compound. For example, in cuprous chloride CuCl copper is monovalent and therefore $\frac{63.6}{1}$ parts copper and $\frac{35.4}{1}$ parts chlorine will be separated from it; from cupric chloride CuCl$_2$, however, in which copper is divalent, $\frac{63.6}{2}$ parts copper and $\frac{70.8}{2}$ parts chlorine will be separated. From stannous chloride SnCl$_2$ there will be separated $\frac{118.5}{2}$ parts tin and $\frac{70.8}{2}$ parts chlorine, from stannic chloride, SnCl$_4$, $\frac{118.5}{4}$ parts tin and $\frac{141.6}{4}$ parts chlorine.

The strength of an electric current can therefore be measured by determining the weight of copper or silver separated at the cathode in a given time from a solution of copper or silver, or by measuring the volume of hydrogen and oxygen produced from water (voltameter).

Since the electric current can be cheaply generated with dynamos, it can be employed for many purposes in chemistry.

Among the practical purposes for which it is used may be mentioned the following:

1. The separation of pure metals from their ores or from the raw products obtained by smelting (Electrometallurgy).

2. The preparation of detachable metal precipitates on the surface of conducting objects, or of objects which have been made conductors (Galvanoplastic), and the coating of the surface of common metals with noble metals (Electroplating).

3. The preparation of valuable chemicals from those which are less valuable; for example, the preparation of soda, chlorine, caustic alkalies, chloride of lime, chlorates, white lead, iodoform, percarbonates, etc.

4. The preparation of certain chemical substances which, as soon as produced, can be employed as oxidizing or reducing agents; viz., the preparation of ozone.

5. The determination of the quantitative composition of compounds (Electrochemical analysis).

6. The production of very high temperatures (up to 3000°)

in order to carry out certain chemical reactions, viz., the reduction of all metal oxides by carbon, the preparation of calcium carbide and other carbides, the preparation of phosphorus, etc.

The electric furnace consists of two blocks of burnt lime (CaO) fitting on one another. The lower block contains a cavity for holding the substance to be heated and has a groove on the edge in which are laid the carbon poles for producing the electric arc. The furnace is closed by the upper block in which there is a concave depression by which the heat of the electric arc is deflected into the cavity below, so that a decomposing action from the electricity itself cannot take place.

2. Theory of the Ions or of Electrolytic Dissociation.

Neither solid electrolytes (p. 74) nor pure water conduct the electric current and therefore neither of them are decomposed by its action, but when electrolytes are dissolved in water the solution becomes a conductor of electricity. The water must therefore cause a change in the electrolytes which makes it possible for them to conduct the electric current, and in fact it has been shown that the electric current is not the cause of the decomposition of the electrolytes, but merely causes the separation of the ions which have already been formed from the molecules of the electrolytes by the action of the water, and which because of this action are already charged with positive and negative electricity (Arrhenius' theory of electrolytic dissociation).

It can be shown by the fact that even the smallest quantity of electricity causes electrolysis; no expenditure of force is therefore necessary for splitting up the electrolyte, and this must have already taken place (however, work is necessary for separation at the electrodes); moreover, only those substances are electrolytes which show, in aqueous solutions, an osmotic pressure which increases with the degree of dilution and a correspondingly greater change in vapor pressure, freezing-point, and boiling-point than corresponds to the molecular weight of the given substance (p. 20).

The positive and negative ions simultaneously formed must bear equal quantities of electricity, since it is a fundamental law of electricity that only equal quantities of the two opposite electricities can result from an originally, electrically neutral substance. Therefore, on the electrolytic dissociation of barium chloride ($BaCl_2$), for example, the barium ion contains the same quantity of positive electricity as the negative electricity of the two chlorine ions together. Simple positively electrified ions or cations are represented by hydrogen and the metals, complex cations by ammonium (NH_4) and its organic derivatives, as well as by the analogous compounds of phosphorus, antimony, and arsenic; other polyvalent elements can also form complex cations with carbon.

Simple negatively electrified ions or anions are formed by the hydroxyl group OH of the bases and by the elements of the chlorine and sulphur group, complex anions are formed by all acids minus the hydrogen which is replaceable by metals on the formation of salts (p. 84).

The degree of dissociation of an electrolyte does not depend only on the degree of dilution of its solution, but also on its temperature and on the nature of the electrolyte. Hydrochloric acid and nitric acid are almost completely dissociated in solutions which contain 0.1 of a gram-molecule in a liter, while in similar solutions carbonic and silicic acids are practically not dissociated at all; of the bases the hydroxides of the alkalies and the alkali earths are almost completely dissociated at this dilution, mercuric chloride, on the contrary, but very little.

If it is assumed that the electrolytes (the acids, bases, and salts) undergo a splitting up into their ions in aqueous solutions, this explains also the increase of the osmotic pressure, etc., on the dilution of such solutions, since by the formation of a greater number of constituents, namely the ions, which is similar in effect to an increase in the number of the molecules, the osmotic pressure, etc., is increased. When the dilution has proceeded to a certain degree, then further dilution causes no further increase in the osmotic pressure, etc., since all the molecules have then been split up into their ions. For example, the molecule $NaCl$, when completely dissociated into its ions $Na+Cl$, acts like two molecules; the molecule Na_2SO_4, when completely dissociated into its ions $Na+Na+SO_4$, acts like three molecules of a non-electrolyte with respect to the osmotic pressure, etc.

An analogous phenomenon is exhibited on the dissociation of gases (p. 72) by heat, which is the cause of the fact that the dissociable gases at a given volume and a given temperature exert a pressure which is greater than that corresponding to the molecular weight of the given gas. The number of the molecules and therefore the pressure of the gas is increased by the dissociation.

The aqueous solutions of the electrolytes show exactly the same behavior with respect to the electric current, since with increasing dilution their conductivity with respect to the quantity of dissolved substance constantly increases until finally a limiting value is attained.

From the knowledge of this limiting value, the degree of dissociation of an electrolyte in more concentrated solution can be calculated; and further, from the increase of the conductivity of chemically pure water on shaking with apparently insoluble electrolytes it can be determined whether these are soluble in water. Many electrolytes when fused conduct themselves with respect to the electric current as they do in aqueous solutions, and therefore, on fusion, a more or less general splitting up of the molecules into ions must take place.

The dissociation of the electrolytes into their ions in aqueous solutions is called *electrolytic dissociation* and differs therein from thermal dissociation in that the compounds do not split up into simpler molecules which can exist free, but into atoms or atomic complexes called ions, which cannot be isolated, but can only exist in solution. Half of the ions, the cations, are electropositive; the other half, the anions, are electronegative.

The power possessed by water to split up a substance into its ions is called dissociating force; besides water certain other liquids also have this force, but none so strongly as water. Formic acid has about six-eighths, methyl alcohol three-eighths, acetone and alcohol about two-eighths of the dissociating force of water. The solutions of electrolytes in liquids which have no dissociating force cannot conduct the electric current and show an osmotic pressure corresponding to the normal molecular weight of the given electrolyte, for which reason such solvents must be used instead of water in determining the molecular weights by the osmotic pressure.

For example, if sal ammoniac NH_4Cl is dissolved in much water, it dissociates into the positive ion NH_4 and the negative ion Cl; if, however, it is brought into the gaseous form by warming, it dissociates into the unelectrified molecules $NH_3 + HCl$, which can be at least partially separated by diffusion. The products of thermal dissociation can be mixed in all proportions, and the separation of the products involves no other work than the components of a gaseous mixture. In the case of the products of electrolytic dissociation, on the contrary, there are always just as many positive as negative ions present as would be necessary for their mutual electrical neutralization, and for separating the ions from one another there is required in addition to the work of moving them to the electrodes also the work of overcoming the force of attraction which the oppositely charged ions exert on one another and which can be brought about only with the help of the electric current. The separation can therefore not be carried out by diffusion, as with thermic dissociation, but can be effected only by electrolysis.

The assumption that the ions formed from the electrolytes are charged partly positively and partly negatively, although no free electricity can be detected in the solutions of the electrolytes, since on dissolving the electrolytes an equal number of oppositely charged ions are simultaneously formed and therefore the given solution can show no electric charge, is explained as follows: The separation of the ions at the electrodes shows that the conduction of the electric current in dissolved electrolytes, in contrast to the conduction in metals, is associated with a transportation of matter, and in fact the cations move to the negatively charged cathode, the anions to the positively charged anode, in order there to separate out. This can be best explained by assuming that the cations, because of their positive electric charge, and the anions, because of their negative electric charge, are attracted by the oppositely charged electrodes.

The fact that many substances which, in the form of atoms and molecules, enter-into chemical reaction with water can exist in it in the form of ions can likewise be explained by the assumption of electrically charged ions. For example, the potassium atoms react with water as follows: $K + HOH = KOH + H$; the electropositively charged potassium ions, on the contrary, act indifferently towards water until on giving up their positive electricity at the cathode they pass over into unelectrified potassium atoms, for which reason the potassium appearing at the cathode (for example, in the electrolysis of KCl) immediately reacts with water. This assumption is further substantiated by the fact that zinc when charged with positive electricity does not dissolve in dilute hydrochloric acid, while the ordinary, non-electrified zinc does dissolve.

The electronegatively charged chlorine ion is entirely different from the free, unelectrified, gaseous chlorine molecule; the former in contrast to the latter has neither color, odor, nor bleaching properties; like all other ions it cannot exist as a gas, but can exist only in solution.

The property of many elements to appear in forms (allotropic modifications) which exhibit an entirely different chemical and physical behavior is called allotropism ($\dot{\alpha}\lambda\lambda o\tau\rho\dot{o}\pi o\varsigma$, in another way).

In addition to the simpler or elementary ions there are also complex ions which have the same composition as other substances which are not ions, but which have. entirely different properties from the latter; for example, the hydroxyl ion OH′ and the hydrogen peroxide molecule O_2H_2, the cyanogen ion CN′ and the dicyanogen molecule C_2N_2.

The property of many compounds of exactly similar qualitative and quantitative composition to appear in forms which show entirely different chemical and physical behavior is called isomerism (p. 30).

Allotropism and isomerism can not only be caused by electrification and non-electrification, but allotropism exists also when (as in the case of ozone sulphur, phosphorus, and carbon, which see) the different modifications of the given element contain a different number of atoms in the molecule. Isomerism, as is known in the case of the carbon compounds (which see), also occurs when the atoms which form the molecules of a compound have a different arrangement (P. 30) or when the molecular weights are whole multiples of one another.

Ions which do not act on water can combine to form free molecules at the electrodes, since after giving up their electric charges they no longer repel each other, and likewise metal ions whose (molecules consist of one atom) can form unelectrified, monatomic, free molecules. For example, two negative chlorine atoms will combine to form one chlorine molecule, two negative hydroxyl ions will combine to form one H_2O_2 molecule, one positive copper ion will pass over into a copper molecule, etc.

. The fact that the electrically oppositely charged ions do not neutralize their electricities in solution can be explained by the assumption that their number is very small as compared to the water molecules which lie between them, so that the latter acts as an insulator, and this action is greater the more the quantity of water (or the quantity of other dissociating liquid) exceeds that of the electrolyte. Therefore in concentrated solutions the possibility that the electrified ions will combine to form non-conducting, and therefore non-dissociated, compounds is much greater, since in this case they are not so widely separated from one another.

The transportation of the electricity supplied to the solution by the electrodes is carried out by the movement of the electrically charged ions to the electrodes. Since, for example, the positive ions are attracted by the negatively charged electrode and therefore move towards this and give up their electricity to it, they can no longer exist as ions, but separate at the electrode as atoms or groups of atoms. The negative ions undergo a similar change at the positively charged electrode. As a result of this giving up of electricity at the electrodes the solution receives at each of these points an excess of electricity with which the electrodes are charged, and this electricity moves with the repelled ions (namely, the ions which carry electricity similar to that of the repelling electrodes) through the liquid to the other electrode.

Since it is exclusively the free ions which conduct the electricity, therefore with an increase in the ions, such as is produced by the increased dissociation caused by dilution, the conductivity of the solution will increase with respect to the quantity of dissolved substance.

The quantity of electricity which the ions transport through the solution, in the manner that a ship carries its cargo, is an unalterable quantity and, for each gram-equivalent of any ion, is equal to 96537 coulombs. From this will be understood the law of Faraday, that equal quantities of current separate equivalent quantities of different substances (p. 75).

Ions which are charged with quantities of electricity equal to that of the hydrogen ion or the chlorine ion are called monovalent ions, those charged with the double quantity of electricity are called divalent ions, etc.

Compounds of monovalent ions when completely dissociated have an osmotic pressure which is twice as great as that which they would have if undissociated, since on dissociation they split up into two ions (for example, KCl into $K+Cl$); compounds of divalent ions have an osmotic pressure three times as great, since they split up into three ions (for example, $BaCl_2 = Ba+Cl+Cl$), etc.

Electropositive ions are denoted, according to their valence, by dots placed after their chemical symbols or formulas, electronegative ions by accents. For example, $K\cdot$, $Ba\cdot\cdot$, Cl', SO_4''.

As already mentioned (p. 77), the movement of the electrolyte in the solution takes place without loss of energy, since an apparently insignificant part of the electricity serves to move the ions to the electrodes (to overcome the friction of the ions on the molecules of the liquid) and appears again as heat (heat-work of the electric current); on the contrary, another part of the electricity is used up in neutralizing the ions which have moved to the electrodes and to separate them as molecules, etc. (chemical work of the electric current); this part is equal to 96537 coulombs.

The ions move to the two electrodes with different velocities and the quantity of electricity carried in the unit time depends not only on the number of the transporting ions, but also on their velocities.

Since equivalent quantities of ions are always separated at the electrodes (P. 77) it might be supposed that the velocities of the ions toward the two electrodes were the same; the contrary, however, can be shown to be the fact, since if the electric current is passed through the solution for some time the solution at one electrode becomes more concentrated, the solution at the other electrode becomes more dilute.

The ions of different substances have a different electroaffinity or intensity of attachment, namely they possess a different power to retain their electric charge and they are therefore distinguished as strong and weak ions; the former form mostly easily soluble and readily dissociating compounds, the latter mostly difficultly soluble and difficultly dissociating compounds.

When a substance of strong electroaffinity which is not in the ion state comes into contact with ions which have a lower electroaffinity, the former seizes the ionic charge of the latter and passes into the form of ions, while the substance with weaker electroaffinity is converted into electrically neutral ions (atoms). For example, on dipping metallic zinc into a solution of a lead salt the zinc passes over into zinc ions and metallic lead is separated, metallic lead separates copper ions as metallic copper, and metallic copper separates mercury ions as metallic mercury (p. 88).

The ion theory not only forms the basis of electrochemistry, but it has furnished an explanation of many other phenomena in physics and chemistry which were formerly not understood. With its assistance many important discoveries in these subjects have been made possible.

The readiness with which electrolytes enter into reactions and the velocity with which their chemical action proceeds, in contrast to the slowness with which non-electrolytes, namely the carbon compounds, enter into reaction, was first explained by the ion theory. The more completely an electrolyte is dissociated the greater is its conductivity and its chemical activity, since it is the free ions only which react. For example, hydrogen chloride when dissolved in chloroform does not act on carbonates; since in this case no dissociation of HCl is possible; on the addition of water the action takes place immediately and the carbonates are decomposed with the evolution of carbon dioxide.

The chemical properties of aqueous solutions of electrolytes depend on the properties of the free ions which they contain; for example, in all solutions of silver salts the silver ion $Ag^{.}$ can be detected by one and the same reagent, namely by compounds which contain the chlorine ion Cl'; likewise in all soluble compounds of chlorine with hydrogen and metals the chlorine ion Cl' can be detected by all compounds which contain the silver ion $Ag^{.}$; all dissolved compounds which contain the ion SO_4'' can be detected by compounds which contain the ion $Ba^{..}$ and *vice versa*.

On the other hand, however, the chlorine ion Cl' does not produce a precipitate in the aqueous solutions of silver salts with potassium cyanide, KCN, since such solutions contain the components of the salt $KAg(CN)_2$, namely the ions $K^{.}$ and $Ag(CN)_2'$, and the latter ion shows different reactions from the $Ag^{.}$ ion.

The aqueous solutions of the copper salts have a blue color which is due to the $Cu^{..}$ ion; the solutions of the salt $K_2Cu_4CN_4$ on the contrary are colorless, since they contain the complex ion CuC_4N_4'; $CuCl_2$ in concentrated aqueous or alcoholic solutions is only slightly dissociated and therefore these solutions have the greenish-yellow color of the $CuCl_2$ molecule; on diluting the solution, because of the increased dissociation, the blue color of the $Cu^{..}$ ions appears.

Those ions which contain constituents that as such are not ions, but which, on the other hand, can exist themselves as ions, are called complexes.

In the solutions of the chlorine containing compounds chloric acid, $HClO_3$, and trichloracetic acid, $HC_2Cl_3O_2$, and in their salts, the silver ion $Ag^{.}$ does not produce a precipitate, since here the chlorine is not present as the Cl' ion, but is combined with other elements as the complex ions ClO_3' and $C_2Cl_3O_2$, respectively.

The iron ion $Fe^{..}$ is precipitated from its solutions by hydrogen sulphide, but is not precipitated by this reagent from the solutions of potassium ferrocyanide, $K_4FeC_6N_6$, since on dissolving this salt the complex ion FeC_6N_6' and not the ion $Fe^{..}$ is formed.

The ion theory is also of importance for physiological chemistry. For example, the poisonous action of certain metallic salts is proportional to their dissociation and therefore alcoholic solutions of mercuric chloride or silver nitrate, which are less dissociated than their aqueous solutions, are not fatal to anthrax bacilli. The salts of many metals which are poisonous, as cations, for example, $Cu^{..}$, $Ag^{.}$, $Hg^{..}$, lose their poisonous properties when they form complex anions (even with the poisonous cyanogen CN'). For example, the silver ion $Ag^{.}$ of the salt $AgNO_3$ has a stronger disinfectant action than the complex ion $Ag(CN)_2'$ of potassium silver cyanide, $KAg(CN)_2$, the anion CN' is a powerful poison, the anion FeC_6N_6' of potassium ferrocyanide is not poisonous.

The group of electrolytes comprising acids, bases, and salts (p. 74) are defined by the ion theory as follows:

Acids are hydrogen compounds which on electrolytic dissociation split up partially or entirely into cations of hydrogen and anions of non-metals or anions consisting of atomic groups; the hydrogen cations produce their acid taste and acid reaction, namely, their property of turning litmus red when it has been colored blue by bases and of decolorizing phenolphthalein solution which has been colored red by bases (see Indicators, p. 87).

Bases are hydroxyl compounds which on electrolytic dissociation split up partially or entirely into hydroxyl ions (OH') and metal ions; the hydroxyl ions cause the alkaline taste and the alkaline (basic) reaction, namely, the property of turning red litmus blue, of turning yellow tumeric brown, and of coloring colorless phenolphthalein red (see Indicators, p. 87).

The strength of acids (avidity) and likewise of bases is determined by the number of H· ions and OH′ ions respectively present in their solutions, also by their degree of dissociation. In completely anhydrous and therefore non-dissociated condition the acids do not exhibit either acid reactions or the ability to form salts (see below). The strongest acids are hydrochloric, hydrobromic, hydriodic, nitric, chloric, and sulphuric; the strongest bases are the hydroxyl compounds (hydroxides) of the alkali and alkaline-earth metals.

Salts are metal compounds which on electrolytic dissociation partially or entirely split up into metal cations (the acid salts also into hydrogen cations) and into acid anions; salts dissociate more readily and more completely than acids and bases and the salts of the monovalent metal cations are the most strongly dissociated. The formation of salts results from the substitution of the hydrogen cation of acids by metal cations, after which the displaced hydrogen cation either escapes in a non-electric form as gas or combines with the hydroxyl ion of the base to form only very slightly dissociated water.

If all of the hydrogen cations of the acid are replaced by metal cations, then a *neutral* (neutral reacting) *salt* is formed, namely, a salt which tastes neither acid nor alkaline, but simply salty, and which does not alter either litmus, tumeric, or phenolphthalein. If the hydrogen cations of a polybasic (i.e., having more than one hydrogen cation) acid are only partially replaced by metal cations then an *acid salt* is formed which still gives hydrogen cations on dissociation and which therefore shows an acid reaction in aqueous solutions.

The formation of salts, which takes place by the action of acids on bases in aqueous solutions, consists merely in a combination of the H· ions of the acid with the OH′ ions of the base to form water, since the latter is only very slightly dissociated (see below), while the metal cations and the acid anions remain to a greater or less extent uncombined in the solution. After the evaporation of the solvent which causes the dissociation these ions combine to form a salt.

That this phenomenon of salt formation depends only on the combination of the H· ions with the OH′ ions is shown by the fact that on the neutralization of equimolecular quantities of acids (i.e., quantities of acids which are neutralized by the same quantity of a given base, viz., $\dfrac{\text{HCl}}{1}$, $\dfrac{\text{H}_2\text{SO}_4}{2}$, $\dfrac{\text{H}_3\text{PO}_4}{3}$, etc.) in very dilute solutions (when the dissociation of the given substance is complete) with bases the heat of neutralization is the same for the different acids. If, however, in the foregoing process a salt was actually formed in the solution, then the heat of neutralization would correspond to the heat of formation of the given salt (which has a different value for different salts). Moreover, when dilute solutions of different neutral salts are mixed (when they do not precipitate one another), there is no thermal effect produced (law of thermal neutrality), which shows that the ions of the different salts remain side by side uncombined in the solution, namely that on mixing no change in condition takes place. A dilute solution of $\text{NaCl} + \text{KNO}_3$ or of $\text{NaNO}_3 + \text{KCl}$ are therefore identical and both contain the ions $\text{Na}^{\cdot} + \text{NO}_3' + \text{K}^{\cdot} + \text{Cl}'$.

Neutral salts of weak acids can show alkaline reactions in aqueous solutions, salts of weak bases can show acid reactions. This is a result of the hydrolytic dissociation, i.e., dissociation under the action of the ions of water, as a result of which the salts of weak acids on the one hand split up into non-dissociating (not acid-reacting) acids and dissociated bases, on the other hand the salts of the weaker bases split up into undissociated (not alkaline-reacting) bases and dissociated acids.

The electrical conductivity of absolutely pure water, which is extremely slight but still measurable, shows that even water is slightly dissociated into its ions $\text{H}^{\cdot}$ and OH' (p. 78). These ions will now combine with the cations and anions produced on dissolving the given salt in water to form undissociated acids or bases, since the slightly dissociating acids and bases require no more $\text{H}^{\cdot}$ or OH' ions to prevent their dissociation than are present in the water. Small quantities of these ions are sufficient to cause a partial re-formation of acid or base, while on the other hand the stronger acids and bases remain dissociated and a corresponding acid or basic reaction must appear.

For example, potassium cyanide, KCN, a salt of the weak hydrocyanic acid HCN and the strong base KOH, when dissolved in water exhibits the alkaline reaction of the OH' ions and the characteristic odor of undissociated hydrocyanic acid, although the $\text{K}^{\cdot}$ ions formed on electrolytic dissociation do not react alkaline and the CN' ions are not volatile. The alkaline reaction is explained by the fact that the alkaline reacting OH' ions of the water are present as well as the $\text{K}^{\cdot}$ ions, but that these cannot combine with one another because KOH being a strong electrolyte is strongly dissociated, while the $\text{H}^{\cdot}$ ions of the water are used up by combining with the CN' ions to form hydrocyanic acid, which is a weak and therefore but slightly dissociated acid.

Ferric chloride, FeCl_3, the yellow salt of the weak base Fe(OH)_3 and the strong acid HCl, on dissolving in water shows a strongly acid reaction and the solution becomes reddish brown, although the $\text{Fe}^{\cdots}$ ion, formed by electrolytic dissociation, is yellow and the Cl' ions do not react acid. The acid reaction of the solution is explained by the fact that acid-reacting $H^{.}$ ions of the water are present in the solution as well as Cl' ions, and these cannot combine with one another because HCl is a strong electrolyte, while the OH' ions of the water are used up in forming the brown base $Fe(OH)_3$ with the $Fe^{...}$ ions, this base being weak and therefore scarcely dissociated.

Since the dissociation of water increases with the temperature, on warming the hydrolytic dissociation increases also.

Hydrolytic dissociation is also called hydrolysis, but this must not be confused with the term applied to the splitting up of the molecules of complicated, non-electrolytic, organic compounds into simpler molecules with the simultaneous addition of the elements of water, which is also called hydrolysis.

The natural or artificial dyes which undergo a change in color by the action of dissolved acids or bases, and which therefore serve for the identification of these substances as well as for demonstrating neutralization (p. 84), are called *indicators*. These indicators are weak acids or bases which have a different color when they are dissociated from that which they have when undissociated. Since only weak acids or bases on solution do not split up into ions, while their neutral salts are very completely dissociated, therefore on the addition of the slightest quantity of an acid or base the formation of a dissociating salt immediately occurs and the color of the indicator immediately undergoes an alteration. Undissociated litmus is a weak, red acid which is colored blue by a trace of alkali (base), since the resulting salt immediately dissociates into metal cations of the base and the blue anions of litmus acid. If an acid which dissociates more readily than litmus be added, then the $H^{.}$ ions of the former convert the blue litmus anions into undissociated, red litmus. Undissociated phenolphthalein is a weak, colorless acid which is colored red by traces of bases, because the resulting salt immediately dissociates into metal cations and red anions of phenolphthalein. The $H.$ ions of a more readily dissociating acid convert the latter into undissociated colorless phenolphthalein again.

3. Transformation of Chemical Energy into Electrical Energy.

The simplest and best known chemical systems in which the energy associated with chemical phenomena is converted into electrical energy are the galvanic elements or voltaic cells. These are combinations of conductors of the first order (metals or carbon) with conductors of the second order (electrolytes) from which an electric current is obtained.

Since it is only in electrolytes that the conduction of the current is associated with chemical reaction, therefore all galvanic elements must contain electrolytes which up to the present have been employed almost exclusively dissolved in water or fused.

It is a characteristic of all galvanic cells that the substances which enter into chemical reaction must be separated from one another and that a conducting connection is necessary in order that the reaction can take place, while without this connection galvanic cells can be kept for a long time without any chemical reaction resulting. From this, however, it cannot be assumed that in a galvanic couple the chemical system is in a state of equilibrium, any more than this can be assumed in the case of a mixture of hydrogen and oxygen, which apparently does not change on standing. The fact is that in both cases the velocity of reaction is extremely smnll (p. 66).

Just as in the transformation of electrical energy into chemical energy a portion of the former passes over into heat, so also in the transformation of chemical energy into electrical energy thermal changes occur. Those cells whose electromotive force decreases with the temperature convert a portion of their chemical energy into heat; on the other hand, those cells whose electromotive force increases with the temperature produce more electrical energy than corresponds to the chemical energy which they consume. They supply the deficit by the absorption of heat from their surroundings and therefore become cooler when they are working.

Just as a salt dissolves in a liquid until the osmotic pressure of its solution is in equilibrium with the particular solution tension (p. 51) of the salt, so every metal has a force, dependent on its chemical nature, sufficient to send positively electrically charged ions (cations) into solution. This force is called the electrolytic solution tension (solution or ionizing pressure) and becomes active when the metal is dipped into the aqueous solution of one of its salts. It is the more active the less cations of the metal there are already in the solution, namely, the smaller the osmotic pressure of the cations already in solution (Nernst's theory of galvanic cells).

The electromotive force obtained in chemical processes therefore depends on the solution tension of the metals, and as shown in the following arrangement, in which hydrogen, which behaves like a metal, is included, decreases from potassium on: K, Na, Mg, Al, Mn, Zn, Cd, Fe, Co, Ni, Sn, Pb, H, Sb, Bi, As, Cu, Hg, Ag, Pd, Pt; Au (electromotive series).

The solution tension of magnesium and zinc, for example, is equal to several millions of atmospheres; of copper, mercury, and silver it is only about one-trillionth of an atmosphere; it is therefore impossible to prepare a solution of a zinc salt which is sufficiently concentrated to prevent the sending out of positive zinc ions into the solution, while in the case of copper and the following metals the osmotic pressure of the metal ions is greater in even extremely dilute solutions than the solution pressure of the metal. Therefore when metallic zinc is immersed in the solution of a zinc salt it becomes negatively electrified by sending out positive zinc ions into the solution, while copper immersed in a solution of a copper salt becomes positively electrified because the positive copper ions pass out of the solution to the copper and leave the solution negatively electrified.

The arrangement of the metals according to their decreasing solution tension corresponds to the arrangement according to their decreasing electroaffinity, so that every metal precipitates those standing to the right of it from their solutions. The tendency of the metals to oxidize also decreases from left to right.

The farther apart two metals stand in the row, the greater is the electromotive force or electric tension of a galvanic element formed by combining them. For this reason this arrangement is also called the electromotive series.

A knowledge of the electromotive series is of practical importance, since in all cases where objects of metal (alloys, combinations of metals in contact, metals with mechanically or galvanically 'prepared metallic coatings) are exposed to the action of the elements an opportunity is afforded for the formation of short-circuited galvanic couples, as a result of which the metal with the highest solution tension dissolves, but the other remains intact. Galvanized iron is therefore not so strongly oxidized at points where the zinc covering has been injured as if it were not galvanized, while tinned iron on an injury to the tin coating oxidizes (rusts) more readily than it would if it were not tinned, because iron has a greater solution tension than tin and a lower solution tension than zinc.

Concentration cells are formed when two rods of the same metal are brought as electrodes into two solutions of a salt of the metal, the solutions being of different concentration but in contact with one another; for example, two silver rods in two solutions of silver nitrate of different strength, or a long tin rod in two solutions of stannous chloride of different strength placed one on the other, in which case the tin rod forms at the same time both electrodes and the metallic circuit between them.

In this case the electricity is produced by the precipitation of the cations of the concentrated solution as metal on the one electrode, to which they give up their electric charge and electrify this electrode positively, while in the dilute solution from the material of the other electrode cations pass out into the solution, which charges this electrode negatively, since positive electricity cannot be produced without the formation of an equal quantity of negative electricity.

The number of cations which are driven out from the given electrode by the solution pressure of the metal can be only very small even in dilute solutions, since a condition of equilibrium is soon brought about from the fact that the negatively charged electrode exerts such an attraction on the positively charged ions which it has sent out that just as many of these metal ions are precipitated again on the negative electrode as metal as are sent out by the electrode into the solution, and on the other hand the attraction exerted by the negatively charged electrolyte on the electrode which it has charged positively is of such a character that just as many metal ions are again dissolved as the electrolyte furnishes to the metal.

As soon, however, as the electricity is allowed to flow away through a conducting wire the metal again drives cations into the solution and the solution drives cations on to the given electrode and this continues until both liquids have become of equal concentration, namely until the solution pressure and the osmotic pressure have become equally great. Therefore from the silver or tin rod which dips into the given dilute solution, silver or tin will be dissolved, while on the silver or tin rod in the given concentrated solution, silver or tin will be precipitated from the solution of the given salt.

Voltaic cells are formed when two rods of different metals dip into solutions of their salts, which must contain the same anion, and the two solutions are in contact. This is the case when the solutions are separated by a porous partition, for example, a porous earthenware cylinder. A zinc sulphate solution in which dips a zinc rod, in contact with a copper sulphate solution in which dips a copper rod, forms a Daniell's cell.

The electricity is produced as follows: The metal with the greater solution tension (the zinc) dispatches its atoms as cations into the electrolyte and becomes the anode, while on the metal with the lower solution tension (the copper) the cations are precipitated from the neighboring solution, so that it becomes cathode. If the electrodes are now connected by a wire an equalization of opposite electricities, namely, a current, is produced, which continues until all the zinc of the anode has dissolved to form zinc sulphate or all of the copper of the copper sulphate solution is precipitated on the cathode. The chemical process here involved is the replacing of the copper in the copper sulphate by the zinc, which therefore itself dissolves in equivalent quantities. Since the cations are discharged on the cathode this is also called the diverting electrode, while the anode, provided that anions separate at it which dissolve the anode metal, is called the solution electrode.

Since the osmotic pressure of the ions of a metal operates to oppose its solution tension, therefore even in the Daniell's element the concentration of the electrolyte must be of importance. If the concentration of the zinc sulphate solution, namely of the zinc ions, is increased then the tendency of the zinc to form zinc ions is decreased and the anode becomes less anodic, i.e., the tension of the cell decreases. If the concentration of the copper sulphate solution is increased, which is equivalent to an increase in the concentration of the copper ions, then the tendency of the copper ions to pass over to the cathode is increased and this becomes more cathodic, i.e., the tension of the element rises.

PHOTOCHEMISTRY

is the study of the relation between chemical energy and radiant energy.

1. Transformation of Chemical Energy into Radiant Energy.

When chemical change is accompanied by the development of light it is called combustion (see Oxygen). In this process a part of the chemical energy of the reacting substances is set free as heat and light, generally the light first appears as heat which raises the temperature of the substances so high that they radiate visible light. Light can also result from the direct transformation of chemical energy (chemical luminosity), as is illustrated by the brilliancy of burning magnesium, the temperature of which is only about 1350° but which would have to be 5000° if the enormous development of light depended on the temperature alone. The light produced by self-luminous organisms (beetles, bacteria) also depends on chemical luminosity, since these show no elevation of temperature.

Many substances burn with a flame, many only with glowing; the flame is gas heated to incandescence by the combustion process and therefore only those substances burn with a flame which are combustible gases or which develop combustible gases from their own combustion.

For example, pure carbon and iron burn only with incandescence since they form no combustible gases. Wood, coal, tallow, etc., burn with a flame because from the effect of the heat they produce gaseous, combustible decomposition products.

Flames can be luminous or non-luminous; the luminosity is caused by the presence of incandescent solid substances in the flame and also by an increase in the temperature and density of the burning gases.

A non-luminous flame can become luminous when solid, non-volatile substances are introduced into it (for example, the Welsbach light) and therefore all flames are luminous when the products of combustion are solid as well as gaseous. Zinc and magnesium burn with luminous flames because the oxides which they form on burning are not volatile and when finely divided are heated to a white heat in the flame. When hydrogen and oxygen are compressed together they burn with a luminous flame because the density of the flame-gases is then greater.

If gases which ordinarily burn with a luminous flame are cooled they lose their light-producing power and, *vice versa*, non-luminous gases become luminous when they are previously warmed. For example, if gases which are not combustible or which do not support combustion (nitrogen, carbon dioxide) are mixed with gases which burn with a luminous flame, then on burning the mixture no light is developed. This is due to the dilution of the combustible gases and to the cooling caused by the non-combustible gases added.

Many substances which burn in the air with a non-luminous flame burn in pure oxygen with a luminous flame, since in the latter case the flame is not cooled by the inert nitrogen of the air and therefore reaches a higher temperature and further because the products of combustion in vessels cannot escape so rapidly with the removal of heat as in the open air.

The luminosity of the ordinary illuminating materials is due to the fact that finely divided carbon separates out in them and is heated to a white heat. This can be demonstrated by holding a cold object in the flame and observing the carbon which is deposited on it as soot.

In the Welsbach light a fabric composed of 99 per cent. thorium and 1 per cent. cerium is heated in a non-luminous flame to incandescence.

The flame of marsh-gas (CH_4) is non-luminous; the flame of acetylene (C_2H_2), on the contrary, is very brilliant because it contains twice as much carbon in the molecule as the former, and this great quantity of carbon, if too much oxygen does not reach the flame, does not burn immediately but separates out in a finely divided state and becomes white hot.

Marsh-gas, hydrogen gas, and carbon monoxide gas burn with non-luminous flames because their products of combustion are exclusively of a gaseous nature.

Flames consist of an envelope of glowing gas, while in the interior of the flame (because of a lack of that gas which surrounds the flame and maintains it) no combustion and accordingly no high temperature can exist; the interior of the flame consists of unburned gas and is cold.

The ordinary, luminous flame consists of three parts: the inner dark part consists of the gases still unburned (hydrocarbons, especially acetylene) which are formed from the decomposition of the wax or tallow, etc., by the heat; then comes a luminous mantle in which incomplete combustion occurs. In this layer ethylene (C_2H_4) splits up into marsh gas (CH_4) and carbon (C); the former burns completely while the separated carbon is heated to incandescence because not enough oxygen is present for its combustion. This part is called the reducing flame because substances which contain oxygen when introduced into it give up their oxygen to the carbon.

In the external, non-luminous, bluish layer, which is surrounded by air, complete combustion of the separated carbon to carbon dioxide takes place. The bluish color is due to the carbon monoxide which burns here to carbon dioxide. This part is called the oxidizing flame, because bodies introduced into it are oxidized.

The temperatures of the different flames are very different and do not depend on the luminosity, as is shown by the hardly visible oxyhydrogen flame (see Water).

If a sufficient quantity of air is introduced into the interior of an illuminating-gas flame, then, because of the complete combustion of the carbon, and also because of the dilution (p. 92), the flame becomes non-luminous; the temperature of the flame, however, becomes noticeably higher. On this depends the construction of the Bunsen burner used in laboratories in which a mixture of gas and air is burned.

The use of the blowpipe, a metal tube through which air is blown into the flame and which directs this sidewise upon the object to be heated, depends on the same principle. This has an oxidizing action on substances held in the outer part of the flame, since an excess of oxygen is here present, and a reducing action of substances held in the inner, luminous flame, since this contains free carbon or reducing hydrocarbons.

Many salts impart a coloration to a non-luminous flame (since the salts are decomposed and reduced by the flame and the metal passes over into the gaseous state), which is characteristic for the metal of the given salt. For example, the salts of sodium color the flame yellow, the salts of potassium color it violet, the salts of barium color it green, etc., so that in this manner many elements can be detected in their compounds. If, however, a number of elements which color the flame are present, then the color of one element can conceal that of another, but in such a case all of the elements present can be detected from the spectra of their gases (p. 44).

Glowing gases send out rays of definite wave length depending on the chemical nature of the given gas, but which are independent of the temperature within wide limits, and which show characteristic lines when split up by a prism (p. 45).

2. Transformation of Radiant Energy into Chemical Energy.

The radiant energy of light can cause chemical changes which are called photochemical reactions. Radiant energy shows considerable analogy to electrical energy; for example, the metals are more or less positively electrified in ultraviolet light, and the action of light converts selenium and phosphorus into modifications which conduct electricity.

Photochemical reactions are exothermic or endothermic; in the former (since no energy is consumed, but is set free as heat) the light appears only to start the chemical process and to accelerate it in the manner of a catalytic agent; in the latter case, at all events, the energy of the radiation changes into chemical energy. When light causes a chemical process it must always be absorbed by the substance on which it acts, since neither the transmitted nor the reflected part of the light produces any chemical action.

Through investigations conducted with the so-called actinometer (see below) it has been found that the chemical action is proportional to the strength of the light, and the photochemical effect of the absorbed light rays is in general proportional to the quantity of light, i.e., the product of the intensity and time of radiation.

Different rays of light have different action; red rays are in many cases inactive, while blue and violet rays, and especially the invisible ultra-violet rays, exert a strong chemical action. The assimilation of carbon dioxide in plants is strongest in red and yellow light, so that the character of ray which can cause the strongest chemical action is very different in different chemical reactions. The observation that the chemical action of light does not generally reach its greatest intensity immediately on absorption, but only after some time, is called photochemical induction.

The photochemical action of light comprises chemical combination (for example, the formation of HCl from $H + Cl$, which is used in actinometric measurements), chemical decomposition (for example, the production of dark-colored halogen compounds of silver from the white or yellow halogen compounds of silver by a partial splitting off of halogen, on which depends their use in photography and actinometric measurements), allotropic transformation (for example, the production of red phosphorus from yellow), and the reduction or oxidation of certain substances.

The most important photochemical process is the assimilation of plants containing chlorophyll under the influence of sunlight, by which the radiant energy of the sun is stored up in the form of chemical energy. In this process the carbon dioxide (CO_2) taken up from the air by the plants is reduced with the evolution of oxygen and the carbohydrates (sugar, starch, etc.) rich in energy are formed.

On the other hand, the sun's rays also retard the vital functions of those plants which do not contain chlorophyll (the fungi) and often are fatal to them (for example, the pathogenic bacteria). Many natural and artificial coloring materials are bleached by light, while on the other hand the action of light causes many other substances to turn darker (for example, white paper made from wood-pulp, the human skin).

PART SECOND.

INORGANIC CHEMISTRY.

DIVISION OF THE ELEMENTS.

ORDINARILY the elements are divided into two groups, the *non-metals* or *metalloids* and the *metals*.

The *metals,* when compact, have the well-known metallic appearance and are good conductors of heat and electricity; they do not as a rule combine with hydrogen, producing non-volatile compounds. Their oxygen compounds generally have the character of basic anhydrides, i.e., they form bases with water (P. 99, *b*). Many metals are soluble in water when in a finely divided state (p. 54), while no metal is soluble in other solvents without undergoing a change. The compounds of the metals with the non-metals are decomposed by the electric current, the metal separating out at the negative pole.

The *non-metals* or *metalloids* are (with the exception of hydrogen) poor conductors of heat and electricity or are non-conductors, and all combine with hydrogen, forming volatile generally gaseous compounds. Most of the oxygen compounds of the non-metals have the character of acidic anhydrides, i.e., they form acids with water (P. 98, *a*). Most of the non-metals are soluble without change in most solvents. Their compounds with metals are always decomposed by the electric current, so that the non-metal separates at the positive pole.

A sharply defined division of the elements into metals and non-metals is not possible because often bodies are separated which have great similarity in all their chemical properties. Thus gaseous hydrogen is closely related in its chemical behavior to the metals, while arsenic and antimony have a metallic appearance, but behave chemically like the metalloids; also various elements like carbon and phosphorus are known in the metallic as well as the non-metallic condition. It is therefore indispensable

95

in the classification to consider also the chemical properties and to divide the elements which are chemically analogous in certain groups. This may be done best according to the periodic system, which classification will be followed in the following description of the elements.

NOMENCLATURE.

1. Radicals or Groups

are those unsaturated complex groups which behave like the elements, that is, they form permanent constituents of a series of compounds, and in these they can be replaced by other equivalent atoms or groups of atoms; thus the residue ^-OH forms a constituent of the following compounds: KOH, $Ca(OH)_2$, $Fe(OH)_3$, $Sn(OH)_4$, and can be transformed unchanged from one compound into another.

The corresponding atomic complexes containing carbon are generally called radicals and all the others groups or residues. The group ^-OH is called *hydroxyl*, $^-NH_2$ *amido* or *amid*, ^-NH *imido* or *imid*, ^-SH *hydrosulphuryl*, $^-NO_2$ *nitro*. The radical ^-CO is called *carbonyl*, $^-CH_3$ *methyl*, ^-CN *cyanogen*, etc.

As the valence of the elements is constant towards hydrogen, the groups which are obtained by removing one or more atoms of hydrogen from the saturated molecule cannot exist in the free state, but, like the atoms, unite together with their free valences, forming complicated bodies; thus from the molecule of water, HOH, of hydrogen sulphide, HSH, of marsh-gas, CH_4, by the removal of an atom of hydrogen we obtain the univalent groups ^-OH, ^-SH, $^-CH_3$, which immediately unite, forming

$$\overset{II}{H}O-\overset{II}{O}H, \qquad \overset{II}{H}S-\overset{II}{S}H, \qquad H_3\overset{IV}{C}-\overset{IV}{C}H_3,$$

Hydrogen peroxide   Hydrogen persulphide.   Dimethyl or ethane.

In regard to other facts about nomenclature see Part III.

2. Binary Compounds.

These are compounds consisting of two elements and are designated by names of the elements following each other with the final word terminating in *ide;* thus all binary compounds of oxygen, sulphur, chlorine, bromine, iodine, fluorine, are called oxides, sulphides, chlorides, bromides, iodides, fluorides, while the metallic compounds of hydrogen, boron, phosphorus, nitrogen, carbon, silicon are called hydrides, borides, phosphides, nitrides, carbides, silicides. Thus:

$$CaO, \qquad ZnS, \qquad AlF_3, \qquad HCl,$$

Calcium oxide.   Zinc sulphide.   Aluminium fluoride.   Hydrogen chloride.

If two different compounds of the same elements are known, we designate that compound having the element with highest valence by the suffix *ic* to the Latin root of the metal, while the one having the element with lower valence terminates in *ous*. Thus:

I	II	I	II	I	II
HgCl,	HgCl$_2$,	Hg$_2$O,	HgO,	Hg$_2$S,	HgS,
Mercurous chloride.	Mercuric chloride.	Mercurous oxide.	Mercuric oxide.	Mercurous sulphide.	Mercuric sulphide.

These compounds used to be indicated by the prefixes *proto* and *per*. Thus:

HgCl,	HgCl$_2$,	FeCl$_2$,	FeCl$_3$,
Mercury protochloride.	Mercury perchloride.	Iron protochloride.	Iron perchloride.

If more than two compounds are known of the same elements, then the number of oxygen, chlorine, bromine, etc., atoms contained in the molecule are indicated by the Greek prefixes. For example:

H$_2$O,	H$_2$O$_2$,	P$_2$O$_3$,	P$_2$O$_4$,	P$_2$O$_5$,
Hydrogen monoxide.	Hydrogen dioxide.	Phosphorus trioxide.	Phosphorus tetroxide.	Phosphorus pentoxide.

Compounds like Cr$_2$O$_3$ have the prefix *sesqui* to differentiate them from the trioxides such as CrO$_3$. The different steps in oxidation are often indicated as follows: *suboxide, oxide, sesquioxide* and *super-* or *peroxide*. Thus: Pb$_2$O, lead suboxide; PbO, lead oxide; Pb$_2$O$_3$, lead sesquioxide; PbO$_2$, lead superoxide (also lead peroxide).

3. Ternary and Higher Compounds.

The most important compounds belonging to this group are the acids, bases, and salts which have the property in common of being electrolytes, which has already been discussed from the standpoint of the theory of ions, p. 84.

A. Acids.

Acids are compounds of hydrogen with non-metals and also with certain metals, this hydrogen being entirely or partly replaceable by the metal on coming in contact with a metal or a metallic hydroxide (a base) or with a metallic oxide (a basic anhydride), producing compounds of the metal called salts. Thus:

$$2\text{HCl} + \overset{\text{II}}{\text{Zn}} = \text{ZnCl}_2 + 2\text{H}$$

Hydrogen chloride.   Zinc.   Zinc chloride.   Hydrogen.

$$\underset{\substack{\text{Sulphuric}\\\text{acid.}}}{\text{H}_2\text{SO}_4} + \overset{\text{II}}{\underset{\substack{\text{Zinc}\\\text{oxide.}}}{\text{ZnO}}} = \underset{\substack{\text{Zinc}\\\text{sulphate.}}}{\text{ZnSO}_4} + \underset{\text{Water.}}{\text{HOH}}$$

$$\underset{\substack{\text{Nitric}\\\text{acid.}}}{\text{HNO}_3} + \underset{\substack{\text{Sodium}\\\text{hydroxide.}}}{\text{NaOH}} = \underset{\substack{\text{Sodium}\\\text{nitrate.}}}{\text{NaNO}_3} + \underset{\text{Water.}}{\text{HOH}}$$

Acids, when soluble in water, have a sour taste and acid reaction (p. 84). According to composition we differentiate between oxygen free acids and those which contain oxygen (oxyacids), and they are called mono-, di-, tri-, etc., basic (mono-, di-, tri-, etc., valent or mono-, di-, tri-, etc., hydric) acids according to whether they contain one, two, three, etc., atoms of hydrogen replaceable by metals.

Acid anhydrides, acidic oxides (also incorrectly called anhydrous acids), are those oxides produced by removing all the hydrogen with the corresponding quantity of oxygen, in the form of water, from one or more molecules of an oxyacid; for example:

$$\underset{\substack{\text{Sulphuric}\\\text{acid.}}}{\text{H}_2\text{SO}_4} = \underset{\substack{\text{Sulphuric}\\\text{anhydride.}}}{\text{SO}_3} + \text{H}_2\text{O} \qquad \underset{\substack{\text{Phosphoric}\\\text{acid.}}}{2\text{H}_3\text{P}\bar{\text{O}}_4} = \underset{\substack{\text{Phosphoric}\\\text{anhydride.}}}{\text{P}_2\text{O}_5} + 3\text{H}_2\text{O}$$

In the same manner the *sulphoacids* yield anhydrides by abstracting H_2S; thus, $2\text{H}_3\text{AsS}_3 = \text{As}_2\text{S}_3 + 3\text{H}_2\text{S}$.

The acid anhydrides do not have any acid reaction and unite with water, forming acids again.

Acid radicals are the groups obtained on the removal of hydroxyl groups OH from a molecule of oxyacids (often do not exist free); for example:

SO_2 acid radical of sulphuric acid $\text{SO}_2(\text{OH})_2$,
NO_2 " " " nitric " $\text{NO}_2(\text{OH})$,
PO " " " phosphoric " $\text{PO}(\text{OH})_3$.

Oxygen free acids are designated by adding the prefix *hydro* to the element or group forming the acid, this element or group ending in *ic;* thus, HCl, hydrochloric acid; HBr, hydrobromic acid; HCN, hydrocyanic acid.

The sulphoacids belonging to this group can be derived from the oxyacids by replacing their oxygen by sulphur and calling them, correspondingly, H_3AsS_4, sulpharsenic acid; HCNS, sulphocyanic acid.

The name of the *oxyacids* is derived by adding the word acid to the name of the element or group forming the acid ending in *ic;*

for example, $HClO_3$, chloric acid; HCNO, cyanic acid; H_2SO_4, sulphuric acid; H_3AsO_4, arsenic acid; H_4SiO_4, silicic acid.

If two acids of the same elements are known we designate the one poorest in oxygen with the termination *ous;* thus, H_2SO_3, sulphurous acid; if a series of acids of the same elements are known, the one poorest in oxygen has the prefix *hypo* (below), while the richest in oxygen has the prefix *hyper* (above) or *per;* thus:

HClO, Hypochlorous acid. $HClO_3$, Chloric acid.
$HClO_2$, Chlorous acid. $HClO_4$, Hyper or Perchloric acid.

If an acid HCl_2O_5 was known it would stand between chlorous acid and chloric acid and would be called hypochloric acid.

The oxyacids of nitrogen are called: hyponitrous acid, HNO; nitrous acid, HNO_2; nitric acid, HNO_3.

B. Bases.

Bases are compounds of hydroxyl groups HO^- with metals, which on coming in contact with an acid replace its hydrogen entirely or partly by the metal contained in it, producing compounds of the metal, called salts; for example:

$$\overset{I}{NaOH} \ + \ \overset{I}{HNO_3} \ = \ NaNO_3 \ + \ HOH$$

Sodium   Nitric   Sodium   Water.
hydroxide.   acid.   nitrate.

When soluble in water the bases have a caustic taste and have an alkaline reaction (basic, p. 84). According as the bases contain one, two, three, etc., hydroxyl groups they are called mono-, di-, tri-, etc., acidic (mono-, di-, tri-, etc., valent or mono-, di-, tri-, hydric bases).

Basic anhydrides are those oxides which are produced when all the hydrogen combined with the corresponding quantity of oxygen as water is removed from one or more molecules of a base; for example:

$$2KOH = K_2O + H_2O; \ Zn(OH)_2 = ZnO + H_2O; \ 2Fe(OH)_3 = Fe_2O_3 + 3H_2O.$$

The name given to the base is obtained by adding the word *hydroxide* to the name of the element forming the base. If an element forms several hydroxides then the designation is the same as the corresponding oxides (p. 97); for example: $Fe(OH)_2$, ferrous hydroxide; $Fe(OH)_3$, ferric hydroxide.

C. Salts.

Salts are the compounds produced by completely or partly replacing the hydrogen of an acid by a metal. This can take place as follows:

By replacing the hydrogen of the acids directly; thus, $Zn + H_2SO_4 = 2H + ZnSO_4$.

By the acid coming in contact with a base or its anhydride, when a double decomposition takes place with the formation of water; thus:

$$KOH + HCl = KCl + HOH;$$
$$Ca(OH)_2 + H_2SO_4 = CaSO_4 + 2HOH;$$
$$CaO + H_2SO_4 = CaSO_4 + HOH.$$

By an acid anhydride combining directly with a basic anhydride; thus, $Fe_2O_3 + 3SO_3 = Fe_2(SO_4)_3$.

Normal or *neutral salts* are those which are obtained when all of the replaceable hydrogen in an acid is replaced by a metal. Most of them have a neutral reaction (p. 85); still, many normal salts and indeed acid salts have an alkaline reaction when they are derived from weak acids; on the other hand many normal salts are acid when they are derived from weak bases (see p. 86).

In giving names to the oxysalts all the salts containing the same salt-forming element are given a generic name by replacing the last syllable of the Latin name of the acid-forming element by the suffix *ate* for those richest in oxygen and the suffix *ite* for an analogous compound poorer in oxygen.

For example:

$MClO$	Hypochlorite
$MClO_2$	Chlorite
$MClO_3$	Chlorate
$MClO_4$	Perchlorate

The special name is formed by placing the name of the metal replacing the hydrogen before the generic name. For example:

$KClO$	Potassium hypochlorite
$NaClO_2$	Sodium chlorite
$NaClO_3$	Sodium chlorate
$AgClO_4$	Silver perchlorate

The special names of two salts which contain the same metal with different valence are derived in the same manner as the oxides, bases, etc. (p. 97), by adding *ic* to the Latin root of the name of the metal replacing the hydrogen when it has the highest valence and *ous* when it has the lowest valence. Thus:

$\overset{\text{II}}{Fe}SO_4$ Ferrous sulphate; $\qquad$ $\overset{\text{III}}{Fe_2}(SO_4)_3$ Ferric sulphate;

$\overset{\text{I}}{Hg}NO_3$ Mercurous nitrate; $\qquad$ $\overset{\text{II}}{Hg}(NO_3)_2$ Mercuric nitrate;

$\overset{\text{I}}{Cu}Cl$ Cuprous chloride; $\qquad$ $\overset{\text{II}}{Cu}Cl_2$ Cupric chloride.

The names of the oxygen free salts are derived by adding *ide* to the name of the acid-forming element; for example, $FeCl_2$, ferrous chloride; $FeCl_3$, ferric chloride; KCN, potassium cyanide. Sulphosalts (see Sulphoacids, p. 98) and complex salts (p. 102) also often terminate with the syllable *ate;* thus, KCNS is called potassium sulphocyanide or potassium sulphocyanate.

Acid salts are those produced when only a part of the replaceable hydrogen in a polybasic acid is replaced by a metal. Ordinarily they have an acid reaction, but still they may be neutral or alkaline in reaction when they are derived from a weak acid (p. 86).

Certain salts, which may be considered as a combination of a neutral salt with an acid anhydride, are often called erroneously acid salts; thus the potassium salt of dichromic acid, $K_2Cr_2O_7 = K_2CrO_4 + CrO_3$, is often called acid potassium chromate.

Acid salts of dibasic acids are named by adding the word *acid* or the prefix *bi* or *hydro* to the acid-forming word; for example, $NaHSO_4$, acid sodium sulphate, sodium bisulphate, sodium hydrosulphate.

Acid salts of tri- and polybasic acids are named according to the number of hydrogen atoms of the acid replaced by metallic atoms as mono-, primary, monobasic; or as di-, secondary, dibasic; or as tri-, tertiary, tribasic salts, etc. Names are also given according to the number of hydrogen atoms left that are replaceable by metals, as follows: simple, double salts, etc. For example:

KH_2PO_4.	$Ca\big\langle\begin{smallmatrix}H_2PO_4.\\ H_2PO_4.\end{smallmatrix}$	Mono-, primary, double acid, monobasic	
K_2HPO_4.	$CaHPO_4$.	Di-, secondary, simple acid, dibasic	Potassium or calcium phosphate.
K_3PO_4.	$Ca_3(PO_4)_2$.	Tri-, tertiary, tribasic, neutral	

Basic salts are those which are obtained when a metal atom replaces all the replaceable hydrogen in an acid with only a part of its valence, while the remaining valences are saturated with basic HO groups or by O atoms. They are designated by adding the word *basic* or the prefix *sub* or *oxy* to the acid-forming element; thus $(HO)_2=Bi-NO_3$, basic bismuth nitrate, bismuth subnitrate, $O=Sn=Cl_2$; tin oxychloride.

Double salts are those salts which are obtained when the replace-able hydrogen of an acid is replaced by different metals—thus, $NaMgPO_4$, sodium magnesium phosphate—or by the combination of several molecules of simple salts; for example, $MgSO_4 + K_2SO_4$, mag-nesium-potassium sulphate.

Complex salts are those compounds which are produced by the combination of simple salts and which differ from the double salts by not having the reactions of their simple constituents (their ions, p. 83), but which have reactions corresponding to the atomic com-plex composed of simple constituents (which functionate as com-plex ions, p. 84). For example, by the union of $2KCl + PtCl_4$ we do not obtain the double salt, but the potassium salt of hydrochlorpla-tinic acid, H_2PtCl_6 (see Platinum and Gold); by the union of $4KCN + Fe(CN)_2$ the potassium salt of ferrocyanic acid, $H_4FeC_6N_6$ (see Cyan-ogen Compounds).

The double salts must not be confounded with the isomorphous mixtures (p. 49) obtained on the common crystallization of isomorphous salts and whose composition changes with the changes in the solution from which they separate out.

I. NON-METALS OR METALLOIDS.

According to the periodic system the elements belonging to this group are classified as follows:

Hydrogen,	Oxygen,	Nitrogen,	Helium,	Carbon,
Fluorine,	Sulphur,	Phosphorus,	Argon,	Silicon,
Chlorine,	Selenium,	Arsenic,	Neon,	Germanium,
Bromine,	Tellurium.	Antimony,	Krypton,	Tin,
Iodine.	———	Bismuth,	Xenon.	Lead.
		Boron.	———	———
———	———			
Univalent.	Bivalent.	Trivalent.		Quadrivalent.

The elements bismuth, germanium, tin, and lead, which have pronounced metallic properties, will be treated of in connection with the metals.

Hydrogen hardly belongs to either of the above groups, as it has both metalloid and metallic characteristics, and forms at the same time the type of all elements.

Hydrogen.

Atomic weight $1.01 = H$.

Occurrence. Free in small quantities in gases of volcanoes and certain petroleum wells, enclosed in the potassium salts of Stassfurt, in the rock salt of Wieliczka, and in the meteoric iron of Lenarto. In the decomposition of organic substances hydrogen is set free; hence it occurs in the intestinal gases of man and certain animals and as traces in the atmosphere. The chief quantity of hydrogen exists in combination with oxygen as water; all plants and animals contain combined hydrogen as a chief constituent. As shown by spectral analysis free hydrogen exists in large quantities in the fixed stars and in the gases surrounding the incandescent solar nucleus.

103

Preparation. 1. Ordinarily hydrogen is prepared by the action of hydrochloric acid or dilute sulphuric acid upon zinc or iron:

$$Zn + H_2SO_4 = ZnSO_4 + H_2.$$
Zinc.   Sulphuric acid   Zinc sulphate   Hydrogen.

On using strong sulphuric acid the generation of hydrogen soon ceases on account of the formation of zinc sulphate, which is not soluble in strong sulphuric acid and which forms a protective coat on the zinc which only dissolves on the addition of water. Pure zinc is only slightly acted upon by dilute acids, but if small amounts of certain salts of the heavy metals (platinum, copper, silver, lead, etc.) are added to the acid, a rapid evolution of hydrogen takes place. The conversion of the H ions of the acid into hydrogen gas takes place on the surface of the zinc with greater difficulty than on the surface of other metals, hence on the addition of salts of other metals the respective metal is precipitated on the surface of the zinc and the hydrogen is therefore evolved with greater ease.

2. By the electrolysis of water (H_2O) to which some acid or base has been added, when the water is apparently directly decomposed (p. 75), 2 vol. hydrogen are simultaneously evolved at the negative pole and 1 vol. oxygen at the positive pole.

3. By the decomposition of water by many metals (p. 113) either at high temperatures or even at ordinary temperatures. The alkali and alkaline-earth metals (which see), such as potassium, sodium, calcium, decompose water even in the cold:

$$2HOH + 2Na = 2NaOH + H_2.$$
Water   Sodium   Sodium hydroxide.   Hydrogen.

Compact zinc, iron, etc., decompose steam only at a red heat (p. 61):

$$3Fe + 4HOH = Fe_3O_4 + 4H_2.$$
Ferrous-ferric oxide.

Red-hot carbon also decomposes water (see Carbon Monoxide), and metallic magnesium decomposes water at its boiling-point.

4. Very pure hydrogen can be prepared by heating sodium formate (see Formic Acid) with sodium hydroxide: $CHNaO_2 + NaOH = Na_2CO_3 + H_2$.

5. Finely powdered zinc, aluminium, iron, evolve hydrogen on warming with caustic alkali: $Zn + 2NaOH = Zn(ONa)_2 + H_2$. $Al + 3KOH = Al(OK)_3 + 3H$.

Technical Preparation. 1. By heating calcium hydroxide, $Ca(OH)_2$, with powdered zinc or iron: $Zn + Ca(OH)_2 = ZnO + CaO + H_2$, or with carbon: $2Ca(OH)_2 + C = CaO + CaCO_3 + 2H_2$.

2. Hydrogen is obtained in the electrolysis of water (see above) or as a by-product in the preparation of potassium hydroxide from a

potassium chloride solution by electrolysis when potassium hydroxide is formed at the negative pole and hydrogen evolved, while chlorine gas is set free at the positive pole:

$$2KCl + 2H_2O = 2KOH + H_2 + Cl_2$$

Properties. Colorless, odorless, and tasteless gas, slightly soluble in water and liquefiable at $-242°$ (p. 41) to a colorless liquid having a sp. gr. 0.07. When allowed to evaporate under the air-pump the temperature sinks to $-252°$ and the remaining liquid solidifies to an ice-like mass. At this lowest obtainable temperature all gases and liquids solidify, hence on introducing them in evaporating liquid hydrogen they liquefy or solidify.

Hydrogen gas is the lightest of all substances; one liter weighing 0.0899 g. at 0° and 760 mm. (p. 42). As a liter of air weighs 1.293 g. at 0° and 760 mm., then hydrogen gas weighs $\dfrac{1.293}{0.0899}$ that of air or is 14.4 times lighter than air, and its sp. gr. relative to air as standard is $\dfrac{0.0899}{1.293}$ or 0.0695. (In regard to the calculation of the absolute weight and specific gravity of gases from their molecular weight see p. 43).

When ignited, hydrogen burns in the air into water with a non-luminous bluish flame which is very hot (p. 112). If a narrow cylinder, open at both ends, be held over a small hydrogen flame, a sound is produced by the vibrations of the heated air (chemical harmonica; also produced by other burning gases).

If hydrogen (or other combustible gas) is passed in the air over finely divided platinum or palladium (spongy platinum, palladium asbestus), it inflames spontaneously (Dobreiner's lamp, gas-lighters) because the metal in the finely divided state has the property of condensing the gases so that their reaction activity is increased; hence they combine and the metal becomes incandescent and ignites the excess of the gaseous mixture (Catalysis, p. 66).

A mixture of hydrogen with oxygen or with air is called "detonating gas," as it explodes on ignition. Hydrogen should therefore be ignited only when all the air has been expelled from the generating apparatus.

Many metals combine with hydrogen, forming so-called hydrides, of which potassium and sodium hydride have a metallic appearance, while the others are white or form gray powders. Palladium and platinum condense hydrogen in themselves without combining therewith and without changing their appearance.

Contrary to the other metalloids, hydrogen shows great conductivity for heat and electricity, a property which is generally ascribed only to the metals.

Because of its lightness hydrogen diffuses (p. 46) readily through animal or vegetable membranes, also through red-hot tubes of iron, platinum, palladium, which do not allow other gases to pass through.

On account of its relationship to oxygen (p. 112) hydrogen abstracts oxygen from many oxygen compounds either in the presence of heat or in the nascent state with the formation of water. This conversion of a compound rich in oxygen into a compound poorer or free from oxygen is called *reduction.*

By reduction we also understand the introduction of hydrogen as well as the replacement of oxygen by hydrogen.

OXYGEN GROUP.

Oxygen, Sulphur, Selenium, Tellurium.

The members of this group are bivalent, although the last three also occur as quadri- and sexivalent elements. They show great similarity to each other. With increase in atomic weight the specific gravity, melting- and boiling-points also increase, and their properties become more and more metallic. All four elements form combinations with 2 atoms of hydrogen which, with the exception of water, H_2O, are gases at the ordinary temperature and have acid-like characters. In regard to the relationship of the members of this group to the elements of the chromium group we refer to this grou⁓

1. Oxygen.

Atomic weight 16 = 0.

Occurrence. Free in the atmosphere (21 vols. per cent.); combined in water in most animal and vegetable tissues, as well as in minerals, so that about one-half of the weight of our planet consists of this element.

Preparation. 1. Ordinarily by heating potassium chlorate ($KClO_3$,) which decomposes into potassium chloride and oxygen: $KClO_3 = KCl + O_3$.

2. By the electrolysis of water to which some acid or base has been added (p. 104) when 1 vol. oxygen is liberated at the positive

pole and as the same time 2 vols. hydrogen are evolved at the negative pole (P. 75).

3. B'y strongly heating many compounds rich in oxygen, i.e., mercuric oxide (HgO) which decomposes into mercury and oxygen: $HgO = Hg + O$; manganese dioxide (MnO_2), which is converted into manganese superoxide and gives off oxygen, $3MnO_2 = Mn_3O_4 + 2O$; and barium dioxide (BaO_2), which decomposes into barium monoxide and oxygen: $BaO_2 = BaO + O$.

4. By heating many substances rich in oxygen, such as manganese dioxide, barium dioxide, potassium persulphate, potassium dichromate with concentrated sulphuric acid, or by heating a solution of chloride of lime with a cobalt salt or copper oxide (see these).

5. It may be prepared in the cold by the action of water upon a mixture of barium dioxide and potassium ferricyanide or by the action of water upon alkali superoxides or percarbonates; of hydrochloric acid upon a mixture of barium dioxide and manganese dioxide; of chloride of lime or potassium permanganate upon hydrogen peroxide, etc. (see these).

Technical Preparation. 1. On heating barium monoxide (BaO) under pressure to 700° in a current of air it is converted into barium dioxide (BaO_2), which at this temperature decomposes into barium monoxide and oxygen when the pressure is diminished. On increasing the pressure and passing air through we again obtain barium dioxide, which decomposes as above stated (p. 72). (Brin process.)

2. Oxygen is evolved on heating calcium plumbate (Ca_2PbO_4) in carbon dioxide gas (CO_2): $Ca_2PbO_4 + 2CO_2 = 2CaCO_3 + Pb + O_2$. On heating the residue in a current of air calcium plumbate is re-formed (Kassner's method).

3. On repeatedly diminishing the pressure on liquefied air (P. 41) a partial evaporation takes place; the liquefied nitrogen evaporating more rapidly than the liquid oxygen on account of its lower boiling-point, l aves a liquid which consists of about 80 per cent. oxygen.

Properties. Colorless, odorless, and tasteless gas, slightly soluble in water, 1.105 times heavier than air, liquefiable at $-182°$ to a light blue liquid having a sp. gr. 1.12 (p. 41) and which solidifies at $-252°$ to an icy mass.

One liter of oxygen gas weighs 1.429 g. at 0° and 760 mm.; hence its specific gravity relative to air is $\frac{1.429}{1.293} = 1.105$.

The atomic weight of oxygen $= 16$ serves as a basis for the determination of the atomic weight of the other elements, and the molecular weight of oxygen = 32 also serves as the basis for the determination of the molecular weight of all other bodies, for the reasons already given on p. 17.

A glowing piece of wood or coal burns in oxygen with great brightness; ignited sulphur burns with a pale blue flame into sulphur dioxide gas, while phosphorus burns with a dazzling white flame into solid phosphorus pentoxide (P_2O_5), etc. Many objects, such as heated iron, which do not burn in the air, burn with scintillations in oxygen into oxides. As oxygen is the constituent of the air which supports combustion, it is natural that bodies should burn more energetically in pure oxygen.

Molten silver absorbs 22 times its volume of oxygen which it gives off on cooling.

All elements, with the exception of fluorine and those of the argon group, combine with oxygen. This process is called *oxidation*, and the resulting compounds are called *oxides*.

By oxidation we also understand the removal of hydrogen by oxygen from a compound or the introduction of oxygen in place of hydrogen with the splitting off of water.

The oxides of the metalloids are nearly always acid anhydrides (p. 98), while the oxides of the metals are nearly always basic anhydrides (p. 99).

We differentiate also between *indifferent oxides* which are derived from metalloids and metals, but which form neither bases nor acids, but still combine directly with acids, forming salts, such as nitrous oxide, nitric oxide, manganese dioxide, lead suboxide, lead peroxide, etc.

These indifferent oxides are those compounds called suboxides and peroxides. These latter readily give off oxygen, hence have a strong oxidizing action. With hydrochloric acid they either evolve hydrogen peroxide or set chlorine free. Oxyacids set free oxygen when they act on peroxides.

Every oxidation is a chemical process connected with the development of heat. If the oxidation of a body takes place very rapidly, it often occurs that such a great development of heat is evolved that the chemical combination takes place with the production of light and the body is popularly said to burn. In ordinary life only such bodies are said to be combustible which burn in atmospheric air because they can combine with the oxygen thereof (combustion in the restricted sense).

If the oxidation takes place very slowly, then the total development of heat is the same as in rapid oxidation; still the temperature cannot rise high enough to produce light. Often the heat cannot be determined on account of the loss of the heat by radiation and conduction.

Decay is called the slow oxidation of organic bodies which takes place with the cooperation of lower organisms. As the final products are the same in this process as in the combustion of organic bodies, we can also consider this process as a slow combustion.

Respiration is also a slow process of combustion when the oxygen taken up by the blood combines with a part of the carbon of the tissues, forming carbon dioxide, which is expired, while the other oxidation products are eliminated by the urine, etc. The body temperature is produced by this oxidation.

The green plants take up from the atmosphere the carbon dioxide produced in respiration of animals, combustion, decay, etc., by means of the stomata of the leaves, and decompose this under the influence of the light into carbon, which serves to build up its tissues, and into oxygen, which is eliminated. The plants therefore perform a reduction process which does not occur in the animal body.

Combustion, in a chemical sense, is any chemical process accompanied by the production of light. As the burning of a body depends upon chemical processes, therefore oxygen cannot burn in the air, as there exists no substance there with which it can combine. Oxygen, on the contrary, burns in hydrogen, ammonia, sulphur vapors, etc., because it combines with these bodies with the development of heat as soon as the necessary temperature, ignition temperature (see below) is sufficient to commence the combination.

As chlorine does not combine directly with oxygen, it does not burn in oxygen, hence also not in the air. Oxygen, on the contrary, does combine with hydrogen and burns therein for the same reason that hydrogen burns in chlorine; coal-gas burns in the air, hence the air (its oxygen) must burn in coal-gas. It follows from this, therefore, that burning or the combustibility of bodies are only relative phenomena.

An ignited body generally continues to burn because of the heat set free in the combination of the particles acting upon each other, heating other particles to their ignition temperature.

On quickly cooling (as by introducing a cold metal in a small flame) every flame may be extinguished. If a wire gauze is held over a tube from which coal-gas is escaping, the gas may be ignited above the wire gauze because the metallic gauze conducts away the heat so well and cools it off so that the gas between the gauze and tube cannot ignite. The Davy safety lamp used in coal-mines as a protection from explosion of fire-damp is constructed on this principle. The oil-lamp is entirely surrounded by wire gauze, and if such a lamp is brought into a mixture of explosive gases they ignite at the flame, but this flame cannot pass through the wire gauze as this cools the gases below their ignition temperature.

Ozone or Active Oxygen.

Oxygen is also known in an allotropic modification (p. 80), which, on account of its powerful oxidizing power, is called "active oxygen," and because of its odor is called "ozone" ($\check{o}\zeta\epsilon\iota\nu$, smell).

Occurrence. Ozone occurs as traces in the air, for instance after a lightning stroke, also in the neighborhood of "graduation houses" (see Common Salt) and on the seashore, as it always forms when large quantities of water quickly evaporate. It is formed to a less extent also in the oxidation and combustion processes in oxygen or in the air. In these processes hydrogen peroxide is nearly always produced, and most of the processes, such as, for instance, "grass-bleaching" which is generally ascribed to the ozone of the air, is due to the hydrogen peroxide (which see).

Turpentine and various other ethereal oils, also triethylphosphine, absorb oxygen with the formation of peroxides, which, like ozone, have an energetic oxidizing power. This property has in the past been ascribed to ozone.

Preparation. 1. By passing the electric spark through oxygen. Ozone may be obtained in larger quantities in a special apparatus, whereby oxygen is exposed to a high tension electric current without the formation of sparks (the dark electric discharge).

2. In the electrolytic decomposition of water ozone is formed, besides oxygen, at the positive pole.

3. By passing oxygen over sticks of moistened phosphorus.

4. By introducing barium dioxide, sodium peroxide, permanganates, persulphates, or percarbonates (best mixed with sand) in cold concentrated sulphuric acid.

5. By passing oxygen over manganese dioxide (MnO_2) or minium (Pb_3O_4), which must not be heated above 400°. By these methods oxygen containing a maximum of 9 per cent. ozone can only be obtained. If this is liquefied by being cooled with liquid air and then

allowed to slowly evaporate, the oxygen goes off at $-182°$, while the ozone which boils at $-120°$ remains as a deep-blue liquid.

Properties. Colorless, in deep layers a bluish gas having a peculiar odor similar to chlorine, when not too dilute causing an irritation of the mucous membranes, liquefiable at $-120°$, forming a deep-blue liquid which readily explodes as it is suddenly transformed in oxygen gas with the development of heat. It is readily soluble in ethereal and fatty oils and only slightly soluble in water. In the solution of ozone in water most of it is converted into oxygen and hydrogen peroxide: $H_2O + O_3 = H_2O_2 + O_2$. Although oxygen generally combines with other bodies only at higher temperatures, ozone has an oxidizing action even at ordinary temperatures (especially when moist). Bright silver is converted by ozone into black silver peroxide, white lead hydroxide into brown lead peroxide, black lead sulphide (PbS) into white lead sulphate ($PbSO_4$), etc. It destroys all vegetable pigments by oxidation (use in bleaching) and oxidizes all organic substances; hence rubber tubes must not be used in its preparation.

If ozone is passed through a glass tube heated above $400°$, it is transformed into oxygen gas, when the volume increases one-half; the specific gravity of ozone is one-half greater than that of oxygen, that is 24 instead of 16. Hence the molecular weight of ozone is 48. From this it follows that a molecule of ozone contains 3 atoms of oxygen; hence 2 volumes of ozone yield 3 volumes of oxygen.

In the formation of ozone a considerable addition of energy takes place in the form of heat, electricity, etc. This explains the great chemical activity of ozone; as on oxidation with ozone 32.4 more calories are set free than with oxygen (p. 69). As an endothermic compound liquid ozone may be suddenly converted into oxygen gas with explosive violence.

Detection. 1. Paper moistened with potassium iodide solution and starch paste turns faintly or deep blue, depending upon the quantity of ozone.

Potassium iodide is not changed by oxygen, but by ozone, on the contrary, the potassium is oxidized and the iodine set free: $O_3 + 2KI + H_2O = O_2 + 2KOH + I_2$. Free iodine can be detected by its property of turning starch deep blue.

Precipitated gum guaiacum turns blue with ozone.

Hydrogen peroxide also gives the reaction with potassium iodide and starch, although only after some time, while chlorine, bromine, and nitric oxide gases also give both reactions.

2. In order to differentiate between ozone and the above-mentioned bodies we make use of paper moistened with an alcoholic solution of tetramethyldiamidodiphenylmethane. In ozone this paper turns violet, with nitric oxide yellow, with chlorine and bromine blue, and does not change with hydrogen peroxide.

a. Compounds with Hydrogen.

Water, H_2O. Hydrogen Peroxide, H_2O_2.

Water, Hydrogen Monoxide, H_2O or $H-O-H$. *Occurrence.* Never chemically pure, but as sea-water, river-water, spring-water, in the form of clouds, fog, rain, ice, snow, hail, dew, as well as invisible vapor of water in the air, as water of crystallization in minerals, and as a constituent of all plants and animals. Water is also one of the products of the combustion of all organic bodies, the processes of respiration of animals, and the union of acids with bases with the formation of salts (p. 100), etc.

Formation. 1. By burning hydrogen in oxygen or air (it is the oxygen of the air which maintains the combustion): $2H + O = H_2O$.

2. By the union of two volumes of hydrogen and one volume of oxygen (synthesis of water). This mixture, which may be kept without combining, is called "detonating gas," as both gases instantly unite with a violent sound and powerful explosion when they are ignited by a burning body or by spongy platinum (p. 105) or by the electric spark; the explosion temperature of detonating gas lies between 500° and 600°, that is, at this temperature the reaction velocity of both gases is accelerated to a remarkable degree (p. 73).

If the water produced is converted into a vapor by heating, then it follows that from three volumes of detonating gas two volumes of vapor of water are produced (p. 12).

The temperature produced on the union of hydrogen with oxygen is about 2000°; the quantity of heat developed is 68 large calories (p. 70).

This high temperature is made use of in the oxyhydrogen blowpipe where both gases first come together at the point of combustion, as otherwise the entire mixture would inflame and explode. Many highly refractory metals, such as platinum, fuse in this flame; when burnt lime (calcium oxide) or zirconia are heated therein to a strong white heat they emit an intense light, which is used for projections, etc. (Drummond light, zircon light).

3. If hydrogen is passed over heated metallic oxides, copper oxide (CuO), iron oxide (Fe_2O_3), these are reduced into metals with the formation of water: $CuO + 2H = Cu + H_2O$. $Fe_2O_3 + 6H = 2Fe + 3H_2O$. If the copper oxide is weighed before and after the experiment, and the water formed also weighed, then the loss in weight of the copper oxide represents the quantity of oxygen present in the water produced.

In explanation of the apparent contradiction that H yields H_2O with iron oxide and that iron with water yields H see p. 61.

Preparation. In order to prepare chemically pure water (*aqua destillata*) on a large scale, ordinary water is distilled, i.e., we convert the water into vapor in a retort by boiling and condensing this vapor again by a cool surface (p. 50). In this procedure all dissolved salts, etc., remain in the retort, as they are not volatilized with the vapor of water, while the dissolved gases (air, carbon dioxide, and ammonia) contained in the water pass off with the vapor produced; hence the first portions of the distillate must be discarded.

Properties. Pure water is a tasteless and odorless fluid, colorless in thin layers and pronouncedly blue in layers of six to eight meters; only slightly compressible, a poor conductor of heat and electricity. It is neutral in reaction, i.e., it has neither acid nor alkaline properties, but forms bases with basic oxides, and acids with acidic oxides (pp. 98 and 99).

The alkali and alkaline-earth metals decompose water even at ordinary temperatures. The other metals (with the exception of lead, bismuth, copper, mercury, silver, platinum, and gold), as well as the metalloid carbon, only decompose water at higher temperatures with the setting free of hydrogen (p. 104).

Water serves as the unit for the determination of specific gravities of solid and liquid bodies, as well as the determination of specific heat of all bodies, as it possesses the greatest specific heat, with the exception of hydrogen (p. 22).

If the electric current is passed through acidulated water, twice as much hydrogen is set free at the negative pole as oxygen at the positive pole.

Water solidifies at 0° (ice formation) and expands at the same

time.   One vol. water at 0° yields 1.07 vols. ice at 0°.   Ice has there‑
fore a sp. gr. 0.93 and floats upon water.   The ice-flowers on the
windows and the snowflakes consist of regularly grouped crystals
of the hexagonal system.

The solidification of water can be prevented, and ice can be melted,
by great pressure (p. 35).   Two pieces of ice pressed together may be
melted on the surface pressed, and be made to adhere to each other,
as the water formed in melting immediately solidifies again as soon
as the pressure is relieved.   The movement of glaciers is dependent
upon this fact, because the masses of ice resting upon the rocks by their
great pressure cause a liquefaction of the under layers of ice and this
allows of the movement of the upper layers.

The expansion of water on freezing is of great mechanical importance
in nature, as the water which has penetrated the rocks ruptures them
on freezing.   By repetition of this process large masses of rocks are
gradually broken up into smaller pieces and then quickly "weather."
Iron bombs filled with water and closed tightly are ruptured on cooling
below zero.

Water has its maximum density at 4°.   Above and below 4° it
again expands; water at 9° has the same density as water at 0°.
The weight of one cubic centimeter of water at 4° serves as the unit
of weight and is called a gram.

The remarkable exception of water to the laws of expansion, although
it is slight, is of considerable importance in the economy of nature.   If
the surface of lakes, ponds, and rivers is cooled, then the water on
the surface becomes heavier and sinks, while warmer, lighter water comes
to the surface, until by degrees the temperature of the total mass of
water becomes 4°.   If now a further cooling takes place, then the colder
water remains on the surface and this only solidifies into ice.   If the
density of water increased to 0°, then the entire mass of water would
be cooled to the freezing-point and converted into ice; the heat of the
summer would then not be sufficient to melt this mass of ice.

Water which contains salts in solution freezes below 0° (p. 19) and
has its maximum density at another temperature; thus with sea-water
it lies below 0°, but this extensive mass is never cooled to its freezing-
point.

When water passes from the solid to the liquid state, there occurs,
besides a diminution in volume, a disappearance of heat (p. 33).

If one kilo of water at 0° is mixed with one kilo of water at 80°, we
obtain two kilos of water at 40°; but if we mix one kilo of ice at 0° and
one kilo of water at 80°, we then obtain two kilos of water at 0°.   The
quantity of heat contained in the warm water has disappeared.   The
heat of fusion of water is, therefore, 80 heat-units (p. 36).

At 100° and a pressure of 760 mm. (p. 37) water is converted into
vapor, and at higher temperatures it decomposes into its elements

(see dissociation, p. 71). Even at ordinary temperatures water and ice evaporate. This conversion into vapor must necessarily be accompanied by absorption of heat; hence the lower temperature of the seacoast depends upon the evaporation of the sea-water. On account of the evaporation of water, the gases generated from watery solutions are always moist (p. 40). Vapor of water is colorless and transparent; one volume of water at 100° yields 1696 volumes of vapor at 100°; one liter of vapor of water weighs at 100° and 760 mm. pressure 0.59 gram. In the passage of water at 100° into steam at 100°, as in the passage of ice into water at 0°, a considerable quantity of heat is absorbed and set free again on the condensation of the steam.

One kilo of vapor of water at 100° warms on conversion into water at 100° 5.36 kilos of water from 0 to 100° or one kilo 536°. The heat of vaporization of water, therefore, amounts to 536 heat-units. Saturated vapors do not follow the law of gases (p. 15). Their vapor tension increases to a much greater degree than their temperature; i.e., the vapor tension (p. 37) of saturated vapor of water at

−10° is	2.1 mm.	100° is	760 mm.
0° "	4.6 "	120.6° "	2 atmospheres.
+20° "	17.4 "	180.3° "	10 "
40° "	54.9 "	365.0° "	194.6 "

Above 365° steam cannot be made liquid by any pressure; 365° is the critical temperature, 194.6 atmospheres the critical pressure of steam (p. 41).

Water of crystallization is that water contained in many crystalline bodies, chemically combined, which stands in certain relationship to the crystalline form.

The quantity of water which a salt at equal temperature takes up in its crystallization is always the same; at different temperatures a salt may unite with different quantities of water and then also has different crystalline forms; thus when magnesium sulphate crystallizes above 20° it forms tetragonal crystals, $MgSO_4 + 4H_2O$, between 7° and 20° triclinic crystals, $MgSO_4 + 5H_2O$, below 6° monoclinic crystals, $MgSO_4 + 7H_2O$.

Water of constitution, or water of hydration, is that portion of the water of crystallization which is firmly united and which, when it is given off, causes a greater change in the properties of the substances to which it belongs than the water of crystallization (see Magnesium Sulphate).

Efflorescence is the change which many crystals undergo when exposed to dry air whereby they lose water of crystallization; they sometimes retain their form, but become dull and non-transparent, and generally are converted into a powder.

Hygroscopic salts are those which attract water from the air, so that they often, when they are very soluble therein, deliquesce.

Natural Water. The water occurring upon the earth is not pure, but it dissolves the earth layers through which it flows to a more or less extent. Besides this it contains carbon dioxide and air. We differentiate between hard water, i.e., those which contain considerable calcium and magnesium salts in solution, and soft water, which contains little solid matter. Because of the formation of insoluble lime soaps, hard water is not suitable for washing (see Soaps); peas and beans when boiled in it do not become tender, as the proteids contained in them form with the lime salts insoluble, hard compounds.

1. Rain- and snow-water (meteoric water) is nearly pure water; it contains only a little air, carbon dioxide, and ammonium nitrate.

2. Spring- or ground-water is generally hard water; with the aid of the absorbed carbon dioxide it dissolves insoluble calcium and magnesium carbonate forming acid carbonates:

$$CaCO_3 + H_2O + CO_2 = Ca(HCO_3)_2.$$
Calcium carbonate.   Acid calcium carbonate.

They also generally contain calcium sulphate (gypsum), etc. On boiling these waters, carbon dioxide is driven off, when the carbonates become insoluble again and precipitate, while the sulphate remains dissolved; the hardness becomes less by boiling.

Absolute or total hardne s is the hardness of waters before boiling; permanent hardness is that which remains after boiling; temporary hardness, that which disappears on boiling. The hardness is determined by the addition of an lcoholic solution of soap whose strength has been previously determined by a calcium salt. It forms an insoluble lime soap, and a fine permanent lather is formed only when all the calcium and magnesium salts have been precipitated. The hardness is measured in degrees of hardness. A degree of hardness corresponds to one part by weight calcium carbonate in 1000 parts by weight of water.

3. River water is soft, although it is often originally hard water; this latter loses its carbon dioxide and hence the carbonates in the act of flowing. Below cities river-waters contain considerable organic matter.

4. Mineral waters are natural waters which contain either large quantities of solid or gaseous bodies or have higher temperature than ordinary water and hence are used for medicinal purposes.

We differentiate chiefly between:

    a. Thermal; these have, as they appear on the surface, a higher temperature than the surrounding atmosphere.

b. Sparkling; these contain especially considerable free carbon dioxide; the alkaline ones besides this sodium carbonate; the saline, sodium chloride; the alkaline-saline, carbon dioxide and sodium sulphate or common salt.

c. Bitter water contains considerable magnesium salts.

d. Sulphur-water contains sulphuretted hydrogen.

e. Salt water contains common salt, also bromine and iodine salts.

f. Chalybeate water contains iron salts.

5. Sea-water differs from all other waters by its containing large amounts of common salt, which amounts on an average to 2.7 per cent., and also contains bromine, iodine, calcium, magnesium compounds, etc., so that the quantity of solid bodies is about 3.5 per cent.

6. Potable water. For this purpose not only is spring-water used, but also water from rivers and lakes. Such waters are purified from impurities and insoluble substances by filtration by passing it through a receptacle which contains above sand, then gravel, then small stones and, below, larger stones. In order to purify potable waters for household use, various filters of carbon or spongy iron are used whereby the water becomes clearer and of a better taste; still the injurious bodies cannot be wholly eliminated by any process of filtration. Good drinking-water must be clear, colorless, and odorless, must have a fresh taste (due to the carbon dioxide), must not contain any ammonia, no nitrous acid, and only small quantities of chlorine, nitrates, or sulphates, organic substances, and must not be too hard; the evaporated residue must not show under the microscope any fungi, infusoria, etc. A drinking-water must not necessarily be discarded because it contains the above compounds, but because their presence may possibly indicate a contamination with animal decomposition products which are not directly detectable.

Hydrogen Peroxide, H_2O_2 or H–O–O–H. *Occurrence.* In very small amounts in the air, in rain and snow; it is formed to a slight extent in the evaporation of oil of turpentine (occurrence in the air of pine forests) and other ethereal oils, and in the slow oxidation (action in grass-bleaching) and combustion in the presence of water as well as in the electrolysis of water.

Preparation. 1. On passing carbon dioxide into water holding barium dioxide (BaO_2) in suspension, or this latter added to dilute cold sulphuric acid (p. 110):

$$BaO_2 + H_2O + CO_2 = BaCO_3 + H_2O_2;$$
$$BaO_2 + H_2SO_4 = BaSO_4 + H_2O_2.$$

The insoluble barium carbonate or barium sulphate which precipitates is filtered off; the obtained dilute solution may be concentrated by evaporation not over 70° until it contains 45 per cent. H_2O_2, this solution is then shaken with ether which dissolves the H_2O_2, and this latter solution on distillation yields H_2O_2, which is further concentrated in a vacuum.

2. By dissolving sodium peroxide in ice-water (or dilute acids) we obtain a dilute solution of hydrogen peroxide besides sodium hydroxide (or the corresponding salt): $Na_2O_2 + 2H_2O = 2NaOH + H_2O_2$.

Properties. Colorless and odorless, bitter, acid-reacting, sirupy liquid readily soluble in water and which irritates the skin, and which spontaneously evaporates in the air. In thick layers it forms a blue liquid having a sp. gr. 1.5 and which is soluble in water and which when strongly cooled forms crystals which melt at $-2°$. If H_2O_2 contains only traces of solid substances of any kind, then it begins even in dilute solutions to slowly decompose at $20°$; with greater heat it effervesces, and decomposes into water and oxygen often with explosive violence. This decomposition may also be brought about by many finely divided metals, such as platinum, gold, silver, or manganese dioxide, carbon, etc., without being changed themselves (Catalysis, p. 66).

In the dilute condition it is rather stable and occurs in commerce as a 3 per cent. by weight (or 10 per cent. by volume) solution in water.

Hydrogen peroxide is a powerful oxidizing agent, because of its ready decomposition with the production of nascent oxygen; it therefore bleaches many pigments (hair, ostrich-feathers), converts dark hair into blonde, transforms black lead sulphide into white lead sulphate (restoration of darkened oil-paintings), arsenious acid into arsenic acid, etc., precipitates brown lead peroxide (PbO_2) from lead acetate solution. This lead peroxide is converted into yellow lead oxide (see below) by an excess of H_2O_2, and red chromium trioxide is oxidized to blue perchromic anhydride.

It also acts as an active reducing agent upon many unstable oxides and peroxides, as well as upon compounds of many metals rich in oxygen, as the loosely combined oxygen atoms in these bodies in contact with H_2O_2 form with the loosely combined oxygen atom a free oxygen molecule.

Thus with ozone it gradually decomposes into water and ordinary oxygen: $O_3 + H_2O_2 = 2O_2 + H_2O$.

Silver oxide (Ag_2O) is transformed with generation of oxygen into metallic silver: $Ag_2O + H_2O_2 = 2Ag + H_2O + O_2$.

Potassium permanganate ($KMnO_4$) is converted in the presence of sulphuric acid into colorless manganous sulphate ($MnSO_4$) whereby one-half of the oxygen is set free: $2KMnO_4 + 5H_2O_2 + 3H_2SO_4 = K_2SO_4 + 2MnSO_4 + 8H_2O + 5O_2$.

Calcium hypochlorite, $Ca(ClO)_2$, is converted into calcium chloride, when one-half of the oxygen is set free: $Ca(ClO)_2 + 2H_2O_2 = CaCl_2 + 2H_2O + 2O_2$.

Detection. 1. Paper moistened with potassium iodide and starch, or with tincture of guaiacum, becomes blue immediately in contact with hydrogen peroxide. Indigo solutions are decolorized after the addition of ferrous sulphate solution (ozone also gives these reactions immediately without ferrous sulphate).

2. By the yellow color produced with a solution of titanic acid in dilute sulphuric acid, or the orange color with vanadic acid in dilute sulphuric acid.

3. If hydrogen peroxide is added to a red watery solution of chromium trioxide (or a chromate treated with sulphuric acid), it oxidizes it into the deep-blue perchromic anhydride, Cr_2O_9, which on carefully shaking with ether is dissolved, but which quickly decomposes into chromium trioxide and oxygen.

2. Sulphur.

Atomic weight $32.06 = S$.

Occurrence. 1. Free, sometimes crystalline, but generally mixed with earthy matter, especially in volcanic regions, principally in Sicily and Iceland, having been formed perhaps by sulphur dioxide (SO_2) and sulphuretted hydrogen (H_2S) coming in contact with each other: $2H_2S + SO_2 = 3S + 2H_2O$.

2. In many minerals it occurs combined with metals, which are divided according to their physical properties into blendes, glance, and pyrites.

3. In the form of sulphuric acid salts, especially as calcium sulphate (gypsum), which forms immense deposits.

4. It is found combined with other elements to a slight extent in animals and plants, especially in the protein bodies, albuminoids, muscles, epidermis, in the bile, and in many algæ and bacteria.

Preparation. 1. The sulphur is separated from the matrix by melting the same, in the locality where it is found, and occurs in commerce in irregular masses called crude sulphur. The crude sulphur is heated in iron vessels, and the sulphur vapors formed are condensed in chambers of masonry (sublimation, p. 36).

If this process takes place slowly, then the temperature of the chambers does not rise above the melting-point of the sulphur and the vaporized sulphur precipitates (like snow from watery vapor) as a fine crystalline powder, and occurs in commerce as flowers of sulphur.

If the sublimation is rapid, or if the sulphur vapors are passed into the chamber for a long time, then the temperature of the chamber becomes so high that the condensed sulphur melts; it is then poured into wooden moulds, forming the rolled sulphur or rolled brimstone of commerce which has a crystalline fracture.

2. In the Leblanc soda manufacture (which see) about 30 per cent: calcium sulphide, CaS, is obtained as a by-product from which the sulphur can be obtained (see Soda).

3 Sulphur may also be obtained by heating iron pyrites in the absence of air: $FeS_2 = FeS + S$.

Properties. A yellow, brittle, crystalline, tasteless and odorless solid which on rubbing becomes electrified, is insoluble in water, but somewhat soluble in alcohol and ether, and readily soluble in turpentine, benzol, carbon disulphide, and rather soluble in fatty and ethereal oils.

It melts at 114°, forming a pale-yellow, môbile liquid which at 160° turns brown and less fluid; at 230° it becomes dark brown and very viscous, so that the vessel may be turned upside down without the liquid flowing out. On heating still higher it becomes again liquid, but not light in color, and at 448° it begins to boil and is converted into a brownish-yellow vapor.

When sulphur is heated in the air or in oxygen to 260° it inflames and burns with a bluish flame with the formation of sulphur dioxide (SO_2), an irritating gas. Sulphur has, next to oxygen, the greatest affinity for other elements, with the exception of fluorine, argon, helium, neon, krypton, and xenon, and unites often in several proportions with the elements. The compounds of sulphur correspond nearly in composition to those of oxygen, and are called sulphides or polysulphides when they contain more than one atom of sulphur in the molecule.

Sublimed sulphur always contains some sulphurous or sulphuric acid, often also arsenic sulphide, which can be removed by treating it with a dilute solution of ammonia and then washing with water.

If sulphur which has been melted at 114° is allowed to cool slowly, thin pliable monoclinic prisms having a specific gravity of 1.96 and melting at 120° are obtained. These crystals on shaking, or on being kept for some time, become non-transparent and brittle and are transformed into rhombic octahedra.

Natural sulphur crystallizes always in rhombic octahedra having a specific gravity of 2.07; from superheated sulphur (p. 35) · rhombic octahedra, which melt at 90°, may separate out. Hence sulphur is dimorphous; both forms separate out always as rhombic octahedra from its solution in carbon disulphide after the evaporation of the same

Detection. 1. When sulphur is heated it melts and volatilizes; it burns with a blue flame, forming sulphur dioxide, which may be detected by its odor.

2. All sulphur compounds when heated with soda upon charcoal yield sodium sulphide, which can be detected by the generation of sulphuretted hydrogen when the fused mass is treated with acids, or when the mass is moistened with water and placed upon a silver coin, when a dark-brown spot of silver sulphide will be obtained.

See also Part III, "Elementary Analysis."

Plastic Sulphur. If sulphur which has been heated above 230° is slowly poured into cold water, we obtain an amorphous, brownish-yellow, transparent, elastic mass, which is called plastic sulphur, having a specific gravity of 1.95 and which gradually solidifies, being converted into rhombic sulphur. Plastic sulphur is a mixture of the two amorphous modifications of sulphur, of which the brown one is soluble in carbon disulphide, while the other remains undissolved as a yellow amorphous powder.

Flowers of sulphur (p. 120) is a mixture of the insoluble amorphous sulphur and of rhombic sulphur.

Milk of sulphur, a third modification, is produced when sulphur is set free from aqueous solutions of metallic polysulphides by acids: $K_2S_5 + 2HCl = 2KCl + H_2S + 4S$. (If, on the contrary, a solution of a metallic polysulphide is added to an excess of acid, hydrogen persulphide separates, see p. 124.) This modification is soluble in carbon disulphide and forms a fine, dirty-white powder which is gradually transformed into rhombic sulphur.

The existence of different allotropic modifications of sulphur (P. 80) can be explained by the fact that the molecule of these different modifications consists of a different number of atoms (see Dissociation, p. 72).

a. *Compounds with Hydrogen.*

Hydrogen Sulphide, H_2S.   Hydrogen Persulphide, H_2S_2.

Hydrogen Sulphide, Sulphuretted Hydrogen, H_2S.   *Occurrence.*
To a slight extent in the gases of volcanoes and in sulphur-waters;
also where organic bodies containing sulphur undergo putrefaction, as
well as in the gases of the intestine of carnivorous animals and in
pathological urine.

Formation. 1. By the action of nascent H upon sulphur dioxide: $SO_2 + 6H = 2H_2O + H_2S$.

2. By heating sulphur and many metallic sulphides in a current
of hydrogen: $Ag_2S + 2H = 2Ag + H_2S$.

Preparation. By the action of dilute hydrochloric or sulphuric
acid upon metallic sulphides, especially ferrous or antimony sulphide:

$$FeS + H_2SO_4 = FeSO_4 + H_2S;$$
$$Sb_2S_3 + 6HCl = 2SbCl_3 + 3H_2S.$$

Properties. Colorless, poisonous, gas having a disagreeable odor
similar to rotten eggs (which contain the same), having a specific
gravity of 1.18; inflames and burns with a blue flame into sulphur
dioxide and water: $H_2S + 3O = H_2O + SO_2$; if the supply of oxygen
is diminished, then sulphur is set free: $2H_2S + 4O = 2H_2O + SO_2 + S$.

It can be liquefied at $-74°$, forming a colorless liquid which
solidifies in crystals at $-85°$. One volume of water at $0°$ dissolves
3.7 volumes of the gas; this solution reddens litmus, and is oxidized
into water with the setting free of sulphur by many compounds containing oxygen, such as chromic acid, nitric acid, etc., as well as by
the oxygen of the air: $H_2S + O = H_2O + S$.

Stronger oxidizing agents, such as fuming nitric acid, lead peroxide,
may ignite the gas, and when mixed with oxygen it explodes when a
light is applied. Sulphuretted hydrogen is, therefore, a powerful
reducing agent. Chlorine, bromine, and iodine decompose it with the
setting free of sulphur, they combining with the hydrogen: $H_2S + 2Cl =$
$2HCl + S$. Sulphur dioxide acts in a similar manner: $SO_2 + 2H_2S =$
$2H_2O + 3S$.

Most metals decompose sulphuretted hydrogen on heating them
together with the formation of sulphides (p. 120), and hydrogen is set
free at the same time. The metallic oxides have a similar action and

form water. With certain of these bodies the combination takes place even in the cold; hence silver, copper, white lead, etc., turn black even in the air, as this often contains small amounts of sulphuretted hydrogen: $2Ag + H_2S = Ag_2S + 2H$.

Sulphuretted hydrogen behaves like the analogous hydrogen acids of the chlorine group, and the sulphides may be considered as salts of hydrosulphuric acid.

The acid sulphides, such as NaHS, have a neutral reaction in aqueous solution because the HS' ion is only slightly dissociated, so that the acid reaction of the H' ions is not sensible, while the neutral sulphides, on the contrary, have an alkaline reaction because the hydroxyl ions OH' are formed by hydrolytic dissociation (p. 86): $Na_2S + HOH = Na\cdot + HS' + Na\cdot + OH'$.

On account of its behavior to metals and their compounds, H_2S, as well as its aqueous solution, is an important reagent and precipitant for the metals.

The sulphides obtained with H_2S may be divided into the following three groups:

1. Sulphides which are not acted upon by dilute acids.

2. Sulphides which are insoluble in water, but are decomposed by acids.

3. Sulphides which are soluble in water.

Because of this behavior sulphuretted hydrogen can be used in chemical analysis in separating the metals into these three groups by passing sulphuretted hydrogen into the solution of the metallic salt to be tested which has previously been acidified with an acid, when the metals of the first group precipitate; if the precipitated sulphides are removed by filtration and the filtrate neutralized, then the sulphides of the second group precipitate, while those of the third group remain in solution.

Many sulphides have a characteristic color, so that H_2S is not only a means of separating the metals into three groups, but it may also serve as a means of identification for certain of the metals. Thus from antimony solutions it precipitates orange-red antimony sulphide, from arsenic solutions yellow arsenic sulphide, from zinc solutions white zinc sulphide, from manganese solutions flesh-colored manganous sulphide, from iron solutions black ferrous sulphide: $FeSO_4 + H_2S = FeS + H_2SO_4$; but as the manganous, ferrous, and zinc sulphides are soluble in acids, the sulphuretted hydrogen only precipitates these when the acid set free is neutralized.

Detection. 1. Sulphides generate sulphuretted hydrogen gas when treated with acids. This gas may be detected by its odor and by its blackening paper moistened with a lead salt solution.

2. Sodium nitroprusside gives a violet color with sulphide solutions.

Hydrogen persulphide, H_2S_2 or $HS-SH$, is obtained when a watery solution of a polysulphide, for instance, of calcium disulphide (CaS_2), is added drop by drop to an excess of dilute hydrochloric acid: $CaS_2 + 2HCl = CaCl_2 + H_2S_2$ (p. 121). It forms a thick, yellow liquid with a disagreeable odor, which bleaches organic pigments, and which gradually decomposes, but quicker on heating, into $H_2S + S$. Hydrogen persulphide probably has the formula H_2S_5, which body is formed first from the unstable hydrogen polysulphide produced: $4H_2S_4 = 3H_2S_5 + H_2S$; $4H_2S_2 = H_2S_5 + 3H_2S$; $4H_2S_3 = 2H_2S_5 + 2H_2S$.

b. Compounds with Oxygen.

—	Hyposulphurous acid,	$H_2S_2O_4$.
Sulphur dioxide, SO_2.	{ Sulphurous acid,	H_2SO_3.
	{ Pyrosulphurous acid,	$H_2S_2O_5$.
Sulphur trioxide, SO_3.	{ Sulphuric acid,	H_2SO_4.
	{ Pyrosulphuric acid,	$H_2S_2O_7$.
Sulphur sesquioxide, S_2O_3	—	
Sulphur heptoxide, S_2O_7		
—	Persulphuric acid,	$H_2S_2O_8$.
—	Oxysulphuric acid,	H_2SO_5.
—	Thiosulphuric acid,	$H_2S_2O_3$.
—	Dithionic acid,	$H_2S_2O_6$.
—	Trithionic acid,	$H_2S_3O_6$.
—	Tetrathionic acid,	$H_2S_4O_6$.
—	Pentathionic acid,	$H_2S_5O_6$.

Of these acids only sulphuric acid and pyrosulphuric acid can be prepared; the other acids are only known in aqueous solution or as salts. Of the last six acids the anhydrides are not known. The last four acids form the group of acids called polythionic acids ($\pi o \lambda \dot{v} s$, many, $\theta \epsilon \hat{\iota} o \nu$, sulphur). Sulphuric acid, pyro- and oxysulphuric acids, as well as their salts, are precipitated by barium salt solutions from their acid solutions.

Sulphur Dioxide, Sulphurous Anhydride, SO_2. *Occurrence.* In the gases of volcanoes.

Preparation. 1. On a large scale by burning sulphur or by roasting sulphides, or from the residue from the Leblanc soda manufacture (p. 127).

2. Ordinarily by heating copper, carbon, or sulphur with concentrated sulphuric acid:

$$2H_2SO_4 + C = 2H_2O + CO_2 + 2SO_2;$$
$$2H_2SO_4 + Cu = 2H_2O + CuSO_4 + SO_2;$$
$$2H_2SO_4 + S = 2H_2O + 2SO_2.$$

3. In smaller quantities by the action of sulphuric acid upon calcium sulphite or upon sodium bisulphite:

$$CaSO_3 + H_2SO_4 = CaSO_4 + H_2O + SO_2;$$
$$2NaHSO_3 + H_2SO_4 = Na_2SO_4 + 2H_2O + 2SO_2.$$

4. By heating sulphur with metallic oxides:

$$2CuO + 3S = 2CuS + SO_2; \qquad MnO_2 + 2S = MnS + SO_2.$$

Properties. Colorless, irritating, neutral gas, 2.21 times heavier than air, liquefiable at $-10°$ (p. 41) to a colorless liquid which solidifies in white flakes at $-76°$. It is not combustible and does not support the combustion of carbon compounds, but many metallic oxides combine when heated in it with the production of flame; thus brown lead dioxide is converted into white lead sulphate:

$$PbO_2 + SO_2 = PbSO_4.$$

It bleaches many organic pigments in the presence of water; this does not depend upon oxidation like the chlorine bleaching, but depends upon the union of the sulphurous acid, H_2SO_3, produced with the pigments. These compounds are unstable and on warming or with bases and acids, etc., decompose under certain circumstances so that the color appears again (use of burning sulphur for bleaching; animal fibres are not decolorized so completely with chlorine as with SO_2). It has rather great affinity for oxygen, hence it removes the same from many oxygen compounds, such as chromic acid, iodic acid (see below), etc. It is, therefore, a powerful reducing agent; it only unites directly with free oxygen in the presence of spongy platinum, etc. (p. 126); it is converted into sulphuric acid by the halogens when in watery solution: $SO_2 + 2H_2O + 2I = H_2SO_4 + 2HI$. It is, therefore, employed to remove the excess of chlorine used in bleaching certain materials which would otherwise destroy the fabric. On the other hand, strong reducing agents (H, H_2S) reduce it again to sulphur (p. 122). It prevents putrefaction and fermentation and serves as a preservative (sulphuring of wine-barrels, etc.). Water dissolves 50 times its volume of . this gas at 15°.

Detection. If a piece of paper moistened with starch and iodic acid (or potassium iodate) is suspended in sulphur dioxide gas, it turns blue, due to the setting free of iodine (p. 111):

$$2HIO_3 + 5SO_2 + 4H_2O = 2I + 5H_2SO_4.$$

An excess of SO_2 decolorizes this paper (process above).

Sulphurous Acid, H_2SO_3 or $HO-SO-OH$. The aqueous solution of sulphur dioxide has an acid reaction and may be considered as dissolved sulphurous acid, although this has never been isolated, as it decomposes again into water and sulphur dioxide on evaporation. If this solution is cooled to $-5°$, then crystals having the formula $H_2SO_3 + 14H_2O$ separate out; on standing in the air the solution is converted into sulphuric acid.

Sulphites. If the aqueous solution of sulphur dioxide is neutralized with bases and then evaporated, we obtain the sulphites: $2NaOH + H_2SO_3 = Na_2SO_3 + 2H_2O$; these are converted in aqueous solution into sulphates on standing in the air. Acids readily decompose the sulphites, setting free sulphurous acid, which decomposes further into water and sulphur dioxide: $NaSO_3 + 2HCl = 2NaCl + H_2O + SO_2$. In regard to the probable existence of sulphites of various constitution, see Taurin.

Detection. SO_2 is set free from the sulphites by the addition of an acid. This can be detected as described above.

Sulphur Trioxide, Sulphuric Anhydride, SO_3. *Preparation.* 1. On a large scale by the so-called contact method, where dry, arsenic-free sulphur dioxide and oxygen (or air) is passed over heated, finely divided platinum on asbestos (platinized asbestos), which acts as the contact substance (p. 125). The temperature must not rise above $450°$, as at higher temperatures the SO_3 formed is again decomposed into $SO_2 + O$ in the presence of the platinum.

2. On heating ferrous sulphate (p. 130) or sodium pyrosulphate (p. 130): $Na_2S_2O_7 = Na_2SO_4 + SO_3$.

3. On the distillation of fuming sulphuric acid (p. 129).

The vapors produced in these three methods are condensed by cold.

Properties. Long, colorless, neutral, caustic prisms which blacken organic substances, melt at $14°$, and which in the presence of traces of water are converted into silky, asbestos-like needles of pyrosulphuric anhydride, S_2O_6, which melt at $50°$, which are less caustic, etc., and which form the commercial product. They fume strongly in the air, as they are somewhat volatile even at ordinary temperatures, and the vapor absorbs water from the air, forming sulphuric acid, which immediately condenses into small visible globules. It dissolves in water with a hissing sound with the production of great heat, forming sulphuric acid: $SO_3 + H_2O = H_2SO_4$.

Sulphuric Acid, Oil of Vitriol, H_2SO_4 or $HO-SO_2-OH$. *Occurrence.* Free to very trivial extent in certain volcanic springs of America, and in the air in regions where large quantities of coal are burnt. It occurs in large quantities combined as sulphates, of which calcium sulphate, $CaSO_4$, gypsum, forms entire geological strata, also as barium sulphate, $BaSO_4$, and strontium sulphate, $SrSO_4$. Animals and plant fluids also contain sulphates, generally of the alkali metals.

Preparation. 1. In small quantities by boiling sulphur with concentrated nitric acid: $S + 2HNO_3 = H_2SO_4 + 2NO$.

2. By dissolving sulphur trioxide prepared by the contact method in water. On account of the ready and less dangerous transportation of SO_3 it is economical to prepare sulphuric acid by dissolving SO_3 in water at the locality where it is to be used. For these reasons the following method of preparation is used to a less extent at the present day.

3. By the oxidation of sulphur dioxide by nitric acid or its decomposition products in the presence of water and air in chambers lined with lead plates (lead chambers). An aqueous solution of sulphur dioxide is only slowly oxidized into sulphuric acid by the oxygen of the air (p. 126); but this may be made to take place more quickly in the presence of contact substances or when compounds which readily give off oxygen are present; such compounds are nitric acid (HNO_3) and nitrogen dioxide (NO_2), which are reduced to nitric oxide (NO) thereby.

The sulphur dioxide is prepared by roasting the residues from the Leblanc soda manufacture, the iron residues used in the purification of illuminating-gas (see Potassium Ferrocyanide), or by roasting metallic sulphides and especially iron pyrites (FeS_2), copper pyrites ($CuS + FeS$), lead pyrites (PbS), zinc blende (ZnS), which are converted into oxides thus: $2FeS_2 + 11O = Fe_2O_3 + 4SO_2$.

The nitric acid is generated from $2NaNO_3 + H_2SO_4$ by introducing this mixture into crucibles placed in the roasting-ovens and the vapors introduced with the sulphur dioxide through the Glover towers into the lead chambers, in which, at the same time, steam and air are conducted.

The acid (chamber acid) which collects contains about 60 per cent. sulphuric acid and is concentrated by heating first in lead pans and contains then about 80 per cent. sulphuric acid (pan acid). As a stronger acid than this attacks lead on boiling, it is further concentrated in glass or platinum vessels until it contains about 92 to 94 per cent. sulphuric acid (crude sulphuric acid). The following transformation takes place in the lead chambers:

    a. $3SO_2 + 2HNO_3 + 2H_2O = 3H_2SO_4 + 2NO$.

    b. A part of the nitric oxide (NO) produced is transformed by the oxygen of the air and by the steam into nitric acid, which again can oxidize a new portion of sulphur dioxide into sulphuric acid:

$$2NO + 3O + H_2O = 2HNO_3.$$

    c. A part of the nitric oxide combines with the atmospheric oxygen, forming nitrogen dioxide (NO_2), which, in the presence of steam, oxidizes the sulphur dioxide into sulphuric acid:

$$SO_2 + H_2O + NO_2 = H_2SO_4 + NO.$$

The regenerated nitric oxide again undergoes the same transformation as described in *b* and *c.*

Besides these, probably other processes may take place, especially the following:

    - *a.* First nitrosylsulphuric acid (nitrosulphonic acid) is produced: $SO_2 + HNO_3 = HO-SO_2-O(NO)$.

    . *b.* Nitrosylsulphuric acid is only stable in certain concentrations and temperatures and is, therefore, in contact with sufficient water or dilute sulphuric acid immediately decomposed:

$$2HO-SO_2-O(NO) + H_2O = 2HO-SO_2-OH + NO + NO_2.$$

    c. Then nitrosylsulphuric acid is again formed, etc.:

$$2SO_2 + NO + NO_2 + H_2O + 2O = 2HO-SO_2-O(NO).$$

If it were possible in this process to prevent any loss, then a small quantity of nitric acid could produce an unlimited quantity of sulphuric acid; but as with one volume of oxygen four volumes of nitrogen are introduced (by the use of air instead of oxygen), the gases are diluted too much and must be continuously resupplied from without. The nitrogen on being removed from the chambers takes a part of the oxides of nitrogen with it, which can be regained again by passing the gases (before they come to the chimney which maintains the draft) through Gay-Lussac towers. In these towers, which are filled with coke, strong sulphuric acid is allowed to trickle, and this dissolves all the oxides of nitrogen. This solution, the so-called nitroso acid, is allowed to trickle through the Glover tower, which is also filled with coke and is placed at the other end of the lead chambers. The sulphur dioxide entering removes from the nitroso acid the oxides of nitrogen and carries them into the lead chamber.

If in these processes water is absent, then nitrosylsulphuric acid deposits in the lead chambers as white crystals, so-called lead-chamber crystals, which decompose in the presence of water (see p. 128, *b*).

Properties. Sulphuric acid is a thick, colorless liquid, which with dust, etc., readily darkens; it has a great affinity for water, hence it is used in drying gases and for filling desiccators. When mixed with water considerable heat is evolved; hence on mixing these the sulphuric acid should always be added to the water, and indeed in a thin stream, otherwise a violent explosion may take place. On account of its great affinity for water, sulphuric acid removes oxygen and hydrogen in the proportion of water from many organic compounds; the preparation of carbon monoxide from oxalic acid, of ethylene from alcohol, the carbonization action of sulphuric acid upon sugar, wood, paper, etc., depends upon this action. If the vapor of sulphuric acid is passed over heated bricks, it decomposes into $SO_2 + H_2O + O$.

Sulphuric acid is a strong bibasic acid, forms, therefore, acid and neutral salts, and expels most other acids from their salts on heating.

Most metals dissolve in cold, dilute sulphuric acid with the generation of hydrogen, or in hot, concentrated sulphuric acid with the generation of sulphur dioxide, forming sulphates. Lead, platinum, gold, and certain rare metals are not attacked by sulphuric acid.

Crude or English sulphuric acid (oil of vitriol) contains 91 to 94 per cent. sulphuric acid, has a specific gravity of 1.830 to 1.837, contains contaminations such as lead, oxides of nitrogen, and often arsenic.

Pure sulphuric acid is obtained by distilling crude sulphuric acid; hereby first a dilute sulphuric acid distils over until the boiling-point rises to 330°, when a pure sulphuric acid distils over which contains only 1.5 per cent. water and has a specific gravity of 1.836 to 1.84 (p. 53).

Anhydrous sulphuric acid is obtained by cooling pure sulphuric acid to −25° or by adding crystals of sulphuric anhydride to strongly cooled sulphuric acid. It forms colorless crystals having a specific gravity of 1.837 which melt at 10°; they cannot be distilled without decomposition. On heating, they first evolve sulphur trioxide, and at 330° a sulphuric acid of 98.5 per cent distils over (Dissociation, p 71).

Fuming sulphuric acid, or *Nordhausen oil of vitriol,* is now prepared by dissolving sulphur trioxide obtained by the contact method in crude sulphuric acid. It used to be obtained by heating roasted ferrous sulphate.

On roasting ferrous sulphate it is transformed into ferric oxide and ferric sulphate, $6FeSO_4 + 3O = 2Fe_2(SO_4)_3 + Fe_2O_3$, which decomposes at a red heat according to the following equation: $Fe_2(SO_4)_3 = Fe_2O_3 + 3SO_3$; the sulphur trioxide which distils off is collected in small quantities of water or sulphuric acid; the iron oxide (Fe_2O_3) which remains in the retort is called colcothar.

It forms a thick liquid which fumes in the air and has a specific gravity of 1.85 to 1.90 and may be considered as a compound of sulphuric anhydride with sulphuric acid as on cooling the following acid separates:

Pyro- or *disulphuric acid,* $H_2S_2O_7$, forms colorless crystals which melt at 35° and is prepared commercially by dissolving SO_3 into concentrated sulphuric acid. On heating it decomposes into H_2SO_4 and SO_3 (p. 126); its salts are obtained by heating the primary sulphates: $2KHSO_4 = H_2O + K_2S_2O_7$, which further decomposes on heating into $K_2SO_4 + SO_3$.

Sulphates are obtained either by dissolving the respective metals in sulphuric acid or by neutralizing a base with sulphuric acid, also by oxidizing the metallic sulphides or metallic sulphites at a gentle heat in the air. In this wise copper and iron sulphates are prepared on a large scale: $CuS + 4O = CuSO_4$.

The sulphates of the alkalies, of the alkaline earths, and of lead are not changed on heating to a high degree; the other sulphates dcompose into sulphur dioxide or sulphur trioxide and metal (or metallic oxides) and oxygen.

Detection. Sulphuric acid and the soluble sulphates give white precipitates of barium sulphate ($BaSO_4$) or lead sulphate ($PbSO_4$) with barium or lead salt solutions. These precipitates are insoluble in acids.

On fusion with sodium carbonate the insoluble sulphates are converted into soluble sodium sulphate in which the sulphuric acid can be detected as above described.

Sulphur sesquioxide, S_2O_3, is produced by introducing powdered sulphur into sulphur trioxide, and forms bluish-green crystals: $S + SO_3 = S_2O_3$; with water it decomposes into $H_2SO_4 + S$.

Hyposulphurous acid, $H_2S_2O_4$, is obtained as a zinc salt by the action of zinc upon aqueous sulphurous acid: $Zn + 2H_2SO_3 = ZnS_2O_4 + 2H_2O$; it is, like its salts, only known in aqueous solution, which is yellow, readily decomposes, and is a strong reducing and bleaching agent.

Sulphur heptoxide, S_2O_7, is produced by the action of dark electrical discharges upon a dry mixture of sulphur dioxide and oxygen, and forms thick drops which solidify at $0°$ and which, on heating, decompose into $2SO_3 + O$, and with water decompose into sulphuric acid and oxygen: $2H_2O + S_2O_7 = 2H_2SO_4 + O$.

Persulphuric acid, $H_2S_2O_8$, is only known dissolved in sulphuric acid or as its salts; it is produced by the electrolysis of 40 per cent. sulphuric acid, whereby this decomposes into its ions (p. 75), $SO_4 + H_2$, and then the ion SO_4 unites with a molecule of H_2SO_4. Its salts or persulphates are obtained by the action of strong bases upon S_2O_7 or by the electrolysis of solutions of sulphates. Persulphates, like the acid, have the properties of hydrogen peroxide, having an oxidizing as well as a reducing action; nevertheless they do not decompose an acid solution of potassium permanganate (p. 118), nor do they react with chromic acid.

Oxysulphuric acid, H_2SO_5 or $HO-SO_2-O-OH$, is obtained by mixing sulphuric acid with hydrogen peroxide, persulphates, sodium or barium dioxide, and is used as one of the most powerful oxidizing agents (Caro's reagent).

3. Selenium.

Atomic weight 79.2 = Se.

Occurs to a slight extent, generally combined, as lead selenide, PbSe, and as sulphur selenide, SSe. It accompanies most of the metallic sulphides to a slight extent, and deposits as a mud in the lead chambers in the sulphuric acid manufacture. It is known in four modifications:

1. As amorphous, black, vitreous masses having a specific gravity of 4.28, soluble in carbon disulphide, and obtained by quickly cooling fused selenium.

2. As a red amorphous powder having a specific gravity of 4.26, soluble in carbon disulphide, and obtained by the reduction of an aqueous solution of selenium dioxide by sulphur dioxide: $H_2SeO_3 + 2SO_2 + H_2O = 2H_2SO_4 + Se$.

3. In dark-red crystals isomorphous with monoclinic sulphur having a specific gravity of 4.5, and obtained on the evaporation of the solutions of modifications 1 and 2 in carbon disulphide.

4. As rhombic, metallic, bluish-gray masses having a specific gravity of 4.8, insoluble in carbon disulphide, and obtained by heating amorphous selenium to $150°$ or by slowly cooling fused selenium. Only this modification conducts electricity.

Selenium melts at $217°$, boils at $680°$, burns when heated in the air with a bluish flame, producing white crystalline selenium dioxide. SeO_2, which has an odor similar to radishes. Selenium dissolves in part unchanged in fuming sulphuric acid, producing a green color. another rt being converted into selenious acid: $Se + 2H_2SO_4 = H_2SeO_3 + H_2SO_3$. ǁaSO_2.

4. Tellurium.

Atomic weight 127.6 = Te.

Occurs only seldom, either free or combined with metals, as graphitic tellurium, $(AuAg_3)Te_3$, tetradymite, $Bi_2Te_3 + Bi_2S_3$, silver telluride,

Ag_2Te, lead telluride, PbTe, etc. All tellurium ores dissolve in hot concentrated acid with a red color. Two modifications of tellurium are known:

1. As amorphous, black powder having a specific gravity of 5.9, obtained on the reduction of an aqueous solution of tellurium dioxide by sulphur dioxide: $H_2TeO_3 + 2SO_2 + H_2O = Te + 2H_2SO_4$.

2. As rhombic, metallic, silvery crystalline masses having a specific gravity of 6.2 and obtained by cooling fused amorphous tellurium.

Tellurium conducts heat and electricity, melts at 450°, boils at about 1390°, and burns when heated in the air with a blue flame, forming white crystalline tellurium dioxide, TeO_2.

HALOGEN GROUP.

Fluorine.   Chlorine.   Bromine.   Iodine.

These monovalent elements differ from all other elements by the fact that their compounds with hydrogen are strongly acid and their compounds with metals are the salts of these acids; hence these elements have also been called halogens, haloids, salt-formers ($\H{a}\lambda s$, salt, $\gamma \epsilon \nu \nu \acute{a} \omega$, to form); their hydrogen acids, halogen acids; and their salts, halogenides or haloid salts.

With an increase in atomic weight their affinity for hydrogen diminishes and for oxygen increases. Chlorine expels bromine and iodine from its hydrogen compounds, bromine expels iodine from these same compounds, while iodine, on the contrary, replaces bromine and chlorine, and bromine replaces chlorine, in their oxygen compounds; for instance, we obtain potassium periodate by treating a solution of potassium perchlorate with iodine, while chlorine is evolved: $HClO_4 + I = HIO_4 + Cl$.

The affinity of fluorine for oxygen is very slight, so that its oxides and oxyacids have not been obtained up to the present time.

The atomic weight and hence also the vapor density of bromine is about the average of that of chlorine and iodine, and bromine stands between these two in all its properties.

1. Chlorine.

Atomic weight 35.45 = Cl.

Occurrence. Only in combination, in small quantities in all parts of animals and in many plants; also as horn-silver, AgCl, as phosgenite, $PbCl_2$, etc.; in larger quantities as carnallite, $MgCl_2 + KCl + 6H_2O$, sylvine, KCl, and rock salt, NaCl; also combined with potassium, sodium, and magnesium (which see), dissolved in sea-water, in salt springs, etc.

Preparation. 1. By heating hydrochloric acid (HCl) with manganese dioxide (MnO_2) or other dioxides:

$$MnO_2 + 4HCl = MnCl_2 + 2H_2O + 2Cl.$$

The manganous chloride ($MnCl_2$) produced is again converted into MnO_2 in technical pursuits (Weldon's process, see Manganous Chloride).

2. By heating sodium chloride with manganese dioxide and sulphuric acid: $MnO_2 + 2NaCl + 2H_2SO_4 = MnSO_4 + Na_2SO_4 + 2H_2O + 2Cl$.

3. By pouring hydrochloric acid or sulphuric acid upon chloride of lime (calcium hypochlorite): $Ca(OCl)_2 + 4HCl = CaCl_2 + 2H_2O + 4Cl$ (see Chloride of Lime).

All other hypochlorites behave like calcium hypochlorite towards hydrochloric acid, also all chlorites and chlorates, as well as many other salts rich in oxygen (potassium dichromate, potassium permanganate). These bodies are therefore used in the preparation of small quantities of chlorine.

4. On the electrolysis of an aqueous solution of hydrochloric acid, HCl (p. 136).

Technical Preparation. 1. Nearly entirely by the electrolysis of fused metallic chlorides as a by-product in the preparation of the respective metals or metallic hydroxides (see Potassium), or by the electrolysis of their aqueous solution.

2. Hydrochloric acid is mixed with air and passed over heated bricks (Deacon's process): $2HCl + O = H_2O + 2Cl$; this decomposition takes place more readily when the bricks are previously impregnated with copper sulphate solution ($CuSO_4$); this cupric salt remains unchanged and can be used for a long time (Catalysis, p. 66).

3. Magnesium chloride, so often obtained as a by-product, decomposes in part on evaporating its watery solution into hydrochloric acid and magnesium oxide: $MgCl_2 + H_2O = MgO + 2HCl$; the magnesium oxychloride, $MgCl_2 + MgO$, which remains behind, decomposes on heating in a current of air into magnesium oxide and chlorine.

4. On heating calcium or magnesium chloride, so extensively obtained as a by-product, with sand in a current of air, we obtain chlorine: $CaCl_2 + SiO_2 + O = CaSiO_3 + 2Cl$.

Properties. A greenish-yellow ($\chi\lambda\omega\rho\delta\varsigma$, greenish yellow), poisonous gas, 2.5 times heavier than air, having an irritating odor, and strongly attacking the respiratory organs, liquefiable at $-40°$, forming a yellow liquid which is heavier than water and not miscible therewith, and which solidifies at $-102°$ into yellow crystals. Liquid chlorine does not attack iron, and is sold in commerce in iron cylinders (p. 41). One volume of water dissolves 2.5 volumes of chlorine at $10°$; on this account, and as it combines with mercury, we collect this gas over hot water or saturated NaCl solution.

Chlorine-water is a solution of chlorine in water which contains 0.4 to 0.5 per cent. by weight; it must be kept in the dark, as daylight slowly and sunlight more rapidly decomposes it with the formation of oxygen and hydrochloric acid: $H_2O + 2Cl = 2HCl + O$. If saturated chlorine-water is cooled to 0°, yellow crystals of chlorine hydrate, $Cl_2 + 10H_2O$, separate out.

It combines directly with all elements with the exception of argon, helium, nitrogen, carbon, oxygen, and certain of the rare platinum metals. Phosphorus, thin sheets of gold and copper, powdered arsenic, antimony, boron, silicon, bismuth, combine in chlorine gas even at ordinary temperatures and burn brightly therein. Chlorine has very strong affinities for hydrogen; a mixture of equal volumes of chlorine and hydrogen combines gradually in diffused daylight, while in sunlight it immediately explodes, producing hydrochloric acid. A hydrogen flame burns in chlorine gas and vice versa (p. 109). Many organic compounds which contain carbon and hydrogen lose their hydrogen with the formation of hydrochloric acid; thus paper moistened with turpentine (a compound of carbon with hydrogen) when introduced into chlorine inflames, and carbon is set free. Chlorine is not combustible in the air, but supports the combustion of a candle or illuminating-gas (a mixture of hydrocarbons) with the setting free of carbon (soot). This phenomenon may be explained by the union of the hydrogen of the respective compound with chlorine and the setting free of heat, which heats the gas to a red heat, while the carbon present separates out.

Chlorine destroys all natural and many artificial coloring matters, as it combines with the same or oxidizes them, since in the presence of water it unites with the hydrogen of the body and sets free nascent oxygen: $H_2O + 2Cl = 2HCl + O$. Its disinfecting action, that is, the destruction of infectious and odoriferous substances by chlorine, also depends upon this fact. (Use of chlorine in the presence of water for bleaching and disinfection.)

Dry chlorine shows much less chemical activity than moist chlorine and does not attack metals; hence liquid chlorine is sold in iron cylinders.

Detection. 1. Free chlorine sets iodine free from potassium iodide, hence it turns potassium iodide and starch paste blue (p. 111).

2. It decolorizes indigo solution and moist litmus paper.

a. Compounds with Hydrogen.

Hydrogen Chloride, Hydrochloric Acid, HCl. *Occurrence.* Free in the gases of volcanoes, and to a slight extent in gastric juice.

Formation. 1. A mixture of equal volumes of hydrogen and chlorine may be kept unchanged in the dark; in diffused daylight they unite gradually; in direct sunlight they combine immediately with explosive violence, forming hydrochloric acid. The volume of the gas remains unchanged.

$$\left\{ \begin{matrix} H \\ H \end{matrix} \right. + \left\{ \begin{matrix} Cl \\ Cl \end{matrix} \right. = \left\{ \begin{matrix} H \\ Cl \end{matrix} \right. + \left\{ \begin{matrix} H \\ Cl \end{matrix} \right.$$

2. By the action of chlorine upon water and many organic compounds.

Preparation. 1. From crude hydrochloric acid (see below) by boiling.

2. It is prepared on a large scale in the Leblanc soda manufacture by heating sodium chloride with sulphuric acid:

$$2NaCl + H_2SO_4 = Na_2SO_4 + 2HCl.$$
Sodium chloride.   Sulphuric acid.   Sodium sulphate.   Hydrochloric acid.

The sodium sulphate obtained as a by-product is used in the manufacutre of glass or is converted into sodium sulphide, Na_2S, which is used in the manufacture of cellulose.

3. On a large scale by passing superheated steam over magnesium chloride (p. 133), which is obtained as a by-product in many chemical processes:

$$2MgCl_2 + H_2O = MgO.MgCl_2 + 2HCl.$$

Properties. Colorless, irritating gas which is non-combustible and which does not support combustion. It is 1.25 times heavier than air, fumes in the air (because it abstracts water therefrom and produces hydrochloric acid, which separates as a vapor), liquefiable at $-81°$, forming a colorless liquid which solidifies at $-116°$.

One volume of water dissolves 450 volumes of this gas at $15°$ and forms therewith a fuming, colorless, strongly acid liquid having a specific gravity of 1.21, which contains 43 per cent. by weight of

HCl. The watery solution of hydrogen chloride is ordinarily called hydrochloric acid or muriatic acid.

Crude hydrochloric acid is contaminated by chlorine. sulphuric acid, arsenic, iron, etc., and hence is yellow in color.   It has a specific gravity of 1.158 to 1.170 and contains from 30 to 33 per cent. HCl.

Pure hydrochloric acid, obtained by the repeated distillation of crude hydrochloric acid, is a colorless liquid of a specific gravity of 1.20 and contains 39.1 per cent. HCl.

Dilute hydrochloric acid consists of equal parts water and pure hydrochloric acid.

On heating concentrated hydrochloric acid it yields HCl until the residue contains 20 per cent. HCl; this hydrochloric acid has a specific gravity of 1.10 and distils without decomposition at 110°; dilute hydrochloric acid on heating gives off water until the residue contains 20 per cent. HCl (p. 53).

If the electric current is passed by means of carbon electrodes (as the chlorine set free would attack metals) through an aqueous solution of hydrogen chloride, equal volumes of hydrogen and chlorine are set free at the negative and the positive pole respectively.

If hydrochloric acid gas is passed over heated potassium or sodium, one-half its volume of hydrogen is set free: $Na + HCl = NaCl + H$.

Hydrochloric acid dissolves most metals with the formation of chlorides and the evolution of hydrogen: $2HCl + Zn = ZnCl_2 + 2H$; only mercury, silver, copper, gold, arsenic, anitmony, bismuth, lead, and the platinum metals are not attacked by this acid, or only to a slight extent.

Chlorides are obtained by the direct union of the elements, but ordinarily by the action of hydrochloric acid upon metals, metallic oxides, or metallic hydroxides.

Detection. 1. When heated with dioxides, hydrochloric acid forms chlorides with the setting free of chlorine (p. 132).

2. Hydrochloric acid and soluble chlorides give with silver nitrate a white, cheesy precipitate of silver chloride (AgCl) which darkens in the light and which is soluble in ammonia, but not soluble in nitric acid: $HCl + AgNO_3 = AgCl + HNO_3$.

3. Lead salts precipitate white lead chloride ($PbCl_2$), soluble in hot water or considerable cold water:

$$Pb(NO_3) + 2HCl = PbCl_2 + 2HNO_3.$$

b. Compounds with Oxygen.

Chlorine monoxide,	Cl_2O.	Hypochlorous acid,	$HClO$.
Chlorine dioxide,	ClO_2.	—	
(Chlorine trioxide,	Cl_2O_3.)	(Chlorous acid,	$HClO_2$.)
Chlorine tetroxide,	Cl_2O_4.	—	
(Chlorine pentoxide,	Cl_2O_5)	Chloric acid,	$HClO_3$.
Chlorine heptoxide,	Cl_2O_7.	Perchloric acid,	$HClO_4$.

Chlorine trioxide and chlorine pentoxide are only known mixed as chlorine tetroxide; chlorous acid is only known in the form of its salts, the chlorites (P. 138).

Chlorine Monoxide, Cl_2O. *Preparation* Cooled chlorine gas is passed over mercury oxide (HgO) and the vapors produced condensed in a cooling mixture: $HgO + 4Cl = HgCl_2 + Cl_2O$.

Properties. It forms a red liquid which boils at $+5°$ and then forms a yellowish-red, poisonous, strongly oxidizing gas with a disagreeable odor. It soon decomposes into its constituents, which can be brought about especially by shock, heating, or in contact with oxidizing sub-stances, often with an explosion.

Hypochlorous Acid, $HClO$. *Preparation.* Chlorine monoxide is very soluble in water, forming hypochlorous acid: $Cl_2O + H_2O = 2HClO$; on concentration this splits into chlorine monoxide again, so that anhydrous $HClO$ is not known.

Properties. The dilute, nearly colorless solution may be distilled without decomposition; the concentrated yellow solution decomposes on heating and in the sunlight with the formation of chloric acid and chlorine: $3HClO = 2HCl + HClO_3$; $2HClO + 2HCl = 4Cl + 2H_2O$. With hydrochloric acid it evolves all the chlorine of both compounds: $HClO + HCl = H_2O + Cl_2$. It is a very weak acid, which does not decompose carbonates, but is a very strong oxidizing agent.

Hypochlorites are also readily decomposable and strong oxidizing agents. They are prepared by saturating $HClO$ with the respective bases. Commercially they are obtained, mixed with chlorides, by passing chlorine (generally electrolytically prepared) into cold and dilute solutions of the strong bases:

$$2NaOH + 2Cl = NaClO + NaCl + H_2O.$$

Sodium   Sodium   Sodium
hydroxide.   hypochlorite.   chloride

Nitric acid sets hypochlorous acid free from the hypochlorites, and this can be separated by distillation: $NaClO + HNO_3 = NaNO_3 + HClO$. Hydrochloric acid sets chlorine free, and certain metallic oxides generate oxygen from hypochlorites (see Chloride of Lime).

Chlorine Dioxide, ClO_2, or Chlorine Tetroxide, Cl_2O_4 (formerly called Hypochloric Acid). *Preparation.* Small quantities of potassium chlorate ($KClO_3$) are gradually and carefully added to concentrated sulphuric acid and then distilled at a temperature not above

30° in order to prevent explosion. The chloric acid first produced decomposes immediately by the dehydrating action of the sulphuric acid: $3ClO_3H = 2ClO_2 + H_2O + ClO_4H$. It may be obtained with less danger by heating potassium chlorate with oxalic acid not above 70°.

Properties. Reddish-yellow, oxidizing gas with a strong odor, liquefiable in an ice mixture forming a red liquid which boils at $+10°$; these bodies are readily decomposed by organic substances, or better by warming above 30° with explosive violence, but not by sunlight. Water dissolves it unchanged, producing a yellow solution; alkali hydroxides decolorize this solution with a formation of chlorates and chlorites: $2ClO_2 + 2KOH = KClO_2 + KClO_3 + H_2O$, so that the compound may be considered as the mixed anhydrides of chlorous and chloric acids $(Cl_2O_5 + Cl_2O_3 = 2Cl_2O_4)$.

The molecule seems at lower temperatures to have the composition Cl_2O_4, and on passage into the gaseous condition it decomposes into two molecules, ClO_2; hence this behavior to alkalies, as well as the analogy with NO_2 and N_2O_4.

Chloric Acid, $HClO_3$. *Preparation.* Only its aqueous solution, containing 40 per cent. $HClO_3$, is known. This is obtained by decomposing barium chlorate with dilute sulphuric acid, filtering off the insoluble barium sulphate, and concentrating the solution under the air-pump receiver: $Ba(ClO_3)_2 + H_2SO_4 = BaSO_4 + 2HClO_3$.

Properties. Thick, odorless and colorless, acid liquid, which on further evaporation under the air-pump, also in light or on warming to 40°, decomposes into chlorine, oxygen, and perchloric acid (which see). It has such a strong oxidizing action that on pouring it upon alcohol, phosphorus, paper, etc., they inflame; with hydrochloric acid it generates chlorine: $HClO_3 + 5HCl = 3H_2O + 6Cl$.

Chlorates are obtained, mixed with chlorides, when chlorine (generally obtained as a by-product in the electrolytic soda manufacture) is passed into hot and concentrated solutions of bases (see Hypochlorites, p. 137): $6KOH + 6Cl = 5KCl + KClO_3 + 3H_2O$.

Chlorates readily give off their oxygen and therefore decompose with explosive violence when they are rubbed or heated with inflammable or oxidizable bodies, such as phosphorus, sulphur, antimony sulphide, sugar, etc. The inflammable mass of the Swedish match consists of antimony sulphide and potassium chlorate, and inflames on friction with the red phosphorus contained on the sides of the box. Silver nitrate does not precipitate the chlorates from their solution; when heated alone they decompose without explosion into per-chlorates (see below) and then into chlorides and oxygen. When hydrochloric acid is poured upon chlorates they generate chlorine besides some ClO_2: $KClO_3 + 6HCl = KCl + 3H_2O + 6Cl$; when heated with sulphuric acid they yield chlorine dioxide (p. 137).

Chlorine heptoxide, Cl_2O_7, is obtained by distilling perchloric acid with phosphorus pentoxide, $2HClO_4 + P_2O_5 = 2HPO_3 + Cl_2O_7$, as a thick, colorless liquid. It has only a slight action upon oxidizable substances, but, on the contrary, when ignited or struck it explodes with violence. It is soluble in water, forming

Perchloric Acid, $HClO_4$. *Preparation.* From chloric acid by ex-posure to light or warmth: $3HClO_3 = HClO_4 + H_2O + 2Cl + 4O$; ordinarily by distilling potassium perchlorate with sulphuric acid: $2KClO_4 + H_2SO_4 = K_2SO_4 + 2HClO_4$.

Properties. Colorless, fuming liquid, which readily oxidizes and therefore inflames carbon, paper, wood, and other organic bodies with explosion; it produces painful sores on the skin, and decomposes with explosion after a few days, even in the dark. On the contrary, its aqueous solution is stable, non-oxidizing, and not decomposable by hydrochloric or sulphuric acids.

Perchlorates are obtained at the positive pole on the electrolysis of an aqueous solution of chlorides hy the oxygen evolved at that pole, or by heating potassium chlorate somewhat above its melting-point, when a part of its oxygen is first given off and it is converted into potassium perchlorate and potassium chloride: $2KClO_3 = KClO_4 + KCl + O_2$. At a higher temperature the potassium perchlorate is completely decomposed into potassium chloride and oxygen. Perchlorates differ from the chlorates in that they are not attacked by hydrochloric acid and do not yield explosive chlorine dioxide with sulphuric acid, as well as by their insolubility in water. Sodium perchlorate is found in Chili saltpeter.

c. *Compounds with Sulphur.*

Sulphur monochloride, S_2Cl_2, and **sulphur dichloride,** SCl_2, are ob-tained by the action of chlorine upon warmed sulphur as reddish-yellow liquids decomposible by water. The first dissolves about 70 per cent. of sulphur and is used in the vulcanization of rubber.

Sulphur tetrachloride, SCl_4, is only stable below $0°$.

2. Bromine.

Atomic weight $79.96 = Br$.

Occurrence. Bromine only occurs combined chiefly with sodium and magnesium in sea-water and in all plants and animals living therein, in many salt springs (Kreuznach, Kissingen) and salt deposits, especially in the Stassfurt "abraum" salts (see Potassium Chloride). It occurs in traces with iodine in the thyroid gland. Silver bromide forms the mineral called bromite.

Preparation. On the partial evaporation of sea-water the more insoluble chlorides separate out first and the readily soluble bromine salts remain in solution; from this mother-liquor (p. 52) or from the mother-liquor from the "abraum" salts (see Potassium) the bromine may be obtained by distilling with manganese dioxide and sulphuric acid: $2KBr + MnO_2 + 2H_2SO_4 = 2Br + K_2SO_4 + MnSO_4 + 2H_2O$, or by passing chlorine therein and heating: $KBr + Cl = KCl + Br$, or by electrolysis when the bromine set free remains in solution and is driven off by distillation.

Properties. Dark reddish-brown, poisonous and caustic liquid (the only liquid element at ordinary temperatures besides mercury) having a specific gravity of 3.18 at 0°, boils at 63° and is very volatile at ordinary temperatures, and solidifies at −8°, forming a crystalline reddish-brown mass.   It has a peculiarly unpleasant odor ($\beta\rho\tilde{\omega}\mu o\varsigma$, stench), attacks the skin, and its yellowish-brown vapors have a powerful action upon the mucous membranes.   Bromine dissolves in 30 parts water (bromine-water), also readily in ether, alcohol, carbon disulphide, and chloroform, with a brownish-red color.   With water it forms at 0° crystalline, yellow bromine hydrate, $Br_2 + 10H_2O$.   Chemically bromine has great similarity to chlorine, but has less affinity for the elements than this; it decomposes water only very slowly, but oxidizes many bodies in the presence of wàter; hence it also has a bleaching action.   It unites with hydrogen first on heating and not in the sunlight, and with sulphur it forms sulphur monobromide, S_2Br_2. Infusorial earth impregnated with bromine is called solid bromine.

Detection. Bromine colors starch paste orange, and dissolves in carbon disulphide or chloroform, producing a brownish-red solution.

a. Compounds with Hydrogen.

Hydrogen Bromide, Hydrobromic Acid, HBr.   *Preparation.* 1. Hydrogen and bromine first unite when their vapors are passed through a gently heated tube (generally filled with spongy platinum in order to increase the yield, p. 126).

2. By distillation of bromides with phosphoric acid:

$$H_3PO_4 + 3NaBr = Na_3PO_4 + 3HBr.$$

On the distillation of bromides with concentrated sulphuric acid we also obtain hydrobromic acid, but this in part is immediately decomposed so that also free bromine and sulphur dioxide are produced: $2HBr + H_2SO_4 = 2H_2O + 2Br + SO_2$.

3. If phosphorus bromide is warmed with water, or if we allow bromine to flow into amorphous phosphorus under water, we obtain hydrobromic acid:

$$PBr_3 + 3H_2O = 3HBr + H_3PO_3.$$

Phosphorus   Phosphorous
bromide.   acid.

4. By passing H_2S into water containing bromine and expelling the excess of H_2S by careful warming, we obtain an aqueous solution of hydrobromic acid: $H_2S + 2Br = 2HBr + S$.

Hydrobromic acid may also be obtained by the action of Br upon organic hydrogen compounds such as naphthalene, $C_{10}H_8$.

Bromides are obtained in a manner analogous to the chlorides (see also Potassium Bromide).

Properties. Colorless gas, irritating odor, fumes in the air, partly decomposes at 800°, liquefiable at −73° and solidifying at −120°, and very readily soluble in water. Its watery solution boils at 125° (p. 53) and then contains 48 per cent. HBr and has a specific gravity of 1.49. This solution turns red in the air, due to the separation of bromine: $2HBr + O = H_2O + 2Br$.

Detection. 1. Silver nitrate precipitates yellowish-white silver bromide, AgBr, from solutions of HBr or of bromides. This precipitate is insoluble in nitric acid, but soluble in considerable ammonia, and darkens in the light.

2. If a solution of HBr or a bromide is treated with chlorine-water, the bromine is set free, which, when shaken with carbon disulphide, gives a reddish-brown color thereto.

b. Compounds with Oxygen.

Hypobromous acid,   HBrO.
Bromic acid, $HBrO_3$.
Perbromic acid, $HBrO_4$.

At the present time we do not know of any oxide of bromine, but only the above-mentioned oxyacids, and these indeed only in watery solution. They are prepared in the same manner as the corresponding chlorine compounds.

Hypobromites, bromates, and perbromates are obtained in the same manner as the chlorine compounds and have great similarity therewith.

3. Iodine.

Atomic weight 126.85 = I.

Occurrence. Only combined with potassium, sodium, calcium, magnesium in certain salt springs and to a less extent in sea-water, from which many sea animals and all algæ remove it; also in Chili saltpeter and in certain rock salts, in many plants, in muscle-tissue, blood, in cow's milk and hen's eggs, in fresh-water animals, and in the thyroid gland. It occurs rarely as silver iodide and lead iodide in certain minerals.

Preparation. 1. Seaweeds are burnt and the ash (kelp or varec) lixiviated with water; the less soluble chlorides are removed from this solution in part by evaporation and the remaining liquid (the mother-liquor) is distilled with manganese dioxide and sulphuric acid or chlorine pass through the solution, when the iodine is obtained (same process as with bromine, p. 140).

2. The mother-liquor of Chili saltpeter contains sodium iodate, from which the iodine can be set free and obtained by heating and passing sulphur dioxide through the solution: $2NaIO_3 + 5SO_2 + 4H_2O = Na_2SO_4 + 4H_2SO_4 + 2I.$

Properties. Steel-gray, poisonous, metallic-like plates, having a specific gravity of 4.94, melting at 116°, and forming violet (ἰώδης, violet) vapors at 183°. Iodine evaporates even at ordinary temperatures, has a peculiar odor, and combines with hydrogen only at a red heat. It does not decompose water, but oxidizes many substances in the presence of water, and colors the skin brown. Only traces of iodine are soluble in water, but it is readily soluble in ether and alcohol, producing a brown solution (tincture of iodine), and in an aqueous potassium iodide solution (Lugol's solution), and in chloroform and carbon disulphide, forming a violet-colored solution. Chemically it behaves very similar to bromine and chlorine, and is separated from its metallic compounds by these bodies.

Iodine unites with chlorine, forming liquid, brown iodine monochloride, ICl, or iodine trichloride, ICl_3, which form orange-yellow needles. It forms brown crystals of IBr with bromine and does not combine with sulphur.

Detection. Iodine in the presence of iodides gives a deep-blue color with starch. paste which, on warming, forms a colorless compound; by this reaction, as well as by its violet solution in carbon disulphide or chloroform, the smallest quantities of iodine may be detected

a. Compounds with Hydrogen.

Hydrogen Iodide, Hydriodic Acid, HI. *Preparation.* In a manner similar to hydrobromic acid (p. 140).

Properties. Colorless gas, readily decomposable, fuming in the air, liquefiable at $-34°$ and solidifying at $-51°$, readily soluble in water. The watery solution shows a constant boiling-point at $127°$ and then contains 37 per cent. HI and has a specific gravity of 1.67 (p. 53). At $518°$ HI begins to decompose; it is decomposed by oxygen or oxidizing substances: $2HI + O = H_2O + 2I$; hence it is a strong reducing agent. With iodic acid all the iodine of both compounds is set free: $5HI + HIO_3 = 6I + 3H_2O$.

Iodides are obtained in the same manner as the chlorides (see also Potassium Iodide).

Detection. 1. Silver nitrate precipitates pale-yellow silver iodide, AgI, from the solution of HI or iodides. This precipitate is insoluble in ammonia (differing from AgCl and AgBr) and nitric acid. When pure AgI is rather stable in sunlight.

2. Chlorine and bromine set iodine free from a solution of HI or iodides. This may be detected by its violet solution in carbon disulphide or chloroform, and by its turning starch paste blue.

b. Compounds with Oxygen.

—	Hypoiodous acid,	HIO
Iodine pentoxide, I_2O_5.	Iodic acid,	HIO_3.
—	Periodic acid,	HIO_4.

Hypoiodous acid, HIO, is only known in the form of its salts, the hypoiodites, which are obtained in addition to iodides on dissolving iodine in dilute cold solutions of bases (p. 137), but which quickly decompose into iodides and iodates: $3NaIO = NaIO_3 + 2NaI$.

Iodine pentoxide, I_2O_5, is produced on warming iodic acid to $170°$ and is a white, crystalline powder, which decomposes into $I_2 + O_5$ at $300°$ and is soluble in water, producing iodic acid, HIO_3.

Iodic acid, HIO_3, is obtained by heating iodine with fuming nitric acid: $3I + 5HNO_3 = 3HIO_3 + 5NO + H_2O$, or from iodine pentoxide (see above). It forms colorless, rhombic crystals which are soluble in water.

Iodates are obtained mixed with iodides by dissolving iodine in bases and evaporating to dryness (see HIO_3). Sulphur dioxide (P. 126), H_2S, HI, and other reducing agents set iodine free from iodic acid and iodates.

Periodic acid, HIO_4, is only known as H_5IO_6 with 2 mol. water, forming colorless crystals. It is prepared by the action of iodine upon a solution of perchloric acid (p. 132).

Periodates are obtained by passing chlorine into a hot alkaline solution of the iodate.

4. Fluorine.

Atomic weight $19 = F$.

Occurrence. Free in certain fluor-spars, but combined especially as fluor-spar, CaF_2, cryolite, $3NaF + AlF_3$, also in phosphorite and apatite (see Phosphorus). It occurs to a slight extent in mineral springs, ashes of plants, and as CaF_2 in bones and teeth.

Preparation. Anhydrous hydrofluoric acid (which contains potassium fluoride in solution in order to make it a conductor) is electrolyzed at $-20°$ in an apparatus of copper or platinum (as free fluorine combines with all other metals); fluorine is set free at the anode, which consists of platinum-iridium alloy: $HF = H + F$.

Properties. A greenish-yellow gas 1.26 times heavier than air, produces irritation of the respiratory organs, liquefies at $-187°$, forming a pale-yellow liquid which at this low temperature has hardly any chemical affinity. Fluorine combines with bromine, iodine, sulphur, arsenic, antimony, boron, phosphorus, silicon, and carbon, even at ordinary temperatures, with the production of flame (not with the other metalloids), also with the metals on gently warming and with gold, platinum, and copper at higher temperatures. It inflames alcohol, ether, turpentine, benzol, and many other organic compounds, as the fluorine combines with the hydrogen contained therein. It combines with hydrogen at $-25°$ even in the dark, and with water it yields HF and oxygen containing ozone; it decomposes the metallic compounds of bromine, chlorine, and iodine; and it does not attack glass (see below).

Compounds of Fluorine.

Hydrogen Fluoride, Hydrofluoric Acid, HF. *Occurrence.* Only as its salts, the fluorides.

Preparation. By heating fluorine compounds with concentrated sulphuric acid in a platinum or lead retort. It is also produced as a by-product in the manufacture of superphosphates (fertilizers) from bones: $CaF_2 + H_2SO_4 = CaSO_4 + 2HF$.

Properties. Colorless, strongly fuming, poisonous, and irritating gas which liquefies on cooling, forming a colorless, volatile liquid, which boils at 19° and solidifies at −102.5°. It dissolves all metals with the evolution of hydrogen, producing fluorides with the exception of gold, platinum, and lead. When anhydrous, or when it contains water, it attacks glass and dissolves the same as it forms with the silicon, the chief constituent of glass, a gaseous compound, silicon fluoride, SiF_4. The solution in water may, therefore, only be kept in lead, platinum, paraffine, or gutta-percha bottles. This solution in water boils constantly at 120° and then contains 35 per cent. HF and has a specific gravity of 1.15. It is used in etching glass and porcelain and to dissolve silicic acid compounds which are not attacked by other acids: $SiO_2 + 4HF = SiF_4 + 2H_2O$. It prevents the acid fermentation of milk and hence is used in fermentive pursuits.

Fluorides are obtained by the action of HF upon the metals They are all soluble in HF; silver fluoride and most of the other fluorides are soluble in water, while the fluorides of the calcium group are insoluble in water.

Detection. Hydrofluoric acid etches glass; the fluorides on being heated in a lead crucible with sulphuric acid evolve hydrofluoric acid vapors which are allowed to act upon a glass plate that has been covered with wax and a portion of the wax removed; after the removal of the wax the marking may be seen upon its surface.

NITROGEN GROUP.

**Nitrogen. Phosphorus. Arsenic. Antimony.
Bismuth. Vanadium. Tantalum. Niobium.**

These elements form similarly constituted compounds, in which they occur either trivalent or quadrivalent. In this group also, as the atomic weight increases the vapor density, the melting- and boiling-point, as well as the metallic character, increase also.

The elements of the second series (Bismuth, etc.) have a perfect metallic character, hence will be discussed in connection with the metals; and they do not combine with hydrogen, while the elements of the first series form gaseous compounds with three atoms of hydrogen.

The basic character of these hydrogen compounds, as well as the acid-forming power of the oxides, diminishes with the increase in metallic properties. Ammonia (NH_3) has strong basic properties and unites with all acids forming salts; phosphureted hydrogen (PH_3) combines only with HBr and HI; arseniureted hydrogen (AsH₃) and antimoniureted hydrogen (SbH₃) have no more basic properties.

While all the oxides of nitrogen, phosphorus, and arsenic have pronounced acidic character, the antimony trioxide shows besides the property of a weak acid anhydride also that of a basic anhydride, and bismuth trioxide has only basic properties. All of these pentoxides are acid anhydrides; the acid derived from bismuth pentoxide is unstable.

1. Nitrogen.

Atomic weight 14.04 = N.

Occurrence. Free to a slight extent in the gases of many springs and of volcanoes, mixed with oxygen in the air where it occurs as 78 volumes per cent.; combined chiefly as potassium nitrate, KNO_3, and sodium nitrate, $NaNO_3$, which latter body is found to a great extent in South America and called Chili saltpeter. It is also found combined in the most important parts of plants and animals (proteids, blood, muscle, nerve substance, etc.), in fossil plants (coal), in guano, and as ammonia (p. 147).

Formation. 1. By burning phosphorus in air contained under a bell-jar over water, the phosphorus combines with the oxygen of the air, forming phosphorus pentoxide (P_2O_5), which dissolves in the water, producing metaphosphoric acid, while the nitrogen of the air remains.

2. In the combustion of all nitrogenous organic compounds with copper oxide in the presence of copper (see Elementary Analysis).

Preparation. 1. By passing pure air over red-hot copper, when copper oxide is formed and nitrogen remains.

The nitrogen obtained from the air always contains small amounts of argon, helium, xenon, krypton, and neon (p. 152), which can only be removed with difficulty.

2. By heating ammonia nitrite, which decomposes into water and nitrogen: $NH_4NO_2 = 2N + 2H_2O$; or by boiling a solution of potassium nitrite (KNO_2) with ammonium chloride (NH_4Cl), when ammonium nitrite is first formed and then this decomposes as above:

$$KNO_2 + NH_4Cl = KCl + NH_4NO_2.$$

3. By passing chlorine into a saturated solution of ammonia (p. 152): $2NH_3 + 6Cl = 2N + 6HCl$.

The HCl formed produces with the still undecomposed ammonia, white vapors of ammonium chloride, NH_4Cl, which dissolve in the water: $NH_3 + HCl = NH_4Cl$ (p. 149).

4. By warming a solution of ammonia with chloride of lime: $3Ca(OCl)_2 + 2NH_3 = 3CaCl_2 + 3H_2O + 2N.$

Properties. Colorless, odorless, and tasteless gas, slightly soluble in water, and liquefiable to a colorless liquid at $-194°$ and solidifying at $-214°$ (p. 41). It does not burn, neither does it support combustion; it has an asphyxiating action, not because it is poisonous, but because of the lack of oxygen (hence the French name *azote* for nitrogen: α, privative, and $\zeta\omega\epsilon\iota\nu$, life). It has a specific gravity compared with air as unit of $\dfrac{1.250}{1.293} = 0.97$. Chemically, nitrogen is very indifferent; in the cold it unites with lithium and at a red heat with calcium, barium, strontium, magnesium, boron, titanium, silicon, and uranium, and the metals of the rare earths forming so-called nitrides. It combines with hydrogen and oxygen only at very high temperatures, as under the influence of the electric spark; nevertheless, nitrogen forms in an indirect way a great number of characteristic compounds.

Both the nitrogen atoms in the molecule are attached very firmly to each other and are only separated from each other at very high temperatures (also by electric discharge). Certain bacteria can bring about a similar separation at ordinary temperatures. These organisms exist in the roots of leguminous plants, hence these plants have the power of taking up atmospheric nitrogen to produce the proteids (nitrification bacteria), while other plants and all animals cannot assimilate atmospheric nitrogen. Pure cultures of these bacilli are used as *nitragin* to inoculate fields planted with leguminous plants, as the above bacteria are not universally distributed.

Detection. Only by the absence of all properties characteristic of other gases.

a. Compounds with Hydrogen.

Ammonia, NH_3.
Hydroxylamine, $NH_2(OH)$.
Hydrazine, N_2H_4.
Hydrazoic acid, N_3H.

Ammonia, NH_3. *Occurrence.* Combined with acids to a slight extent in the air, in rain-water, in the soil, and also in many spring-waters. Ammonium salts (p. 150) are found in the gases of volcanoes, in carnallite, in guano and ammonium sulphate forms the mineral mascagine.

Formation. Ammonium nitrite and nitrate are produced by the action of the electric spark upon moist air, and in the evaporation of water and in every combustion in the air: $2H_2O + 2N = NH_4NO_2$.

Nitrogen and hydrogen combine to form ammonia only with the dark electric discharge; nevertheless, they combine in the nascent state: thus when zinc is dissolved in 5 to 6 per cent. nitric acid (p. 160). In this process no hydrogen is generated, but the hydrogen atoms act upon the nitric acid and reduce it to nitrogen, which combines with the hydrogen, forming ammonia: $3Zn + 6HNO_3 = 3Zn(NO_3)_2 + 6H$; $2HNO_3 + 10H = 6H_2O + 2N$; $2N + 6H = 2NH_3$. This reduction of nitric acid in watery solution can be done more readily by finely powdered aluminium, zinc, or iron in the presence of strong bases (p. 160).

Ammonia is formed in a similar manner in the putrefaction of organic nitrogenous substances, or by heating them in the absence of air (in their dry distillation), or by fusing them with strong bases, or on heating them with concentrated sulphuric acid, when the nascent nitrogen and hydrogen unite (see Elementary Analysis).

In the past, ammonium chloride, a salt of ammonia, used to be obtained by the dry distillation of camel-dung in the oasis of Jupiter Ammon, hence the name, sal ammoniacum.

Preparation. Ammonia is chiefly prepared by heating ammonium salts with stronger bases such as KOH, NaOH or cheaper with calcium hydroxide, $Ca(OH)_2$:

$$(NH_4)_2SO_4 + Ca(OH)_2 = CaSO_4 + 2H_2O + 2NH_3.$$
Ammonium   Calcium   Calcium
sulphate   hydroxide.   sulphate.

The starting-point in this preparation is the ammonia obtained as a by-product in the dry distillation of coal (coke and gas manufacture). Coal contains about 1.5 per cent. nitrogen which on heating, in great part, combines with the hydrogen which is also present, forming ammonia. On passing the generated gases (illuminating-gases) through water, the ammonia is absorbed, the solution (ammoniacal liquor) is neutralized with sulphuric acid: $2NH_3 + H_2SO_4 = (NH_4)_2SO_4$ (p. 150), and evaporated to dryness. The ammonium sulphate which remains is purified by sublimation and then decomposed by heating with calcium hydroxide.

Properties. Colorless gas having when moist an alkaline reaction and whose peculiar odor may be detected in the smallest quantity; it is 0.59 times lighter than air and must be collected over mercury, as one volume of water at $0°$ dissolves 1148 volumes ($=0.875$ parts by weight) of the gas and at $20°$ 739 volumes ($=0.526$ parts by weight). At $-34°$ it liquefies to a colorless liquid, which solidifies at $-75°$. Liquid ammonia boils at $-34°$ and absorbs considerable heat on evaporation. This is made use of in the production of artificial ice and cold, as it may be liquefied at $10°$ by a pressure of 7 atmospheres.

It is a body having very great activity, as it has powerful action upon many metalloids and metals, on the latter especially in the presence of air and water. On passing dry ammonia over heated potassium, potassium nitride is produced: $3K + NH_3 = NK_3 + 3H$; over heated magnesium, magnesium nitride is obtained: $3Mg + 2NH_3 = Mg_3N_2 + 6H$.

Ammonia does not burn in the air, but it burns, on the contrary, in pure oxygen with a yellowish flame, forming water and nitrogen; oxygen may also be made to burn in ammonia (p. 109): $2NH_3 + 3O = 3H_2O + 2N$.

Chlorine gas inflames in ammonia gas with the formation of white vapors of ammonium chloride (p. 146).

When mixed with oxygen and passed over heated spongy platinum it is converted into nitric acid: $NH_3 + 4O = HNO_3 + H_2O$.

Hypochlorites or hypobromites, for instance solutions of chloride of lime or sodium hypobromite, decompose ammonia, setting free all the nitrogen (p. 147).

Caustic ammonia is the name given to the watery solution of ammonia, whose specific gravity is less according to the amount of ammonia it contains. It has the odor of ammonia and an alkaline reaction, and precipitates the heavy metals as hydroxides from their solutions: $FeSO_4 + 2NH_3 + 2H_2O = Fe(OH)_2 + (NH_4)_2SO_4$. It behaves, therefore, as a strong base and may be considered as a watery solution of $NH_4{-}OH$ (corresponding to $K{-}OH$). This compound is indeed not known free, but organic derivatives of the same, the so-called ammonium bases (see Part III), which behave like aqueous solutions of the bases KOH or NaOH. Caustic ammonia having a specific gravity 0.96 contains 10 per cent. by weight, and is ordinarily obtained by diluting the commercial caustic ammonia with water.

Commercial ammonia has a specific gravity 0.90 and contains 28 per cent NH_3.

Salts of Ammonia. Moist gaseous ammonia, as well as its solution, has strong basic properties and yields salts with acids by direct addition; these salts have great similarity to the alkali salts, therefore they will be treated of in connection with these.

$$NH_3 + HCl \quad = NH_4Cl \qquad \text{Ammonium chloride;}$$
$$2NH_3 + H_2SO_4 = (NH_4)_2SO_4 \qquad \text{Ammonium sulphate;}$$
$$NH_3 + H_2S \quad = (NH_4)HS \qquad \text{Ammonium hydrosulphide;}$$
$$2NH_3 + H_2S \quad = (NH_4)_2S \qquad \text{Ammonium sulphide.}$$

In the ammonia compounds, the group NH_4 behaves like a univalent metal, in that it replaces the hydrogen of acids with the formation of salts; This group NH_4 is, therefore, called *ammonium*, and the salts of ammonia are called ammonium salts.

Detection. 1. Free ammonia is readily detected by its odor, as well as by the brown coloration produced on turmeric paper and the blue color produced upon moist red litmus paper, when introduced in an atmosphere containing ammonia gas. On allowing the paper to be exposed to air, the original color returns, due to the evaporation of the ammonia.

2. A glass rod moistened with hydrochloric acid forms visible vapors of ammonium chloride when introduced in an atmosphere containing ammonia.

3. The merest traces of ammonia in watery solution can be detected by the brown coloration or precipitation produced by Nessler's reagent (see this).

4. In regard to the detection of ammonia in its compounds see ammonium salts.

Hydroxylamine, Oxyammonia, $NH_2(OH)$, may be considered as NH_3, in which one atom of hydrogen is replaced by the univalent hydroxyl group, OH.

Preparation. It is produced by the action of dilute nitric acid upon tin (p. 160): $HNO_3 + 6H = NH_2(OH) + 2H_2O$, or by the decomposition of mercury fulminate (see this) by HCl, and is obtained from its salts by decomposition by bases. Hydroxylamine sulphate is obtained from the potassium hydroxylamine disulphonate, produced by the action of alkali bisulphites upon potassium nitrite, by heating with water: $2KHSO_3 + KNO_2 = KOH + N(SO_3K)_2(OH)$.

Properties. It forms explosive, colorless, deliquescent crystals, which melt at 33°, and which boil at 58° at a pressure of 20 mm. It is without odor, combustible, readily soluble in water, having a strong reducing action and is a weak base. With the aldehydes it forms oximes (see these).

Hydroxylamine salts are produced by the direct union of a hydroxylamine molecule with one hydrogen of the acid. Hydroxylamine chloride, $NH_2(OH)HCl$, forms colorless, poisonous crystals, which are readily soluble and which have a reducing action but not explosive.

Hydrazine, Diamid, H_2N-NH_2. *Preparation.* By the action of sulphur dioxide upon potassium nitrite (KNO_2) we obtain nitric oxide-potassium sulphite, $K_2SO_3 \cdot N_2O_2$, which yields hydrazine when treated in aqueous solution with nascent hydrogen (sodium amalgam): $K_2SO_3 \cdot N_2O_2 + 3H_2 = N_2H_4 + K_2SO_4 + H_2O$. On fractional distillation of the solution we obtain hydrazine hydrate, $N_2H_4 \cdot H_2O$, which yields anhydrous hydrazine by distilling in a vacuum with barium oxide.

Properties. Caustic, basic liquid having a peculiar odor, volatile at ordinary temperatures and boiling at 113° and forming colorless crystals at 1° and stable even at 300°. Hydrazine and its salts have a strong reducing power.

Hydrazine Salts. Hydrazine combines with acids by addition, forming two series of salts which we may consider as being formed by the bivalent radical diammonium, $-H_3N-NH_3-$, replacing 2 H atoms of the acid thus: $N_2H_4 + H_2SO_4 = (N_2H_6)SO_4$, or that the univalent radical, H_4N-NH-, replaces 1 H atom of the acid, thus: H_4N-NHI, $(H_4N-NH_2)SO_4$; these latter salts are the most stable. Diammonium sulphate, $(N_2H_6)_2SO_4$, separates on account of its insolubility from solutions of all the other hydrazine salts by treating them with sulphuric acid.

Hydrazoic Acid, $HN\langle\!\!\begin{smallmatrix}N\\\|\\N\end{smallmatrix}\!\!$. *Preparation.* 1. An ice-cold solution of nitrous acid (HNO_2) is poured into an ice-cold solution of hydrazine: $HNO_2 + N_2H_4 = N_3H + 2H_2O$.

2. Ammonia (NH_3), when heated with metallic sodium, yields sodium amid, NH_2Na, which when heated with nitrous oxide, N_2O, yields the sodium salt of hydrazoic acid: $2NH_2Na + N_2O = NaN_3 + NaOH + NH_3$. On distilling this salt with dilute sulphuric acid, we obtain an aqueous solution of hydrazoic acid.

3. By the action of a solution of nitrogen trichloride in benzol (see opposite page) upon hydrazine: $N_2H_4 + NCl_3 = N_3H + 3HCl$. The hydrazoic acid thus obtained can be made anhydrous by careful distillation and treatment with calcium chloride.

Properties. Colorless, poisonous, caustic, strongly acid liquid, boiling at 37°, and having an intolerable odor. With glowing bodies, but often also spontaneously, it explodes with great violence. Miscible with alcohol and water, and the aqueous solution can only be kept with safety; it dissolves all metals which are soluble in HCl.

Hydrazoates, or azides, are similar in every respect to the salts of hydrochloric acid with the exception that they all readily explode. Silver nitrate precipitates white silver hydrazoate, N_3Ag; mercurous nitrate precipitates white mercurous hydrazoate, N_3Hg.

b. Compounds with the Halogens.

Nitrogen Chloride, NCl_3. *Preparation.* If chlorine is passed in a saturated ammonia solution, ammonium chloride and nitrogen are evolved (p. 146). If, on the contrary, chlorine is passed in excess, then the ammonium chloride is still further decomposed with the formation of nitrogen chloride:

$$NH_4Cl + 6Cl = NCl_3 + 4HCl.$$

Properties. Nitrogen chloride is a thick, pale-yellow liquid having an irritating odor and a specific gravity of 1.7, which on shaking, warming, or in contact with many bodies, especially of organic kind, decomposes into its constituents. NCl_3 is rather soluble in water, and this solution often spontaneously decomposes with explosive violence on standing with the generation of nitrogen; the solution in carbon disulphide, chloroform, benzol, etc., is, on the contrary, rather stable and may be handled with safety.

Nitrogen bromide, NBr_3, is quite similar to the above.

Nitrogen iodide, NI_3 or NHI_2, NH_2I, $N_2H_3I_3 = (NH_3 + NI_3)$ are produced by the action of considerable iodine upon ammonia solutions according to conditions. They form black powders which violently explode. Triazoiodide, N_3I, is produced from iodine and N_3Ag (see Hydrazoates) as colorless crystals which are extremely explosive.

c. Mixture of Nitrogen and Oxygen.

Atmospheric Air. The air consists of 78.1 volumes per cent. of nitrogen, 0.9 volume per cent. of argon, and 21 volumes per cent. of oxygen, or 75.5 per cent. by weight of nitrogen, 1.3 per cent. by weight of argon, and 23.2 per cent. by weight of oxygen. Besides these, the air always contains some vapor of water, carbon dioxide, traces of ammonia, hydrogen, nitrous and nitric acid, as well as the elements helium, krypton, neon, and xenon, and where considerable coal is burnt we also have sulphurous and sulphuric acids. The air also contains numerous microorganisms which cause fermentation, putrefaction, and many diseases.

Properties. Air, especially when dry, is a poor conductor of heat and electricity; one liter of dry air weighs at 0° and 760 mm. pressure 1.293 grams; the air is, therefore, 773.4 times lighter than water. Its density relative to $O_2 = 32$ (i.e., its molecular weight if it were a compound) is 28.95. The air is liquefiable at $-191°$, forming a colorless liquid having a specific gravity of 0.995 (p. 41). The air is very readily liquefied by the use of the apparatus as suggested by Linde and by Hampton, and is extensively used for technical purposes.

Nearly all liquid and gaseous bodies become solid when cooled in liquid air, and chemical transformations are retarded or arrested at these temperatures.

In regard to the calculation of the specific gravity of gases from their molecular weights using air as unit see p. 44.

The proportion of nitrogen and oxygen in the air is nearly un-changeable; still the air is to be considered as a mixture of these gases for the following reasons:

1. If oxygen and nitrogen are mixed in the proportion that they exist in air, and the electric spark is passed through the mixture, we find no temperature or volume changes, and the mixture shows the same properties as before, namely, the properties of the air.

2. The proportion by weight between oxygen and nitrogen in the air does not correspond to their atomic weights.

3. If the air were a chemical combination, then its composition must be retained when it is dissolved in water. If air is shaken with water, more oxygen dissolves than nitrogen; if the air is expelled by boiling and its composition determined, we find that it consists of 35 volumes per cent. oxygen (important for aquatic animals).

4. Liquid air yields first nearly all of its nitrogen on evaporation in open vessels (p. 107).

The composition of the air is the same because of the diffusion of the gases (p. 46).

Considerable oxygen is removed from the air by the processes of respiration, combustion, putrefaction, decay, and returns to the air as carbon dioxide, but the green parts of the plant absorb this carbon dioxide and decompose it with the evolution of free oxygen, the carbon going to form the structure of the plant.

The amount of carbon dioxide in the air amounts on an average to 0.04 volume per cent., but may rise to 1 or 2 per cent. in rooms occupied by many persons, or by the burning of many gas-flames; experience has shown that it is not advisable to have more than 0.1 volume per cent. in a room for habitation. The purity of the air can be determined by estimating the amount of carbon dioxide contained therein.

The vapor of water in the air (humidity) is dependent upon the temperature of the air and corresponds nearly to the vapor tension of the water in millimeters at this temperature (p. 115). 1000 liters of air saturated with vapor of water contain at −10° 2.4 grams water, at 0° 4.9 g. water, at +20° 17.2 g. water, at +40° 55 g. water.

Ordinarily the vapor of water amounts to only from 50 to 70 per cent. of the quantity which is necessary completely to saturate the air; if the quantity is greater, then the air appears close and damp; when less, it is dry.

1. *Determination of Nitrogen and Oxygen. a.* According to volume, by taking a certain volume of air, placing it in a tube closed at one end and graduated in millimeters (eudiometer), placing it over mercury, and adding a certain amount of hydrogen. If an electric spark is passed between two platinum wires placed in the upper part of the tube, an explosion takes place and water is formed, and the volume of the gases diminishes on the cooling of the same. As example, 100 volumes of air + 50 volumes of hydrogen on explosion leave 87 volumes of gas. Hence 63 volumes of gas have been condensed into water, and the third o this = 21 volumes is the quantity of oxygen contained in the 100 volumes of air.

b. A measured volume of air is passed over a weighed amount of copper heated to red heat, and the increase of weight determined (= the oxygen), and the volume of nitrogen remaining also measured.

2. *Determination of Vapor of Water. a.* Physically by the hygrometer or psychrometer.

b. Chemically, at the same time as the carbon dioxide, by passing a known volume of air through weighed calcium chloride tubes, and then through tubes filled with potassium hydroxide. The increase in weight of the calcium chloride tubes gives the amount of water in the volume of air passed through; the increase in weight of the potassium hydroxide tubes gives the carbon dioxide.

3. *Determination of Carbon Dioxide. a.* See above, 2 *b.*

b. A large glass vessel whose volume has been carefully determined is filled with the air to be tested and a measured quantity of barium hydroxide solution added, the vessel tightly stoppered and shaken. All the carbon dioxide is absorbed and is precipitated as insoluble barium carbonate: $Ba(OH)_2 + CO_2 = BaCO_3 + H_2O$. The quantity of barium hydroxide contained in the solution is, after the experiment, less than before and the quantity of carbon dioxide is determined by the difference.

d. Compounds with Oxygen.

Nitrous oxide,	N_2O.	Hyponitrous acid,	HNO.
Nitric oxide,	NO.	—	
Nitrogen trioxide,	N_2O_3.	Nitrous acid,	HNO_2.
Nitrogen dioxide,	NO_2.	—	
Nitrogen tetroxide,	N_2O_4.		
Nitrogen pentoxide,	N_2O_5.	Nitric acid,	HNO_3.

Oxygen combines directly with nitrogen only when the electric spark is passed for a long time through a dry mixture of these gases. Red nitrogen dioxide gas, NO_2, is in part produced. If water is present, ammonium nitrite is produced (p. 148).

Nitrous Oxide, Laughing-gas, N_2O or $O{<}^{N}_{N}{>}$. *Preparation.* It is produced in addition to nitric oxide by the action of 15 to 18 per cent. nitric acid upon zinc or tin (see p. 160): $4Zn + 10HNO_3 = 4Zn(NO_3)_2 + 5H_2O + N_2O$. It is obtained pure by heating ammonium nitrate: $NH_4NO_3 = N_2O + 2H_2O$.

Properties. Colorless and odorless, neutral gas, 1.52 times heavier than air, with a sweetish taste; it liquefies at $-88°$, forming a colorless liquid which solidifies to an ice-like mass at $-103°$. Glowing carbon or ignited phosphorus burns in nitrous oxide the same as in oxygen gas; feebly burning sulphur is extinguished, as there is not sufficient heat generated to decompose the gases into $N_2 + O$. With an equal volume of hydrogen it explodes when ignited: $N_2O + 2H = H_2O + 2N$; it does not combine with oxygen. One volume of water dissolves at $10°$ 1.1 volumes of this gas; on account of this property and because it does not yield red nitrogen dioxide with nitric oxide, it can be readily differentiated from oxygen. When nitrous oxide is inspired it produces insensibility, and on account of the condition thus produced it has been called laughing-gas.

If metallic sodium is heated with an enclosed volume of nitrous oxide, it burns into sodium oxide, and a volume of pure nitrogen remains corresponding to the volume of N_2O used. As 2 volumes of nitrous oxide weigh 44 parts by weight, and 2 volumes of nitrogen $= 28$ parts by weight, then the combined oxygen must weigh $16 = 1$ volume, and hence nitrous oxide must have the formula N_2O.

Hyponitrous Acid, Nitrosylic Acid, HNO, more correctly $H_2N_2O_2$ or $HO-N=N-OH$. It is not produced from $N_2O + H_2O$, but, on the contrary, decomposes into these (see below). It is obtained in solution by the action of nitrous acid upon hydroxylamine: $HNO_2 + NH_2OH = 2HNO + H_2O$, and in the anhydrous form by treating silver hyponitrite, $Ag_2(NO)_2$ (see below), with a solution of HCl in ether, when $Ag_2(NO)_2$ decomposes into AgCl and free $H_2N_2O_2$. The liquid obtained on the evaporation of the ether is readily explosive, thick, and solidifies on cooling into colorless crystals.

Hyponitrites are produced by the action of potassium amalgam upon the aqueous solutions of the nitrate or nitrite; by precipitating the alkali salts with silver nitrate we obtain pale-yellow silver hyponitrite. On decomposing the salts with sulphuric acid, the hyponitrous acid set free immediately decomposes in part into $H_2O + N_2O$ and in part according to the equation $3H_2N_2O_2 = 2N_2O_3 + 2NH_3$.

Nitric Oxide, NO. *Preparation.* By the action of 30 to 35 per cent. nitric acid (p. 160) upon copper, mercury, silver, and many other metals: $3Cu + 8HNO_3 = 3Cu(NO_3)_2 + 4H_2O + 2NO$.

Properties. Colorless, neutral gas, 1.04 times heavier than air, liquefiable to a colorless liquid at $-154°$. It is decomposed into nitrogen and oxygen by burning substances, but it requires a higher temperature than nitrous oxide; burning sulphur, ignited carbon or wood, are extinguished therein, while burning magnesium or ignited phosphorus burns brightly in this gas. When mixed with hydrogen, nitric oxide burns on ignition without explosion: $NO + H_2 = H_2O + N$; with carbon disulphide it burns with an intense blue flame which is very rich in chemical rays: $CS_2 + 6NO = CO_2 + 2SO_2 + 6N$. It is slightly soluble in water ($\frac{1}{20}$ volume), but, on the contrary, is soluble in an aqueous solution of ferrous salts, producing a brown solution; on warming this solution, it is driven off (method of purification).

Nitric oxide differs from all other gases by the fact that with the air or in the presence of oxygen it forms brownish-yellow nitrogen dioxide gas; hence traces of free oxygen in gaseous mixtures can be determined by this brownish-yellow coloration.

On heating sodium in a closed volume NO, one-half of the volume of gas used remains as nitrogen; hence the formula for this gas can be derived in the same way as that for nitrous oxide.

Nitrogen Trioxide, Nitrous Anhydride, N_2O_3. *Preparation.* 1. Equal volumes of nitric oxide and nitrogen dioxide or 4 volumes nitric oxide and 1 volume of oxygen are condensed in a tube cooled to $-21°$.

$$NO + NO_2 = N_2O_3. \qquad 2NO + O = N_2O_3.$$

2. By the distillation of arsenic trioxide with nitric acid and cooling the vapors produced to $-21°$.

$$As_2O_3 + 2HNO_3 + 2H_2O = 2H_3AsO_4 + NO + NO_2.$$

3. By mixing a sodium nitrite solution with sulphuric acid and cooling the dried vapors set free to $-21°$ (see Nitrites).

Properties. Deep-blue liquid at $-21°$ which gradually decomposes into $NO_2 + NO$, but much quicker at its boiling-point (3.5°). These two gases on cooling unite again, producing N_2O_3; hence it only exists as a liquid.

Sulphuric acid absorbs N_2O_3, respectively $NO_2 + NO$, with the formation of nitrosylsuphuric acid (lead chamber crystals, p. 128).

$$2SO_2\!\!<^{OH}_{OH} + N_2O_3 = 2SO_2\!\!<^{OH}_{O(NO)} + H_2O.$$

This solution of nitrosylsulphuric acid in sulphuric acid is called nitrated acid and when undiluted, is very stable.

Nitrous Acid, HNO_2 or $O\!\!=\!\!N^-OH$. *Occurrence.* As ammonium nitrite, $NH_4^-NO_2$ to a trivial extent in the air, in rain-water, in many spring-waters; it is also formed to a slight extent on the evaporation of water in the air, by the action of the electric spark upon moist air (in rain after thunder-storms), in all combustion processes in the air, and in the slow oxidation of phosphorus in the air, on rusting of iron, and in the electrolysis of water containing air: $2N + 2H_2O = NH_4^-NO_2$. It occurs as nitrites in many plant juices, nasal mucus, and saliva.

Preparation and Properties. Liquid nitrogen trioxide mixes with a small amount of ice-cold water, producing a blue liquid, which may be considered as HNO_2 in solution, but which decomposes at ordinary temperatures or in the presence of considerable water: $3HNO_2 = HNO_3 + 2NO + H_2O$.

Nitrites are more stable; they are produced by saturating the watery solution of HNO_2 with bases, oftener by carefully heating the corresponding nitrate. With sulphuric acid, they develop reddish-brown nitric dioxide gas mixed with colorless nitric oxide gas, as the sulphuric acid removes the water from the nitrous acid set free: $2HNO_2 = H_2O + NO + NO_2$. Weak acids set nitrous acid free, but this decomposes immediately into $HNO_3 + 2NO + H_2O$. Nitrous acid and its decomposition products have an oxidizing action, therefore bleach plant pigments and set iodine free from iodine salts: $2KI + 2HNO_2 = 2KOH + 2I + 2NO$. On the other hand, the nitrites, on account of their tendency to be converted into nitrates, have a reducing action; that is, they decolorize acidified solutions of potassium permanganate: $5HNO_2 + 2KMnO_4 + 3H_2SO_4 = 5HNO_3 + K_2SO_4 + 2MnSO_4 + 3H_2O$.

Detection. 1. The aqueous solution of nitrites on acidification with sulphuric acid turns starch paste and potassium iodide blue.

2. Metaphenylendiamine gives an intense yellow with nitrite solutions after acidification.

3. Diphenylamine gives the same reaction as with nitric acid.

Nitrogen Dioxide, NO_2. *Preparation.* 1. By mixing 2 volumes of nitric oxide and 1 volume of oxygen.

2. By heating lead nitrate and cooling the vapors produced: $Pb(NO_3)_2 = PbO + 2NO_2 + O$.

3. By passing the electric spark for a long time through a dry mixture of 1 volume of nitrogen and 2 volumes of oxygen (p. 155).

Properties. Reddish-brown, suffocating, caustic, and poisonous gas having a strong oxidizing action; hence it sets iodine free from iodine compounds and supports the combustion of many substances. It is 1.5 times heavier than air, decomposes with water into nitric acid and nitric oxide, $3NO_2 + H_2O = 2HNO_3 + NO$; hence it used to be considered as an acid and called hyponitric acid. With decreas_ing temperatures, it becomes lighter in color, when its volume weight steadily increases, as more molecules NO_2 are converted into N_2O_4 and condense finally into pure

Nitrogen tetroxide, N_2O_4, which at $0°$ is a nearly colorless liquid, which finally solidifies at $-20°$ into colorless crystals, which melt at $-10°$. Above $0°$ the liquid becomes more and more yellow, as with the increased temperature NO_2 is produced and at $26°$ it boils and then contains 20 per cent. NO_2 (Dissociation).

N_2O_4 is the mixed anhydride of nitric acid and nitrous acid ($N_2O_5 + N_2O_3 = 2N_2O_4$), as it forms with ice-cold water (or with bases), nitric acid and blue nitrous acid (or their salts): $N_2O_4 + H_2O = HNO_3 + HNO_2$. With water at ordinary temperature it forms nitric acid and nitric oxide, as the N_2O_4 is first split into $2NO_2 : 3NO_2 + H_2O = 2HNO_3 + NO$.

Nitrogen pentoxide, nitric anhydride, N_2O_5, is obtained by passing dry chlorine gas over dried silver nitrate at $60°$: $2AgNO_3 + 2Cl = 2AgCl + N_2O_5 + O$, or by distilling nitric acid with phosphorus pentoxide: $2HNO_3 + P_2O_5 = N_2O_5 + 2HPO_3$. The phosphorus pentoxide removes the elements of water from the nitric acid and is converted into metaphosphoric acid (HPO_3), which is not volatile, while nitrogen pentoxide distils over and is obtained on cooling the distillate as colorless prismatic crystals. These melt at $30°$, forming a yellow liquid, and slowly decompose at a slightly higher temperature; on rapidly heating, as well as by keeping, it decomposes with explosive violence into nitrogen dioxide and oxygen. With water it forms nitric acid.

Nitric Acid, HNO_3 or $O_2N(OH)$. *Occurrence.* It is formed to a slight extent with ammonium nitrate and ammonium nitrite (p. 148) by passing the electric spark through moist air; otherwise it is found, only as its salts, the nitrates, especially sodium nitrate (Chili salt peter, $NaNO_3$), in Peru in large quantities. It is also found as potas sium nitrate very widely diffused, although not in large quantities, on the surface of the earth, especially in Egypt and India. It is found as calcium nitrate, $Ca(NO_3)_2$, or so-called wall saltpeter, on the walls of stables and urinals. Nitrates are formed in all decomposition processes of organic nitrogenous substances (see Nitrates); also many plants contain nitrates.

Preparation. By distilling alkali nitrates, chiefly Chili saltpeter, with sulphuric acid at a temperature not above 220°:

$$NaNO_3 + H_2SO_4 = NaHSO_4 + HNO_3.$$
$$\text{Sodium nitrate.} \qquad \text{Sodium bisulphate.}$$

The crude nitric acid thus obtained contains 61–65 per cent. HNO_3 and has a specific gravity of 1.36–1.40, and the pure nitric acid of commerce is obtained from this by redistillation and contains 68 per cent. HNO_3 (see below).

If, in the distillation the temperature rises above 220°, then the sodium bisulphate decomposes a second molecule of the nitrate: $NaHSO_4 + NaNO_3 = Na_2SO_4 + HNO_3$; the temperature in this case becomes so high that a part of the nitric acid decomposes, $2HNO_3 = H_2O + 2NO_2 + O$, and red, fuming nitric acid is obtained.

Anhydrous nitric acid is obtained by distilling pure nitric acid with concentrated sulphuric acid, and removing the nitrogen oxides produced by passing air through the distillate.

Properties. Anhydrous nitric acid is a colorless, fuming liquid having an irritating odor and a specific gravity of 1.54, forms a crystalline solid at $-40°$, and begins to boil at 86° with a partial decomposition into $H_2O + 2NO_2 + O$, and miscible with water in all proportions. If dilute nitric acid is distilled, the boiling-point gradually rises until it is 121° (p. 53), when nitric acid distils over which contains 68 per cent. HNO_3 and has a specific gravity of 1.41. Stronger nitric acids on heating give off HNO_3 until a 68 per cent. acid is produced.

Nitric acid vapors, on being passed through a red-hot tube, decompose into nitrogen dioxide, water, and oxygen (see above); this decomposition takes place in part on heating and on standing in the sunlight; hence nitric acid gradually becomes yellow. It contains 76.1 per cent. oxygen, which is in part readily given off to oxidizable bodies; hence nitric acid is one of the strongest oxidizing agents and destroys organic pigments (indigo solution) and other organic compounds, or the hydrogen atoms of many organic compounds may be replaced by NO_2 (nitration).

Nitric acid is also called *aqua fortis*, as it is used in the separation of silver from gold. It dissolves or oxidizes all metals with the exception of gold, platinum, rhodium, iridium, ruthenium, and oxidizes all metalloids with the exception of fluorine, chlorine, nitrogen, and the members of the argon group. In this change it is in part reduced to nitrogen oxides (N_2O, .NO, NO_2), and indeed higher oxides are produced the more concentrated the acid is. Very dilute acid is reduced by zinc to NH_3, and by tin indeed into NH_3 and $NH_2(OH)$ (p. 150). This reduction into NH_3 takes place still more readily in alkaline solution (p. 148).

With 6 per cent. nitric acid NH_3 is produced: $4Zn + 9HNO_3 = 4Zn(NO_3)_2 + 3H_2O + NH_3$; with 18 per cent. nitric acid N_2O is produced: $4Zn + 10HNO_3 = 4Zn(NO_3)_2 + 5H_2O + N_2O$; with 30 per cent. nitric acid NO is obtained: $3Zn + 8HNO_3 = 3Zn(NO_3)_2 + 4H_2O + 2NO$; and with stronger nitric acid NO_2 is produced.

Red fuming nitric acid is a reddish-brown liquid which gives off red vapors in the air and has a specific gravity of 1.48 to 1.50. It consists of a solution of about 8 parts nitrogen dioxide dissolved in 100 parts 68 per cent. nitric acid. It has a still stronger oxidizing power than colorless nitric acid; 'when mixed with water it becomes first blue, then green, and finally colorless, because nitrous acid and then nitric acid are produced from the nitrogen dioxide. When this acid acts upon combustible materials they may be made to take fire.

Nitro-hydrochloric acid, *aqua regia*, is the name given to a strong oxidizing mixture of 3 parts hydrochloric acid with 1 part nitric acid, because it dissolves nearly all metals, including platinum and gold (called the king of metals by the alchemists).

This action takes place especially on heating and depends upon the formation of chlorine and the following compounds:

$HNO_3 + 3HCl = 2Cl + 2H_2O + NOCl$ (nitrosylchloride).
$HNO_3 + HCl = H_2O + NO_2Cl$ (nitrylchloride, existence questionable).

These two chlorine compounds may be considered at the chloranhydrides of nitric acid and nitrous acid.
(Acid chlorides, acid chloranhydrides are those acids whose ^-OH groups are entirely or in part replaced by Cl).

Nitrates are produced where organic nitrogenous substances undergo decay in the presence of stronger bases and air. It has been shown that in alkaline soils certain bacteria accumulate, which have the property of oxidizing the ammonia produced in the decay into nitric acid (so-called nitrification germs): $KOH + NH_3 + 4O = KNO_3 + 2H_2O$.

These conditions for nitric acid formation are present in nearly all soils, and this is the reason why soluble nitrates are found in spring-water and the upper layers of the earth, especially in the neighborhood of stables. On the other hand, a series of bacteria are known which can reduce the nitric acid formed into nitrous acid, ammonia, and indeed into nitrogen. These so-called saltpeter-destroyers are the cause of the loss of nitrogen in the decomposition of manure on the manure-heap.

Nitrates are nearly soluble in water; on strongly heating the alkali nitrates are decomposed into nitrites with the setting free of oxygen; the other nitrates decompose into the oxide of the metal and develop $NO_2 + O$. The nitrates are also strong oxidizing agents; placed on red-hot carbon they all explode. By nascent hydrogen these soluble nitrates are reduced to nitrites or ammonia (p. 148).

Detection. 1. Copper is dissolved by nitric acid with the production of blue copper nitrate and the development of red nitrogen dioxide fumes. The nitric acid is first set free from the nitrates by the addition of sulphuric acid.

2. If a solution of nitric acid or nitrates is mixed with concentrated sulphuric acid and a cold solution of ferrous sulphate floated thereon, a dark ring forms between the two solutions.

The nitric acid oxidizes a part of the ferrous sulphate into ferric sulphate, and nitric oxide is formed, which gives a brown color with the still undecomposed ferrous sulphate (p. 156):

$$6FeSO_4 + 3H_2SO_4 + 2HNO_3 = 3Fe_2(SO_4)_3 + 4H_2O + 2NO.$$

3. A colorless solution of diphenylamine in sulphuric acid is colored deep blue by traces of nitric acid, and also by NO_2, HNO_2, and other oxidizing substances.

2. Phosphorus.

Atomic weight $31 = P$.

Occurrence. Only in the form of phosphates, that is, salts of phosphoric acid, H_3PO_4. These occur in the mineral kingdom especially as calcium phosphate, $Ca_3(PO_4)_2$; in many varieties of rocks which are called phosphate rock or phosphorites, as well as in the coprolites and certain varieties of guano (both fossil excrements); in the minerals phosphorite and apatite, $Ca_3(PO_4)_2+CaClF$, vivianite, $Fe_3(PO_4)_2$; as cerium phosphate in monozite sand; as aluminum phosphate in wavelite, turquois, osteoliths, etc. Phosphorus occurs in the plant kingdom not only as the widely distributed phosphates, but also as a constituent of the proteids. In the animal kingdom calcium phosphate, $Ca_3(PO_4)_2$, forms two-thirds of the dried skeleton; also in the nerve substance, in the yoke of the egg, urine, brain, and blood. In the iron industry the phosphorus contained in certain iron ores collects in the so-called Thomas-slag (which see), which contains about 50 per cent. $Ca_3(PO_4)_2$. Phosphorus was first (1669) obtained by heating evaporated urine, whereby the phosphates contained therein were reduced on heating to a red heat with the carbon formed (see below).

Preparation. 1. From mineral phosphates or bones. These latter are first purified from gelatine and fat; they are then burned, when the organic substance is destroyed and the inorganic remains as bone-ash, which contains about 85 per cent. calcium phosphate.

a. The mineral phosphates or bone-ash are treated with two-thirds their weight of sulphuric acid, when insoluble calcium sulphate and soluble primary calcium phosphate are produced:

$$\overset{II}{Ca_3}\!\left\{\!{PO_4 \atop PO_4}\right.\!+2\overset{}{H}\!\left\{\!{H \atop H}\right.\!SO_4 = 2CaSO_4 + \overset{II}{CaH_4}\!\left\{\!{PO_4 \atop PO_4}\right..$$

b. The primary calcium phosphate is decanted from the precipitate and evaporated to sirupy consistency, mixed with charcoal powder, dried, and heated to a faint-red heat; in this process the primary calcium phosphate is transformed into calcium metaphosphate and water is given off:

$$CaH_4(PO_4)_2 = Ca(PO_3)_2 + 2H_2O.$$

c. The residue after heating is placed in clay retorts after being mixed with sand (SiO_2) and heated to a white heat, when all the phosphorus contained in the calcium metaphosphate is set free:

$$Ca(PO_3)_2 + SiO_2 + 5C = CaSiO_3 + P_2 + 5CO.$$

2. By heating mineral phosphates with carbon and a flux in the electric oven.

The phosphorus form as a vapor, which is condensed by passing it into water; it is purified by redistillation, and then melted under water and generally moulded in the form of sticks in glass moulds.

Properties. Phosphorus is a transparent, crystalline, faintly yellowish, very poisonous body which is brittle when cold, and at ordinary temperatures as soft as wax and having a faint garlic-like odor (see below). It has a specific gravity of 1.83, melts at 44°, boils at 288°, and then forms colorless vapors whose vapor density is 62, from which it follows that its molecular weight is 124 (p. 21). As its atomic weight is only 31, then the phosphorus molecule in the vaporous state must consist of 4 atoms.

Phosphorus is insoluble in water, slightly soluble in ether and alcohol, readily soluble in fatty oils (*oleum phosphoratum*) and carbon disulphide. On evaporation of the solution (in the absence of air, as it will inflame otherwise) we obtain the phosphorus in regular rhombododecahedra; it shines in the dark ($\phi\tilde{\omega}\varsigma$, light, and $\phi\acute{o}\rho o\varsigma$, carry); it smokes in moist air (vapors of P_2O_5) with the generation of an ozone odor, whereby it is oxidized and forms phosphorous acid (H_3PO_3) and phosphoric acid (H_3PO_4), and at the same time ammonium nitrite, ozone, and hydrogen peroxide are formed. Even at 45° it inflames and burns with a dazzling white flame into phosphorus pentoxide (P_2O_5); the ignition may take place even on lying in the air or by friction; hence it must be kept under water and should be cut only under water. It must also be kept in the dark, as in light and under water it becomes covered with a layer of phosphorus suboxide, P_4O.

It is oxidized to phosphoric acid (p. 167) on warming with nitric acid or aqua regia, and with caustic alkalies it forms, on warming, PH_3 besides hypophosphorous acid (p. 165).

Phosphorus combines with the halogens even at ordinary temperatures with the development of flame, and with most metals on warming, producing phosphides. It has, especially when finely divided, a strong reducing action and separates many metals (or their phosphides) from their salt solution; hence paper moistened with silver nitrate and exposed to an atmosphere containing traces of phosphorus turns black, due to the formation of PAg_3. If hydrogen is passed over gently heated phosphorus, the gas burns with a pale-green flame, due to the phosphorus it contains.

Detection. If the substance which is to be tested for yellow phosphorus is placed in a flask which is connected with a long tube surrounded with cold water, and if the contents of the flask are heated, then the phosphorus volatilizes with the steam and we see in the dark a prominent generally ring-formed light where the vapor condenses. If the quantity of phosphorus is not too small, small masses of phosphorus may be found in the condenser (Mitscherlich's method).

Red Phosphorus (Amorphous Phosphorus). *Preparation.* If yellow phosphorus is heated in closed tubes with a gas that has no action upon it (carbon dioxide or nitrogen) for a few minutes to 300° or for a longer time to at least 250°, then it is converted into a deep-red modification having entirely different properties. On treatment with carbon disulphide or caustic alkali any unchanged yellow phosphorus present may be removed.

Properties. Reddish-brown, odorless and tasteless, non-poisonous, hexagonal microcrystalline powder having a specific gravity of 2.2, insoluble in carbon disulphide, fatty oils, etc., does not shine in the dark, does not change in the air, does not ignite on friction, but first at 260°, and combines with the halogens only on warming. It slowly vaporizes without decomposition at 100°; when heated to 260° in a gas with which it is not active it vaporizes without previously melting, and if the vapor can expand, it passes over again into yellow phosphorus.

From the depression of the freezing-point (p. 20) of a dilute solution of red phosphorus in PBr_3 it follows that the molecule consists of eight or more atoms of P, and also that it is a polymolecular modification of ordinary phosphorus.

The low chemical energy of red phosphorus seems to be dependent upon the diminished motion of its atom; in its formation from yellow phosphorus a considerable development of heat takes place, hence the red phosphorus contains less energy.

Black or metallic phosphorus can be obtained by heating ordinary or red phosphorus in vacuous tubes to 530°, or more readily on heating with lead, whereby the phosphorus dissolves in the fused lead and separates on cooling as metallic-like, dark rhombohedra which have a specific gravity of 2.34 and are even less active than red phosphorus. This modification is perhaps only better crystalline red phosphorus.

a. Compounds with Hydrogen.

Gaseous phosphureted hydrogen, PH_3.
Liquid phosphureted hydrogen, P_2H_4.
Solid phosphureted hydrogen, P_4H_2.

Gaseous Phosphureted Hydrogen, Hydrogen Phosphide, Phosphine, PH_3. *Preparation.* 1. Analogous to arseniureted hydrogen, AsH_3, from dilute sulphuric acid, zinc, and phosphorus.

2. By heating phosphorous acid (H_3PO_3) or hypophosphorous acid (H_3PO_2):

$$4H_3PO_3 = PH_3 + 3H_3PO_4 \text{ (phosphoric acid)};$$
$$2H_3PO_2 = PH_3 + H_3PO_4$$

3. By decomposing calcium phosphide with water or dilute hydrochloric acid: $Ca_3P_2 + 6HCl = 3CaCl_2 + 2PH_3$.

4. Ordinarily by boiling phosphorus with alkali hydroxide solution: $3KOH + 4P + 3H_2O = PH_3 + 3KH_2PO_2$.

5. From phosphonium iodide (see below) by heating with alkali hydrate solution: $PH_4I + KOH = PH_3 + KI + HOH$.

Properties. The phosphureted hydrogen gas obtained according to methods 1 to 4 always contains small amounts of hydrogen and liquid phosphureted hydrogen, P_2H_4, which is spontaneously inflammable and which condenses when the gas is passed into a tube surrounded by an ice mixture, or is decomposed by passing the gas through hydrochloric acid or alcohol. Pure PH_3 is a colorless, poisonous, neutral gas which is insoluble in water, has a garlic-like odor, and liquefies at $-86°$ and solidifies at $-134°$. It does not inflame spontaneously, but inflames with chlorine and bromine vapors, and with oxidizing agents it yields spontaneously inflammable P_2H_4. When ignited, PH_3 burns into water and phosphoric anhydride (P_2O_5), forming a white smoke which often forms rings that rise. In contradistinction to NH_3 phosphureted hydrogen has only weak basic properties, as it unites directly only with HBr and HI, forming so-called phosphonium compounds for instance phosphonium iodide, PH_4I; with HCl it unites only at $-35°$. It has a reducing action and hence separates the pure metal or the metallic phosphide from many metallic salt solutions.

If the electric spark is passed through a tube filled with PH_3, amorphous phosphorus separates, the volume of the gas increases one-half and consists of pure hydrogen: 2 vols. $PH_3 = 34$ parts by weight give **3 vols.** $H = 3$ parts by weight, the phosphorus separated weighing 31 parts.

Liquid phosphuretted hydrogen, P_2H_4, is obtained by cooling self-inflammable PH_3, which forms a colorless, refractive liquid insoluble in water, boiling at 57°, and spontaneously inflammable in the air and burning to $P_2O_5 + 2H_2O$. Its presence in inflammable gases (marsh-gas), etc., makes these spontaneously inflammable, and this is the reason why PH_3 when it has not been passed through cool tubes inflames spontaneously in the air.

Solid Phosphureted Hydrogen, P_4H_2. Liquid phosphureted hydrogen in contact with carbon, sulphur, concentrated HCl, or by sunlight, decomposes into gaseous and solid phosphureted hydrogen: $5P_2H_4 = 6PH_3 + P_4H_2$. P_4H_2 is also obtained by dissolving Ca_3P_2 in warm concentrated sulphuric acid. It is a yellow, odorless, and tasteless powder which inflames at 160° or by shock and burning into $2P_2O_5 + H_2O$.

b. *Compounds with the Halogens.*

Phosphorus Trichloride, PCl_3. On passing dry chlorine over gently warmed phosphorus it inflames and forms PCl_3, which may be collected in a receiver. It is a colorless liquid which boils at 76° and fumes in the air, as it decomposes with the water contained in the air into phosphorous acid and hydrochloric acid: $PCl_3 + 3H_2O = H_3PO_3 + 3HCl$.

Phosphorus pentachloride, PCl_5, is obtained by the action of chlorine upon PCl_3 as a crystalline, yellowish mass which vaporizes at 148°, and decomposes with considerable water into phosphoric acid: $PCl_5 + 4H_2O = H_3PO_4 + 5HCl$; with little water it decomposes into hydrochloric acid and

Phosphorus oxychloride, $POCl_3$, a colorless fuming liquid which boils at 107°: $PCl_5 + H_2O = POCl_3 + 2HCl$.

Phosphorus bromides and **iodides** are entirely analogous to the chlorine compounds. They are obtained by bringing together the constituents in the proportion represented by their formulæ.

c. *Compounds with Oxygen.*

Phosphorus monoxide,	P_2O.	Hypophosphorous acid,	H_3PO_2.
Phosphorus trioxide,	P_2O_3.	Phosphorous acid,	H_3PO_3.
		Orthophosphoric acid,	H_3PO_4.
Phosphorus pentoxide,	P_2O_5.	Metaphosphoric acid,	HPO_3.
		Pyrophosphoric acid,	$H_4P_2O_7$.
Phosphorus tetroxide,	P_2O_4.	Hypophosphoric acid,	$H_4P_2O_6$.

Phosphorus monoxide, P_2O, obtained from H_3PO_3 with PCl_3, is a yellowish-red, amorphous body which does not combine with water and alkalies: $2H_3PO_3 + 2PCl_3 = 6HCl + P_2O_5 + P_2O$.

Hypophosphorous acid, H_3PO_2, is produced by decomposing barium hypophosphite (p. 165, 4) with sulphuric acid. The solution obtained after filtering off the barium sulphate is concentrated by evaporation under the air-pump receiver and forms a colorless, sirupy liquid which solidifies into large white plates at 0°:

$$\overset{II}{Ba} < \genfrac{}{}{0pt}{}{H_2PO_2}{H_2PO_2} + \genfrac{}{}{0pt}{}{H}{H} > SO_4 = \genfrac{}{}{0pt}{}{H_3PO_2}{H_3PO_2} + BaSO_4.$$

On warming, it decomposes into phosphorous acid and PH_3 (p. 165); on taking up oxygen it is readily converted into phosphoric acid and is, therefore, a strong reducing agent. It reduces sulphuric acid to sulphur dioxide and even into sulphur and precipitates gold and silver from their solutions, etc. Only one of its hydrogen atoms can be replaced by metals; it is, therefore, a monobasic acid, having the formula $H_2PO(OH)$.

Hypophosphites are produced on warming an aqueous solution of the bases with phosphorus; they have a reducing action and hence are readily converted into phosphates.

Phosphorus trioxide, P_2O_3 (according to its vapor density, phosphorus hexoxide, P_4O_6, formerly erroneously called phosphorous anhydride), is produced on passing a slow current of dried air over gently heated phosphorus and forms a volatile, white amorphous powder which is readily oxidized into P_2O_5 and which dissolves in water with the separation of a yellow amorphous powder (perhaps P_2O). On heating to 400° it decomposes into phosphorus and

, Phosphorus tetroxide, P_2O_4, the anhydride of hypophosphoric acid, which forms colorless needles.

Hypophosphoric acid, $H_4P_2O_6$, is only known in aqueous solution and is produced when moist phosphorus is exposed for a long time to the air. It is characterized by its difficultly soluble acid sodium salt, $Na_2H_2P_2O_6$, which is used in the separation of this acid from the phosphorous acid and phosphoric acid produced at the same time.

Phosphorous acid, H_3PO_3, is formed beside phosphoric acid and hypophosphoric acid in the slow oxidation of phosphorus in moist air, and may be obtained pure by decomposing phosphorus trichloride with water: $PCl_3 + 3H_2O = H_3PO_3 + 3HCl$. If the solution obtained is evaporated under the air-pump receiver, phosphorous acid separates in colorless, deliquescent crystals which melt at 70°, and on further heating decompose into phosphoric acid and PH_3 (p. 165). Analogous to H_3PO_2 it takes up oxygen and is readily converted into phosphoric acid, and hence acts like a very strong reducing body. Only two of its hydrogen atoms can be replaced by metals, therefore it is a bibasic acid having the formula $HPO(OH)_2$.

Phosphites are also strong reducing agents, but are not oxidized in the air.

Phosphorus pentoxide, phosphoric anhydride, P_2O_5, is produced when phosphorus is burnt in a strong current of dry oxygen or air It forms white, neutral, flocculent masses which have a greenish phosphorescence in the dark and which readily absorb moisture and deliquesce, forming metaphosphoric acid (process, p. 168). Because of this relationship to water it is used in drying gases, as well as in removing water from many substances.

Orthophosphoric Acid, Phosphoric Acid, H_3PO_4 or $OP(OH)_3$. *Occurrence.* See Phosphorus.

Preparation. 1. On dissolving phosphorus pentoxide in cold water metaphosphoric acid (HPO_3) is produced, which, on boiling the solution, is converted into orthophosphoric acid:

$$P_2O_5 + H_2O = 2HPO_3. \qquad HPO_3 + H_2O = H_3PO_4.$$

2. By treating bone-ash with the corresponding quantity of sulphuric acid (p. 162) and evaporating the liquid obtained on decantation from the calcium sulphate:

$$Ca_3(PO_4)_2 + 3H_2SO_4 = 3CaSO_4 + 2H_3PO_4.$$

3. By heating phosphorus or, better, amorphous phosphorus with nitric acid; this gradually dissolves with the development of red vapors of the oxides of nitrogen:

$$3HNO_3 + P = H_3PO_4 + 2NO_2 + NO.$$

This solution is evaporated so as to drive off the excessive nitric acid, and a colorless aqueous solution of phosphoric acid is obtained.

Properties. On further evaporation of the aqueous solution obtained in the various methods of preparation, colorless, rhombic crystals which melt at 38° and which deliquesce in the air separate out. It is readily soluble in water. It is a weaker acid than sulphuric or nitric acid, but H_3PO_4 sets these acids free from their compounds on heating, because it is less volatile. Commercial phosphoric acid is a 25 per cent. watery solution having a specific gravity of 1.154.

On heating H_3PO_4 it loses water and is converted into pyrophosphoric acid and then into metaphosphoric acid. These three phosphoric acids, which differ from each other very markedly in composition and properties, may be considered as combinations of P_2O_5 with 3, 2, and 1 molecule water:

$$P_2O_5 + 3H_2O = 2H_3PO_4, \text{ orthophosphoric acid};$$
$$P_2O_5 + 2H_2O = H_4P_2O_7, \text{ pyrophosphoric acid};$$
$$P_2O_5 + H_2O = 2HPO_3, \text{ metaphosphoric acid}.$$

Phosphates. Orthophosphoric acid forms three series of salts according as 1, 2, or 3 atoms of hydrogen are replaced by metals.

The phosphates used to be divided, according to their behavior towards litmus, into acid salts, MH_2PO_4, neutral salts, M_2HPO_4, and basic salts, M_3PO_4. According to the present view of salts, the compounds having the formula MH_2PO_4 as well as M_2HPO_4 may be considered as acid salts. In regard to the nomenclature of such salts and the polybasic acids see p. 101.

Detection. 1. Orthophosphoric acid and its salts give a yellow crystalline precipitate of ammonium phosphomolybdate, $12MoO_3 +$ $(NH_4)_3PO_4 + 6H_2O$, when their solutions are treated with nitric acid and an excess of ammonium molybdate solution.

2. Silver nitrate produces a yellow precipitate of silver phosphate, Ag_3PO_4, in neutral phosphate solutions.

3. A solution of magnesium salt treated with ammonium salts and ammonia (magnesia mixture, see Magnesium, *c*) gives a white precipitate of ammonium magnesium phosphate, $Mg(NH_4)PO_4 + 6H_2O$, with phosphate solutions.

Pyrophosphoric Acid, $H_4P_2O_7$. *Preparation.* By heating orthophosphoric acid to $250°–300°$: $2H_3PO_4 = H_4P_2O_7 + H_2O$.

Properties. White crystalline masses, readily soluble in water and slowly converted into orthophosphoric acid on standing, but more quickly on heating.

Detection. Ammonium molybdate and magnesia mixture give a precipitate in aqueous solutions only when a transformation into H_3PO_4 has taken place.

Pyrophosphates are obtained by heating the secondary phosphates to a red heat: $2K_2HPO_4 = K_4P_2O_7 + H_2O$. They are stable on boiling their watery solution, but on boiling with dilute acids they are transformed into orthophosphates. With silver nitrate they give a white precipitate of silver pyrophosphate, $Ag_4P_2O_7$. The sodium salt readily dissolves iron salts (removal of iron-stains and ink-spots from linen).

Metaphosphoric Acid, HPO_3. *Preparation.* 1. By dissolving phosphoric acid anhydride in cold water.

2. By continuously heating ortho- or pyrophosphoric acid until no more water is driven off:

$$H_3PO_4 = HPO_3 + H_2O; \quad H_4P_2O_7 = 2HPO_3 + H_2O.$$

Properties. Colorless transparent masses (glacial phosphoric acid) which melt on heating, volatilize at a red heat (hence free phosphoric acid is not used in the preparation of phosphorus). It deliquesces in moist air, dissolves in water, and on standing gradually yields orthophosphoric acid, but more quickly on boiling.

Detection. It differs from ortho- and pyrophosphoric acid by its property of precipitating proteid solutions (use in the detection of proteids in urine, etc.) and by being precipitated by barium chloride solution. Magnesia mixture and ammonium molybdate only yield a precipitate in aqueous solution when a transformation into orthophosphoric acid has taken place.

Metaphosphates are prepared by heating the primary phosphates to a red heat: $KH_2PO_4 = KPO_3 + H_2O$; on boiling their aqueous solution they are converted into orthophosphates; with silver nitrate they give a white precipitate of silver metaphosphate, $AgPO_3$.

d. Compounds with Sulphur.

Phosphorus trisulphide, P_2S_3, and *phosphorus pentasulphide,* P_2S_5, are produced by carefully melting together the corresponding weights of sulphur and amorphous phosphorus. They are yellowish, crystalline bodies which are decomposed by water:

$$P_2S_3 + 6HOH = 2H_3PO_3 + 3H_2S;$$
$$P_2S_5 + 8HOH = 2H_3PO_4 + 5H_2S.$$

3. Arsenic.

Atomic weight 75 =As.

Occurrence. Arsenic occurs native in crystalline masses as arsenic. Combined it occurs as orpiment, As_2S_3, and realgar, As_2S_2, in pyrargyrite and tennanite (p. 181) as arsenopyrite or mispickel, FeSAs, as cobalt glance or cobaltite, CoSAs, as smaltite, $CoAs_2$, gersdorffite, NiSAs, as niccolite, NiAs, as löllingite, $FeAs_2$, and as arsenolite, As_2O_3. It occurs in small quantities in many ores, coal, shale, in sulphur, etc.; so that bodies prepared from such materials contain arsenic. Arsenical mineral springs (Levico, Roncegno) are also known. Many alloys, such as speculum metal, brass, lead shot, often contain arsenic.

Preparation. 1. By heating mispickel, FeSAs, in earthenware tubes, when it decomposes into iron sulphide and arsenic; this latter vaporizes and is passed through tubes, where it condenses in crystalline masses.

2. By heating arsenic trioxide (flowers of arsenic) with carbon: $As_2O_3 + 3C = 2As + 3CO$.

Properties. Grayish-white, metallic-like, shining, opaque, brittle, crystalline masses, sometimes well defined rhombohedra, having a specific gravity of 5.7. It is readily oxidizable, and not soluble without change in any solution, is extremely poisonous, as well as its compounds, and on heating (in the absence of air or in an indifferent gas) it volatilizes at about 450° without melting and forms a colorless and odorless vapor.

The molecular weight of arsenic vapors at 450° is 300.   As the atomic weight of arsenic is 75, it follows that its molecule consists of 4 atoms, analogous to that of phosphorus.   At 1700°, on the contrary, the molecular weight falls to 150, so that its molecule consists of only 2 atoms at this temperature (dissociation, p. 72).

Arsenic is insoluble in hydrochloric acid and dilute sulphuric acid; it dissolves in dilute nitric acid or in concentrated sulphuric acid, forming arsenious acid (or arsenic trioxide, process, p. 174); in concentrated nitric acid or in aqua regia it forms arsenic acid (p. 175).   On heating it combines directly with sulphur, chlorine, bromine, iodine, and most of the metals, and when finely powdered it burns in chlorine gas with the formation of arsenic trichloride.   With the exception of the sulphur and haloid compounds, no arsenic salts are known (see Antimony).   The compounds of arsenic with the metals (the arsenides) are isomorphous with the metallic sulphides and have an analogous constitution; in these sulphur and arsenic can replace each other in atomic proportions.

Arsenic does not change in dry air, but in the presence of water it gradually oxidizes into arsenic trioxide which is soluble in water. On heating in the air it burns into arsenic trioxide with a garlic-like odor.

Brown arsenic (corresponding to red phosphorus) is produced on heating arseniureted hydrogen (p. 172) or ordinary arsenic in a current of hydrogen.   It forms brownish-black, metallic-like masses, and in thin layers it is reddish brown and transparent and consists of microscopic rhombohedra.   It is odorless and tasteless, oxidizable with difficulty, insoluble in all solvents, has a specific gravity of 4.7, and vaporizes above 280° without melting, and is then transformed into the white modification.

Yellow arsenic (corresponding to yellow phosphorus) is produced by subliming ordinary arsenic in carbon dioxide ; it condenses on the cold part of the apparatus as sulphur-like, crystalline masses, readily oxidizable, having a specific gravity of 3.9 and an odor similar to garlic. It dissolves readily in carbon disulphide and separates from this solution on evaporation in regular rhombodecahedra.   At ordinary temperatures, even in the dark, it is quickly transformed into ordinary arsenic.

a. *Compounds with Hydrogen.*

Gaseous arseniureted hydrogen, AsH_3.
Solid arseniureted hydrogen, As_4H_2.
Liquid arseniureted hydrogen, As_2H_4,

is not known free, but derivatives of the same, containing hydrocarbon groups such as cacodyl, $As_2(CH_3)_4$ (see Part III), are known.

Gaseous Arseniureted Hydrogen, Arsenic Hydride or Arsine, AsH_3.

Preparation. 1. It is obtained pure by the action of dilute sulphuric acid or hydrochloric acid upon an alloy of arsenic and zinc: $As_2Zn_3 + 6HCl = 2AsH_3 + 3ZnCl_2$.

2. Mixed with hydrogen it may be obtained by the action of nascent hydrogen upon dissolved arsenic compounds. If a solution containing arsenic, but free from nitric acid (p. 173), is introduced into a flask in which zinc and dilute sulphuric acid are present, we obtain the gas which may be dried by passing over solid calcium chloride (Marsh apparatus).

Properties. Colorless, very poisonous gas, having a garlic-like odor, liquefiable at $-56°$ and solid at $-114°$ and which does not combine with acids, but itself shows acid-like characters (p. 173). When ignited it burns into arsenic trioxide and water: $2AsH_3 + 6O = As_2O_3 + 3H_2O$; if a cold object, for instance a porcelain dish, is held in the flame, this is then cooled down below the combustion (oxidation) temperature of the arsenic, which does not burn into its oxide, but deposits as brownish-black, shining spots, so-called "arsenic-stains," upon the porcelain.

Arseniureted hydrogen is decomposed into arsenic and hydrogen by the electric spark or by a faint white heat; hence if the gas is passed through a heated glass tube, the arsenic deposits as a black, shining coating, so-called "arsenic mirror," after having passed the heated portion of the tube.

Arsenic stains and arsenic mirrors consist of brown arsenic (p. 171); they volatilize on heating without melting; they readily dissolve in sodium hypochlorite solution; when touched with nitric acid they dissolve, forming arsenic acid or arsenious acid; if this solution is neutralized with ammonia and treated with silver nitrate solution, brownish-red silver arsenate or yellow silver arsenite are produced (differing from antimony-stains, p. 179).

Arseniureted hydrogen has reducing properties and precipitates various metals from their solution, and arsenious acid, which is produced, remains in solution:

$$AsH_3 + 6AgNO_3 + 3H_2O = 6Ag + H_3AsO_3 + 6HNO_3.$$

Many metals are precipitated as arsenides by AsH_3:

$$2AsH_3 + 3HgCl_2 = Hg_3As_2 + 6HCl.$$

If AsH_3 is passed over paper moistened with dilute silver nitrate or mercuric chloride solution, these become dark, due to the setting free of metallic silver or the formation of mercury arsenide (light and the presence of other reducing gases being excluded). If a very concentrated solution of silver nitrate is used, then a yellow compound of $AsAg_3 + 3AgNO_3$ is produced which, when moistened with water, turns black.

Solid **arseniureted hydrogen**, As_4H_2, is obtained as a reddish-brown powder when nascent hydrogen acts upon arsenic compounds in the presence of nitric acid. (Hence nitric acid must be absent in the Marsh test for arsenic.)

b. Compounds with the Halogens.

These are obtained by the direct union of the respective elements and have similar properties to the corresponding phosphorus compounds. They are decomposed by water.

Arsenic trichloride, $AsCl_3$, is also produced by warming As_2O_3 or As_2O_5 (which see) with hydrochloric acid. It forms a colorless, thick liquid, boiling at 130°.

Arsenic penta-iodide, AsI_5, the only arsenic penta compound of the halogens known in the free state. It forms red crystals.

Arsenic tribromide, AsB_3, forms colorless crystals.

Arsenic triiodide, AsI_3, forms reddish-yellow crystals.

c. Compounds with Oxygen.

Arsenic trioxide, As_2O_3.	{ (Arsenious acid, H_3AsO_3.) ((Metarsenious acid, $HAsO_2$.)
Arsenic pentoxide, As_2O_5.	{ Orthoarsenic acid, H_3AsO_4. { Metarsenic acid, $HAsO_3$. { Pyroarsenic acid, $H_4As_2O_7$.
Arsenic tetroxide, As_2O_4.	

Arsenic Trioxide, As_2O_3 (white arsenic?), arsenious anhydride, or the arsenious acid of the druggist.

Occurrence. In the mineral kingdom as arsenolite.

Preparation. 1. By burning arsenical ores (as a by-product in the roasting of ores) or by burning arsenic in the air. The arsenic trioxide produced vaporizes and is passed through chambers where it condenses as a white crystalline (octahedra) powder (white arsenic or arsenic meal). It is purified by sublimation, and is obtained as a transparent amorphous colorless mass (vitreous arsenic).

2. On boiling arsenic with sulphuric acid or dilute nitric acid a solution of arsenious acid is obtained which on evaporation yields octahedral crystals of arsenic trioxide: $As + HNO_3 + H_2O = H_3AsO_3 + NO$; $2As + 3H_2SO_4 = 2H_3AsO_3 + 3SO_2$; $2H_3AsO_3 = As_2O_3 + 3H_2O$.

Properties. It volatilizes at 220° without melting, forming colorless vapors which at 700° have a density corresponding to the formula As_4O_6, and at 1800° to the formula As_2O_3 (dissociation, p. 71). On quickly cooling these vapors regular octahedra having a specific gravity of 3.69 are obtained, while on slowly cooling monoclinic prisms having a specific gravity of 4.0 are obtained; hence it is dimorphous. If As_2O_3 is heated under pressure or for a long time in the neighborhood of 220°, it becomes amorphous and fusible (formation of amorphous As_2O_3 in its purification) and has a specific gravity of 3.74; this gradually becomes opaque, white, porcelain-like, being transformed into octahedral As_2O_3. Both modifications dissolve in hot hydrochloric acid without forming compounds, and separate on cooling in colorless, regular octahedra.

If its solution in hot concentrated hydrochloric acid is boiled, then $AsCl_3$, which volatilizes, is produced. On heating As_2O_3 with carbon (p. 170), or many metals, or nascent hydrogen, it is reduced to arsenic; on the other hand, it has a reducing action itself (see below) because of its aptitude to be converted into As_2O_5. It dissolves in water with difficulty, the amorphous variety somewhat more easily and in larger quantities than the crystalline form. The solutions have a faint acid reaction and may be considered as dissolved

Arsenious Acid, $H_3AsO_3 (A_2O_3 + 3H_2O = 2H_3AsO_3)$. This acid is not known free.

Arsenites are, with the exception of the alkali salts, insoluble in water; hence freshly prepared ferric hydroxide, $Fe(OH)_3$, is used as an antidote in arsenical poisoning, as the insoluble ferric arsenite, $FeAsO_3$, is formed, and at the same time the acid of the gastric juice, which would partly dissolve the iron arsenide, is neutralized. Soluble arsenites are strong reducing agents, as they are readily converted into arsenates; with silver nitrate they give a yellow precipitate of silver arsenite; Ag_3AsO_3; with cupric salts a green precipitate of cupric arsenite, $Cu_3(AsO_3)_2$; both these compounds are soluble in nitric acid and ammonia.

Metarsenious acid, $HAsO_3$ (behaves towards arsenious acid like meta-phosphoric acid to phosphoric acid, $H_3AsO_3 = HAsO_2 + H_2O$), is not known free, but its salts are known.

Arsenic pentoxide, arsenic anhydride, As_2O_5, is obtained by gently heating (see Arsenic Tetroxide) arsenic acids (which see) as a white porous mass, which on boiling with concentrated hydrochloric acid forms volatile arsenic trichloride and free chlorine, as the arsenic pentachloride formed decomposes immediately:

$$As_2O_5 + 10HCl = 2AsCl_5 + 5H_2O;$$
$$2AsCl_5 = 2AsCl_3 + 4Cl.$$

It dissolves in water and is slowly converted into orthoarsenic acid:

$$As_2O_5 + 3H_2O = 2H_3AsO_4.$$

Arsenic Tetroxide, As_2O_4. Arsenic pentoxide melts on strongly heat-ing with the evolution of oxygen and forms a viscous, yellow liquid, As_2O_4, which on cooling forms an amorphous, vitreous mass that decomposes into $As_2O_3 + O$ on stronger heating.

Orthoarsenic acid, H_3AsO_4, arsenic acid, is prepared by warming arsenic or arsenic trioxide with aqua regia or with concentrated nitric acid: $3As + 5HNO_3 + 2H_2O = 3H_3AsO_4 + 5NO;$

$$As_2O_3 + 2HNO_3 + 2H_2O = 2H_3AsO_4 + NO + NO_2.$$

On evaporating the solution, colorless rhombic crystals having the composition $2H_3AsO_4 + H_2O$ separate out.

Arsenates are analogous to the salts of orthophosphoric acid and are also isomorphous therewith. Silver nitrate precipitates reddish-brown silver arsenate, Ag_3AsO_4, from their solution, and this precipi-tate is soluble in ammonia and nitric acid. Precipitates analogous to the phosphates are also obtained on warming with ammonium molybdate or with magnesia mixture.

Pyroarsenic acid, $H_4As_2O_7$, is obtained as crystals by heating arsenic acid to 180°: $2H_3AsO_4 = H_4As_2O_7 + H_2O$. .

Metarsenic acid, $HAsO_3$, is produced as a crystalline mass by heating ortho- or pyroarsenic acids to 200°.

On dissolving both of these acids in water they yield orthoarsenic acid. The salts of both of these acids are produced from the corresponding arsenates in a manner analogous to the corresponding phosphates (pp. 169, 170).

d. Compounds with Sulphur.

The following three compounds are known, which, like the sulphides of phosphorus, are obtained by fusing together the corresponding quantities of arsenic and sulphur.

Arsenic disulphide, As_2S_2, is found native as realgar in readily fusible ruby-red crystals.

Arsenic trisulphide, arsenious sulphide, As_2S_3, occurs native as orpiment in yellow crystalline masses. It is obtained as a yellow amorphous powder by passing H_2S into a solution of arsenious acid containing hydrochloric acid, or into a solution of arsenite or of arsenic acid (see below): $As_2O_3 + 3H_2S = As_2S_3 + 3H_2O$.

Neutral solutions of arsenious acid are only colored yellow by H_2S, as colloidal arsenic trisulphide (p. 53) is produced. This formation, which is analogous to the formation of metallic sulphides, shows the metallic character of arsenic.

As_2S_3 is insoluble in water and acids, but soluble in ammonia, caustic and fixed alkalies.

Arsenic pentasulphide, arsenic sulphide, A_2S_5, is prepared by passing H_2S into potassium arsenate solution, when dissolved potassium sulpho-arsenate, K_3AsS_4 (see below), is produced, from which hydrochloric acid precipitates arsenic pentasulphide as a pale-yellow powder: $K_3AsO_4 + 4H_2S = K_3AsS_4 + 4H_2O$; $2K_3AsS_4 + 6HCl = 6KCl + 3H_2S + As_2S_5$. If H_2S is passed into a solution containing free arsenic acid and acidified with acid, sulphur first precipitates out, and after some time arsenic trisulphide precipitates because the H_2S first reduces the arsenic acid into arsenious acid: $As_2O_5 + 2H_2S = As_2O_3 + 2H_2O + 2S$; $As_2O_3 + 3H_2S = As_2S_3 + 3H_2O$.
If H_2S is rapidly passed into a warmed, faintly hydrochloric acid solution of arsenic acid As_2S_5 is nevertheless precipitated.

Sulphoacids and sulphosalts of arsenic. Just as we have sulphides which are analogous to the oxides, so we also have sulphur arsenic salts which correspond to the oxygen arsenic salts. These salts are derived from the following unknown acids:

Sulphoarsenious acid, H_3AsS_3.
Sulpharsenic acid, H_3AsS_4.

The salts of these acids are obtained by dissolving the corresponding sulphide in potassium or ammonium sulphide solution and evaporating, etc., thus:

$As_2S_3 + 3K_2S = 2K_3AsS_3$, Potassium sulphoarsenite.
$As_2S_5 + 3K_2S = 2K_3AsS_4$, Potassium sulphoarsenate.
$As_2S_5 + 3(NH_4)_2S = 2(NH_4)_2AsS_4$, Ammonium sulphoarsenate.

Antimony, tin, gold, platinum form similar sulphosalts (p. 181).

e. Detection of Arsenic Compounds.

1. The garlic-like odor of burning arsenic, which is produced when all its compounds are heated with soda upon charcoal, is characteristic.

2. Sulphuretted hydrogen immediately precipitates yellow arsenic trisulphide from acidified solutions of arsenious acid or arsenites, and from solutions of arsenic acid or arsenates only after passing the gas for a longer time. This precipitate is soluble in alkali and ammonium sulphides (see above), and differs from all other yellow sulphides (Sb_2S_3, SnS_2, CdS) by its insolubility in hot hydrochloric acid and its solubility in ammonia.

3. All arsenic compounds when heated in a glass tube with sodium carbonate and potassium cyanide yield arsenic, which condenses in the upper cold portion of the tube as an arsenical mirror (p. 172).

4. Even traces of arsenic and arsenic compounds (with the exception of arsenic sulphide) may be detected by converting them into AsH_3 and decomposing this by heat (Marsh test for arsenic, p. 172).

5. Stannous chloride dissolved in concentrated hydrochloric acid causes a precipitate of arsenic from the HCl solution of many arsenical compounds, even when they contain only traces of arsenic (nitric acid being absent). This takes place gradually in the cold, and is shown by the brown coloration in the previously colorless solution (Bettendorf's arsenic test): $3SnCl_2 + 6HCl + As_2O_3 = 3SnCl_4 + 3H_2O + 2As$.

4. Antimony (Stibium).

Atomic weight $120.2 = Sb$.

Occurrence. Seldom free, generally as antimony glance or stibnite, Sb_2S_3, as senarmontite, Sb_2O_3, and as kermesite, Sb_2OS_2. Also in many nickel, copper, lead, and silver ores, combined with sulphur (p. 181).

Preparation. From stibnite by heating with iron: $Sb_2S_3 + 3Fe = 3FeS + 2Sb$, or by roasting with air, when sulphur dioxide is developed and antimony trioxide remains: $Sb_2S_3 + 9O = Sb_2O_3 + 3SO_2$; the antimony oxide is then reduced to metal by heating with carbon: $Sb_2O_3 + 3C = 2Sb + 3CO$.

Properties. Bluish-white, metallic-like, shining, very brittle masses having a specific gravity of 6.7, melting at 630°, and on solidifying forming rhombohedral crystals similar to arsenic. On heating above 1450° it vaporizes, and does not change in the air at ordinary temperatures, but burns on heating in the air into white odorless antimony trioxide, Sb_2O_3. Like phosphorus and arsenic, it combines directly on warming with the halogens, and when powdered it inflames in chlorine gas. It is nearly insoluble in hydrochloric acid and dilute sulphuric acid; in hot concentrated sulphuric acid it dissolves with the formation of antimony sulphate: $2Sb + 6H_2SO_4 = Sb_2(SO_4)_3 + 3SO_2 + 6H_2O$; nitric acid oxidizes it into antimony trioxide or orthoantimonic acid, according to the concentration. Both of these bodies are insoluble in nitric acid. Aqua regia dissolves antimony, forming antimony trichloride or antimony pentachloride dependent upon the length of action. The molecule of antimony consists of 2 atoms (p. 171).

A silver-white modification of antimony, having a specific gravity of 5.8, is obtained on the electrolysis of a concentrated solution of antimony trichloride in HCl.

The alloys are type-metal (one part antimony, four parts lead) and Britannia metal (one part antimony and four parts tin).

a. Compounds with Hydrogen.

Antimoniureted hydrogen, antimony hydride or stibine, SbH_3, the only compound of antimony with hydrogen (p. 172), is obtained in the same manner as the analogous arsenic compound.

Properties. It is a colorless gas without basic properties, has a characteristic odor, and becomes liquid at low temperatures and then solid. It burns when ignited into water and antimony trioxide, and is more readily decomposable by heat than AsH_3, and deposits in spots or as a mirror, similar to arsenic. The antimony mirror is produced in front as well as behind the portion heated, is grayish black and less volatile than the arsenic mirror, and melts before it volatilizes. Antimony-stains are, contrary to the arsenic-stains, dull and nearly black, insoluble in sodium hypochlorite solution, and give after oxidation with nitric acid and neutralization with ammonia a black stain of antimony silver, $SbAg_3$, with silver nitrate solution. On passing SbH_3 into silver nitrate solution all the antimony precipitates as $SbAg_3$.

b. Compounds with the Halogens

are prepared like the corresponding phosphorus compounds. They form with many metallic chlorides readily soluble, crystalline double salts, which are used as mordants in dyeing: thus, $NaCl + SbCl_3$.

Antimony Trichloride, Antimonous Chloride, $SbCl_3$. *Preparation.* By dissolving antimony oxide or antimony sulphide in hydrochloric acid: $Sb_2S_3 + 6HCl = 2SbCl_3 + 3H_2S$. This is then distilled, when first H_2S, then the excess of hydrochloric acid, and finally the antimony trichloride pass over.

Properties. Colorless, fuming, crystalline, caustic, soft masses (hence formerly called butter of antimony), which melt at 73° and boil at 223° and which absorb water from the air and deliquesce. Antimony chloride is soluble in hydrochloric acid; if this solution or solid antimony chloride is treated with considerable water, a white crystalline precipitate of

Antimony oyxchloride, $SbOCl$, is obtained. This was formerly called Algarot powder. It contains various amounts of antimony trioxide according to the method of preparation:

$$2SbCl_3 + 3H_2O = Sb_2O_3 + 6HCl;$$
$$SbCl_3 + H_2O = SbOCl + 2HCl.$$

Antimony pentachloride, antimonic chloride, $SbCl_5$, is a yellowish fuming liquid which crystallizes at −6° and which with a little water solidifies into crystals having the composition $SbCl_5 + H_2O$ or $SbCl_5 + 4H_2O$, and with more water deposits ortho- or pyroantimonic acids (p. 180).

c. Compounds with Oxygen.

Antimony trioxide, Sb_2O_3.	Antimonous acid,	H_3SbO_3.
	Metantimonous acid,	$HSbO_2$.
Antimony pentoxide, Sb_2O_5.	Orthoantimonic acid,	H_3SbO_4.
	Metantimonic acid,	$HSbO_3$.
Antimony tetroxide, Sb_2O_4.	Pyroantimonic acid,	$H_4Sb_2O_7$.

Antimony trioxide, Sb_2O_3 (according to vapor density Sb_4O_6), antimonous anhydride, occurs as valentinite in rhombic prisms, as senarmontite in regular octahedra (isomorphous with both forms of arsenic trioxide). It is formed on the burning of antimony or on treating the same with dilute nitric acid, also by the careful heating of antimonous and metantimonous acids. It forms a white crystalline powder which fuses and sublimes in the absence of air (see Antimony Tetroxide), insoluble in water, and acts like metantimonous acid towards acids and alkalies.

Antimonous acid, H_3SbO_3, is precipitated as a white precipitate.by treating tartar emetic (see below) with dilute sulphuric acid.

Metantimonous acid, $HSbO_2$, is obtained from H_3SbO_3 as a white powder by the removal of water, or by treating antimony trichloride solution with alkali hydroxide or alkali carbonate: $2SbCl_3 + 3Na_2CO_3 + H_2O = 2HSbO_2 + 6NaCl + 3CO_2$. Both acids dissolve in an excess of alkali hydrates, forming metantimonites, thus: $NaSbO_2$; the solutions decompose on evaporation with the setting free of Sb_2O_3. They act like bases towards acids; $_{th}e_y$ dissolve in hydrochloric acid, forming antimony trichloride, and in sulphuric acid, producing antimony sulphate, $Sb_2(SO_4)_3$, and are insoluble in nitric acid.

The monovalent group SbO, called antimonyl, which is contained in the metantimonous acid, SbO(OH), can at the same time replace the hydrogen of acids with the formation of salts; for instance, in antimonyl sulphate, $(SbO)_2SO_4$, in tartar emetic, $C_4H_4K(SbO)O_6$.

Antimony pentoxide, antimonic anhydride, Sb_2O_5, is obtained by heating the antimonic acids to 400°. It forms an infusible, amorphous, yellow powder, insoluble in water and nitric acid, but soluble in hydrochloric acid, forming $SbCl_5$, and soluble in alkalies, forming antimonates.

Orthoantimonic acid, H_3SbO_4, is produced on warming antimony with concentrated nitric acid or on mixing antimony pentachloride with cold water: $2SbCl_5 + 8H_2O = 2H_3SbO_4 + 10HCl$. It is a white powder, insoluble in water, ammonia, nitric acid, but soluble in hydrochloric acid, forming $SbCl_5$, also in caustic alkalies, forming antimonates which decompose even on evaporation.

Pyroantimonic acid, $H_4Sb_2O_7$, is produced from H_3SbO_4 on heating to 200°, as well as by mixing $SbCl_5$ with warm water. It is a white powder, soluble in large quantities of pure water, in caustic alkalies, ammonia, and hydrochloric acid: $2SbCl_5 + 7H_2O = H_4Sb_2O_7 + 10HCl$.

The most important salt of this acid is the secondary sodium pyroantimonate, $Na_2H_2Sb_2O_7 + 6H_2O$, as it is the only sodium salt insoluble in water.

Metantimonic acid, $HSbO_3$, is produced from ortho- or pyroantimonic acid by heating to 300°, and forms a white powder.

Antimony tetroxide, Sb_2O_4, also considered as antimony antimonate, $(Sb)SbO_4$, is produced when any oxygen compound of antimony is heated in the air (to about 800°). It is a white amorphous powder which turns yellow when heated and which does not melt nor volatilize. It is insoluble in water, but soluble in caustic alkalies and hydrochloric acid.

d. Compounds with Sulphur.

Antimony trisulphide, Sb_2S_3, occurs in gray, brittle, crystalline masses as stibnite. It is a constituent of many minerals, such as tetrahedrite, pyrargyrite, bournonite, etc. (see below). It is readily fusible and volatile at higher temperatures.

It may be obtained as an amorphous orange-red powder by passing sulphuretted hydrogen into antimony trichloride solution: $2SbCl_3 + 3H_2S = Sb_2S_3 + 6HCl$. It is soluble in alkali sulphides (see below) and becomes gray and crystalline on heating in the absence of air.

Antimony oxysulphide, Sb_2OS_2, is found in nature as kermesite, and is produced on melting stibnite with an insufficient supply of air. It forms a brownish-red vitreous mass (antimony vermilion). A mixture of antimony trisulphide and antimony trioxide is called *kermes*.

Antimony pentasulphide, Sb_2S_5, gold sulphur, is obtained by passing sulphuretted hydrogen into a solution of antimony pentachloride: $2SbCl_5 + 5H_2S = Sb_2S_5 + 10HCl$, or by decomposing sodium sulphoantimonate, Na_3SbS_4 (which see), with acids. It is an orange-red powder, soluble in alkali sulphides (p. 182), and decomposes on heating into $Sb_2S_3 + S_2$.

Sulphoacids and Sulphosalts of Antimony. Like the sulphides of arsenic, the sulphides of antimony form compounds with metallic sulphides which are derivatives of the following acids unknown in the free state:

$$\text{Sulphantimonous acid, } H_3SbS_3;$$
$$\text{Sulphantimonic acid, } H_3SbS_4.$$

The salts of these acids are obtained in a manner analogous to that of arsenic (p. 177); these salts are also found in the following minerals:

$$\overset{I}{\text{Pyrargyrite, }} \overset{}{Ag_3SbS_3}, \qquad \overset{II\ I}{\text{Bournonite, }} (PbCu)SbS_3,$$

$$\text{Tetrahedrite group or Fahl ores, } \overset{II}{M_3}(SbS_3) + \overset{II}{M}\overset{II}{S} = \overset{II}{M_4}Sb_2S_7,$$

where

$$\overset{II}{M} = \overset{I}{Cu_2}, \text{ in part } \overset{I}{Ag_2}, \overset{II}{Fe}, \overset{II}{Zn}, \overset{II}{Hg}; \text{ thus, } (\overset{I}{Cu_2})_3(\overset{I}{Ag_2})Sb_2S_7.$$

e. Detection of Antimony Compounds.

1. By the obtainment of an antimony mirror on heating the antimoniureted hydrogen gas produced (p. 178).

2. Soluble antimony salts give with water a white precipitate of basic antimony salts, which are soluble in tartaric acid (differing from basic bismuth salts).

3. Solutions of antimony compounds acidified with hydrochloric acid form an orange-red precipitate of antimony trisulphide with sulphuretted hydrogen. This precipitate is soluble in alkali or ammonium sulphide, and precipitable from this solution by acids as orange-red antimony pentasulphide.

4. If the substance to be tested is placed upon a piece of platinum-foil with a small piece of zinc and then a little hydrochloric acid added, a black spot of metallic antimony will be obtained upon the platinum. This spot cannot be washed off with water.

ARGON GROUP.

Argon. Helium. Neon. Krypton. Xenon.

These elements seem to belong to the nitrogen group, as they always accompany the free-occurring nitrogen and replace nitrogen in many minerals. These minerals are of such a complicated composition that the valence of argon, etc., has not been determined, and up to the present time no compound of argon, etc., has been obtained, as these elements chemically are still more indifferent than nitrogen (ά and ἐργός, *carrier*). From certain physical properties it follows that their molecule consists only of one atom. They possess characteristic spectra and may be obtained on the fractional distillation of liquid air.

1. Argon.

Atomic weight 39.9 = A.

Occurrence. Free in the air (0.9 volumes per cent.) and to a less extent in the gases of many mineral springs, combined to a somewhat greater extent in rare minerals, such as clevite, broggerite, and uranite, nearly always accompanied with helium.

Preparation. The nitrogen prepared from the air is passed over heated magnesium or lithium, which combines only with the nitrogen. The small quantities of helium which are nearly always present cannot at the present time be separated.

Properties. Colorless, odorless, and tasteless gas, liquefiable at $-185°$ and crystalline at $-190°$, soluble in 25 parts water. It is 1.4 times heavier than air, 20 times heavier than hydrogen; 1 liter weighs 1.78 grams (p. 44). The characteristic spectrum consists of numerous blue, red, and green lines.

2. Helium.

Atomic weight 4 = He.

Occurrence. It occurs free as traces in the air and in the gases of certain mineral springs; it is found combined to a greater extent in certain rare earths which contain vanadium, tantalum, niobium, thorium, yttrium, uranium, especially in cleveite, euxenite fergusonite, monazite, äschynite, and to an enormous extent free in the atmosphere of the sun (hence its name) and of the fixed stars.

Preparation. The powdered minerals are heated in order to remove water and gases, which are less firmly combined than the helium. This product is then heated with potassium dichromate in a vacuum, when helium and argon are evolved; in order to separate these two gases, they are exposed in a Geissler tube with magnesium electrodes to a strong electric current, when the argon is absorbed by the magnesium.

Properties. Colorless, odorless, and tasteless gas, liquefiable at $-250°$ and nearly insoluble in water. Helium is 0.14 times lighter than the air and twice as heavy as hydrogen. Its characteristic spectrum consists of bright lines of which six lie in the red, green, blue, and violet; the seventh, which is especially characterized, occurring in the yellow, and indeed to the right of the yellow sodium lines.

BORON.

Atomic weight $11 = B$.

Boron occurs only trivalent and belongs, from the constitution of its compounds, to the nitrogen group, and from its position in the periodic system, as well as the behavior of certain of its compounds, it belongs to aluminum. Most of the boron compounds show great similarity to the corresponding silicon compounds. When free it has great similarity to carbon and to silicon.

Occurrence. Only combined as boric acid, H_3BO_3 (which see), and in the tetraborates, such as tinkal or borax, $Na_2B_4O_7 + 10H_2O$, in India. Boracite, $4MgB_4O_7 + 2MgO + MgCl_2$ and borocalcite, $CaB_4O_7 + 6H_2O$, occur in Stassfurt salt. Boron is also found to a trivial extent in many plants.

Preparation. 1. By heating boron trioxide with magnesium when amorphous boron is obtained: $B_2O_3 + 3Mg = 3MgO + 2B$; the magnesium oxide produced is dissolved by hydrochloric acid.

2. Crystalline boron is obtained by heating amorphous boron or boron trioxide with aluminium: $B_2O_3 + 2Al = Al_2O_3 + 2B$. The boron dissolves in the aluminium and crystallizes out on cooling; the aluminium is dissolved by hydrochloric acid.

Properties. *Amorphous boron* is a brown powder having a specific gravity of 2.45, burns on heating into boron trioxide, combines at higher temperatures with chlorine, bromine, sulphur, nitrogen, also with platinum and silver with the production of so-called borides. It has a reducing action, hence explodes when rubbed with lead peroxides, decolorizes potassium permanganate solution, precipitates metallic silver from silver salt solutions, etc. It is oxidized into boric acid by boiling with nitric or sulphuric acid: $2B + 3H_2SO_4 = 2H_3BO_3 + 3SO_2$; on boiling with caustic potash solutions it dissolves with the formation of potassium metaborate: $2KOH + 2B + 2H_2O = 2KBO_2 + 6H$.

Crystalline boron forms colorless, transparent quadratic crystals having a specific gravity of 2.63, and being next to the diamond in refractive power and in hardness. It is not oxidized on heating and is not attacked by acids and caustic potash solutions. When fused with potassium hydroxide both modifications yield potassium metaborate.

Compounds of Boron.

Gaseous boron hydride, BH_3, is produced on decomposing a fused mixture of boron and magnesium with hydrochloric acid. It is colorless, has a disagreeable odor, burns with a green flame into $B_2O_3 + 3H_2O$; it precipitates black silver boride, Ag_3B, from silver salt solutions, and on heating decomposes into its constituents.

Solid boron hydride has probably the formula B_4H_2.

Boron trichloride, BCl_3, and also **boron trifluoride,** BF_3, are obtained in the same way as the similar si icon compounds.

Boron nitride, BN, is obtained on heating amorphous boron in nitrogen, as a white amorphous powder which is insoluble and infusible. If steam is passed over boron nitride at 200°, boric acid and ammonia are formed: $BN + 3H_2O = H_3BO_3 + NH_3$.

Boron carbide, B_6C, produced on heating boron with carbon in the electric furnace, forms black crystals which are very stable and next to the diamond in hardness.

Boron trioxide, boric anhydride, B_2O_3, is obtained by heating the boric acids. It forms colorless, fusible, vitreous masses which are only slightly volatile at a high white heat and which dissolve in water, forming boric acid.

Orthoboric Acid, H_3BO_3 or $B(OH)_3$, Boracic Acid. *Occurrence.* Free as the mineral sassolite, and to a slight extent in many mineral waters (Wiesbaden, Aachen), and in the steam which is evolved in the fumaroles of Tuscany and in California, and streaming from the earth in the Volcano Islands. Salts of this are not found in nature (see below).

Preparation. 1. In Tuscany the vapors from the "fumaroles" or "soffioni" are condensed in water placed in basins. These solutions are evaporated in flat pans, which are heated by the vapors, until the boric acid crystallizes out and which is then purified by recrystallization.

2. In Stassfurt boracite or borocalcite are treated with hydrochloric acid: $CaB_4O_7 + 2HCl + 5H_2O = CaCl_2 + 4H_3BO_3$.

3. Chemically pure boric acid is obtained by treating a hot, saturated solution of borax with hydrochloric acid gas: $Na_2B_4O_7 + 2HCl + 5H_2O = 2NaCl + 4H_3BO_3$; the boric acid which separates from the solution on cooling is purified by recrystallization.

Properties. Colorless, shining laminæ having a specific gravity of 1.43, possessing a fatty touch, and which are soluble in 3 parts boiling and 25 parts cold water, as well as in alcohol. Boric acid is a very weak acid, but on account of the slight volatility of its anhydride it drives out most acids from their salts on heating therewith.

Detection. The lighted alcoholic solution burns with a green flame, and on boiling the aqueous solution the boric acid volatilizes with the steam. Boric acid solutions color blue litmus paper faint red, turmeric paper reddish brown on drying. (Alkalies turn turmeric paper brown immediately, and acids change this brown into yellow again. Boric acid only produces a brown stain after drying, which is not changed by acids and becomes greenish black with alkalies).

Borates. These are all derived from tetraboric acid; only organic salts of orthoboric acid are known. The salts of metaboric acid are very unstable. All borates give the reactions mentioned for boric acid if they are treated with hydrochloric acid for the turmeric reaction and with sulphuric acid for the flame reaction.

Metaboric acid, HBO_2, is obtained by heating orthoboric acid to 100°: $H_3BO_3 = HBO_2 + H_2O$.

Pyro- or Tetraboric Acid, $H_2B_4O_7$, is produced on heating ortho- or metaboric acids to 140°: $4H_3BO_3 = H_2B_4O_7 + 5H_2O$;
$$4HBO_2 = H_2B_4O_7 + H_2O.$$

Perboric acid, HBO_3, is known only in the form of salts, which are obtained from solutions of the tetraborates by H_2O_2. They are energetic oxidizing agents.

CARBON GROUP.
Carbon. Silicon.
Germanium. Tin. Lead.
Titanium. Zirconium. Cerium. Thorium.

The elements of the first two series are di- and tetravalent; lead generally occurs divalent; the elements of the last series are tetravalent only.

Germanium, tin, lead, bear the same relationship to carbon and silicon that arsenic, antimony and bismuth do to nitrogen and phosphorus. As the atomic weight increases, the character of the elements becomes more metallic:

$$\begin{array}{ccccc} +17 & +44 & +45 & +88 \\ N=14 & P=31 & As=75 & Sb=120 & Bi=208 \\ +16 & +44 & +46 & +89 \\ C=12 & Si=28 & Ge=72 & Sn=118 & Pb=207 \end{array}$$

The elements of the second and third series show even more metallic character than arsenic and antimony, and will be discussed with the metals. They do not combine with hydrogen, but, on the contrary, combine with the halogens, forming volatile compounds. Their monoxides are bases, while their dioxides are acid anhydrides.

1. Carbon.

Atomic weight 12=C.

Occurrence. Free in three allotropic modifications as diamond, graphite, and amorphous carbon, which in their properties show the greatest differences and are only correspondent with each other through the fact that they all yield carbon dioxide, CO_2, on burning. It occurs combined as the chief constituent of all animal and plant substances and the products, such as peat and various coals, produced by the slow decomposition of plant substances. Combined with hydrogen it forms rock-oil (petroleum) and asphalt; as carbon dioxide it occurs combined with oxygen in the air and in the carbonates, such as marble, limestone, chalk, dolomite, often forming entire mountains.

Preparation. See the individual modifications.

Properties. Solid, odorless, and tasteless, soluble only in molten iron, not fusible, only volatile at about 3500° in the electric arc. At ordinary temperatures it remains unchanged; at higher temperatures it burns into carbon dioxide with the development of light and heat and leaves an ash which consists of admixed inorganic substances. At a white heat amorphous carbon abstracts the oxygen from most bodies and is, therefore, a powerful reducing agent.

Carbon combines with fluorine even at ordinary temperatures; with oxygen, sulphur, and iron at a red heat; and with hydrogen, boron, silicon, and most metals, on the contrary, at about 3000°. Its compounds with silicon, boron, and the metals are called *carbides* and are characterized by their stability and non-fusibility at the highest temperatures which have been obtained up to the present time. These carbides generally form beautiful crystalline masses which are decomposed by acids and, with the exception of iron carbide, chromium carbide, and titanium carbide, are decomposed by water.

Diamond occurs in regular rhombodecahedra, seldom as octahedra, generally with curved surfaces and edges. It has a specific gravity of 3.5, is transparent, generally colorless, sometimes red, green, blue, and black, readily pulverizable, is a non-conductor of electricity and a poor conductor of heat. The diamond possesses the greatest brilliancy and of all bodies has the highest refractive power, as well as the greatest hardness, as it can only be polished with its own powder (bort). When heated in oxygen it burns into carbon dioxide; when strongly heated in the absence of oxygen it is converted into graphite; even the strongest oxidizing agents do not attack it. Diamonds may be obtained in microscopic crystals by dissolving carbon in molten iron and cooling the mixture under great pressure.

Graphite, plumbago, black lead, occurs chiefly as amorphous, scaly grayish-black masses which, when rubbed on paper, leave a mark, and is used in the lead-pencil manufacture and as iron paint (stove-polish). It seldom occurs in the crystalline state except as gray hexagonal plates. It has a specific gravity of 2.25, conducts heat and electricity well, is more difficult of combustion in an oxygen current than the diamond, but is oxidized by strong oxidizing agents into graphitic acid, $C_{11}H_4O_5$, or mellitic acid, $C_{12}H_6O_{12}$.

It is artificially obtained by melting amorphous carbon with iron, when the carbon dissolves and separates out in black plates on cooling the iron.

Amorphous carbon is obtained by heating many organic substances (p. 4) in the absence of air, when the volatile carbon compounds are driven off and a part of the carbon mixed with the inorganic constituents remains behind. It forms a black, voluminous powder, or gray to black opaque compact masses which have differing specific gravities.

   a. Lampblack is obtained by burning substances such as turpentine, rosin, etc., which are very rich in carbon with an insufficient supply of air. It is the purest form of amorphous carbon and is used in the preparation of India ink and printing-inks.

   b. Wood charcoal is obtained by the carbonization of wood in heaps or closed vessels (in the dry distillation of wood). It shows the structure of the wood and is porous, hence a poor conductor of heat and electricity. It absorbs gases, pigments, bitter principles, alkaloids, and many metallic salts from their solutions. One volume of wood charcoal absorbs 90 volumes of ammonia or 9 volumes of oxygen. Because of this property it is used in the filtration of water, in the absorption of putrefactive gases, and in the defusilization of alcohol. Freshly heated wood charcoal inflames in the air spontaneously, as in the absorption of oxygen heat is set free.

   c. Animal charcoal, bone-black, ivory-black, is obtained on the carbonization of animal bodies (blood-carbon, bone-carbon). It absorbs gases, coloring matter, etc., with greater activity than wood charcoal, as the carbon is mixed with considerable mineral matter and hence is very finely divided.

   d. Gas-carbon, retort graphite, deposits, in the manufacture of illuminating-gas, on the walls of the retorts. It is very hard and dense, a good conductor of electricity, and is used in galvanic batteries.

e. Coke remains in the retorts in the manufacture of illuminating-gas. It conducts heat and electricity.

f. Fossil carbon has been produced from prehistoric plants by a form of decay similar to carbonization, where the hydrogen and oxygen go off in great part as water, and the longer the process goes on the greater the amount of carbon is left, until finally the plant structure disappears completely. Peat contains 60 per cent. carbon, brown coal 70 per cent., soft coal 75–90 per cent., and anthracite coal 95–98 per cent. carbon.

a. Compounds with Hydrogen.

Carbon and hydrogen unite directly and indeed at high temperatures, forming only acetylene, C_2H_2, methane, CH_4, ethane, C_2H_6; still the number of known compounds of carbon and hydrogen is very great, as will be seen later in Part III.

All compounds of carbon may be derived from the hydrocarbons, as their hydrogen can be replaced completely or in part by other atoms or by atomic groups. For these reasons the chemistry of the hydrocarbons and their derivatives is treated of in a special part, and the old term "organic chemistry" has been retained for this part.

b. Compounds with the Halogens

are obtained indirectly by the action of the halogens upon the hydrocarbons and will be considered in this connection.

c. Compounds with Oxygen.

Carbon monoxide, CO.

Carbon dioxide, CO_2.   Carbonic acid, H_2CO_3.

Carbon Monoxide, CO. *Formation.* 1. In the combustion of carbon with an insufficient supply of air, as in furnaces or stoves with closed dampers, or by burning charcoal in braziers, in illuminating-gas and also in the so-called generator gases.

These latter are produced for technical purposes in specially constructed furnaces (generators) by the insufficient combustion of deep layers of carbon with air, and consist chiefly of carbon monoxide and nitrogen. Dowson's gas is a mixture of generator-gas and water-gas (see 189).

Preparation. 1. By passing carbon dioxide over heated carbon when the volume of the gas is doubled:

$$\underbrace{1 \text{ mol.} = 2 \text{ vol.}}_{CO_2} + C = \underbrace{2 \text{ mol.} = 4 \text{ vol.}}_{2CO}.$$

By passing CO_2 over red-hot zinc-dust or by heating zinc-dust with magnesium or calcium carbonate, when the latter decomposes into calcium or magnesium oxide and carbon dioxide and the CO_2 is reduced by the zinc powder, thus: $Zn + CaCO_3 = ZnO + CaO + CO$.

2. Ordinarily by heating oxalic acid $(C_2H_2O_4)$ with sulphuric acid: $C_2H_2O_4 = H_2O + CO_2 + CO$. The gaseous mixture obtained is passed through a watery solution of potassium hydroxide which absorbs all the CO_2, while the CO passes through unchanged.

Carbon monoxide may also be obtained by warming many other carbon compounds, such as citric acid, malic acid, formic acid, potassium ferrocyanide (which see) with sulphuric acid. Carbon dioxide is also produced in these changes.

3. By heating an excess of carbon with various metallic oxides, thus: $CuO + C = Cu + CO$.

4. By passing steam over heated carbon a mixture of carbon monoxide and hydrogen (water-gas) is obtained: $C + H_2O = CO + 2H$. This mixture is also obtained when the electric arc passes between carbon poles under water.

Properties. Colorless and odorless, neutral gas 0.967 times lighter than the air and which liquefies at $-190°$. It is nearly insoluble in water, but is quickly absorbed by a solution of cuprous chloride in hydrochloric acid or ammonia. When lighted it burns in the air with a blue flame (characteristic) into carbon dioxide; it does not support the combustion of bodies, but on account of its tendency to form CO_2 it is a strong reducing agent, especially at higher temperatures. When inspired even in small quantities it is poisonous, as it forms carbon monoxide hæmoglobin with the blood. When mixed with oxygen it explodes on ignition; it combines in the sunlight with chlorine and bromine, forming $COCl_2$ or $COBr_2$.

It combines with finely divided iron and nickel at 38° to 40°, forming iron or nickel carbonyl, $Fe(CO)_4$ or $Ni(CO)_4$, both colorless refractive liquids, and with potassium (which see) it forms at 80° solid carbon monoxide potassium $(COK)_6$, all of which explode on heating. $Fe(CO)_5$ and $Fe_2(CO)_7$ are also known.

When exposed with H to the dark electric discharge it forms formaldehyde: $CO + H_2 = CH_2O$; and with water it forms formic acid: $CO + H_2O = CH_2O_2$. With fused caustic alkali it forms alkali formates.

Detection. 1. Because of its reducing powers CO darkens paper moistened with palladium chloride, when metallic palladium separates out: $PdCl_2 + CO + H_2O = Pd + 2HCl + CO_2$. In order to test for CO in air it is passed through a palladium chloride solution.

2. Carbon monoxide hæmoglobin possesses a characteristic absorption spectrum consisting of two dark bands which are not changed by reducing agents (differing from oxyhæmoglobin spectrum). The merest traces of carbon monoxide can be detected by passing the gas to be tested through dilute blood, and investigating this with a spectroscope.

3. If blood containing CO is treated with a solution of potassium ferrocyanide and some dilute acetic acid, a cherry-red coagulum is produced, while normal blood with the same treatment gives a dark-brown coagulum.

Carbon oxychloride, carbonyl chloride, phosgene gas, $COCl_2$, is obtained as a colorless irritating gas by mixing equal volumes of carbon monoxide and chlorine in the sunlight (hence the name). This gas on cooling condenses to a colorless liquid which boils at 8° and is characterized by great chemical activity. Water decomposes it into hydrochloric acid and carbon dioxide: $COCl_2 + H_2O = CO_2 + 2HCl$.

Carbon dioxide, carbonic anhydride (erroneously called carbonic acid), CO_2. *Occurrence.* It is produced on the complete oxidation of all carbon compounds, also in their combustion, in respiration, in decay, and is hence found to a slight extent in the air (p. 152) and in every natural water (p. 116), in greater quantities in the so-called acid waters. It streams up from the earth in large quantities in volcanic regions, such as in the poisonous valley of Java, in the dog-grotto of Naples, in Pyrmont, and in the crater of Eifel. It is often found in deep wells and mines (choke-damp), as well as absorbed by many eruptive rocks. Carbon dioxide is found combined in the carbonates, such as limestone, $CaCO_3$, and dolomite, $CaCO_3 + MgCO_3$, forming entire mountain ranges. It occurs enclosed in certain minerals in the liquid state.

Formation. 1. In alcoholic fermentation (cause of accidents in fermentation-vats).

2. By burning carbon with an excess of air or oxygen: $C + 2O = CO_2$.

3. By heating carbon with an excess of metallic oxides (p. 189): $2CuO + C = 2Cu + CO_2$.

Preparation. 1. Ordinarily by pouring acids upon carbonates. Marble and hydrochloric acid are generally used: $CaCO_3 + 2HCl = CaCl_2 + H_2O + CO_2$.

2. By heating carbonates, such as calcium carbonate (limestone) or magnesium carbonate: $CaCO_3 = CaO + CO_2$.

Properties. Colorless and odorless gas having an acid taste, 1.5 times heavier than air, and as a product of complete combustion is not combustible nor does it support combustion. The presence of a few per cent. of carbon dioxide in the air has an asphyxiating action since the elimination of carbon dioxide from the lungs is very much slower because of the diminished diffusion. If potassium or magnesium is heated in CO_2, they are oxidized and the carbon separates as amorphous carbon. At $-80°$ or at $0°$ and a pressure of 39 atmospheres (p. 41) it is converted into a colorless neutral liquid which is not miscible with water and has a specific gravity 0.92 and occurs in commerce in wrought-iron cylinders.

If liquid carbon dioxide is allowed to flow out in a thin stream, a part becomes gaseous immediately and takes up so much heat that another part solidifies into snow-like flakes which melt at $-65°$ under a pressure of 3.5 atmospheres; hence solid CO_2 passes immediately into the gaseous state at ordinary temperatures (p. 35).

Solid carbon dioxide volatilizes only slowly on lying in the air, when its temperature sinks to $-78°$. A mixture of the same with ether produces, on evaporation, a temperature of about $-90°$; if this mixture is evaporated under the air-pump receiver, its temperature falls to $-140°$.

One volume of water dissolves at $15°$ 1 volume carbon dioxide. The watery solution has a faint acid reaction, the dry gas itself being neutral. Carbon dioxide is readily absorbed by caustic alkalies, forming carbonates: $CO_2 + 2KOH = K_2CO_3 + H_2O$.

With the dark electric discharge in the presence of water or hydrogen CO_2 is converted into formic acid: $CO_2 + 2H = CH_2O_2$; $CO_2 + H_2O = CH_2O_2 + O$.

Its constitution can be determined by burning pure carbon in a known volume of oxygen, when on cooling, in place of the oxygen used, an equal volume of carbon dioxide is found. Two volumes of carbon dioxide, which weigh 44 parts, contain 2 volumes or 32 parts by weight of oxygen, and hence $44 - 32 = 12$ parts by weight of carbon.

Carbonic acid, H_2CO_3 or $O = C = (OH)_2$, is not known free, but the aqueous solution of carbon dioxide may be considered as a solution of carbonic acid: $CO_2 + H_2O = H_2CO_3$. As H_2CO_3 is a weak acid, hence neutral salts formed with strong bases still have a basic reaction.

Carbonates may be obtained by saturating bases with carbon dioxide: $2KOH + CO_2 = K_2CO_3 + H_2O$; $K_2CO_3 + H_2O + CO_2 = 2KHCO_3$. They are decomposed, with effervescence, by every stronger acid, as the carbonic acid formed immediately decomposes: $K_2CO_3 + H_2SO_4 = K_2SO_4 + H_2CO_3$; $H_2CO_3 = H_2O + CO_2$.

Detection of Carbon Dioxide and Carbonates. If a glass rod moistened with a clear solution of barium or calcium hydrate is introduced into a vessel containing carbon dioxide, the solution becomes cloudy with the formation of insoluble barium or calcium carbonate.

The carbon dioxide is set free from the carbonates by the addition of an acid and the moistened glass rod held above the liquid. Smaller quantities are detected by the cloudiness produced in the above solutions, when the gas to be tested is passed through these solutions for a longer time.

Percarbonic acid, $H_2C_2O_6$ or HO-COO-OOC-OH, is known only in the form of salts, the percarbonates. These are produced when an aqueous solution of a carbonate is exposed to electrolysis below $-10°$. In aqueous solution they have an oxidizing action like H_2O_2, which is also generated by dilute acids. $K_2C_2O_6 + 2H_2SO_4 = 2KHSO_4 + H_2O_2 + 2CO_2$. Water or bases at ordinary temperatures develop oxygen from these:

$$H_2C_2O_6 + 2KOH = 2KHCO_3 + H_2O + O.$$

d. Compounds with Sulphur.

Carbon monosulphide, CS.
Carbon disulphide, CS_2.   Sulphocarbonic acid, H_2CS_3.

Carbon monosulphide, CS, is produced as dark yellow masses when carbon disulphide and hydrogen or carbon disulphide and carbon monoxide are exposed to the dark electric discharge:

$$CS_2 + H_2 = CS + H_2S; \; CS_2 + CO = CS + COS.$$

Carbon Disulphide, CS_2. *Preparation.* By passing sulphur vapors over heated carbon and condensing the carbon disulphide vapors.

Properties. Colorless, highly refracting liquid, having a specific gravity of 1.27, boiling at 46°, very easily inflamed, and burning with a bluish flame: $CS_2 + 6O = CO_2 + 2SO_2$. When pure it has an ethereal odor, but the commercial carbon disulphide has a very unpleasant odor. If a strong current of air is blown over CS_2, it rapidly evaporates and absorbs so much heat that a part solidifies into a white crystalline mass. When the vapor is inhaled it produces poisonous symptoms; when mixed with oxygen it explodes even by a spark. It readily dissolves sulphur, phosphorus, bromine, iodine, resins, rubber, and fatty oils, and mixes with alcohol and ether in all proportions, but not with water.

Sulphocarbonic acid, H_2CS_3, is prepared from the sulphocarbonates by acids as a thick, nauseating, reddish-brown, very unstable liquid.

Sulphocarbonates. As carbon dioxide unites with oxides, forming carbonates, so carbon disulphide combines with sulphides, forming sulphocarbonates; thus sodium sulphocarbonate is produced by dissolving carbon disulphide in sodium sulphide solution: $CS_2 + Na_2S = Na_2CS_{1.}$

Carbon oxysulphide, COS, is found in mineral springs and is produced by passing sulphur vapors and carbon monoxide through a red-hot tube, or, better, by pouring sulphuric acid upon potassium sulphocyanide (KCNS), when sulphocyanic acid is set free which takes up water, forming carbon oxysulphide and ammonia: $HCNS + H_2O = COS + NH_3$. It is a colorless gas having an odor similar to sulphuretted hydrogen, readily inflammable, and burns into carbon dioxide and sulphur dioxide.

Monosulphocarbonic acid, H_2CO_2S ($= H_2O + COS$) and
Disulphocarbonic acid, H_2COS_2 ($= H_2S + COS$) are known only as salts.

e. Compounds with Nitrogen.

One compound called dicyanogen, $NC-CN$, is known which forms numerous other compounds. Dicyanogen cannot be obtained by direct union of the elements. Dicyanogen and its compounds will be treated of in organic chemistry.

2. Silicon, or Silicium.

Atomic weight 28.4 = Si.

Occurrence. Silicon is, next to oxygen, the most widely diffused of the elements, but does not occur free. Combined with oxygen, it is found as silicon dioxide or silica in the three natural kingdoms; silicates form many minerals and nearly all crystalline varieties of rocks (see Silicon Dioxide and Silicates).

Preparation. 1. By heating sodium fluosilicate with metallic sodium: $Na_2SiF_6 + 4Na = 6NaF + Si$; this product is treated with water, which dissolves the sodium fluoride and leaves the amorphous silicon.

2. Alloyed with some magnesium it may be obtained by fusing sand (SiO_2) and magnesium powder together.

3. If zinc be added in this method of preparation, then the silicon dissolves in the molten zinc and separates on cooling as crystalline silicon, which remains on dissolving the zinc in hydrochloric acid.

Properties. Crystalline silicon forms black octahedra of a specific gravity 2.5, which scratch glass; it is not oxidized on heating in the air, and melts at about 1500°.

Amorphous silicon is a reddish-brown powder with the specific gravity 2.35 and burning in the air, forming SiO_2. Both modifications are insoluble in acids; on heating in chlorine gas they burn into silicon chloride; on boiling with caustic potash solution they dissolve, forming alkali silicates: $Si + 4KOH = K_4SiO_4 + 4H$.

The compounds of silicon ·with the metals are called *silicides*—for example, barium silicide, $BaSi_2$—and are obtained by heating the respective metallic oxide wth carbon and silicic acid to about 3000°. They form white crystalline masses which are decomposed by water.

a. Compounds of Silicon.

Silicon hydride, SiH_4, is prepared in a similar manner to arsenic hydride by dissolving an alloy of silicon and magnesium in dilute hydrochloric acid, $SiMg_2 + 4HCl = SiH_4 + 2MgCl_2$, as a colorless gas. Even on gently warming, it inflames and burns into water and silicon dioxide, the latter forming a ring-like cloud. If the gas is diluted with hydrogen, it is spontaneously inflammable.

Silicon chloride, $SiCl_4$, is produced on heating silicon or a mixture of silicon dioxide and carbon in chlorine gas: $SiO_2 + 2C + 4Cl = SiCl_4 + 2CO$. It is a colorless, fuming, irritating liquid which boils at 75° and which decomposes with water, forming silicic acid: $SiCl_4 + 4H_2O = H_4SiO_4 + 4HCl$.

Silicon chloroform, $SiHCl_3$, so called because it is similarly constituted to chloroform, $CHCl_3$, is obtained with silicon chloride on heating silicon in hydrochloric gas. It is a colorless, fuming liquid which boils at 36°, and which is readily inflammable.

Besides these compounds, other silicon compounds having an analogous constitution to the corresponding carbon compounds are known (see compounds of the alcohol radicals with metalloids in Part III).

Silicon carbide, SiC, which is next to the diamond and boron carbide in hardness, is not attacked by any acid, and is more stable with heat than the diamond. It forms colorless needles which are prepared by the reduction of sand (SiO_2) with carbon in the electric furnace. It is used under the name of carborundum as a substitute for emery and diamond-powder.

Silicon Fluoride, SiF_4. *Preparation.* By the action of hydrofluoric acid upon silicon dioxide or a silicate.

Calcium fluoride and sand (SiO_2) or glass powder is warmed with sulphuric acid. First hydrofluoric acid is formed, which then acts upon the silicon dioxide: $2CaF_2 + H_2SO_4 = CaSO_4 + 4HF$; $4HF + SiO_2 = 2H_2O + SiF_4$. The water formed unites with the sulphuric acid, and silicon fluoride is evolved.

Properties. Colorless, irritating, fuming gas which is neither combustible nor supports combustion, and is liquefiable at −160°. It is decomposed by water into silicic acid and hydrofluosilicic acid: $3SiF_4 + 4H_2O = 2H_2SiF_6 + H_4SiO_4$; the gelatinous silicic acid is removed by filtration, and the watery solution of

Hydrofluosilicic acid, H_2SiF_6, is obtained. This is an acid, fuming, colorless liquid; it is not known in the anhydrous state; on evaporation in a platinum vessel it leaves no residue, as it decomposes into volatile SiF_4 and $2HF$. It forms salts, the silicofluorides, with bases. K_2SiF_6 and $BaSiF_6$ are insoluble in water, hence they are used in the quantitative estimation of potassium and barium.

Silicon sulphide, SiS_2, is obtained by heating amorphous silicon with sulphur, and forms silky prisms which are decomposed by water: $SiS_2 + 4H_2O = H_4SiO_4 + 2H_2S$.

Silicon Dioxide, Silica, Silicic Anhydride, SiO_2. *Occurrence.* It forms as silicates the chief constituent of the earth's crust and is a constituent of all plants and animals (see Silicic Acids). It occurs free in the crystalline as well as in the amorphous state.

Crystalline silicon dioxide occurs as rock crystals in transparent, colorless, hexagonal columns, and as quartz as colorless, opaque, granular masses. Both of these occur in many modifications, especially as

Citrine, gold topaz, when transparent and pale yellow.
Amethyst when transparent and violet by manganese.
Smoky topaz when transparent and colored brown or black by bituminous matter.
Tridymite, in hexagonal twin crystals.
Common quartz when opaque, gray, or yellowish.
Milky quartz when opaque, milky white.
Rose-quartz when opaque, rose-color.
Agate, aventurine, chalcedony, chrysoprase, flint, heliotrope, jasper, carnelian, cat's-eye, onyx, tiger-eye, are mixtures of amorphous and crystalline silicon dioxide having various colors and occurring as minerals.
Quartz is the chief constituent of granite, syenite, and gneiss. In boulders and grains (sand) quartz covers a great part of the earth's surface; sandstone consists of individual quartz granules which are united together by some other substance.

Amorphous silicon dioxide is found as opal in colorless or colored vitreous masses, and also in many vitrifications; tripolite or infusorial earth consists of the siliceous shells of infusoria.

Preparation. Amorphous silicon dioxide is obtained by burning amorphous silicon or by heating silicic acid. It forms white, soft powder. Crystalline silicon dioxide is obtained by heating dialyzed silicic acid (see below) under pressure for a long time or by strongly heating amorphous silicon dioxide for a long time.

Properties. It is only soluble in hydrofluoric acid. When boiled with alkali hydroxide solution the amorphous SiO_2 dissolves, forming the corresponding silicate (p. 197). The amorphous form has a specific gravity of 2.2, the crystallized a specific gravity of 2.6. It melts in the oxyhydrogen blowpipe-flame.

Detection. Similar to the silicates (p. 197).

Silicic Acids. *Occurrence.* The silicates form the chief constituent of most rocks (see Aluminium) and many minerals and are found in the animal kingdom, especially in the shell of infusoria, in feathers, hair, and quills. They occur in the plant kingdom, especially in many grasses, in straw, and in rattan. Dissolved silicic acid occurs free in many mineral waters, especially in the hot springs of Iceland and New Zealand, which deposit in the air as compact masses (siliceous sinter).

1. *Orthosilicic Acid,* H_4SiO_4. If fine sand is fused with sodium carbonate, it forms a vitreous mass, soluble in hot water, consisting of sodium silicate (sodium water-glass): $2Na_2CO_3 + SiO_2 = Na_4SiO_4 + 2CO_2$. If hydrochloric acid is added to this solution, orthosilicic acid is set free: $Na_4SiO_4 + 4HCl = 4NaCl + H_4SiO_4$, which separates in part as a gelatinous mass containing water, while another part remains in solution, as it is somewhat soluble in water, but more soluble in dilute HCl.

A solution of pure silicic acid can be obtained from the above solution, containing NaCl and HCl, by means of dialysis, as silicic acid is a colloid, while the impurities are crystalloid and diffuse through the membrane of the dialyzer (p. 47).

The silicic acid solution obtained by dialysis is colorless, faintly acid, and may be concentrated by evaporation, but soon solidifies into a transparent jelly (silicic acid gele, p. 54) of silicic acid, which can be precipitated from the silicic acid solution by very small amounts of sodium carbonate or other salts. If the evaporation is continued to dryness, we obtain a fine white amorphous powder having the composition $H_2SiO_3 + xSiO_2$.

Orthosilicic acid has not been obtained pure because on drying it gives off water and is then mixed with metasilicic acid or polysilicic acids:

$$H_4SiO_4 = H_2SiO_3 + H_2O;$$
$$2H_4SiO_4 = H_6SiO_7 + H_2O.$$

2. *Metasilicic acid,* H_2SiO_3, corresponding approximately to the preceding formula, is obtained when orthosilicic acid solution is evaporated under the air-pump receiver at 15° and the glass-like residue dried over sulphuric acid.

3. *Polysilicic Acids.* As the polybasic sulphuric acid, phosphoric acid, and arsenic acid may, by condensation of several molecules with the elimination of water, form anhydro- or poly-acids, so also does silicic acid, and in fact to a much greater extent. There are a great number of polysilicic acids which are only known mixed with each other. According to the number of silicon atoms contained in the molecule, we call the respective acids di-, tri-, tetra-, pentasilicic acid, etc., and the corresponding salts di-, tri-, tetra-, pentasilicates, etc.

$$H_2Si_2O_5 = H_2SiO_3 + SiO_2 \quad \text{or} \quad = 2H_4SiO_4 - 3H_2O.$$
$$H_6Si_2O_7 = H_4SiO_4 + H_2SiO_3 \quad \text{``} \quad = 2H_4SiO_4 - H_2O.$$
$$H_2Si_3O_7 = H_2SiO_3 + 2SiO_2 \quad \text{``} \quad = 3H_4SiO_4 - 5H_2O.$$
$$H_4Si_3O_8 = 2H_2SiO_3 + SiO_2 \quad \text{``} \quad = 3H_4SiO_4 - 4H_2O.$$

On heating, all silicic acids yield SiO_2.

Silicates. The silicates occurring in nature are nearly all derived from the polysilicic acids and only a few are derived from ortho- or metasilicic acid. Artificially silicates are obtained as amorphous, vitreous masses by fusing silicic anhydride (p. 195) with bases or metallic carbonates. They are insoluble in water (with the exception of the alkali silicates) and are decomposed by acids with the separation of silicic acid.

Of the natural silicates, only a few are decomposed by acids, most of them not being attacked at all. Such silicates, in order to be able to determine the metals contained therein, must be first made decomposable by acids, which is done by fusing the finely powdered substance with dried sodium carbonate.

Detection. If silicic anhydride, silicic acid, or a silicate is melted on a platinum loop with phosphorus salt (which see), then the bases dissolve in the metaphosphate produced, while silicic acid separates and causes the otherwise clear bead to become opaque, forming the so-called silica skeleton.

II. METALS.

For reasons given on page 95 the elements which are called metals are divided into the following groups:

1. Potassium, sodium, cæsium, rubidium, lithium (ammonium).
2. Calcium, barium, strontium.
3. Beryllium, magnesium, zinc, cadmium.
4. Copper, silver, mercury.
5. Aluminium, gallium, indium, thallium, scandium, yttrium, lanthanum, cerium, praseodymium, neodymium, samarium, gadolinium, erbium, thulium, ytterbium.
6. Tin, zirconium, titanium, thorium.
7. Bismuth, vanadium, tantalum, niobium.
8. Chromium, molybdenum, tungsten, uranium.
9. Iron, manganese, cobalt, nickel.
10. Gold, platinum, osmium, iridium, ruthenium, rhodium, palladium.

Properties. The metals have more properties in common than the non-metals. The metals are non-transparent, and only certain of them are transparent in very thin layers. With the exception of mercury they are all solids at ordinary temperatures.

When compact, especially when the surface is polished, they have a peculiar shine, which has been called metallic lustre. When finely divided they form dark powders. They are good conductors of heat and electricity. The color of most metals is white to bluish gray. Copper is red; gold, barium, and strontium are yellow. Most metals are crystalline in regular systems. Only a few, having a metalloid character, are not regular in their crystallization; thus bismuth crystallizes in the hexagonal system, and tin in the quadratic system.

The specific gravity of the metals is very different and variable, from 0.59, the specific gravity of lithium, to 22.5, the specific gravity of osmium. Metals whose specific gravity is below 5 are called light metals, and those above are called heavy metals. Most metals are malleable and tough and may be converted into foil and wire. Only bismuth and tin, which have metalloid character, are brittle.

All metals are fusible. Mercury has the lowest melting-point, namely, $-40°$; potassium melts at 63°, zinc at 243°, copper at 1100°, platinum at 1770°, iridium at 1950°, chromium at 2100°, osmium at 2500°. All metals can be converted into a vaporous state; their volatility corresponds to their fusibility. Mercury vaporizes at 360°, zinc at about 1000°; platinum and the other difficultly fusible metals can be vaporized in the electric furnace.

No metal in the compact condition is soluble as such. If the soltion of a metal in acids or bases is evaporated, we obtain a salt or oxide. On the contrary, numerous metals are soluble in water at the moment when they are precipitated from the solution of their salts in a very finely divided state (see Colloidal Solution, p. 53).

The metals are as a rule more active chemically than the metalloids, as the molecule of the metals consists only of an atom, while in the metalloids the molecule must be first split into atoms. The metalloids are therefore more active in the nascent state than in the ordinary condition, where the atoms have already united to form molecules (p. 16).

The combinations of the metals between themselves according to no certain proportion by weight are called *alloys* (p. 49). These generally have the average properties of the metals of which they are composed, so that it is possible to obtain alloys by the selection of suitable metals which have properties of technical importance. The alloys with mercury are called *amalgams*.

The color of the alloys is different according to the constituents; still it is independent of the proportion of the metals; thus an alloy of copper with 30 per cent. of tin is white, while with 30 per cent. zinc it is yellow. The hardness and toughness are generally greater than those of the individual metals, while the melting-point is lower, and often lower than that of the metal having the lowest melting-point; thus a mixture of 2 parts bismuth, 1 part tin, and 1 part lead (Wood's metal) melts at 94°, while pure bismuth melts at 270°, tin at 235°, and lead at 334°. The melting-point of amalgams lies always above that of mercury. Acids often attack alloys with greater difficulty than the constituents alone; still many alloys are dissolved by acids which would not dissolve the metals contained in the mixture.

The compounds of the metals with the non-metals do not have the properties of the metals. While the oxides of the non-metals are nearly all acid-forming, only a few of the higher oxides of the metals are acid, while most of the metallic oxides are basic in character. The compounds with hydrogen are colorless powders with the exception of palladium, sodium, and potassium, which are metallic in appearance; gaseous combinations of hydrogen with the metals are not known. The halogen compounds of the metals are mostly very stable, while those of the metalloids, with the exception of those with carbon, are readily decomposed by water.

The combinations with boron, silicon, and carbon (borides, silicides, carbides) are not volatile even at the temperature of the electric furnace. The carbides (p. 186) have great technical importance. We differentiate between the metals according to their chemical behavior into—

1. *Non-noble Metals.* Certain of these oxidize even on being exposed to the air, but all oxidize on being heated. They decompose water either at ordinary temperatures (alkali metals, alkaline earths) or at higher temperatures (with the exception of lead, bismuth, and copper).

2. *Noble Metals.* These show slight affinity for oxygen and do not change in the air. They do not decompose water even at high temperatures, and their indirectly obtained oxides decompose on heating into the metal and oxygen. To this group belong silver, gold, platinum, also mercury and certain rare platinum metals which indeed oxidize when heated in the air, but decompose again into metal and oxygen when heated to a higher degree.

Occurrence. 1. Mercury, silver, copper, arsenic, antimony, bismuth, lead, rarely iron are found combined as well as free (native). Gold and the platinum metals occur nearly entirely free, as they only have a slight affinity for oxygen and are not at all or only slightly changed by atmospheric influences.

2. Most of the heavy metals do not exist in a free state. The naturally occurring compounds of the heavy metals are called ores.

3. The light metals to which the elements of the potassium, calcium, and aluminium groups, and magnesium and certain rare metals belong do not occur native, but are found chiefly as silicates, these forming the chief mass of the rocks of the earth's crust (p. 6). As the metals can be obtained from these silicates only with difficulty,

they are not used for the preparation of the metals, but other compounds, occurring to a less extent, are used for this purpose.

Preparation. 1. The native metals are separated from the accompanying rock formation by fusion (bismuth), by washing (gold), or by distillation (mercury).

2. The metals are obtained from their ores by various complicated methods.

The metallic oxides are mostly reduced by heating with carbon. This can be done in the case of the oxides of the rare metals by the action of the electric arc or by aluminium (which see).

Metallic sulphides are either converted into oxygen compounds by heating them in the air (roasting) and then reduced by carbon, or they are heated with some cheaper metal which sets the contained metal free (precipitation).

Many ores are dissolved and then precipitated from their solution by cheaper metals or by the electric current. Thus metallic copper is precipitated from a solution of copper sulphate by iron, while the iron is dissolved as ferrous sulphate.

3. The light metals (see above), as they can only be separated from their widely distributed silicates with difficulty, are obtained on a large scale from their chlorides by electrolysis, or to a less extent by heating the chlorides with sodium, or from their oxides by heating with aluminium.

ALKALI METAL GROUP.

Lithium. Sodium. Potassium. Rubidium. Cæsium (Ammonium).

Monovalent metals, soft at ordinary temperatures, readily fusible and volatile on heating strongly. They oxidize even at ordinary temperatures, decompose water rapidly in the cold, and then form hydroxides (the alkalies) which are very soluble in water and volatile without decomposition at high temperatures and are the strongest bases. Their carbonates, sulphates, phosphates, and sulphides are soluble in water, while the carbonates and phosphates of all other metals are insoluble in water.

1. Potassium (Kalium).

Atomic weight 39.15 = K.

Occurrence. Only combined. Potassium chloride and potassium sulphate occur in sea-water, and besides this they form large deposits, generally above rock salt, in North Germany and Galicia. These "abraum salts" (so called from the German word *abraumen,* to remove, because it must first be removed to get at the rock salt), which were formerly obtained only at Stassfurt (Germany), contain especially the potassium salts carnallite and sylvine (p. 204) and kainite and schönite (p. 206), which are valuable fertilizers (potash fertilizers), and form the original substance from which many other potassium salts are prepared.

Potassium is also widely distributed as a constituent of many varieties of rock, especially as potassium-aluminium silicate; thus as feldspar, leucite, and mica, which undergo weathering and supply potassium compounds to the soil, these being taken up by the plants and remain in the ash, after burning of the plants. The potassium compounds are introduced into the animal organism by means of the plants and are found especially in the muscles, blood corpuscles, eggs, and in milk.

Preparation. 1. By the electrolysis of fused potassium hydroxide, potassium chloride, or potassium cyanide, when the potassium separates at the negative pole, or by the electrolysis of an aqueous solution of potassium chloride, using mercury as the negative pole, when the potassium which is set free does not decompose the water but forms an amalgam with the mercury from which it is separated by heating to 360°.

2. By heating an intimate mixture of potassium carbonate with carbon: $K_2CO_3 + 2C = 2K + 3CO$. The potassium volatilizes and the vapors are condensed in flat iron boxes, which when filled are cooled in petroleum.

The potassium vapors used to be condensed by conducting the vapors directly into petroleum. In this case a part of the potassium combined with the carbon monoxide, forming a black and very explosive compound $(COK)_6$ (p. 189).

Properties. Shining, silver-white metal which is as soft as wax and brittle at 0°. It has a specific gravity 0.86, melts at 62.5° and distils as a greenish-blue vapor at about 670°. It quickly oxidizes in moist air, the surface being covered with potassium hydroxide; hence it is kept in petroleum. When fused in the air it inflames and burns with a violet flame into not well-known oxides. It decomposes water with the formation of potassium hydroxide and hydrogen, whereby considerable heat is evolved so that it inflames the hydrogen which burns with a violet flame due to the volatilized potassium, $2K + 2H_2O = 2KOH + 2H$; when alloyed with mercury it decomposes water without the production of flame. It combines directly with the halogens, sulphur, and phosphorus, with the production of flame, and when heated with hydrogen to 400° it combines, forming silver-like brittle potassium hydride, K_2H, which spontaneously inflames in the air and which decomposes at about 420°. On account of its relationship to oxygen and chlorine potassium is used in the setting free of metals from their oxygen and chlorine compounds.

a. Compounds of Potassium.

Potassium oxide, K_2O, is not known with certainty.

Potassium Hydroxide, Caustic Potash, KOH.

Formation. 1. By the action of potassium on water (see above).

Preparation. 1. On a large scale by the electrolysis of an aqueous solution of potassium chloride: $KCl + H_2O = KOH + H + Cl$.

2. In small quantities by boiling a solution of potassium carbonate with slaked lime (calcium hydroxide); insoluble calcium carbonate settles out and the solution contains potassium hydroxide: $K_2CO_3 + Ca(OH)_2 = CaCO_3 + 2KOH$. The solutions obtained in 1 and 2 are evaporated in silver dishes, as the KOH acts upon iron and porcelain.

Properties. White, crystalline substance, the strongest of all bases, fusible at a red heat, and volatile without decomposition at higher temperatures. When exposed to the air it absorbs moisture and deliquesces and at the same time absorbing carbon dioxide. It destroys most plant and animal substances and is therefore used as a caustic agent. It is very soluble in water and alcohol; the alcoholic solution is called alcoholic potash. As these solutions contain one of the strongest bases they precipitate most metals as hydroxides or oxides from their solution; thus, $FeSO_4 + 2KOH = Fe(OH)_2 + K_2SO_4$.

Potassium sulphide, potassium sulphuret, K_2S, is obtained by fusing potassium sulphate with carbon, as an anhydrous, dark-red, crystalline mass: $K_2SO_4 + 2C = K_2S + 2CO_2$. It is very hygroscopic, has an alkaline reaction, absorbs oxygen and water from the air, and is converted into potassium thiosulphate and potassium hydroxide:

$$2K_2S + H_2O + 4O = K_2S_2O_3 + 2KOH.$$

On mixing a watery solution of potassium hydrosulphide with a solution of potassium hydroxide we obtain a solution of the sulphide from which colorless crystals of $K_2S + 5H_2O$ separate on evaporation: $KSH + KOH = K_2S + H_2O$. With acids it develops H_2S:

$$K_2S + H_2SO_4 = K_2SO_4 + H_2S.$$

Potassium hydrosulphide, potassium sulphydrate, KSH, is obtained in solution by saturating caustic potash with H_2S: $KOH + H_2S = KSH + H_2O$. On carefully evaporating this solution we obtain alkaline, colorless crystals having the formula $2KSH + H_2O$, which on heating lose their water and anhydrous KSH is obtained as a yellow mass. With acids it generates H_2S, with sulphoacids or their anhydrides it combines (like its anhydride K_2S), producing sulphosalts; thus, $6KSH + As_2S_3 = 2K_3AsS_3 + 3H_2S$:

$$K_2S + CS_2 = K_2CS_3 \text{ (p. 193).}$$

Potassium Polysulphides. Besides the sulphides K_2S and KSH we have the polysulphides or polysulphurets, K_2S_2, K_2S_3, K_2S_4, K_2S_5, which are obtained by fusing potassium monosulphide, K_2S, with a corresponding amount of sulphur. They form red or yellow masses which are readily soluble in water and are decomposed by acids with the development of H_2S and at the same time finely divided, nearly white sulphur, so-called milk of sulphur (p. 121) separates:

$$K_2S_3 + H_2SO_4 = K_2SO_4 + H_2S + 2S.$$

Liver of sulphur, hepar sulfuris, is the mixture of potassium polysulphides with potassium thiosulphate or potassium sulphate obtained by heating potassium carbonate with sulphur. It is deliquescent, readily soluble in water, and forms masses which have a brown color (hence the name).

Potassium Chloride, KCl. Occurs a ssylvine, KCl, as carnallite $(MgCl_2 + KCl + 6H_2O)$, also in sea-water, in certain salt springs, and in the ash of plants, in animal fluids and tissues.

Preparation. It crystallizes directly from a hot saturated solution of carnallite on cooling and is also formed by the action of hydrochloric acid upon potassium hydroxide or carbonate: $K_2CO_3 + 2HCl = 2KCl + H_2O + CO_2$. It forms colorless, shining cubes which melt at a red heat and are volatile at a white heat. These crystals dissolve in 3 parts water.

Potassium Bromide, KBr. *Preparation.* By dissolving bromine in caustic potash, when potassium bromide and potassium bromate are formed: $6Br + 6KOH = 5KBr + KBrO_3 + 3H_2O$. This is evaporated to dryness and heated with carbon when the potassium bromate is reduced to bromide: $KBrO_3 + 3C = KBr + 3CO$. On dissolving this in water, filtering and evaporating to crystallization, we obtain the crystals.

Properties. Colorless, shining cubes which are fusible and volatile and soluble in three parts of water.

Potassium iodide, KI, is obtained in an analogous manner to potassium bromide.

Iodine is rubbed with considerable powdered iron under water, when ferrous iodide is formed and enough iodine is added to this solution until ferrous-ferric iodide, $Fe_2I_8(2FeI_3 + FeI_2)$, is formed. This solution is now treated with an equivalent quantity of potassium carbonate and heated to boiling, when ferrous-ferric oxide precipitates and the dissolved KI obtained by evaporation to crystallization:

$$Fe_3I_8 + 4K_2CO_3 = Fe_3O_4 + 8KI + 4CO_2.$$

Properties. It forms colorless cubes which are fusible and volatile and soluble in 0.75 part water. Aqueous solutions of potassium iodide dissolve iodine readily, and many iodine compounds which are insoluble in water, such as HgI_2, are also soluble therein.

Potassium Chlorate, $KClO_3$. *Preparation.* 1. Besides KCl by passing chlorine into a hot concentrated caustic potash solution and by evaporating this to crystallization when the more insoluble potassium chlorate separates out first (p. 138).

2. It is prepared on a large scale as follows: Warm milk of lime, $Ca(OH)_2$, is saturated with chlorine and the calcium chlorate formed is treated with potassium chloride, $Ca(ClO_3)_2 + 2KCl = CaCl_2 + 2KClO_3$, or by the electrolysis of potassium chloride (p. 104), when the chlorine set free at the anode is passed directly into the potassium hydroxide formed at the cathode.

Properties. Colorless plates having a characteristic cooling taste, soluble in 16 parts cold water and melting at 334°. On heating above this point it decomposes into potassium chloride and potassium perchlorate, and oxygen is given off (p. 139). On heating still higher all the oxygen is expelled and it is converted into potassium chloride. In regard to other properties of the chlorate see p. 138.

Potassium Perchlorate, $KClO_4$. It is prepared as described on p. 139, and occurs often to a slight extent in Chili saltpeter, and forms colorless crystals which are very insoluble in water and hence is readily freed from the potassium chloride produced in its preparation by washing with water.

Potassium hypochlorite, KClO, is known only as a watery solution (eau de Javelle) and is used as a bleaching agent, especially for wine-

and fruit-stains. See p. 137 for further information in regard to hypo-chlorites.

Potassium sulphate, secondary potassium sulphate, K_2SO_4, occurs in the lava of Vesuvius and in schönite $(K_2SO_4 + MgSO_4 + 6H_2O)$ and kainite $(K_2SO_4 + MgSO_4 + MgCl_2 + 6H_2O)$; also in most plants and to a slight extent in the urine and blood. It is obtained by heating potassium chloride with sulphuric acid, $2KCl + H_2SO_4 = K_2SO_4 + 2HCl$, and also commercially from kainite and schönite. It forms white, hard crystals or crystalline crusts which melt at a red heat without decomposition and are soluble in 10 parts cold water.

Potassium bisulphate, potassium acid sulphate, primary potassium sulphate, $KHSO_4$, occurs as misenite and is obtained as a by-product in the chemical industries or by dissolving potassium sulphate in sulphuric acid: $K_2SO_4 + H_2SO_4 = 2KHSO_4$.

It forms colorless acid crystals, readily soluble in water and readily fusible. On heating somewhat above its melting-point (197°) it forms potassium pyrosulphate, $K_2S_2O_7$, (p. 130): $2KHSO_4 = K_2S_2O_7 + H_2O$. On heating higher it decomposes into potassium sulphate and sulphur tri-oxide: $K_2S_2O_7 = K_2SO_4 + SO_3$. As this decomposition takes place at about 600° acid potassium sulphate is used in decomposing minerals which are not attacked by sulphuric acid at its melting-point (338°).

Potassium persulphate, $K_2S_2O_8$, is produced in the electrolysis of a watery solution of potassium bisulphate and forms a white crystalline powder. (Properties, see p. 131.)

Potassium Nitrate, Saltpeter, KNO_3. *Occurrence.* To a con-siderable extent in the soil in warm climates (Bengal, India) and to a less extent in nearly every arable soil. Its formation depends upon the fact that when nitrogenous organic matter suffers decay in the air and in the presence of strong bases the micro-organisms oxi-dize the nitrogen into nitric acid (p. 161). So-called wall saltpeter, which forms in stables and closets, is calcium nitrate.

Preparation. 1. It was formerly prepared according to the fol-lowing process: Animal refuse is piled up with rubbish, wood ashes, or other bodies rich in potassium and then moistened with urine and liquid manure (saltpeter plantation). After 2 or 3 years the saltpeter effloresces out and is scraped off, dissolved in water, and treated with potassium carbonate in order to decompose any cal-cium nitrate present. It is obtained pure by repeated crystallization.

2. By purifying the natural potassium nitrate by repeated recrys-tallization.

3. Hot concentrated solutions of Chili saltpeter ($NaNO_3$, p. 212) and the potassium chloride obtained from Stassfurt salt are mixed: $NaNO_3 + KCl = NaCl + KNO_3$. The sodium chloride is about as solu-ble in cold water as in warm, so it remains in solution on cooling the

same, while the less soluble potassium nitrate separates out on cooling (conversion saltpeter).

Properties. Colorless prisms or crystalline powder, soluble in 4 parts cold water and in less than one-half its weight of hot water. It is nearly insoluble in alcohol, melts at 340°, and on further heating it yields oxygen and forms potassium nitrite, KNO_2, and finally decomposes into potassium oxide, nitrogen, and oxygen and hence has an active oxidizing action (p. 159). Paper impregnated with potassium nitrate is called touch-paper (charta nitrata) and is used in medicine.

Gunpowder is a granular mixture of about 75 per cent. saltpeter and 12.5 per cent. each of sulphur and wood charcoal. One gram powder yields 260 c.c. explosion gases, measured at 0° and 760 mm. pressure, which with heat set free at the time of the explosion expands to 2100 c.c. The decomposition takes place theoretically as follows: $2KNO_3 + S + 3C = K_2S + 2N + 3CO_2$. As a matter of fact the decomposition is more complicated.

Potassium nitrite, KNO_2, is obtained by heating saltpeter (which see) or by melting the same with lead. It forms a colorless deliquescent salt, which is readily soluble.

Potassium arsenite, K_3AsO_3, is obtained by neutralizing a solution of arsenious acid with K_2CO_3. A watery solution containing 1 per cent. As_2O_3 as K_3AsO_3 is called Fowler's solution.

Potassium Carbonate, Potash, K_2CO_3. *Preparation.* 1. In countries rich in forests the ashes of the trees are lixiviated with water and the filtered solution evaporated to dryness and the brown residue heated until white.

Land plants contain sodium, calcium, magnesium, and especially potassium, which are combined with organic acids, sulphuric acid, phosphoric acid, and chlorine. On burning the organic salts are converted into carbonates.

2. From the fat-free wash-water from wool by evaporation, ashing, lixiviation, etc., as described in method 1.

3. From the residue remaining after the obtainment of alcohol from beet-root molasses, which, unlike the residue left after the preparation of alcohol from potatoes, has no value as a food for animals. The residue is ashed and lixiviated and treated as given in method 1.

The crude potash (calcined potash) obtained by methods 1, 2, and 3 contain up to 10 per cent. foreign salts, especially potassium chloride.

4. It is prepared in greatest quantities from potassium chloride of "abraum" salts in an analogous manner to sodium carbonate from sodium chloride (p. 213) or by saturating a mixture of magnesium

carbonate and potassium chloride solution with CO_2 when the insoluble double salt $MgCO_3 + KHCO_3 + 4H_2O$ separates out. On heating this with water under pressure it decomposes into insoluble magnesium carbonate and potassium carbonate, which remains in solution and which is obtained therefrom by evaporation.

The purified potash thus obtained still contains a small percentage of foreign salts. Such potash is also obtained by dissolving crude potash (pearlash) in a small quantity of water, which leaves the impurities in part undissolved, and then evaporating the solution, when at first the less soluble foreign salts separate and on further evaporation the pure potash results.

Potash was formerly obtained by heating tartar (see Tartaric acid), when potassium carbonate and carbon were derived; hence pure potash is sometimes called salt of tartar or sal. tartari. It can also be obtained by heating potassium bicarbonate, which can be readily prepared pure: $2KHCO_3 + K_2CO_3 + H_2O + CO_2$.

Properties. White, granular, strongly alkaline powder which melts at a high temperature without decomposition and is soluble in equal parts of water, and can be obtained from this solution as colorless crystals, $2K_2CO_3 + 3H_2O$, on evaporation. It deliquesces in the air to a thick liquid, due to its absorbing water.

Primary Potassium Carbonate, $KHCO_3$, Potassium Bicarbonate or Acid Carbonate. If carbon dioxide is passed into a concentrated solution of potassium carbonate the acid salt crystallizes out as colorless transparent crystals because of the less solubility of this salt: $K_2CO_3 + H_2O + CO_2 = 2KHCO_3$. This salt is soluble in 4 parts water and is alkaline in reaction.

Potassium silicate, potassium water-glass, is obtained by fusing potassium carbonate with sand (SiO_2, p. 196) as a vitreous mass, which has no constant constitution but is a mixture of various polysilicates (p. 197). When powdered it is soluble in water on boiling for a long time; the concentrated aqueous solution soon solidifies in the air and after a time dries to a non-transparent mass.

b. Detection of Potassium Compounds.

1. They give a violet coloration to the non-luminous flame. The spectrum of this flame is characterized by a red and a violet line.

2. Platinum chloride produces a yellow crystalline precipitate of potassium-platinum chloride, K_2PtCl_6.

3. Tartaric acid produces a gradual precipitate of white crystalline potassium acid tartrate, $C_4H_5KO_6$ (cream of tartar).

4. Sodium picrate precipitates yellow crystalline potassium picrate, $C_6H_2(NO_2)_3(OK)$, from potassium salt solutions.

Reactions 2, 3, and 4 are also given by ammonium salts, but these volatilize when gently heated and hence can be separated in this way.

2. Sodium (Natrium).

Atomic weight 23.05 = Na.

Occurrence. Only combined, abundantly and widely distributed, especially as sodium chloride, as enormous deposits (rock salt), as well as in solution in sea-water, in various salt seas or lakes and springs. Sodium nitrate occurs deposited in South America as Chili saltpeter, while sodium silicate is a constituent of many minerals and crystalline rocks. Traces of sodium salts are always found in the atmospheric dust.

Sodium is very widely distributed in the plant kingdom; still potassium compounds exist to a greater extent in land plants. The sodium salts are extensively distributed in the animal kingdom, and especially in the fluid parts of the body, while the potassium salts exist to a greater extent in the solid parts.

Preparation. Sodium is prepared in the same way as potassium. Sodium does not form an explosive compound with carbon monoxide in its preparation from sodium carbonate and carbon (see p. 202).

Properties. Shining white, soft metal having a specific gravity of 0.97, melting at 95.6°, and volatile at 742°, forming a colorless vapor. It quickly oxidizes in the air, hence it is kept beneath petroleum. It burns on warming with a yellow flame into Na_2O_2 and decomposes water like potassium; still the heat produced is not sufficient to ignite the hydrogen. With hydrogen it forms silvery-white sodium hydride, Na_2H, which does not spontaneously inflame.

a. Compounds of Sodium.

Sodium oxides and **Sodium sulphides** have analogous constitution and similar properties to the corresponding potassium compounds and are prepared in the same way.

Sodium hydroxide, NaOH, caustic soda, is prepared in a similar manner to potassium hydroxide (see also Sodium Carbonate) and has the same properties. When impure it occurs in commerce as white, bluish, or reddish pieces.

Sodium peroxide, Na_2O_2, produced by passing air over heated metallic sodium. It forms white or yellowish crystalline masses which with ice-cold water forms NaOH and H_2O_2, and with water at ordinary temperature generates oxygen and NaOH. It has energetic oxidizing properties, often with explosive violence.

Sodium Chloride, Common Salt, NaCl. *Occurrence.* In enormous layers as rock salt, dissolved in sea-water (up to 3 per cent.), and in many springs, the salt springs. It is a constituent of plant-ashes and occurs in all fluids of the animal body, especially in the blood and urine.

Preparation. 1. From rock salt, which is often found very pure, sometimes in regular cubes, but mostly as transparent masses. If it contains foreign salts (clay, gypsum), it is dissolved in water in the mine and the solution (salt brine) pumped to the surface of the earth and evaporated until crystallization occurs.

2. From sea-water. In hot climes the sea-water is allowed to evaporate to crystallization in flat basins ("salterns") by means of the summer heat. In cold climes the sea-water is allowed to freeze in the "salterns"; the ice formed consists only of water, and on removal a concentrated brine solution is obtained, which is evaporated to crystallization.

3. In Germany, where the climate is not suitable for the evaporation of sea-water and as most of the brine springs would require considerable fuel for evaporation, the salt solutions are concentrated by spontaneous evaporation. The solutions are allowed to trickle repeatedly drop by drop through bundles of fagots piled up together and exposed to the prevailing winds (graduation-house). During the "graduation" the insoluble impurities, such as calcium sulphate, calcium and magnesium carbonate, attach themselves to the twigs. The concentrated brine is then evaporated, when more of the above-mentioned insoluble salts with NaCl separate out, called pan-stone (lick-salt for animals). On further evaporation the pure salt then deposits. The mother-liquor contains the more soluble salts, such as calcium and magnesium bromides.

Properties. Large, colorless cubes or smaller, hollow, step-like pyramids, or hopper-shaped crystals, or a crystalline powder. They melt at red heat and are volatile at white heat. Hardly more soluble in hot water than in cold water; 100 parts water at 0° dissolve 36 parts and at 100° 39 parts salt. At various temperatures during evaporation of the brine we obtain fine salt, coarse salt, or medium salt. If salt becomes moist in the air, it contains magnesium salts.

Sodium iodide, NaI, and

Sodium bromide, NaBr, are obtained in the same manner as the corresponding potassium salt and have the same properties.

Sodium hypochlorite, NaClO, is known only in solution (eau de Labarraque).

Sodium sulphite, $Na_2SO_3 + 7H_2O$, is obtained by the action of SO_2 upon sodium hydrate solution (see Sulphites, p. 126).

Sodium bisulphite, $NaHSO_3$, gives off SO_2 in the air and is oxidized to Na_2SO_4.

Sodium Sulphate, Na_2SO_4, Glauber Salts. *Occurrence.* In many mineral waters (Carlsbad, Marienbad), salt springs, sea-water, and in Spain as enormous deposits.

Preparation. 1. On an extensive scale by heating sodium chloride with sulphuric acid in the manufacture of soda (p. 213) and hydrochloric acid (p. 135):

$$2NaCl + H_2SO_4 = Na_2SO_4 + 2HCl.$$

2. By decomposing dissolved magnesium sulphate (kieserite) with sodium chloride: $MgSO_4 + 2NaCl = MgCl_2 + Na_2SO_4.$

This process is only accomplished at low temperatures, and is performed at Stassfurt during winter. The $MgCl_2$ produced remains in solution, while Na_2SO_4 crystallizes out.

3. By passing sulphur dioxide, air, and steam over heated sodium chloride (Hargreave's process): $2NaCl + SO_2 + O + H_2O = Na_2SO_4 + 2HCl.$

Properties. Colorless prisms of the composition $Na_2SO_4 + 10H_2O$ or containing 56 per cent. water. On lying in the air they lose a part of their water of crystallization and then become cloudy and non-transparent.

Sodium sulphate melts at 33° and decomposes into a saturated solution and into anhydrous Na_2SO_4. On further heating it loses its water of crystallization completely and is converted into anhydrous Na_2SO_4.

100 parts water at 18° dissolves 20 parts, at 30° 200 parts, and at 33° 354 parts of the salt $Na_2SO_4 + 10H_2O$. Above 33° the solubility diminishes, at 50° only 263 parts are dissolved, at 100° 238 parts of the salt. This depends upon the fact that at 33° the anhydrous salt Na_2SO_4 is formed and separates out, and at the same time a part of it as such goes again into solution, so that a condition is obtained where its solubility diminishes in increasing temperatures. If a solution which is saturated at 33° is cooled to a lower temperature, no salt separates out, although the salt is less soluble therein (supersaturated solution, p. 52).

Dry sodium sulphate, $Na_2SO_4 + H_2O$, is obtained by drying sodium sulphate.

Artificial Carlsbad salt is a mixture of sodium sulphate, sodium carbonate, sodium chloride, and potassium sulphate.

Sodium thiosulphate, $Na_2S_2O_3$, sodium hyposulphite, is obtained from the soda residues or by boiling an aqueous solution of sodium sulphite with flowers of sulphur: $Na_2SO_3 + S = Na_2S_2O_3$. It crystallizes with 5 molecules of H_2O in prisms which are stable in the air and are readily soluble. The solution is decomposed by acids with the setting free of sulphur (milk of sulphur, p. 121) and sulphur dioxide: $Na_2S_2O_3 + 2HCl = 2NaCl + SO_2 + S + H_2O$. It has a reducing action converting chlorine, bromine, and iodine into sodium salts: $2Na_2S_2O_3 + 2I = 2NaI + Na_2S_4O_6$ (sodium tetrathionate, p. 124) and hence serves as "antichlor" in chlorine bleaching in order to remove the excess of chlorine. It dissolves the halogen salts of silver and is used in photography as a "fixing salt," removing the silver compounds not acted upon by light.

Sodium Nitrate, Chili saltpeter, $NaNO_3$, occurs in large deposits in Chili, from whence the world receives its supply, using it to a great extent as a fertilizer. It is purified by recrystallization and forms colorless rhombohedric anhydrous crystals soluble in 1.2 parts water. The crystals are cubical, hence the salt is called cubical saltpeter. It deliquesces in the air, hence it cannot be used in the manufacture of gunpowder.

Tertiary sodium phosphate, Na_3PO_4, trisodium phosphate, so-called basic sodium phosphate (p. 169), is obtained by treating the secondary sodium phosphate with NaOH and evaporating to crystallization. It forms colorless, strongly alkaline prisms, $Na_3PO_4 + 12H_2O$.

Secondary sodium phosphate, Na_2HPO_4, disodium phosphate, so-called neutral sodium phosphate (P. 169), occurs in carnivorous urine and other animal fluids, and is obtained by saturating phosphoric acid with NaOH until a faint alkaline reaction is obtained and then evaporating to crystallization. It forms colorless, readily efflorescing, faintly alkaline prisms, $Na_2HPO_4 + 12H_2O$, which are soluble in 5.8 parts water. On heating it fuses in its water of crystallization, then loses this and is converted into sodium pyrophosphate: $2Na_2HPO_4 = Na_4P_2O_7 + H_2O$.

Primary sodium phosphate, NaH_2PO_4, monosodium phosphate, so-called acid sodium phosphate (P. 169), occurs in carnivorous urine and gives the acid reaction to the same. It is obtained when phosphoric acid is treated with a calculated amount of NaOH and evaporated. It forms colorless acid crystals, $NaH_2PO_4 + H_2O$. It loses its water of crystallization at 100°, and on heating more intensely acid sodium pyrophosphate, $Na_2H_2P_2O_7$, and finally sodium metaphosphate, $NaPO_3$, is obtained:

$$2NaH_2PO_4 = Na_2H_2P_2O_7 + H_2O; \quad Na_2H_2P_2O_7 = 2NaPO_3 + H_2O.$$

This last is used in blowpipe analysis like borax (see below).

Sodium sulphantimonate, $Na_3SbS_4 + 9H_2O$ (Schlippe's salt), is obtained by boiling antimony trisulphide with sulphur and caustic soda, or, instead of the latter, with soda solution and lime. It crystallizes in colorless tetrahedra, soon becoming pale yellow, and is used in the preparation of gold sulphur (p. 181):

$$2Na_3SbS_4 + 6HCl = 6NaCl + Sb_2S_5 + 3H_2S.$$

Sodium Tetraborate, $Na_2B_4O_7 + 10H_2O$, sodium biborate, borax, occurs dissolved in certain lakes of California and Thibet, and occurs in commerce as tinkal. It is obtained at the present time by saturating a solution of boracic acid with soda and recrystallizing as monoclinic colorless prisms of alkaline reaction (turning red litmus blue; in regard to the reaction of borax with turmeric see p. 185). Borax is soluble in 17 parts water, and on heating borax it loses its water of crystallization and yields vitreous, fused borax, $Na_2B_4O_7$;

which is used in blowpipe analysis, as it gives characteristic colors with many metallic oxides.

Sodium Carbonate, Na_2CO_3, Soda. *Occurrence.* Dissolved in certain lakes in Asia, Africa, North America, and in certain mineral waters (Carlsbad, Vichy). It occurs as an efflorescence on the soil in Egypt, Hungary, and South America. It is contained to a great extent in the ashes of certain sea-plants (especially in the varieties of Salsola and Salicornia), as well as from seaweeds, from which it was formerly prepared in the same manner as potassium carbonate from the ashes of land-plants.

Preparation. 1. Leblanc's Method (1794). Common salt is heated with sulphuric acid (or SO_2, air, and steam are passed over NaCl, p. 211), when sodium sulphate and hydrochloric acid are produced. This last is allowed to pass off and is absorbed by water.

The sodium sulphate is fused at a red heat with chalk ($CaCO_3$) and hard coal, when the sodium sulphate is first reduced by the carbon into sodium sulphide: $Na_2SO_4 + 2C = Na_2S + 2CO_2$. The sodium sulphide is then transformed by the calcium carbonate into sodium carbonate and calcium sulphide: $Na_2S + CaCO_3 = Na_2CO_3 + CaS$.

The black· mass thus obtained, called black ash, contains 30–45 per cent. sodium carbonate, 30 per cent. calcium sulphide, besides lime (CaO), undecomposed calcium carbonate, coal, and sand. It is lixiviated with cold water, which dissolves the sodium carbonate.

The soda solution (lye) is evaporated, while a continuous flow of new lye is added, until sodium carbonate, as $Na_2CO_3 + H_2O$, separates out, and this removed as soon as formed. These crystals are heated to remove water and any organic matter mixed with them, and occurs in commerce as calcined soda, or the crystals are recrystallized from watery solution and obtained as $Na_2CO_3 + 10H_2O$ (p. 214).

For every kilogram of soda there remains from 1 to 1½ kilograms of residue, which contains all the sulphur derived from the sulphuric acid used. This residue by the action of the air develops sulphuretted hydrogen, while the polysulphides produced dissolve when exposed to rain, so that the air and water in the neighborhood of a Leblanc soda factory are contaminated to a considerable extent. In more recent times the fresh residue is treated with carbon dioxide, $CaS + H_2O + CO_2 = CaCO_3 + H_2S$; the sulphuretted hydrogen is either burnt with a diminished supply of air, when most of the sulphur separates out (P. 122), or it is completely burnt into sulphur dioxide, which is used again in the preparation of sulphuric acid (Chance-Claus method of regenerating sulphur).

2. Solvay or Ammonia Method (1863). A solution of ammonium bicarbonate reacts with a solution of sodium chloride, forming ammonium chloride and sodium bicarbonate, which is less soluble and separates out: $NaCl + NH_4.HCO_3 = NH_4Cl + NaHCO_3$.

Sodium bicarbonate yields calcined soda on heating:
$$2NaHCO_3 = Na_2CO_3 + H_2O + CO_2.$$

The ammonia is reobtained from the ammonium chloride by heating with lime or magnesia: $2NH_4Cl + CaO = CaCl_2 + 2NH_3 + H_2O$.

On passing carbon dioxide and ammonia into a solution of salt the process described above takes place again:
$$NaCl + NH_3 + CO_2 + H_2O = NaHCO_3 + NH_4Cl.$$

The decomposition takes place even at the ordinary temperature. In practice the carbon dioxide is passed into the liquid under pressure not below 40°. The residue of $MgCl_2$ or $CaCl_2$ obtained in this process is used in the preparation of chlorine (p. 133).

The Solvay process requires much less coal, does not yield any residue which contaminates the neighborhood, and yields a purer soda than the Leblanc method. It would have entirely supplanted this last method if hydrochloric acid, obtained by this process, were not an important by-product.

3. Electrolytic Process. Sodium chloride solutions are decomposed by the electric current, when chlorine appears at the carbon positive pole and sodium at the iron negative pole. This sodium forms NaOH with the water, and this is then transformed into the difficultly soluble $NaHCO_3$ (see 2) by passing CO_2 into the solution, which yields calcined soda on heating (see below).

4. For preparation of soda from cryolite see Aluminium Hydroxide, p. 250.

Properties. Anhydrous or calcined soda forms a white mass or a white powder which melts at a red heat and has an alkaline reaction. If a not too concentrated, hot, aqueous solution is allowed to cool in the air, large colorless monoclinic prisms, $Na_2CO_3 + 10H_2O$, sodium carbonate (crude washing soda), separate out. By repeated recrystallization purified soda is obtained, $Na_2CO_3 + 10H_2O$. Crystalline soda quickly loses a part of its water of crystallization in the air and changes into a white powder: $Na_2CO_3 + 2H_2O$, sodium carbonate (siccum).

100 parts water at 15° dissolve 55 parts soda, at 38° 138 parts, at 100° 100 parts. Hence on heating a solution saturated at 38° soda separates out and precipitates at 50° $Na_2CO_3 + 7H_2O$, at 100° $Na_2CO_3 + H_2O$, and below 38° $Na_2CO_3 + 10H_2O$.

Sodium bicarbonate, $NaHCO_3$, primary or acid sodium carbonate, sodium hydrocarbonate (baking-soda), separates out on passing carbon dioxide into a concentrated solution of sodium carbonate: $Na_2CO_3 + H_2O + CO_2 = 2NaHCO_3$. The cheaper varieties are prepared according to the Solvay process (p. 214) as an intermediary product. It forms crystalline crusts or a white crystalline powder having a faint alkaline reaction and soluble in 12 parts cold water. The moist powder or the powder exposed to moist air loses carbon dioxide.

Sodium silicate is prepared in a manner similar to that of potassium silicate (p. 208). Its watery solution is called "water-glass."

b. Detection of Sodium Compounds.

1. They color a non-luminous flame intensely yellow. The spectrum of this flame shows a single bright-yellow line.

2. All sodium salts are soluble in water with the exception of disodium pyroantimonate ($Na_2H_2Sb_2O_7$), which is obtained as a white granular precipitate when a solution of a sodium compound is treated with a solution of dipotassium pyroantimonate.

3. Cæsium.   4. Rubidium.
Atomic weight 132.9 = Cs.   Atomic weight 85.4 = Rb.

Both of these elements occur only in combination, rather widely distributed, always together, but always in small amounts; thus in many plant-ashes, in several minerals, such as lepidolite, pollucite, triphylite, and many mineral springs, especially of Nauheim and Dürkheim, also in the mother-liquors of carnallite. They give characteristic spectral lines and hence their names.

They are both obtained by the electrolysis of their chlorides, and inflame spontaneously in the air when in large pieces, hence must be kept under petroleum. Cæsium has a specific gravity of 1.84, is soft, silvery in appearance, melts at 26.4°, boils at 270°, and the metal as well as its compounds gives a violet color to a non-luminous flame. The spectrum of this flame show two intense blue lines.

Rubidium, sp. gr. 1.52, is soft, silvery in appearance, melts at 38.5, boils at about 500°, and the metal and its compounds give a violet color to a non-luminous flame. The spectrum of this flame shows two dark red and two violet lines.

5. Lithium,
Atomic weight 7.03 = Li,

occurs only combined, very widely distributed, but always in very small quantities, as in many mineral waters (Baden-Baden, Carlsbad, Marienbad), and in the ashes of many plants, especially in tobacco and the common beet. Certain minerals, such as triphylite, lepidolite, petalite, amblygonite, contain as much as 4.5 per cent. lithium. It is obtained by the electrolysis of the chloride as a silver-white metal which melts at 186° and which boils above 800°. The metal has a specific gravity of 0.59 and hence is the lightest of all metals In other regards it behaves like sodium. It forms the connecting link between the alkali metals and the alkaline earths, as its carbonate and phosphate are difficultly soluble in water. Lithium and its compounds give a beautiful red color to the non-luminous flame, and the spectrum of this flame gives one red line.

Lithium carbonate, Li_2CO_3, is obtained as a white crystalline precipitate by treating a concentrated solution of $LiCl_2$ with ammonium carbonate. It is soluble in 80 parts water, giving an alkaline reaction to the solution.

6. Ammonium.

Ammonia combines directly with all acids, forming salts which on account of their similarity to the potassium compounds will be treated of at this place. The hydrogen of the acid is replaced by the monovalent radical ammonium NH_4, which does not exist free. All ammonium salts are isomorphous with the analogous potassium salts.

The metallic character of the NH_4 group is shown by the existence of ammonium amalgam, which has the same character as potassium amalgam. The ammonium amalgam is obtained as a voluminous metallic mass when sodium amalgam is treated with a solution of ammonium chloride: $(Hg.Na) + NH_4Cl = (Hg.NH_4) + NaCl.$ It quickly decomposes into hydrogen, mercury, and ammonia.

a. Compounds of Ammonium.

Ammonium oxide $(NH_4)_2O$, and **ammonium hydroxide,** $(NH_4)OH$, have not been isolated (see Ammonia).

Ammonium peroxide, $(NH_4)_2O_2$, separates on mixing solutions of $2NH_3$ and H_2O_2 cooled to $-20°$, and forms colorless cubes which decompose at ordinary temperatures into $2NH_3 + H_2O + O$.

Ammonium sulphide, $(NH_4)_2S$, separates as colorless crystals on bringing 1 vol. H_2S gas and 2 vols. NH_3 together at $-20°$.

Ammonium hydrosulphide, ammonium sulphydrate, $(NH_4)SH$, forms colorless crystals on cooling equal volumes of NH_3 and H_2S gas to 0°. This body is obtained in solution by saturating a solution of ammonia with sulphuretted hydrogen: $NH_3 + H_2S = NH_4SH$. It forms a colorless solution which becomes yellow in the air, whereby ammonium polysulphide and water are produced (ammonium sulphide used in the laboratory).

Ammonium chloride, sal ammoniac, NH_4Cl, occurs in volcanic and coal deposits, is obtained by neutralizing the ammoniacal liquor of the gas-works (p. 148) with hydrochloric acid, evaporating the solution to dryness and subliming the residue, or by the sublimation of sodium chloride with ammonium sulphate:

$$2NaCl + (NH_4)_2SO_4 = 2NH_4Cl + Na_2SO_4.$$

When sublimed it forms fibrous crystalline masses, and when crystallized from water it forms a crystalline powder. It is colorless and odorless, volatile on heating, without melting, dissolves in 4 parts water, and is nearly insoluble in alcohol.

Ammonium bromide, NH_4Br, is obtained on the sublimation of potassium bromide with ammonium sulphate, and forms white cubes or a white crystalline powder which is readily soluble in water and difficultly soluble in alcohol.

Ammonium iodide, NH_4I, is prepared from potassium iodide and ammonium sulphate and forms white deliquescent cubes which are readily soluble in alcohol. It splits off iodine in moist air.

Ammonium sulphate, $(NH_4)_2SO_4$, occurs as the mineral mascagnine. It is obtained by saturating the ammoniacal liquor of the gas-works with sulphuric acid and forms colorless and odorless crystals which decompose at high temperatures. It is used as a valuable fertilizer.

Ammonium persulphate, $(NH_4)_2S_2O_8$, forms colorless prisms which are readily soluble. It is prepared on a large scale being used as an oxidizing agent (p. 131) and is also used in photography.

Ammonium nitrate, $(NH_4)NO_3$, produced by saturating a solution of ammonia with nitric acid and evaporating this product. It forms colorless crystals which are soluble in water and which decompose into water and nitrous oxide (p. 155) when heated. It is used as an explosive when mixed with saltpeter and resins.

Ammonium nitrite, $(NH_4)NO_2$. In regard to formation and occurrence see Nitrous Acid. It is obtained by mixing a solution of silver nitrite with an ammonium chloride solution, filtering off the silver chloride, and carefully evaporating the filtrate. It forms colorless crystalline masses which on heating yield water and nitrogen (p. 146).

Ammonium phosphates. Tertiary ammonium phosphate $(NH_4)_3PO_4$, secondary ammonium phosphate, $(NH_4)_2HPO_4$, and primary ammonium phosphate, $(NH_4)H_2PO_4$, are obtained in a manner similar to that of the corresponding sodium phosphates. On careful heating they leave a residue of metaphosphoric acid thus:

$$(NH_4)H_2PO_4 = HPO_3 + H_2O + NH_3.$$

Sodium-ammonium phosphate, $Na(NH_4)HPO_4 + 4H_2O$ (microcosmic salt), occurs in guano and stale urine (hence formerly called Sal urinæ fixum). If 5 parts disodium phosphate and 2 parts diammonium phosphate are dissolved in hot water and allowed to cool, colorless prisms separate out:

$$Na_2HPO_4 + (NH_4)_2HPO_4 = 2Na(NH_4)HPO_4.$$

·It is used in blowpipe analysis, as on heating it is converted into sodium metaphosphate, which dissolves many metallic compounds with the production of characteristic colors:

$$Na(NH_4)HPO_4 = NaPO_3 + H_2O + NH_3.$$

Ammonium carbonate, $(NH_4)_2CO_3 + H_2O$, precipitates when ammonia is passed into a concentrated solution of commercial ammonium carbonate (see below). On allowing to lie in the air it is converted into

Ammonium bicarbonate, $(NH_4)HCO_3$, which precipitates on saturating an ammonia solution with carbon dioxide. Both of ·these bodies form colorless crystals which decompose at 60°· into carbon dioxide and ammonia.

Commercial ammonium carbonate is a combination of primary ammonium carbonate with ammonium carbamate (which see):

$$HO-CO-O(NH)_4 + H_2N-CO-O-(NH_4).$$

It is produced in the decay of many organic nitrogenous bodies and was formerly obtained by the dry distillation of such bodies, e.g., horn, hoofs, claws, bones, and leather refuse, etc.

It was then strongly contaminated with inflammable oils and was called in pharmacy Sal cornu cervi volatile, Hartshorn salt, ammonium carbonicum- pyroleosum. It is now obtained by subliming calcium carbonate (chalk) with ammonium sulphate or ammonium chloride. It forms white, fibrous, crystalline masses which are soluble in 5 parts water. On exposure to the air it develops NH_3 and CO_2 and is transformed into ammonium bicarbonate which is less soluble.

b. Detection of Ammonium Compounds.

1. The hydroxides of the alkali metals and the alkaline earthy metals set ammonia free from ammonium compounds, which can be readily detected by the odor as well as by turning turmeric paper brown (p. 150).

2. Platinum chloride and also tartaric acid produce precipitates in ammonium salt solutions analogous in composition to the potassium compounds, namely, ammonium-platinum chloride, $(NH_4)_2PtCl_6$, and ammonium bitartrate, $C_4H_5(NH_4)O_6$.

3. All ammonium salts are volatile on heating.

4. In solution traces of ammonium salts produce a brown coloration or precipitate with Nessler's reagent (see Mercuric Iodide).

GROUP OF ALKALINE-EARTH METALS.
Calcium. Strontium. Barium.

These are divalent metals, which combine at ordinary temperatures with oxygen and also decompose water, forming strong basic oxides (the alkaline earths) and hydroxides which are less soluble in water than the alkalies. The oxides are not reduced into the metals by carbon nor by hydrogen. Their normal phosphates, sulphates, and carbonates are soluble with very great difficulty in water, or are insoluble therein, while their sulphides, like those of the alkalies, are readily soluble. The solubility of the hydroxides increases from calcium to barium, and the solubility of the sulphate diminishes from calcium to barium.

They are called alkaline-earth metals because they have properties similar to the alkali metals as well as to the earthy metals.

Their specific gravity, their melting-point, their volatility and chemical energy increase with their atomic weight. Thus calcium hydroxide and carbonate are readily decomposed on heating, while barium hydroxide is not at all decomposed, and barium carbonate only with difficulty, and the corresponding strontium compounds are intermediate in their behavior. .

Their salts are not decomposed by ammonia solution, in contradistinction to the salts of the groups which follow.

With hydrogen they form hydrides, CaH_2, SrH_2, BaH_2, which are colorless solids.

These metals combine with nitrogen at a red heat, forming nitrides, Ca_3N_2, Sr_3N_2, Ba_3N_2, which with water suffer the following decomposition:

$$Ca_3N_2 + 6HOH = 3Ca(OH)_2 + 2NH_3.$$

With the heat of the electric furnace they combine with carbon, forming carbides, CaC_2, SrC_2, BaC_2, which is the reason that the alkaline-earth metals are not obtained on heating their oxides with carbon as the metal which is set free immediately combines with the carbon. .

1. Calcium.

Atomic weight $= 40.1 = Ca$.

Occurrence. Never free, but its compounds are found extensively, often in large quantities. Calcium carbonate and calcium sulphate (which see) occur as very large deposits and form the chief constituents of the solids of spring- and river-waters. Calcium silicate is a constituent of nearly all siliceous minerals, and calcium phosphate occurs in apatite and phosphorite (p. 162). Calcium salts always occur in the plant and animal organisms.

Preparation and Properties. Calcium is obtained by the electrolysis of fused calcium chloride or by heating calcium iodide with sodium. It is a silvery-white soft metal, harder than lead and having a specific gravity of 1.58 and melting at 760°, but, as it slowly oxidizes in dry air, it must like potassium be kept beneath petroleum.

a. Compounds of Calcium.

Calcium oxide, CaO, caustic lime, lime, burnt lime, is obtained by heating to a red heat pure calcium carbonate (white marble); but on a larger scale by heating impure calcium carbonate (the limestones) in so-called limekilns. It forms hard, white, amorphous, alkaline masses which become crystalline at 2500° and fusible at about 3000°, absorb moisture and carbon dioxide from the air, and are transformed into $CaCO_3$ and $Ca(OH)_2$. On account of its high refractory power calcium oxide is used in the construction of crucibles for the oxyhydrogen blowpipe, as well as for the Drummond lime-light (p. 113). It combines with water with the evolution of great heat (slaking of lime), producing

Calcium hydroxide, $Ca(OH)_2$, slaked lime, which is a white porous powder that forms a pasty mass with little water, and with more water the so-called milk of lime. It is soluble in 700 parts cold water and in 1300 parts hot water; hence a cold saturated solution becomes cloudy on heating.

The saturated watery solution, called lime-water, aqua calcis, becomes turbid when exposed to the air because of the formation and precipitation of calcium carbonate.

Mortar. Calcium hydroxide forms calcium carbonate with the carbon dioxide of the air and gradually becomes crystalline and hence solid; and upon this fact depends the hardening of air-mortar, which is a pasty mixture of sand, slaked lime, and water. The equation $Ca(OH)_2 + CO_2 = CaCO_3 + H_2O$ explains the great abundance of moisture in newly constructed buildings.

Cement. If a limestone contains considerable aluminium silicate (clay), then calcium silicate is formed on burning the same. This calcium silicate cannot be further slaked with water, i.e., calcium hydroxide cannot be produced. Nevertheless such a product hardens when mixed with water and remains hard even under the surface of water; hence it is called hydraulic lime or cement, and is used especially for hydraulic use. The pasty mixture of sand, cement, and water is called water-mortar. The hardening is caused by the formation of hydrated aluminium-calcium silicates and basic calcium aluminate (see Aluminium).

Calcium sulphide, CaS, is obtained pure in the same manner as K_2S, and impure from the residues in soda manufacture. It is insoluble in cold water, but is slowly decomposed by it into difficultly soluble calcium hydroxide and readily soluble calcium hydrosulphide:

$$2CaS + 2H_2O = Ca(OH)_2 + Ca(SH)_2.$$

Pure calcium, strontium, and barium sulphides shine in the dark after exposure to sunlight, but lose this property in moist air (Bolognian stone, used as a phosphorescent paint).

Calcium hydrosulphide, calcium sulphydrate, $Ca(SH)_2$, is produced in purifying illuminating-gas (gas-lime), and is obtained as a gray mass by the action of H_2S upon $Ca(OH)_2$; thus, $Ca(OH)_2 + 2H_2S = Ca(SH)_2 + 2H_2O$. It may also be prepared from calcium sulphide (see above). Its watery paste destroys hair (Böttger's depilatory, rhusma) and is used to remove wool from   ee ' hides. On boiling with water it decomposes: $Ca(SH)_2 = CaS + H_2S$h p s

Calcium polysulphides, CaS_2, CaS_3, CaS_5, etc., are obtained mixed with calcium sulphate on heating calcium oxide with the corresponding amounts of sulphur. They behave like the corresponding potassium polysulphides and are used in the preparation of precipitated sulphur (milk of sulphur) and H_2S_2.

Calcium chloride, $CaCl_2$, occurs in tachhydrite, $CaCl_2+MgCl_2+12H_2O$, and is obtained by dissolving calcium carbonate (marble, chalk) in hydrochloric acid and is also procured in large quantities in the manufacture of ammonia as well as in the Solvay soda process. It is obtained from its solutions by evaporation as colorless crystals, $CaCl_2+6H_2O$, which dissolve in water with great lowering of temperature. The crystals melt at 29° in their water of crystallization, and at 200° white porous anhydrous $CaCl_2$ is obtained. This dissolves in water with generation of heat and melts at 719°, forming a crystalline mass which absorbs water with activity and deliquesces in the air, and hence is used as a drying agent. It also absorbs ammonia and forms a white powder therewith: $CaCl_2+8NH_3$. The large quantities of calcium chloride obtained as a by-product in technical processes are used in the preparation of chlorine (which see).

Calcium fluoride, CaF_2, occurs as colorless, yellow, green, or violet crystals or masses called fluor-spar, and in small amounts in plant-ashes, bones, and the enamel of the teeth. It is insoluble in water and phosphoresces on heating or when exposed to the sunlight. On account of its ready fusibility it is used as a flux in metallurgical processes.

Calcium hypochlorite, $Ca(OCl)_2$, is obtained as thin crystals by strongly cooling a concentrated aqueous solution of chloride of lime.

Chloride of lime, bleaching-powder, consists of calcium chloride and calcium hypochlorite. $Ca{<}^{OCl}_{OCl}+CaCl_2$, or of $2Ca{<}^{OCl}_{Cl}$.

Calcium chloride cannot be isolated from the chloride of lime, hence the last formula is probably correct.

Preparation. Chlorine is passed over layers of dry calcium hydroxide at a temperature which must not rise above 25°, so that no calcium chlorate is produced:

$$2Ca(OH)_2+4Cl=2Ca(OCl)Cl+2H_2O.$$

It is not possible to have all the calcium hydroxide combine with chlorine, so that the product only contains 25 and 30 per cent. available chlorine (p. 222) and always contains calcium hydroxide.

Properties. White powder having a chlorine-like odor. The calcium chloride and hypochlorite dissolve in water, while the calcium hydroxide remains in great part undissolved. The watery solutions have a bleaching action. Hydrochloric acid and sulphuric acid evolve chlorine from chloride of lime, hence it serves in the preparation of chlorine for the destruction of organic pigments, disagreeable odors, and for disinfection. Chloride of lime should contain at least 25 per cent. of active chlorine. By active chlorine we mean the quantity of chlorine set free on adding the above-mentioned acids to chloride of lime. It amounts to double the quantity contained in the calcium hypochlorite:

$$CaCl(OCl) + 2HCl = CaCl_2 + H_2O + 2Cl;$$
$$CaCl(OCl) + H_2SO_4 = CaSO_4 + H_2O + 2Cl.$$

Chloride of lime evolves hypochlorous acid on standing in the air, being set free by the action of the carbon dioxide of the air. On heating or in sunlight it slowly decomposes into $CaCl_2 + O_2$, and the generation of oxygen may take place with explosive violence; hence the generation of oxygen may take place with explosive violence; hence chloride of lime must be kept cool and dark and not in completely closed vessels.

Many metallic oxides, such as Co_2O_3, CuO, develop oxygen from chloride of lime on warming. Probably a part of the oxygen of the metal combines with the chloride of lime, setting free oxygen: $CaCl(OCl) + 2Co_2O_3 = CaCl_2 + 2CoO + 2O$. The lower metallic oxide formed is then again transformed by the chloride of lime into the higher metallic oxide, which in turn acts upon the chloride of lime.

Ammonia when heated with chloride of lime generates nitrogen (p. 147).
Nitric acid sets hypochlorous acid free (p. 137).
Hydrogen peroxide develops oxygen (p. 119).

Calcium Sulphate, $CaSO_4$. *Occurrence.* 1. In most spring-waters (permanently hard waters, p. 116).

2. As anhydrite in anhydrous rhombic crystals.

3. As gypsum, $CaSO_4 + 2H_2O$, it forms dense masses; as crystalline masses it is called alabaster, and as monoclinic prisms it forms selenite, etc.

Preparation. By treating a concentrated solution of a calcium salt with concentrated .sulphuric acid a white crystalline precipitate of calcium sulphate, $CaSO_4 + 2H_2O$, is obtained.

Properties. It is soluble in 400 parts water and loses 1½ molecules of its water of crystallization on being heated to 120°; the product, $2CaSO_4 + H_2O$, is called burnt gypsum or plaster of Paris. If this is mixed into a paste with water, it unites with this with the development of heat and quickly hardens. The use of plaster of Paris in the preparation of casts, moulds, plaster bandages, etc., is dependent upon this property. If gypsum is heated to 160°, it loses all its water of crystallization and no longer combines with water, being then called dead burnt plaster. Anhydrite does not combine with water.

Tertiary or Tricalcium phosphate, $Ca_3(PO_4)_2$ (also neutral calcium phosphate), occurs as apatite and phosphorite (p. 162) and constitutes two-thirds of the bones of animals; it forms the chief mass of the coprolites, phosphorites, osteolites, and certain varieties of guano, and is found in the ash of all animal and plant organs. It is insoluble in pure water but, on the contrary, is partly soluble in water containing CO_2 or certain salts.

A basic calcium phosphate, $Ca_3(PO_4)_2 + CaO$, is the chief constituent of the Thomas slag obtained in the dephosphorization of iron and which is an excellent fertilizer. Its value is dependent upon the fact that it does not have to be converted into soluble superphosphate like the other calcium phosphate, but is ground finely (Thomas phosphate), and is readily decomposed by the carbonic acid and moisture of the soil.

Pure calcium phosphate is obtained by mixing a solution of disodium phosphate treated with ammonia with a solution of calcium chloride. This precipitate is gelatinous and when dry forms a white, amorphous, odorless and tasteless mass.

Secondary calcium phosphate, $CaHPO_4 + 2H_2O$ (dicalcium phosphate), occurs often in urinary calculi and sediments as microscopic crystals. It is obtained as a white crystalline powder insoluble in water by mixing a solution of disodium phosphate with a calcium chloride solution.

Primary calcium phosphate, $CaH_4(PO_4)_2$ (monocalcium phosphate), is obtained by treating tertiary or secondary calcium phosphate with sulphuric acid (P. 162, *a*), and separates out on the evaporation of the solution as colorless, deliquescent, acid-reacting scales. It is readily soluble in water and hence is better suited as a fertilizer than the tertiary salt. When mixed with gypsum it forms superphosphate, which is used as a fertilizer.

Calcium carbide, CaC_2, is obtained by heating CaO with carbon in the electric furnace. When pure it forms gold-like crystals, and in commerce it forms grayish-black masses which with water and acids generate acetylene gas, C_2H_2 (which see).

Calcium Carbonate, $CaCO_3$. *Occurrence.* 1. Amorphous or not markedly crystalline as limestone; in crystalline granules as marble; in amorphous or crystalline granules combined with magnesium carbonate as dolomite, $CaCO_3 + MgCO_3$; amorphous as chalk; schistous-like as lithographic stone, etc.

2. As calc-spar (calcite) in hexagonal crystals (rhombohedrons), and as aragonite in rhombic columns.

3. In all plant-ashes and in all animals, especially in the bones; in the urine of herbivorous animals, and in many pathological concretions, as in urinary calculi, etc. It forms the chief constituent of the corals, oyster-, mussel-, snail-, and egg-shells, and pearls.

4. It also occurs in fertile soil and in most natural waters.

Preparation. By mixing a solution of a calcium salt with alkali carbonate as a white crystalline precipitate (precipitated chalk).

Properties. On heating to white heat it decomposes into carbon dioxide and calcium oxide: $CaCO_3 = CaO + CO_2$. It is insoluble in pure water, but it gradually dissolves in the presence of CO_2, forming the primary carbonate: $CaCO_3 + H_2O + CO_2 = CaH_2(CO_3)_2$; hence all waters coming from lime soil contain primary calcium carbonate (temporary hardness of water, p. 116). On allowing such water to stand in the air it loses CO_2 and calcium carbonate precipitates, and this accounts for the formation of thermal tuff, stalactites, lime-tuff, etc. The same change takes place on boiling, when the calcium carbonate separates as a crystalline crust on the vessel and forms the boiler incrustations or scales.

Calcium silicate, $CaSiO_3$, is a constituent of many siliceous minerals and is found pure in crystalline masses as wollastonite. It is obtained as a white crystalline mass by fusing silicon dioxide (sand) with calcium carbonate.

Glass. Calcium silicate is opaque and insoluble in water; alkali silicates are transparent and soluble in water. The calcium as well as the alkali silicates are decomposable by acids. If, on the contrary, calcium silicate is fused with the proper proportion of alkali silicate, we obtain a transparent amorphous compound which is neither attacked by water nor by acids and is called glass. In the preparation of common glass there is fused together a mixture of sand, limestone, and soda. Bohemian glass is made with potash instead of with soda; the constituents of flint glass are sand, potash, and lead oxide.

1. Common white glass is sodium-calcium silicate and is used in the preparation of drinking glasses, window glass, etc.

2. Crude green glass, or bottle glass, is sodium-calcium silicate, which is made from impure material in which the ferrous silicate gives a green color to the glass and the ferric silicate gives a yellow color.

3. Bohemian glass, crown glass, is potassium-calcium silicate; it is less fusible than the sodium glass, and is especially used in the preparation of chemical utensils which must withstand high temperatures.

4. English crystal glass, flint glass or paste, is potassium-lead silicate, which fuses readily, is highly refractive, takes a high polish, and is used for optical purposes and decorative objects.

5. Jena glass is characterized by its refractive and dispersive power and hence is used almost exclusively in optical apparatus. It is prepared by partly replacing the silicates by boric acid, phosphoric acid, or fluorine compounds.

6. Colored glass is obtained by dissolving small amounts of metallic oxides in the fused glass. With cobaltous oxide a blue glass is obtained, with ferric oxide a yellow, with cupric or chromic oxide an emerald-green, with uranium oxide a fluorescent greenish yellow, with manganese oxide a violet, with sodium sulphide a brown, and with metallic gold or copper a ruby-red glass. Black glass is simply deeply colored violet, brown, or blue glass. Green or common glass (see 2) is decolorized on the addition of manganese oxide (brownstone) or didymium salts or traces of selenium, as the violet color produced is complementary to the green. Milk-glass is obtained by adding bone-ash or stannic oxide to the glass.

b. Detection of Calcium Compounds.

1. They color the non-luminous flame yellowish red, and the spectrum of this flame shows an intense green and a violet line.

2. Ammonium oxalate causes a white precipitate of calcium oxalate even in very dilute solutions of calcium salts. This precipitate is insoluble in acetic acid and oxalic acid. Barium and strontium salts are only precipitated by this reagent in very concentrated solutions.

3. Sulphuric acid precipitates calcium salts only from concentrated solutions and not from dilute solutions, as the white precipitate of calcium sulphate, $CaSO_4$, is soluble to a slight extent in water.

2. Strontium.

Atomic weight 87.6 = Sr.

Occurs only combined and then as strontianite, $SrCO_3$, and as cölestine, $SrSO_4$. The yellow metal has a specific gravity of 2.5, melts at 600°, is obtained by the electrolysis of fused strontium chloride, and is similar to calcium in properties.

a. Compounds of Strontium.

Strontium oxide, SrO, is produced by heating strontium nitrate. It is a gray mass which with water generates heat and forms strontium hydroxide.

Strontium hydroxide, $Sr(OH)_2$, which is soluble in 50 parts water, can be obtained as crystalline $Sr(OH)_2 + 8H_2O$ from this solution. On heating it decomposes into water and strontium oxide.

Strontium salts are prepared from strontium carbonate by treating it with the respective acid. In regard to their use in the manufacture of sugar see Cane-sugar.

b. Detection of Strontium Compounds.

1. They give a beautiful crimson-red color to the non-luminous flame (use in fireworks), and the spectrum of this flame shows one orange and one blue line.

2. Sulphuric acid precipitates white crystalline $SrSO_4$ from even very dilute solutions of strontium salts.

3. Barium.

Atomic weight 137.4 = Ba.

Occurs only combined as witherite, $BaCO_3$, and as barite or heavy spar, $BaSO_4$. The pale-yellow metal, having a specific gravity of 3.7 and melting at 475°, is obtained by the electrolysis of fused barium chloride. It behaves like Ca and Sr, but is more readily oxidized and decomposes water more readily.

a. Compounds of Barium.

Barium oxide, BaO, forms a pale-gray amorphous mass.

Barium hydroxide, $Ba(OH)_2$, crystallizes from its aqueous solution as $Ba(OH)_2 + 8H_2O$ in tetragonal prisms.

Both of these are obtained like the corresponding strontium compounds and have the same properties. Barium hydroxide is soluble in 20 parts water (baryta-water) and fuses without decomposition at a red heat.

Barium dioxide, BaO_2, is produced by heating barium oxide in a current of oxygen to about 350°. On heating to 700° it decomposes again into barium oxide and oxygen. It is a white powder insoluble in water and has, like all peroxides, no basic properties. With dilute acids BaO_2 yields H_2O_2 (p. 117), which decomposes with MnO_2 into water and oxygen: $MnO_2 + BaO_2 + 2H_2SO_4 = MnSO_4 + BaSO_4 + 2H_2O + 2O$. With concentrated sulphuric acid BaO_2 produces oxygen containing ozone (p. 110). Potassium ferricyanide (which see), as well as all salts of the heavy metals, develops oxygen from BaO_2 in the presence of water.

Barium salts are obtained from barium carbonate by decomposition with the respective acids, evaporating and recrystallizing.

Barium chloride, $BaCl_2 + 2H_2O$, forms colorless rhombic crystals which are soluble in water.

Barium sulphate, $BaSO_4$, is used as a paint (permanent white), also mixed with basic lead carbonate (Venetian, Hamburg, Dutch white).

b. Detection of Barium Compounds.

1. They give a yellowish-green color to the non-luminous flame, and the spectrum of this flame, which is rich in lines, gives one prominent pale-green line.

2. Sulphuric acid precipitates from even very dilute solutions of barium salts white amorphous barium sulphate, $BaSO_4$, insoluble in acids.

3. Potassium chromate precipitates yellow barium chromate, $BaCrO_4$; hydrofluosilicic acid precipitates white $BaSiF_6$ from barium salt solutions. Both are insoluble in acetic acid. Ca and Sr salts are not precipitated by these reagents.

MAGNESIUM GROUP.
Beryllium. Magnesium. Zinc. Cadmium.

These are divalent metals which are volatile on heating and which burn with a flame into oxides when air is supplied. Their specific gravity, fusibility, and volatility increase with their atomic weights. They do not oxidize on being exposed to dry air. Zinc and cadmium do not decompose water at the boiling temperature, beryllium and compact magnesium only with difficulty, while they all decompose water at a red heat. Their oxides, hydroxides, carbonates, and phosphates are insoluble in water, and their sulphates are readily soluble in water. The sulphides of beryllium and magnesium are soluble in water, while zinc and cadmium sulphide are insoluble therein. The oxides of beryllium and magnesium are not reduced to the metallic state by carbon. The carbonates and chlorides are readily decomposed to the basic salts on warming.

Their sulphates (with the exception of beryllium sulphate) form isomorphous double salts, which are readily soluble, with the alkali sulphates: $MgSO_4 + K_2SO_4 + 6H_2O$. Their hydroxides are soluble in ammonia or ammonium salts.

1. Beryllium (Glucinum).
Atomic weight 9.1 = Be.

Occurs only combined in certain rare minerals; thus in phenacite, Be_2SiO_4, in chrysoberyl, $Be(AlO_2)_2$, and especially in beryl, $(Be_3Al_2)Si_6O_{18}$, which when colored green with chromium oxide forms the precious stone called the emerald.

The metal is prepared by heating its chloride with sodium or its oxide with aluminium. It forms a silver-white metal having a specific gravity of 1.8 and melting at about 1000°, and the metal with its salts gives no coloration to the flame and no spectrum. The copper alloy of beryllium gives a beautiful tone when struck and is used for technical purposes.

Ammonia or caustic alkali precipitates $Be(OH)_2$ from its sweet salt solution, this precipitate being soluble in an excess of caustic alkali. Ammonium carbonate precipitates $BeCO_3$, which is soluble in an excess of $(NH_4)_2CO_3$.

2. Magnesium.
Atomic weight 24.36 = Mg.

Occurrence. Only combined: 1. As carbonate in magnesite, $MgCO_3$, and dolomite, $MgCO_3 + CaCO_3$.

2. As silicate, forming the minerals olivine, talc, soapstone, serpentine, meerschaum. Powdered talc, a white greasy powder, is used as a vulnerary powder. Magnesium-calcium silicate forms the minerals augite, hornblendes, and asbestus (amianthus); magnesium-aluminium silicate, chlorite and magnesium mica.

3. As sulphate and chloride in sea-water and in certain mineral waters, giving a bitter taste thereto. In the "abraum" salts (p. 202)

they occur as carnallite, $MgCl_2 + KCl + 6H_2O$, kainite, $K_2SO_4 +$ $MgSO_4 + MgCl_2 + 6H_2O$, kieserite, $MgSO_4 + H_2O$, schönite, $MgSO_4 +$ $K_2SO_4 + 6H_2O$, boracite, $4MgB_4O_7 + 2MgO + MgCl_2$.

4. As phosphate and carbonate in the vegetable and animal kingdoms, especially in the seeds and bones. Many animal concretions, such as urinary and intestinal calculi, consist of ammonium-magnesium phosphate, $Mg(NH_4)PO_4$, which also occurs in guano.

Preparation. By heating magnesium chloride with sodium or by the electrolysis of fused magnesium chloride or carnallite.

Properties. Silver-white metal, non-oxidizable in dry air, having a specific gravity of 1.75 and which can be hammered, drawn into wire, and cast into various shapes. Magnesium melts at 800° and vaporizes at about 1100°. When heated with air it burns with a dazzling white light which is rich in actinic rays. This light is produced by the non-volatile MgO, becoming incandescent (used in fireworks and photography). Dilute acids dissolve magnesium readily with the generation of hydrogen. On boiling magnesium with water hydrogen is slowly generated. Alkalies do not attack this metal. It is a strong reducing agent and removes the oxygen from most metallic oxides and many acid anhydrides on being heated therewith. On being heated white-hot it absorbs nitrogen and eventually also argon (p. 182).

a. Alloys with Magnesium.

The alloy with aluminium, magnalium, is used for technical purposes, and with zinc for fireworks.

b. Compounds of Magnesium.

Magnesium oxide, MgO, magnesia, bitter-earth, is obtained by heating magnesium carbonate or magnesium hydroxide. It is a white, amorphous, infusible, very light powder which is insoluble in water. When gently heated with water it forms $Mg(OH)_2$.

At higher temperatures it conducts electricity and shines with an intense light (Nernst light). This light is produced by passing the electric current through pencils of MgO which are previously heated.

Magnesium hydroxide, $Mg(OH)_2$, is formed on treating a magnesium salt solution with caustic alkali (preparation of the metallic hydroxides). It forms a white amorphous powder which is nearly insoluble in water and which decomposes on heating into $MgO + H_2O$.

Magnesium chloride, MgCl₂. It occurs as stated on pages 221 and 228 and is obtained as a by-product in many chemical processes. It crystallizes with 6H₂O, forming deliquescent crystals which on evaporation decompose and are used in the manufacture of hydrochloric acid and chlorine (p. 133). It may be obtained in an anhydrous form by heating the magnesium chloride containing water in the presence of NH₄Cl or in a current of hydrochloric acid.

Magnesium sulphate, MgSO₄. *Occurrence.* As kieserite, also in kainite and schönite (p. 228); in sea-water and many mineral waters, to which it gives a bitter taste.

Preparation. By dissolving magnesium carbonate in dilute sulphuric acid, or by boiling kieserite for a long time with water, which gradually dissolves the same.

Properties. On evaporating its aqueous solution it is obtained as neutral, colorless, rhombic crystals having the composition $MgSO_4 + 7H_2O$. It has a disagreeable, bitter taste and is soluble in 1.5 parts by weight of cold water (Epsom salts).

At 70° $MgSO_4 + 6H_2O$ crystallizes from its saturated solution, and at 0° $MgSO_4 + 12H_2O$. Kieserite, $MgSO_4 + H_2O$, is soluble only in 400 parts water, but it gradually dissolves on boiling with water and then crystallizes as $MgSO_4 + 7H_2O$.

On heating to 100° Epsom salts lose 5 molecules of water, forming $MgSO_4 + 2H_2O$, and at 150° they lose still another molecule, the last molecule being driven off only at 260° (water of constitution, p. 115).

With the sulphates of the alkali metals magnesium sulphate forms double salts (p. 102), which crystallize in monoclinic prisms, whereby the firmly combined molecule of water of crystallization is replaced thus:

Potassium-magnesium sulphate, $MgSO_4 + K_2SO_4 + 6H_2O$;
Ammonium-magnesium sulphate, $MgSO_4 + (NH_4)_2SO_4 + 6H_2O$.

The sulphates of zinc, cadmium, manganese, iron, cobalt, nickel, which are constituted similarly to the magnesium sulphate, are all isomorphous, and also, like cupric sulphate, which likewise crystallizes with 5 molecules H₂O, contain 1 molecule H₂O firmly combined. These all, like mercuric sulphate, form double salts which have an analogous composition and are isomorphous with double magnesium sulphates:

$NiSO_4 + K_2SO_4 + 6H_2O$; $CuSO_4 + K_2SO_4 + 6H_2O$, etc.

Magnesium carbonate, MgCO₃, occurs as rhombohedra in magnesite (isomorphous with calcite, smithsonite, rhodochrosite, siderite, etc.), and in crystalline masses as magnesite, also in dolomite (p. 223).

If carbon dioxide is passed into water containing suspended basic magnesium carbonate (see below), we obtain from the filtered solution, on standing, neutral magnesium carbonate, $MgCO_3 + 3H_2O$, as colorless crystals. On boiling this with water we again obtain basic magnesium carbonate.

If, on the contrary, a magnesium salt solution is precipitated with alkali carbonate, we do not obtain the normal carbonate, but the basic carbonate, which has a varying composition, dependent upon the concentration and temperature of the liquids. This forms the magnesium carbonate of commerce (magnesia alba) and is generally given the following formula: $4MgCO_3 + Mg(OH)_2 + 4H_2O$. It forms a white porous powder which ordinarily occurs in cubes, and which is insoluble in water and on heating decomposes into $MgO + CO_2 + H_2O$.

c. Detection of Magnesium Compounds.

1. They are colored red when moistened with a cobalt salt solution and heated on charcoal.

2. They differ essentially from the compounds of the alkaline earths by the solubility of their sulphate, their precipitation by ammonia, as well as by their behavior in the presence of ammonium salts (see 3).

3. The tendency of the magnesium salts to form soluble double compounds with ammonium salts is the reason why they are not precipitated by carbonates, nor by the alkali hydroxides, nor by ammonia in the presence of sufficient amounts of an ammonium salt.

4. A mixture of sodium phosphate and ammonia produces a white crystalline precipitate of ammonium magnesium phosphate, $Mg(NH_4)PO_4 + 6H_2O$, in magnesium salt solutions containing ammonium salts in solution. This double salt is the most insoluble salt of magnesium (p. 169).

3. Zinc.

Atomic weight 65.4 = Zn.

Occurrence. Only combined, especially as zinc spar or smithsonite, $ZnCO_3$, as zinc silicate or willemite, $Zn_2SiO_4 + H_2O$, as ordinary calamine (zinc silicate and carbonate), as zinc blend or sphalerite, ZnS, and as zincite, ZnO. Nearly all these ores contain some cadmium besides zinc.

Preparation. By heating the zinc ore in the air (roasting) zinc oxide is obtained, and this on being heated with carbon is reduced ($ZnO + C = Zn + CO$). The zinc volatilizes and is collected in the receivers.

At first the receivers are so cold that the zinc separates as a powder, mixed with some zinc oxide (which is produced by the air present at the beginning of the operation), and with the more volatile cadmium. The powder, which is called zinc-aust, is used as a reducing agent and also in the preparation of cadmium.

Properties. Bluish-white crystalline metal having a specific gravity of 7.1 and brittle at ordinary temperatures, malleable and rollable at 100°, again brittle at 200°, fusible at 433°, and vaporizable at 920°. The vapor of zinc inflames in the air and burns into zinc oxide with a greenish flame.

Zinc becomes coated in moist air with basic zinc carbonate, and on account of its stability in the air it is used as rolled sheet zinc or as a coating to sheet iron (galvanized iron). Zinc in pieces decomposes water only at a red heat, while zinc-dust gradually decomposes water at ordinary temperatures.

Dilute acids dissolve zinc with rapidity, according to the amount of impurities it contains (p. 104); thus pure zinc is hardly attacked. With nitric acid no generation of hydrogen takes place, as the hydrogen produced reduces the nitric acid (p. 160). On heating zinc with concentrated sulphuric acid sulphur dioxide is produced, and on heating with caustic alkalies it dissolves with the development of hydrogen (p. 104). Zinc precipitates the metal as a powder, and often in a spongy form, from the solutions of salts of copper, lead, tin, silver, gold, platinum, etc.

a. Alloys of Zinc.

See Copper and Nickel.

b. Compounds of Zinc.

Zinc oxide, ZnO, is formed on burning zinc in the air (zinc-white, flowers of zinc) or by heating zinc carbonate. It is a white amorphous powder which is insoluble in water and which becomes momentarily yellow on being heated.

Zinc hydroxide, zinc hydrate, $Zn(OH)_2$, is precipitated from zinc salt solutions by caustic alkalies or ammonia. It forms a white precipitate which is soluble in an excess of the precipitant: $ZnSO_4 + 2KOH = Zn(OH)_2 + K_2SO_4$. On heating it decomposes into zinc oxide and water.

Zinc chloride, $ZnCl_2$, is produced by dissolving zinc in hydrochloric acid and carefully evaporating the solution. It forms a white caustic mass which deliquesces in the air and is readily soluble in water and alcohol. It also occurs in commerce cast in sticks.

On evaporation it behaves like $MgCl_2$ (p. 229). It crystallizes from strong hydrochloric acid solution as $ZnCl_2 + H_2O$, and with zinc oxide it forms a plastic mass which soon hardens and is used by dentists in filling teeth. It absorbs ammonia with the formation of $ZnCl_2.NH_3$.

Zinc sulphate, $ZnSO_4$, white vitriol, with $7H_2O$, is chiefly prepared by gently roasting zinc blende (ZnS) and extracting the residue with water and evaporating or, in a purer form, by dissolving zinc in dilute sulphuric acid.   It forms white acid-reacting crystals which are isomorphous with magnesium sulphate and like these form double salts with the alkali sulphates. Zinc sulphate dissolves in 0.6 part by weight of water.

Zinc carbonate, $ZnCO_3$, occurs as smithsonite, isomorphous with magnesium carbonate, etc. (p. 229).   On precipitating zinc salt solutions with alkali carbonates a white precipitate of basic zinc carbonate is formed analogous to the magnesium salts. Its composition is dependent upon the concentration and temperature of the solutions; thus it may be $ZnCO_3 + 2Zn(OH)_2$ or $ZnCO_3 + 3Zn(OH)_2$.

c. Detection of Zinc Compounds.

1. On being moistened with a cobaltous salt solution and heated on charcoal in a blowpipe flame they give a beautiful green infusible mass ($ZnO.CoO$), which is also used as an artist's pigment (Rinmann's green, Saxony green).

2. Ammonium sulphide precipitates white zinc sulphide from zinc salt solutions: $ZnSO_4 + (NH_4)_2S = ZnS + (NH_4)_2SO_4$. This ZnS is insoluble in acetic acid, but is soluble in dilute mineral acids and is used mixed with $BaSO_4$ as a white paint (lithopone, zincolith).

3. The precipitate of $Zn(OH)_2$ produced by alkali hydroxides is soluble in an excess of the alkali hydroxide.

4. The spark-spectrum shows a series of lines, amongst which those in the red and blue are most striking.

4. Cadmium.

Atomic weight $112.4 = Cd$.

Occurs only in combination in small amounts in the zinc ores, and rarely as greenockite, CdS.   It is obtained by distilling zinc-dust (p. 231) with charcoal and then redistilling the metal obtained.   It forms a white tough metal having a specific gravity of 8.6, melting at 315°, and boiling at 778° (zinc at 1040°).

a. Alloys of Cadmium.

See mercury and lead.

b. Compounds of Cadmium.

These are obtained in the same manner as the corresponding zinc compound.

Cadmium oxide, CdO, is obtained by heating cadmium nitrate. It is a brown microcrystalline powder.

Cadmium hydroxide, $Cd(OH)_2$ is precipitated from cadmium salt solutions by caustic alkalies or ammonia as a white powder which is soluble in ammonia but not in caustic alkali, and when heated decomposes into $CdO + H_2O$.

Cadmium sulphate, $3CdSO_4 + 8H_2O$, crystallizes only at $-20°$ with $7H_2O$, but forms double salts with alkali sulphates, corresponding to the magnesium, zinc, etc., salts (p. 229).

c. Detection of Cadmium Compounds.

1. On passing H_2S into a solution of cadimum salt a beautiful yellow precipitate of CdS is obtained. This is insoluble in $(NH_4)_2S$, which differs from the yellow sulphides of tin and arsenic. This sulphide is used as a pigment (cadmium yellow).

2. When heated with soda on charcoal the cadmium compounds give a brown crust of cadmium oxide, CdO.

3. The spark-spectrum gives characteristic bright lines.

SILVER GROUP.

Copper. Silver. Mercury.

These metals appear as monovalent, copper and mercury also as divalent. They do not decompose water even at a red heat, and are not dissolved by hydrochloric acid or dilute sulphuric acid, but are readily dissolved by nitric acid or concentrated sulphuric acid. Their sulphides are insoluble in dilute acids, and the halogen compounds of the monovalent series are insoluble in water. The hydroxides of copper decompose even at 100° in the presence of water, while the hydroxides of silver and mercury are not known. With alkali sulphates the sulphates of the metals of the divalent series form double salts which are isomorphous and correspond with those of the magnesium group.

1. Copper (Cuprum).

Atomic weight 63.6 = Cu.

Occurrence. Native in large quantities, especially in America and Siberia, often crystallized in cubes or octahedra. The most common copper ores are red copper ore (cuprite), Cu_2O, copper glance (chalcocite), Cu_2S, azurite, $2CuCO_3 + Cu(OH)_2$, malachite, $CuCO_3 + Cu(OH)_2$, chalcopyrite or copper pyrites, $Cu_2S + Fe_2S_3$, and the tetrahedrite group (Fahl ores) (p. 181). Traces of copper are found in most plants and animals.

Preparation. 1. Native copper is obtained from the ores by stamping them and then washing with water. The granular masses obtained are refined by melting in the presence of small quantities of carbon.

2. From its oxides and carbonates by reduction with carbon:
$$CuCO_3 + 2C = Cu + 3CO.$$

3. Copper can be obtained from the sulphides only with difficulty, as the iron and other impurities must first be removed.

a. The sulphides are roasted with air until a great part of the foreign metallic sulphides (p. 127) are transformed into oxides, while a part of the copper sulphide is also oxidized.

b. The mass thus obtained is fused with carbon and silicates, when the copper is reduced on account of its low affinity for oxygen, while the other oxides are in great part dissolved by the fused silicates, which are removed as a slag. The copper combines with the metallic sulphides still present and collects on the bottom of the furnace as coarse metal (containing about 32 per cent. Cu).

c. This is roasted in turn and again fused with carbon and silicates. A part of the foreign metallic oxides is here reduced to the metallic state; it no longer forms a slag, but unites with the metallic copper, forming the so-called black copper (containing about 95 per cent. Cu).

d. This black copper is fused on a blast-hearth, when the foreign metals oxidize more readily than the copper, separate on the surface, and are removed.

e. Or the black copper is suspended as the anode in a cupric sulphate solution and the copper dissolved by the electric current and deposited on the cathode, consisting of a pure copper plate, while the impurities separate out as a slime.

4. In the wet way, by converting the copper ore into cupric sulphate (which see) by roasting, or into cupric chloride by roasting with alkali chlorides and then precipitating the copper from these watery solutions by means of iron; thus, $CuSO_4 + Fe = FeSO_4 + Cu$; $CuCl_2 + Fe = FeCl_2 + Cu$.

5. Chemically pure copper can be obtained by reducing heated copper oxide in a current of hydrogen or by decomposing a copper salt solution by the electric current.

Properties. Red, very ductile and tough metal having a specific gravity of 8.9 and unacted upon in dry air, but becoming in moist air, coated with green basic carbonate (verdigris). On being heated in the air it is coated with a black layer of cupric oxide and melts at 1065°. It is insoluble in hydrochloric acid and dilute sulphuric acid, but gradually dissolves when moistened therewith and exposed to the air, when oxygen is absorbed. Copper dissolves in hot concentrated sulphuric acid, forming cupric sulphate with the production of sulphur dioxide (p. 124) and in dilute nitric acid with the generation of nitric oxide (p. 156). Metallic copper can be separated from its aqueous solutions by iron, zinc, and phosphorus.

Univalent copper forms cuprous compounds, and bivalent copper forms cupric compounds. All cupric compounds are blue when dilute, as they all have the same cation, Cu″, while the cation Cu′ is colorless as cuprous salts.

a. Alloys of Copper.

Copper cannot be cast, as on cooling it contracts irregularly and hence does not fill the mould. With zinc or tin copper forms alloys which can be cast.

Alloys of copper with zinc. Brass is a golden-yellow alloy of 2 parts copper and 1 part zinc. The more copper it contains the redder is the color, and the more zinc the whiter is the product. Such alloys are called pinchbeck or red metal, talmi-gold, Muntz metal, impure gold-leaf, so-called gold foam, etc.

Alloys of copper with tin are called bronzes. We differentiate between gun-metal, bell-metal, speculum-metal, antique bronze. Modern bronze consists of copper, tin, and zinc. Copper coins contain 95 per cent. copper, 4 per cent. tin, and 1 per cent. zinc. Phosphorus bronze contains 0.5–0.8 per cent. P and is very hard and stable. Silicium bronze contains 0.5–0.8 per cent. Si, is very hard, and is a good conductor of electricity.

Alloys of copper with aluminium, silver, nickel, manganese, and gold will be treated of in connection with these metals.

b. Cuprous Compounds.

Cuprous oxide, copper suboxide, Cu_2O, occurs as cuprite and may be obtained by heating a cupric salt solution with glucose in the presence of caustic alkali. It forms a red crystalline powder which is not changed in the air and is insoluble in water. See also cuprous salts.

Cuprous hydroxide, $Cu(OH)$, is obtained on gently heating an alkaline cupric salt solution (see below) with glucose, or by treating cuprous chloride with caustic alkali. It forms a yellow powder which quickly oxidizes to cupric hydroxide.

Cuprous sulphide, Cu_2S, occurs as copper glance, copper e , etc., and is produced on burning copper in sulphur vapors, as well as by heating cupric sulphide in the absence of air ($2CuS = Cu_2S + S$), or in a current of hydrogen.

Cuprous salts are known only of oxygen-free acids, as Cu_2O dissolves in oxygen acids, forming cupric salts, with the separation of metallic copper: $Cu_2O + H_2SO_4 = CuSO_4 + H_2O + Cu$. The cuprous salts are colorless, but quickly turn green or blue in the air, due to the absorption of oxygen.

The haloid cuprous salts are insoluble in water. Cuprous chloride dissolves readily in NH_3 or HCl, and both of these solutions absorb CO (p. 189).

c. Cupric Compounds.

Cupric oxide, copper oxide, CuO, obtained by heating cupric nitrate or cupric carbonate, or by precipitating a boiling cupric salt solution with hot caustic alkali or alkali carbonate solution. It is a black amorphous powder which does not change on heating, but which on heating with carbon, hydrogen, or organic substances gives off its oxygen; hence it is used in the analysis of organic bodies (Part III).

Cupric hydroxide, copper hydrate, $Cu(OH)_2$, is precipitated from cupric salt solutions by caustic alkalies as a blue amorphous mass which on heating, even under water, decomposes into black cupric oxide and water.

The presence of many organic substances prevents the precipitation of cupric salts by alkalies. A clear mixture of cupric sulphate with tartrates and caustic alkali is used as an alkaline copper solution, or Fehling's solution, for the detection of sugar (which see). Cupric oxide and cupric hydroxide are soluble in ammonia, the solution becoming deep blue. This solution is the only solvent for cellulose (Schweitzer's reagent).

Cupric sulphide, copper sulphide, CuS, is precipitated from solutions of cupric salts by H_2S as an amorphous brownish-black powder insoluble in dilute acids. When heated in the absence of air it decomposes into cuprous sulphide and sulphur, and when moist it oxidizes into cupric sulphate.

Cupric sulphate, Copper Vitriol, Blue Vitriol, $CuSO_4 + 5H_2O$. *Preparation.* 1. On a large scale by carefully roasting the natural copper sulphides, which are thereby converted into cupric sulphate, while the iron sulphide is in great part converted into iron oxide. On lixiviating with water a cupric sulphate solution comparatively free from iron is obtained (crude). This is purified by repeated recrystallization.

2. It may also be obtained pure by dissolving copper in concentrated hot sulphuric acid and evaporating, etc.:

$$Cu + 2H_2SO_4 = CuSO_4 + SO_2 + H_2O.$$

Properties. Large blue triclinic crystals which dissolve in 2.5 parts water. At 100° they lose 4 molecules of water of crystallization, and at 200° they become anhydrous and form a white powder, which becomes blue again with the slightest amount of water (detection of water in alcohol). On highly heating cupric sulphate it decomposes into cupric oxide, sulphur dioxide, and oxygen: $CuSO_4 = CuO + SO_2 + O$. With alkali sulphates cupric sulphate yields double salts which are analogous in composition and isomorphous with the magnesium salts, etc.:

$$CuSO_4 + K_2SO_4 + 6H_2O \quad (p.\ 229).$$

Basic cupric sulphate forms the color called Casselmann's green.

Copper alum is a fused mass of cupric sulphate, alum, camphor, and potassium nitrate.

Cupric ammonium salts, e.g.,

$$Cu{<}{NH_3 \atop NH_3}{>}SO_4 + xNH_3, \qquad Cu{<}{NH_3Cl \atop NH_3Cl} + xNH_3,$$

may be considered as ammonium salts in which 2H atoms are replaced by a divalent Cu atom. They contain a complex ion (p. 84) with a cupric atom (see Cobalt and Platinum) instead of a cupric ion.

If a cupric sulphate solution is treated with ammonia, a blue basic cupric sulphate is precipitated. This dissolves in an excess of ammonia, giving a deep-blue liquid from which deep-blue prisms of $(NH_3)_2CuSO_4 + 2NH_3 + H_2O$ are precipitated by alcohol.

On heating this salt to 150° it loses water and ammonia and a green cupric ammonium sulphate, $(NH_3)_2CuSO_4$, is obtained.

Cupric chloride, $CuCl_2$, forms pale-green needles with 2 mols. water, and is yellowish brown when anhydrous. It forms double salts with alkali chlorides.

Cupric arsenite, $CuHAsO_3$, is obtained by precipitating a cupric salt with potassium arsenite, as a beautiful green precipitate which is used as a pigment, called Scheele's green.

Cupric Carbonate, $CuCO_3$, is not known. If a cupric salt solution is treated with sodium carbonate, a precipitate of the green basic cupric carbonate, $3CuCO_3 + 3Cu(OH)_2 + H_2O$ or $CuCO_3 + Cu(OH)_2$, is obtained, dependent upon the temperature and concentration of the liquid. This compound forms the color Brunswick green and occurs as the beautiful green mineral malachite. Verdigris, patina, copper, rust, which is produced by the action of air and water upon copper or bronze, has the same composition. $2CuCO_3 + Cu(OH)_2$ occurs as the beautiful blue mineral azurite, which when powdered forms the artist's pigment known by that name.

d. Detection of Copper Compounds.

1. They give a green or blue color to the non-luminous flame. The spectrum of this flame contains many lines, but the blue and green lines are characteristic.

2. Sulphuretted hydrogen precipitates from solutions of copper brownish-black CuS, which is insoluble in dilute acids.

3. Ammonia precipitates greenish-blue basic salts which are soluble in excess (see above) from copper solutions. All copper com-

pounds with the exception of the sulphide are soluble with blue coloration in an excess of ammonia. With cuprous compounds the color appears only after standing.

4. Zinc and iron deposit metallic copper, which appears as a red coating on the metal.

5. Cuprous compounds are characterized by the insolubility of their halogen compounds.

2. Silver (Argentum).

Atomic weight 107.93 = Ag.

Occurrence. Often native in large pieces, but especially in combination as silver glance, argentite, Ag_2S, as silver-copper glance, stromeyerite, $Cu_2S + Ag_2S$, as stephanite, $Ag_2S + Ag_3SbS_3$, as proustite, Ag_3SbS_3, horn-silver, AgCl, and in the Fahl ores (p. 181). Lead and copper pyrites often contain small amounts of silver.

Preparation. Silver glance and similar ores, which contain no other metal, need only to be roasted and melted. All other ores containing silver require complicated methods of separation. The silver obtained from them always contains several per cent. of foreign metals.

1. Amalgamation Process. This is used in Mexico when there is lack of fuel.

   a. The ore is rubbed up with common salt, cupric sulphate, and water and roasted, when the silver is converted into silver chloride.

   b. The mass is shaken in barrels with water, iron filings, and mercury. The metallic iron converts the silver chloride into metallic silver: $2AgCl + Fe = 2Ag + FeCl_2$. The mercury dissolves the separated silver (silver amalgam) and collects at the bottom of the barrel, where it is drawn off. The mercury is distilled, leaving the silver as a residue.

2. The extraction method is used for copper ores containing silver. The ore or the copper-stone containing silver (p. 233) is roasted in the air, when first ferric sulphate and then cupric sulphate are formed, and finally at a higher temperature, when all the ferric sulphate and a part of the cupric sulphate are decomposed into oxides, silver sulphate is formed. The product thus obtained is extracted with hot water, which dissolves the silver sulphate and any cupric sulphate present, and the silver then precipitated from this solution by metallic copper, when cupric sulphate is obtained as a by-product: $Ag_2SO_4 + Cu = 2Ag + CuSO_4$.

3. The lead method is chiefly applied to obtain the silver that is always found in lead pyrites, the lead containing silver being first prepared from the ore (see Lead). Other ores poor in silver are melted with lead or lead glance, and the argentiferous lead prepared therefrom.

   If the lead obtained is rich in silver ($\frac{1}{2}$ per cent.), it is directly exposed to cupellation, which is based upon the fact that the molten lead oxidizes when exposed to the air, but the molten silver does not. The lead is fused and air is blown in by a blast; the lead is converted into oxide (litharge) which melts and passes off, leaving finally the unoxidized silver covered only with a thin layer of lead oxide, which is finally ruptured, and the shining silver remains behind.

If the lead contains only small amounts of silver, it is concentrated by Pattinson's or Parke's process. Pattinson's process is based upon the fact that a molten alloy of silver and lead solidifies slower than pure lead. Hence if this alloy is allowed to cool slowly, pure lead first crystallizes out, which can be removed by perforated ladles, and finally a eutectic mixture containing 2.2 per cent. silver remains.

Parke's process is based from the fact that on the addition of zinc to molten lead poor in silver the two metals do not mix well together, but the silver leaves the lead and goes over to the zinc, forming a difficultly fusible alloy, which on cooling separates on the surface. The zinc is separated therefrom by distillation.

Pure silver can be obtained from the crude silver by melting it with lead and then driving this off, or by dissolving the crude silver in HNO_3 and precipitating all the silver from solution as $AgCl$ by means of hydrochloric acid, and reducing this as described on p. 240.

Pure silver may moreover be obtained by dissolving the crude silver in sulphuric acid and precipitating the silver with copper or iron: $Ag_2SO_4 + Fe = FeSO_4 + Ag_2$; also by electrolysis, where the crude silver is suspended as the anode in a concentrated solution of silver nitrate, when pure silver deposits at the cathode, which consists of a sheet of silver.

Properties. White, shining, rather soft metal having a specific gravity of 10.5 and not oxidizing when molten, but absorbing 22 times its volume of oxygen, which it gives up on cooling: this causes the sputtering that takes place when fluid silver cools Ozone causes a superficial coating of silver peroxide (p. 240). Silver melts at 960° and vaporizes at about 2000°, forming a pale-blue vapor. Concentrated hydrochloric acid and dilute sulphuric acid do not attack silver, but hot concentrated sulphuric acid dissolves it, forming silver sulphate: $2Ag + 2H_2SO_4 = Ag_2SO_4 + SO_2 + 2H_2O$. Dilute nitric acid, even in the cold, dissolves silver, producing silver nitrate: $3Ag + 4HNO_3 = 3AgNO_3 + NO + 2H_2O$. Silver unites directly with the halogens; with sulphur it combines very readily; hence silver objects turn black when exposed to air containing H_2S. Silver is very ductile and is the best conductor of heat and electricity. Thin hammered sheets are called silver-foil.

Silver only occurs univalent, hence the designation argentous and argentic salts is not used. The silver compounds have the same composition as the cuprous compounds.

There are also certain allotropic modifications of silver, namely, a bluish-green and a golden-yellow form, which are very quickly transformed into ordinary silver.   On the reduction of silver chloride we obtain a fine gray powder, so-called molecular silver, while on heating silver citrate we obtain the so-called colloidal silver as a brownish-black powder which forms a deep-red pseudo-solution with water (p. 53)

a. Alloys of Silver.

Silver is too soft to be worked as such.   A small addition of copper makes it harder without changing its white color.   Ordinary silverware contains 12 parts silver in every 16 parts, or 75 per cent. At the present time we generally express the amount of silver in the alloy in parts per thousand.   German, French, Austrian, and United States silver coins have a fineness of $\frac{900}{1000}$, and the English coins $\frac{925}{1000}$ or sterling.

b. Compounds of Silver.

Silver **oxide**, Ag_2O, is obtained by treating a silver salt solution with caustic alkali, producing a dark-brown amorphous precipitate, which when moist is a strong base and behaves like AgOH.   On heating it splits into silver and oxygen.

Silver **hydroxide**, $Ag(OH)$, is obtained as a white powder on mixing an alcoholic solution of a silver salt with alcoholic caustic alkali at $-40°$. At ordinary temperatures it decomposes into $Ag_2O + H_2O$.

Silver **peroxide**, Ag_2O_2, is formed by the action of ozone upon Ag_2O. It is a black crystalline powder.

Silver **sulphide**, Ag_2S, occurs as octahedrons in silver glance and is produced by passing H_2S into a silver salt solution.   It is an amorphous black precipitate which is insoluble in dilute acids.

Silver chloride, chloride of silver, AgCl, occurs in octahedra as hornsilver, and is obtained by precipitating a silver salt solution with hydrochloric acid or a soluble chloride.   It forms an amorphous, white, cheese-like precipitate:

$$AgNO_3 + HCl = AgCl + HNO_3;$$
$$AgNO_3 + KCl = AgCl + KNO_3.$$

When exposed to light it turns violet and then black, when it decomposes into chlorine and silver subchloride, Ag_2Cl.   Silver chloride is insoluble in acids, but readily soluble in ammonia, potassium cyanide, and sodium hyposulphite.   Nascent hydrogen (zinc and HCl) readily reduces it into the metal, and fusion with alkali carbonate produces the same change.   The silver carbonate first formed is decomposed by the heat with separation of metallic silver: $2AgCl + Na_2CO_3 = Ag_2CO_3 + 2NaCl$; $Ag_2CO_3 = 2Ag + CO_2 + O$.

Silver bromide, AgBr, is obtained by treating a silver salt solution with HBr or a bromide, producing a yellowish-white precipitate having the same properties as silver chloride, although not as soluble in ammonia.

Silver iodide, AgI, is obtained by treating a silver salt solution with **HI or an** iodide, producing a yellowish precipitate having similar properties as silver chloride, but insoluble in ammonia, and is sensitive to light in the presence of silver nitrate.

The art of photography depends upon the fact that light causes a transformation of the haloid salts of silver into silver subhaloid salts, i.e., Ag_2Cl, which by certain reducing agents (developers) are readily reduced to silver, while the unchanged haloid silver is not acted upon by the developer. After development of the picture, in order to make it permanent it must be "fixed" by placing it in a solution of compounds (generally sodium hyposulphite), which dissolves the undecomposed and still sensitive haloid silver compounds. At the present time plates covered with silver bromide (by the aid of gelatine) are chiefly used, as this is more stable and more sensitive to light than plates covered with silver iodide and chloride. Silver iodide can be used only when wet; while silver chloride is generally used for printing-paper.

Silver hydrazoate, N_3Ag, and

Silver amid, $2NH_2Ag + H_2O$, silver fulminate, produced when freshly precipitated silver oxide is treated with a concentrated solution of ammonia, both explode when dry with slight friction.

Silver nitrate, $AgNO_3$, is prepared by dissolving silver in nitric acid (p. 239) and evaporating the solution. It forms colorless rhombic crystals which dissolve in 1 part by weight of water or 4 parts of alcohol. It has a caustic action and blackens in contact with organic substances (skin, linen, etc.). Silver nitrate melts at 200°, and on further heating it decomposes into silver nitrite, $AgNO_2$, and oxygen, and finally into silver, nitrogen, and oxygen.

It is moulded when hot into sticks which are called lunar caustic, *Lapis infernalis.* On fusion with two parts of potassium nitrate we obtain mitigated caustic.

c. Detection of Silver Compounds.

1. When fused on charcoal with soda they yield a silver globule.

2. All silver compounds with the exception of the iodide and sulphide are readily soluble in ammonia.

3. Hydrochloric acid or a soluble chloride precipitates white, cheese-like silver chloride from solutions of silver. This precipitate is soluble in ammonia, insoluble in acids, and darkens in the light.

4. Metallic zinc, iron, copper, and mercury precipitate metallic silver from its solutions.

3. Mercury (Hydrargyrum or Quicksilver).

Atomic weight $200 = Hg$.

Occurrence. Native to ·a slight extent, but chiefly as cinnabar (HgS) in Almaden, Idria, California, etc. It is also found in many Fahl ores (p. 181).

Preparation. Generally by roasting cinnabar with access of air: $HgS + 2O = Hg + SO_2$. The vapors of mercury formed are passed into a cooled chamber, where they condense.

Pure mercury is obtained by treating the commercial quicksilver with a ferric chloride solution or cold dilute nitric acid, which dissolve the contaminating metals more readily than the mercury. The mercury may also be purified by squeezing it through leather and then distilling in vacuum. When pure it flows over smooth surfaces as shining, round drops, but when impure it forms mat, elongated drops.

Properties. Mercury is the only liquid metal; at ordinary temperatures it is silver-white, has a specific gravity of 13.5, solidifies at ⋅ $-39°$, and boils at $357°$, but volatilizes even at ordinary temperatures, yielding vapors which are poisonous. It is unchanged in the air, but if it is heated in the neighborhood of its boiling-point it turns into red crystalline mercuric oxide (p. 244). Mercury is insoluble in hydrochloric acid and cold sulphuric acid; moderately warm nitric acid dissolves an excess of mercury, forming mercurous nitrate: $3Hg + 4HNO_3 = 3HgNO_3 + 2H_2O + NO$; white hot nitric acid in excess dissolves mercury, forming mercuric nitrate: $3Hg + 8HNO_3 = 3Hg(NO_3)_2 + 4H_2O + 2NO$. Sulphuric acid also forms mercurous or mercuric sulphate under the same conditions:

$$2Hg + 2H_2SO_4 = Hg_2SO_4 + 2H_2O + SO_2;$$
$$Hg + 2H_2SO_4 = HgSO_4 + 2H_2O + SO_2.$$

On shaking with air, water, etc., or by rubbing with sugar, fat, etc., mercury can be obtained finely divided, so that it forms a gray powder which prevents the coalescing of the layers of air, water, fat, etc., between the drops. On rubbing mercury with two parts fat we obtain gray mercury salve. Mercury is also known as a black powder which gives a brown pseudo-solution with water (p. 53) and which is used in medicine as hyrgol, or colloidal mercury.

The compounds of mercury are mostly poisonous and have an analogous composition and nomenclature as the copper compounds.

a.·Alloys of Mercury

are called *amalgams* and are obtained by the direct union with the metals. Most metals dissolve in mercury even in the cold, the alkali metals dissolving with the production of a flash of light. Amalgams are also obtained when mercury is added to a metallic salt solution or a metal placed in a mercuric nitrate solution. The amalgams richest in mercury are liquid, while those poor in mercury are solid and often crystalline, many retaining mercury even when heated to 450°.

Tin amalgam is used in the preparation of mirrors; tin-zinc amalgam, for the frictional-electric machine; cadmium, platinum, or copper amalgams, as filling for teeth. Ammonium amalgam, see p. 216. The amalgams of the alkali metals, of aluminium or magnesium evolve a constant current of H with water and hence are used as reducing agents.

b. Mercurous Compounds.

Mercurous oxide, mercury suboxide, Hg_2O, is obtained as a brownish-black powder when a mercurous salt solution is treated with caustic alkali: $2HgCl + 2KOH = Hg_2O + 2KCl + H_2O$. It decomposes in the light into $HgO + Hg$.

Black wash is prepared by mixing calomel with lime-water and consists, therefore, of finely divided mercurous oxide suspended in lime-water

Mercurous hydroxide, $Hg(OH)$, is not known, as it decomposes into water and mercurous oxide the moment it is produced from mercurous salts.

Mercurous sulphide, Hg_2S, is precipitated from mercurous salt solutions by H_2S as a black powder which quickly decomposes into mercuric sulphide and mercury: $2HgNO_3 + H_2S = Hg_2S + 2HNO_3$; $Hg_2S = HgS + Hg$.

Mercurous Chloride, Calomel, Mercury Sub- or Protochloride, HgCl. *Preparation.* 1. By treating a mercurous nitrate solution with hydrochloric acid or a solution of a chloride we obtain mercurous chloride as a pure, white amorphous precipitate: $HgNO_3 + HCl = HgCl + HNO_3$.

2. By the sublimation of mercuric chloride with mercury: $HgCl_2 + Hg = 2HgCl$. The calomel is obtained by this method as a radiated crystalline yellowish-white mass which when rubbed forms a yellowish powder, and when scratched gives a yellow mark.

If in the sublimation the mercurous chloride vapors are quickly cooled by passing them into a cool chamber or by condensing them by means of steam, we obtain a very fine mercurous chloride which appears as a pure white microcrystalline powder.

3. By reducing a warm solution of mercuric chloride by sulphur dioxide we obtain a white crystalline precipitate of mercurous chloride:

$$2HgCl_2 + SO_2 + 2H_2O = 2HgCl + 2HCl + H_2SO_4.$$

Properties. Mercurous chloride is insoluble in water, alcohol, and dilute acids, but soluble in concentrated acids, forming mercuric salts. On boiling with hydrochloric acid or with solutions of chlorides of the alkali group it is transformed into the very poisonous mercuric chloride and mercury separates out. On heating it vaporizes without melting.

By caustic alkalies it is converted into the brownish-black mercurous oxide, and when treated with ammonia it becomes black (hence its name, καλός, beautiful, μέλας, black), when probably **Mercurous ammonium chloride,** $NH_2(Hg_2)Cl$, is formed:

$$2HgCl + 2NH_3 = NH_2(Hg_2)Cl + NH_4Cl.$$

Mercurous Iodide, HgI. *Preparation.* By treating a dilute mercurous nitrate solution with potassium iodide solution, or ordinarily by rubbing mercury and iodine together in atomic proportions.

Properties. Amorphous greenish-yellow powder insoluble in water and alcohol and gradually decomposing in the light into mercuric iodide and mercury, and quickly on heating: $2HgI = HgI_2 + Hg$. An aqueous solution of potassium iodide produces the same change.

It is produced as crystalline yellow leaves by heating mercurous nitrate solution with iodine: $2HgNO_3 + I = HgI + Hg(NO_3)_2$. On heating the amorphous and crystalline salt it sublimes into red, needle-shaped crystals which quickly turn yellow again.

Mercurous sulphate, Hg_2SO_4 (preparation. p. 242), precipitates also from concentrated mercurous nitrate solutions by the addition of H_2SO_4 as colorless crystals which are difficultly soluble and which decompose on heating into $Hg_2 + SO_2 + O_2$, and with cold water form greenish-yellow insoluble basic mercurous sulphate, $Hg_2O + Hg_2SO_4 + H_2O$.

Mercurous nitrate, $HgNO_3$, is produced by the action of cold or slightly warmed nitric acid upon an excess of mercury (process, p. 242). It forms colorless monoclinic crystals, $HgNO_3 + H_2O$, which are soluble in a small amount of warm water, but which decompose on the addition of more water into the soluble acid salt and into the pale-yellow insoluble basic salt, $HgOH.HgNO_3$. If the aqueous solution is treated with ammonia, a black precipitate of **Mercurous ammonium nitrate,** $NH_2(Hg_2)NO_3$, mixed with mercurous oxide is obtained. This precipitate is used in pharmacy.

c. Mercuric Compounds.

Mercuric Oxide, Mercury Oxide, HgO. *Preparation.* 1. Mercury is heated nearly to its boiling-point for a long time in the air. On a large scale a mixture of mercuric nitrate and mercury are heated until no more red vapors of nitrogen oxides are evolved: $Hg(NO_3)_2 + 3Hg = 4HgO + 2NO$. Thus produced it forms a red crystalline powder (red precipitate) which does not combine with oxalic acid.

2. If a mercuric salt solution is treated with caustic alkali, an orange-yellow amorphous powder is obtained which when shaken with an oxalic acid solution forms white mercuric oxalate: $HgCl_2 + 2KOH = HgO + 2KCl + H_2O$.

Properties. It is insoluble in water, readily soluble in acids. When carefully heated it becomes red, then black, and on cooling red again. At a red heat it decomposes into its elements, and with ammonia it combines, forming white $2HgO + NH_3$, which explodes on heating.

Yellow wash is prepared by mixing mercuric chloride with lime-water and hence consists of finely divided mercuric oxide suspended in lime-water.

Mercuric hydroxide, $Hg(OH)_2$, is not known because as soon as it is formed it decomposes into mercuric oxide and water:

$HgCl_2 + 2KOH = Hg(OH)_2 + 2KCl; \quad Hg(OH)_2 = HgO + H_2O$.

Mercuric Sulphide, HgS. *Occurrence.* As cinnabar in dark-red crystals or masses.

Preparation. 1. Black microcrystalline mercuric sulphide can be obtained by passing H_2S into a mercuric salt solution, $HgCl_2 + H_2S = HgS + 2HCl$, or as a black powder by the continuous rubbing of quicksilver and moist sulphur together.

2. Red crystalline mercuric sulphide is produced by prolonged heating of black HgS with alkali sulphides and water, forming a beautiful scarlet-red powder. On the sublimation of black HgS in the presence of air a dark-red crystalline mass is obtained which is similar to natural cinnabar and which on being ground gives a scarlet-red non-poisonous powder (vermilion, Chinese red).

Properties. Both modifications are insoluble in water, alcohol, hydrochloric or nitric acids, but readily soluble in aqua regia, forming mercuric chloride. On heating in the air they decompose into sulphur dioxide and mercury (p. 242).

Mercuric Chloride, Corrosive Sublimate, Mercury Bichloride, $HgCl_2$. *Preparation.* 1. On heating mercury in chlorine gas or by dissolving mercury in aqua regia and evaporating to crystallization.

2. By heating sodium chloride with mercuric sulphate, when the mercuric chloride sublimes, while the sodium sulphate remains: $2NaCl + HgSO_4 = Na_2SO_4 + HgCl_2$.

Properties. Sublimed it forms a white crystalline mass, and when ground a white powder (mercurous chloride gives a yellowish powder). It is soluble in 16 parts of water, 3 parts of alcohol, 14 parts of ether, and crystallizes from these solutions in rhombic prisms. Reducing bodies (SO_2, p. 243, $SnCl_2$, p. 258) convert it into mercurous chloride and then into mercury. With alkali chlorides it forms stable neutral double salts, $HgCl_2 + 2KCl + H_2O$, which are soluble in water. It is very poisonous and prevents putrefaction of organic bodies by forming insoluble compounds with the proteids thereof. As mercuric chloride is precipitated by proteid solutions, these bodies are used as antidotes in mercurial poisoning.

Tablets composed of equal parts $HgCl_2$ and NaCl are used in surgery.

Mercuric ammonium chloride, NH_2HgCl (NH_4Cl in which 2 H atoms are replaced by the divalent Hg atom), is formed by precipitating mercuric chloride solution with an excess of ammonia solution: $HgCl_2 + 2NH_3 = NH_4Cl + NH_2HgCl$. It forms a white mass or an amorphous powder which is insoluble in water and alcohol, but readily soluble in acids.

On heating it volatilizes without melting, but decomposes; hence it is called also white infusible precipitate, in distinction from the fusible precipitate $ClH_3N-Hg-NH_3Cl$.

Mercuric Iodide, Biniodide of Mercury, Red Iodide, HgI. *Preparation.* 1. In a manner analogous to that of mercurous iodide by rubbing atomic amounts of mercury and iodine.

2. Ordinarily by precipitating mercuric chloride with a potassium iodide solution: $HgCl_2 + 2KI = HgI_2 + 2KCl$.

Properties. An amorphous powder which at first is yellow, then turns a beautiful red; it is insoluble in water, readily soluble in alcohol, potassium iodide, and mercuric chloride. It crystallizes from alcoholic solutions as red quadrioctahedra.

On sublimation or on heating to 126° it is converted into a yellow rhombic modification which quickly, especially in the light, is transformed again into the red modification.

Nessler's reagent is a solution of mercuric iodide in potassium iodide made alkaline with caustic alkali. It is used in the detection of the smallest traces of ammonia or its salts, as it gives a brown precipitate or coloration of dimercuric ammonium iodide, $NHg_2I + H_2O$.

Mercuric sulphate, $HgSO_4$, (preparation, p. 242), forms colorless crystals which by boiling water are converted into a basic mercuric sulphate, $HgSO_4 + 2HgO$. With alkali sulphates it forms double salts, $HgSO_4 + K_2SO_4 + 6H_2O$, which are isomorphous with those of the magnesium group.

Mercuric nitrate, $Hg(NO_3)_2$, is obtained by dissolving mercury in an excess of hot nitric acid (process, p. 242). It forms colorless crystals, $2Hg(NO_3)_2 + H_2O$, which with large quantities of water form a white basic salt, $Hg(NO_3)_2 + 2HgO + H_2O$. It is used in the estimation of urea, according to Liebig-Pflüger, as well as a reagent for proteids (Millon's reagent).

d. Detection of Mercury Compounds.

1. If they are heated in a glass tube with Na_2CO_3, a gray sublimate of metallic mercury deposits on the cold part of the tube.

2. When placed on metallic copper mercury deposits as a gray coating which becomes bright on being rubbed, and which disappears on heating.

3. Sulphuretted hydrogen precipitates black mercuric sulphide from solutions containing mercury. This precipitate is characterized from all other sulphides by being insoluble in hot nitric acid.

4. Stannous chloride precipitates white mercurous chloride or black (finely divided) mercury (P. 258) from mercurial solutions.

5. Mercurous compounds turn black with caustic alkalies, black with ammonia (p. 244), yellowish green with potassium iodide, and white with hydrochloric acid.

6. Mercuric compounds become yellow with caustic alkalies, white with ammonia (P. 246), red with potassium iodide, and are not changed with hydrochloric acid.

GROUP OF EARTHY METALS.

ALUMINIUM. GALLIUM. INDIUM. THALLIUM.
Scandium. Yttrium. Lanthanum. Cerium.
Praseodymium. Neodymium. Samarium. Gadolinium.
Erbium. Thulium. Ytterbium.
Decipium. Dysprosium. Holmium. Philippium. Terbium.

The earthy metals occur only combined in nature and, with the exception of gallium, thallium, and cerium, exist as trivalent elements.

The preparation of pure earthy metals is very difficult, as in their chemical behavior they are extremely similar. They are especially characterized by their spectrum, and many of their oxides (especially cerium dioxide, CeO_2, see thorium, p. 263) emit an intense white light on being heated in a non-luminous flame.

Their salts and their solution are partly colored, and each metal gives a characteristic absorption spectrum.

They differ from the alkali and alkaline-earth metals in that their hydroxides are precipitated by ammonia and that they decompose water only at high temperatures. They are not, with the exception of indium, thallium, and gallium, in common with these two groups, precipitated by H_2S, as their sulphides are soluble or unstable.

Aluminium, gallium, indium form soluble salts with oxalic acid. Their sulphates form double salts having the formula $MK(SO_4)_2$, called alums, with the alkali sulphates and ammonium sulphate. These are soluble' in water and form octahedral crystals.

Thallium shows a different behavior (P. 255). The elements of the second and following series are called rare earthy metals and occur in certain rare minerals found in Sweden, Norway, and America, such as cerite, euxenite, gadolinite, orthite, æschynite, fergusonite, monazite, polycrase, yttrotantalite, samarskite, and especially in monazite sand.

They form difficultly soluble or insoluble salts with oxalic acid, and their soluble sulphates yield double salts with potassium sulphate, having the formula $MK_3(SO_4)_3$, of which those of scandium, samarium, thulium, decipium, erbium, holmium, philippium, and terbium are insoluble in a saturated potassium sulphate solution. The elements of the last series consist perhaps of a mixture of still unknown elements.

1. Aluminium.

Atomic weight 27.1 = Al.

Occurrence. Only combined. To a slight extent as aluminium oxide, sulphate, and hydroxide (which see), but in large quantities as aluminium silicate; thus feldspars, micas, zeolites, chlorites, and many other minerals are compounds of aluminium silicate (which see) with other metallic silicates, forming granite, porphyry, gneiss, mica schist, slate, clay, the chief constituents of the earth's crust. Andalusite, obsidian (pumice-stone), garnets, lapis-lazuli, topaz, tourmalin, etc., consist chiefly of aluminium silicates. Cryolite, $AlF_3 + 3NaF$, forms enormous deposits in Greenland. Despite this wide distribution of aluminium compounds they occur in only a few plants and not at all in the animal kingdom.

Preparation. 1. By heating aluminium chloride or cryolite with sodium: $AlCl_3 + 3Na = Al + 3NaCl$.

2. By the electrolysis of aluminium oxide (alumina) which is dissolved in fused cryolite.

Properties. Silver-white ductile metal of a specific gravity of 2.6 and melting at 700° without oxidation. On heating higher it burns without volatilizing, producing aluminium oxide. When finely powdered or in thin leaves it burns readily in the air with a bright light and decomposes boiling water with the generation of hydrogen. When compact it does not decompose water even at a white heat; at its melting-point it is an energetic reducing agent; therefore it is used in the obtainment of metals which can otherwise only be reduced from their oxides in the electric furnace.

The heat of reaction set free in its union with oxygen is used in order to obtain temperatures up to 3000° (by the compound called thermite, which is a mixture of powdered aluminium with iron oxide), and is also used in welding and melting processes, as well as in reduction processes which do not take place at lower temperatures. This reduction is brought about according to Goldschmidt's method, which consists in mixing the oxide of the metal to be obtained with aluminium powder, and to ignite this mixture (reduction mass) by means of a mixture of aluminium and BaO_2 (ignition mass). By this method numerous metals which were formerly fused and reduced with difficulty are now obtained in a very pure and fused state.

Aluminium dissolves readily in hydrochloric acid or caustic alkali with the development of hydrogen: $Al + KOH + H_2O = KAlO_2 + 3H$. It dissolves in hot sulphuric acid with the generation of SO_2, and it is only slowly acted upon by dilute sulphuric acid and nitric acid, as it becomes covered with a thin layer of Al_2O_3 besides H or NO. Under the air-pump receiver it dissolves readily in all acids, as the protective layer of gases is removed. On account of its lightness and its stability in the air it is very extensively used in the preparation of many useful objects.

a. Alloys of Aluminium.

The copper alloy, aluminium bronze, is characterized by being very hard and stable and by its golden color.

Magnalium is an alloy with magnesium and has properties similar to brass.

Partinium is an alloy with tungsten and is much more stable than aluminium.

b. Compounds of Aluminium.

Aluminium Oxide, Alumina; Al_2O_3. *Occurrence.* In colorless, transparent hexagonal crystals it is corundum; colored red with chromium it is called ruby; colored blue by cobalt it is known as sapphire; and the bluish-gray crystalline masses are called emery.

Preparation. Amorphous by heating aluminium hydroxide; crystalline by heating boron trioxide with aluminium fluoride, $2AlF_3 + B_2O_3 = Al_2O_3 + 2BF_3$; also by passing chlorine over red-hot sodium aluminate, $2NaAlO_2 + 2Cl = Al_2O_3 + 2NaCl + O$, as well as in Goldschmidt's process (see above).

Properties. Next to diamond and boron crystalline alumina is the hardest of all bodies and forms colorless hexagonal prisms. Amorphous alumina is a white odorless and tasteless powder which melts in the oxyhydrogen flame and then forms a very hard mass similar to burnt clay. When calcined it is, like crystalline alumina, insoluble in acids and can only be converted into a soluble compound by fusion with caustic potash or potassium bisulphate (p. 206).

Aluminium Hydroxide, $Al(OH)_3$. *Occurrence.* It occurs as hydrargillite, combined with $AlK(SO_4)_2$ as alum-stone (p. 251); as bauxite, $Al_2O(OH)_4$, mixed with iron oxide; and as $AlO(OH)$, diaspore.

Preparation. 1. $Al(OH)_3$ is precipitated from aluminium salt solutions by ammonia. Sodium carbonate also precipitates $Al(OH)_3$ from aluminium salt solutions, as the aluminium carbonate first formed decomposes immediately:

$$2AlCl_3 + 3Na_2CO_3 = 6NaCl + Al_2(CO_3)_3;$$
$$Al_2(CO_3)_3 + 3H_2O = 2Al(OH)_3 + 3CO_2.$$

2. On a large scale from cryolite or bauxite.

Cryolite is heated with limestone, when soluble sodium aluminate and insoluble calcium fluoride are formed: $AlF_3.NaF + 3CaCO_3 = Na_3AlO_3 + 3CaF_2 + 3CO_2$. Bauxite is heated with sodium carbonate: $Al_2O(OH)_4 + 3NaCO_3 = 2Na_3AlO_3 + 2H_2O + 3CO_2$. The Na_3AlO_3 is dissolved in water, and the carbon dioxide generated passed into the solution, which forms sodium carbonate and precipitates aluminium hydroxide: $2Na_3AlO_3 + 3CO_2 + 3H_2O = 3Na_2CO_3 + 2Al(OH)_3$. On treatment with water $Al(OH)_3$ remains undissolved and the sodium carbonate is obtained from the aqueous solution by evaporation.

3. As a by-product in the preparation of potassium alum from alum-stone (p. 252).

Properties. White gelatinous precipitate which when dry forms a white powder and which is insoluble in water; still by dialysis its solution in aluminium chloride and also in water can be obtained. It is soluble in acids as well as in caustic alkalies: $Al(OH)_3 + KOH = KAlO_2 + 2H_2O$. It acts like other weak bases, being an acid towards stronger bases, forming salts that are called aluminates (see below). On careful warming the $Al(OH)_3$ can be transformed into the above-mentioned hydrate, $Al_2O(OH)_4 = 2Al(OH)_3 - H_2O$ and $AlO(OH) = Al(OH)_3 - H_2O$.

When freshly precipitated it has the property of carrying down with it dissolved inorganic and organic bodies. This is made use of in the clarification of many liquids, and in the precipitation of organic pigments from their solution (preparation of lake colors), as well as in dyeing, where it is employed as a mordant (fixing of colors in plant-fibres), whereby the $Al(OH)_3$ is deposited directly upon the tissues by dipping first in aluminium acetate or sodium aluminate and then exposing to steam, when $Al(OH)_2$ is formed.

Aluminates. On treating solutions of aluminium hydroxide in alkali hydroxides with alcohol a precipitate having the composition $KAlO_2$ or $NaAlO_2$ is obtained. The acid-forming aluminium hydrate, $HAlO_2$, corresponding thereto occurs as diaspore, $Mg(AlO_2)_2$ as spinel, $Be(AlO_2)_2$ as chrysoberyl, $Zn(AlO_2)_2$ as gahnite, and $Fe(AlO_2)_2$ as pleonaste.

Sodium aluminate is prepared on a large scale (see Aluminium Hydroxide, p. 250).

The aluminates soluble in water are immediately decomposed by carbon dioxide into aluminium hydroxide.

Aluminium chloride, $AlCl_3$, is obtained by dissolving aluminium hydroxide in hydrochloric acid and evaporating. It forms white deliquescent crystals, $AlCl_3 + 6H_2O$, which decompose on heating: $2AlCl_3 + 3H_2O = Al_2O_3 + 6HCl$. With metallic chlorides it forms double salts whose solution can be evaporated without decomposition: $AlCl_3 + 3KCl$. On a large scale the anhydrous aluminium chloride can be prepared by strongly heating a mixture of aluminium oxide and carbon in chlorine gas. This forms a white deliquescent crystalline mass: $Al_2O_3 + 3C + 6Cl = 2AlCl_3 + 3CO$.

Aluminium fluoride, AlF_3, occurs in cryolite (p. 248), and with $9H_2O$ forms colorless crystals which are soluble in hot water.

Aluminium sulphide, Al_2S_3, is obtained by heating aluminium with sulphur, which gives a yellow mass decomposable by water.

Aluminium Sulphate, $Al_2(SO_4)_3 + 18H_2O$, also called concentrated alum, occurs as alunogen, feather-alum, aluminite, websterite, as well as alum-stone (p. 252).

Preparation. 1. By dissolving aluminium hydroxide (obtained on a large scale from bauxite or cryolite, p. 250) in sulphuric acid and then evaporating.

2. Alum-slate or alum-earth (aluminium silicate which contains carbon and iron pyrites) or ordinary clay (p. 252) are transformed into aluminium sulphate by means of sulphuric acid, then lixiviated with water, which leaves the silicic acid behind. The solution is evaporated when most of the difficultly soluble iron sulphate first crystallizes out and then the aluminium sulphate.

Properties. Aluminium sulphate forms colorless monoclinic crystals or crystalline masses which are readily soluble and have an acid reaction.

If a solution of aluminum sulphate is treated with a solution of a sulphate of the alkali metals, of ammonium, silver, or thallium, we obtain on evaporation the double salts, which crystallize in regular octahedra, are much less soluble in water than the aluminium sulphate, and have the composition $MAl(SO_4)_2 + 12H_2O$, where M may be K, Na, Cs, Rb, Ag, Tl, or NH_4, or an organic derivative of NH_4. These double salts are called alums and, according to the monovalent metal they contain, are designated potassium alum, sodium alum, ammonium alum, silver alum, etc.

Ferric, manganic, and chromic sulphates and those of indium and gallium also form double salts with the sulphates of the monovalent metals. These double salts have the same shape and composition, hence the term alums is also applied to these double salts.

Alums are hence isomorphous double salts having the formula $\overset{I}{M}\overset{III}{M}(SO_4)_2 + 12H_2O$, where $\overset{I}{M}$ may be K, Na, Cs, Rb, Ag, Tl, or NH_4; and $\overset{III}{M}$ may be Al, Fe, Mn, Cr, In, or Ga:

> $FeNa(SO_4)_2 + 12H_2O$, Sodium iron alum.
> $CrNH_4(SO_4)_2 + 12H_2O$, Ammonium chromium alum.
> $MnK.(SO_4)_2 + 12H_2O$, Potassium manganese alum.

Ammonium alum, $Al(NH_4)(SO_4)_2 + 12H_2O$, and

Potassium alum, alum, $AlK(SO_4)_2 + 12H_2O$, are the most important of these double salts and are at the present time more and more replaced in technology by aluminium sulphate or sodium aluminate (preparation, see Alumiunium Hydroxide, p. 250).

Preparation. 1. From alum-stone (alunite), $AlK(SO_4)_2 + 2Al(OH)_3$, by heating and then extraction with water, when the aluminium hydroxide remains undissolved and the potassium alum is obtained from the solution by evaporation and crystallization.

2. On a large scale from the aluminium sulphate obtained from alum slate, bauxite, or cryolite, by treating its solution with potassium or ammonium sulphate and evaporating to crystallization.

Properties. Large colorless octahedral crystals soluble in 10 parts water, giving an acid reaction to the solution, and which on heating fuse in their water of crystallization, and on further heating are converted into anhydrous white porous alum (burnt alum). From its hot solution in water treated with some alkali carbonate alum crystallizes on evaporation in cubes as so-called neutral or cubical alum, which is soluble in water with neutral reaction.

Aluminium Silicates. *Occurrence.* As constituent of the most important rocks of the earth (see Aluminium, p. 248), also in the impure form as alum-slate, alum-stone, and as ordinary clay, and in the pure state as white clay (kaolin, porcelain earth, argilla): $H_2Al_2(SiO_4)_2 + H_2O$. Ordinary clay forms immense layers and is produced by the weathering of feldspathic rocks, and is a mixture of white clay with other silicates, calcium carbonate, ferric hydroxide, etc.

By the influence of water and carbon dioxide the feldspars, $MAl_2(SiO_4)_2$ or $MAl_2Si_6O_{16}$ ($M = Na_2$, K_2, Mg, Ca), are so decomposed that the alkaline earths are converted into soluble bicarbonates and the alkalies into soluble silicates (which are further decomposed into carbonate and free silicic acid), while the insoluble aluminium silicate remains as clay:

$$K_2Al_2Si_6O_{16} + H_2O = K_2Si_4O_9 + H_2Al_2(SiO_4)_2;$$
$$CaAl_2(SiO_4)_2 + 2H_2O + 2CO_2 = CaH_2(CO_3)_2 + H_2Al_2(SiO_4)_2.$$

Ochre, Sienna earth, Naples red, are natural mixtures of clay with considerable ferric oxide. Loam is a mixture of clay, sand, and ferric oxide, while marl is a mixture with considerable calcium carbonate.

Properties. Aluminium silicates are partly decomposable by acids, while some must first be decomposed (p. 197). The clays are gray, brown, or yellow in color and absorb water and then form plastic masses which shrink on heating, when their hardness increases, so that they yield sparks when struck with steel. Clays lose the property of being plastic with water after heating and become porous and allow water to pass through. The purer the clay the more infusible it is, and mixtures of clay and lime, iron oxide, lead oxide, alkali salts are more or less fusible and impervious to water.

Porcelain and earthenware consist of burnt clay to which have been added bodies (feldspar, for example) which accelerate fusion, or the clay may be covered with a thin layer of readily fusible silicates which produce an impervious covering (glaze).

According to the purity of the clay and the temperature used in its preparation we differentiate between:

a. Compact earthenware with glass-like fracture and more or less transparent. Although these are impervious to water by the flux added before burning, they are still covered with a glaze in order to make the rough surface smooth and polished. Porcelain is white and transparent, while earthenware (stoneware) is white, gray, yellow, or brown, and non-transparent. Unglazed porcelain is called bisque.

b. Porous earthenware, which is earthy on fracture and completely non-transparent. It absorbs water and sticks to the tongue. To this group belong Fayence ware, crockery, Majolica ware, Delft ware, ordinary pottery, terra cotta, bricks, Hessian crucibles, etc.

Cement (p. 220) is also artificially made by heating clay with limestone and is a calcium-aluminium silicate.

Ultramarine. By heating a mixture of porcelain clay, wood charcoal, soda, and sulphur in the absence of air, we obtain green ultramarine, which forms blue ultramarine on heating again with sulphur in the air.

Violet and red ultramarine are produced when dry hydrochloric acid gas and air are passed over blue ultramarine at 150°.

Dilute acids decolorize ultramarine with the setting free of H_2S, sulphur, and gelatinous silicic acid. It probably consists of sodium-aluminium silicate, $Na_2Al_2(SiO_4)_2$, and sodium polysulphides.
The rare blue mineral lapis-lazuli has perhaps a similar composition.

c. Detection of Aluminium Compounds.

1. When moistened with a cobaltous salt solution and heated upon charcoal in the blowpipe flame they give a beautiful blue cobalt aluminate, $Co(AlO_2)_2$ (Thenard's blue, cobalt blue, Leyden blue, and cobalt ultramarine).

2. Ammonia or ammonium sulphide precipitates aluminium hydroxide from its solutions. The precipitate is insoluble in an excess of the precipitant.

3. Caustic alkalies precipitate aluminium hydroxide, which is soluble in an excess of the precipitant.

2. Gallium.

Atomic weight 70=Ga.

Occurs only combined to a very trivial extent in certain zinc blendes as gallium sulphide, Ga_2S_3. It is obtained by the electrolysis of its sulphate. It is a hard white metal having a specific gravity of 5.9 and fusing at 30°, volatile at about 900°, and stable in the air.

Gallium compounds do not give any color to the colorless flame. On volatilization by the electric spark a spectrum consisting of two bright violet lines is obtained.

Sulphuretted hydrogen precipitates white gallium sulphide, Ga_2S_3, from neutral or acetic acid solutions of gallium only in the presence of other precipitable metallic salts. This precipitate is soluble in mineral acids.

3. Indium.

Atomic weight 114=In.

Occurs only combined to a very slight extent in many zinc blendes as indium sulphide, In_2S_3. It is obtained by the electrolysis of its chloride as a soft white metal which is stable in the air, has a specific gravity of 7.4, melts at 176°, and vaporizes at about 1200°.

Indium compounds give a bluish-violet color to the non-luminous flame, and the spectrum of this flame consists of one indigo-blue line (hence the name indium) and a violet line.

Sulphuretted hydrogen precipitates indium from its solution as yellow indium sulphide, In_2S_3, which is soluble in alkali sulphides, forming sulpho salts, and insoluble in dilute acids.

4. Thallium.

Atomic weight 204.1=Tl.

Thallium occurs as a univalent and as a trivalent element. According to behavior it belongs on one side to the alkali metals, as it forms when monovalent a soluble hydroxide, sulphate, carbonate, and silicate, also because it may replace the alkali metals in the alums. On the other hand it has resemblances to lead on account of its chloride and iodide, which are soluble with difficulty, and its insoluble sulphide, and also on account of its physical properties.

Thallium occurs only in combination, up to 17 per cent. in crookesite, and to a less extent in many sulphur and copper ores, in certain brines, especially those of Nauheim, and in sylvine and carnallite. It is obtained by the electrolysis of its chloride as a very soft white metal which melts at 290° and vaporizes at about 1600°, has a specific gravity of 11.8, and which oxidizes in moist air. Thallium compounds give a beautiful green color to the colorless flame, and the spectrum of this flame consists of a single bright-green line ($\vartheta\alpha'\lambda\lambda o\varsigma$, green bud). It is precipitated from its solutions by ammonium sulphide as black thallium sulphide, Tl_2S, which is soluble in dilute acids.

TIN GROUP.

Tin. Germanium. Lead.
Titanium. Zirconium. Thorium.

These elements with carbon and silicon form a group (p. 185) and exist as divalent and tetravalent elements.

Like silicon they form tetravalent compounds, such as volatile tetrachlorides, etc., and their tetrafluorides combine with other metallic fluorides, forming salts corresponding to and isomorphous with the silicofluorides; thus potassium fluostannate, K_2SnFl_6.

On strongly heating they burn into dioxides (lead into PbO or Pb_3O_4) which are acid anhydrides.

They decompose water only at high temperatures.

Titanium, zirconium, and thorium compounds are not precipitated from their neutral or acid solution by H_2S. Zirconium and thorium sulphates give double salts with potassium sulphate, $Zr(SO_4)_2+K_2SO_4+2H_2O$, which are insoluble in saturated solution of potassium sulphate.

1. Tin (Stannum).

Atomic weight 119 = Sn.

Occurrence. Very seldom native, to a slight extent as SnS_2, but very widely distributed as cassiterite or tin-stone, SnO_2.

Preparation. Crushed tin-stone is purified from foreign materials by means of water, roasted, and then heated with coal. The tin thus obtained contains other metals and is therefore slowly heated again to fusion, when the more readily fusible tin flows off, while the more infusible metals remain behind in the unmolten state.

Properties. Silver-white, soft, malleable, not tough metal having a specific gravity of 7.3. At 200° it becomes so brittle that it can be pulverized, at 231° it melts and is covered with white tin oxide, and at about 1600° it is volatile and burns, when air is supplied, into tin dioxide (tin-ash). It is stable in the air, hence it is used in covering (tinning) iron and copper objects. It has a crystalline fracture and a peculiar crackling sound on bending, because the crystals rub against each other (cry of tin). It decomposes water at a red heat, and hydrochloric acid dissolves it with the liberation of hydrogen and forms stannous chloride; concentrated sulphuric acid liberates SO_2 and forms stannous sulphate, while dilute nitric acid evolves NO, forming stannous nitrate. Concentrated nitric acid oxidizes it into insoluble metastannic acid, while anhydrous nitric acid does not attack tin at all. When boiled with caustic alkali tin dissolves with the generation of hydrogen and forms alkali stannate (p. 257): $Sn + 2KOH + H_2O = K_2SnO_3 + 4H$. When beaten or rolled into thin leaves it is called tin-foil.

Gray tin, an allotropic modification having a specific gravity of 5.8, is produced by the action of low temperatures ($-20°$) upon pieces of tin which gradually change into gray tin, consisting of small quadratic crystals which on fusion are transformed into normal tin. Even on contact with gray tin the ordinary white tin can be transformed into the first variety (tin-disease).

Tin occurs divalent in the stannous compounds. These readily take up oxygen and are therefore strong reducing agents. Stannous hydroxide has strong basic properties. Tetravalent tin occurs in the stannic compounds. Stannic hydroxide behaves sometimes as a weak acid and again as a weak base.

a. Alloys of Tin.

Cooking utensils made of tin always contain some lead, and the German law prescribes that such utensils must not contain more than 10 per cent. of lead. Sheet iron covered with a layer of tin is called tin-plate or sheet tin. Soft-solder contains 30–60 per cent. lead. For alloys with copper see p. 235; with antimony, p. 178; with bismuth, p. 264; with mercury, p. 243.

b. Stannous Compounds.

Stannous oxide, SnO, obtained by heating stannous hydroxide in the air, is a stable black powder insoluble in caustic alkali and which on heating in the air ignites and burns into stannic oxide. On slowly evaporating a solution of $Sn(OH)_2$ in caustic alkali it is obtained as dark-green crystals.

Stannous hydroxide, $Sn(OH)_2$, is produced by precipitating a stannous salt with caustic alkali. It forms a white precipitate which oxidizes in the air into stannic hydroxide and dissolves in an excess of the caustic alkali. Stannous hydroxide forms salts on solution in acids.

Stannous chloride, tin protochloride, $SnCl_2$, is ordinarily prepared by dissolving tin in hydrochloric acid and evaporating the solution. The product thus obtained, the tin salt of commerce, $SnCl_2 + 2H_2O$, forms colorless monoclinic crystals which become anhydrous at $1C0°$ and which volatilize without decomposition at a red heat. $SnCl_2$ is soluble in little water, but when considerable water is added thereto it decomposes and the solution becomes cloudy, due to the formation of basic stannous chloride, $Sn(OH)Cl$, which goes into solution again on the addition of acids. The same precipitate is obtained when the clear aqueous solution is allowed to stand in the air: $3SnCl_2 + O + H_2O = 2Sn(OH)Cl + SnCl_4$. This tendency to oxidize is so strong that even dry stannous chloride, on exposure to the air, is transformed into tin oxychloride, $SnOCl_2$, and hence is an energetic reducing agent. In regard to its use in the detection of arsenic see p. 177.

Stannous sulphide, SnS, is obtained by mixing a stannous salt solution with H_2S, which forms an amorphous brownish-black powder, or by melting tin with sulphur, when a bluish-gray crystalline mass is produced. It is fusible, insoluble in dilute acids and alkali monosulphides. It is soluble in alkali polysulphides, forming alkali sulphostannates: $SnS + K_2S_2 = K_2SnS_3$.

c. Stannic Compounds.

Stannic oxide, tin oxide, stannic anhydride, SnO_2, occurs as tin-stone in quadratic crystals or compact masses, seldom white, but generally colored. It is produced as fine needles by heating tin in the air, or as an amorphous white or yellowish powder by heating the two stannic acids. This powder is insoluble in water and in acids, but soluble in alkali hydrates, forming stannates (see below).

Stannic hydroxide has the formula $Sn(OH)_4$ or $SnO(OH)_2$ according to the method of drying. Both these hydroxides are known as ortho- and metastannic acids.

Orthostannic acid is produced on boiling a watery solution of stannic chloride or by treating this solution with ammonia. It forms a gelatinous precipitate which when dried forms a vitreous mass. It dissolves readily in acids, forming stannic salts, and in alkali hydrates, forming stannic acid salts or stannates; thus, Na_2SnO_3, which, on evaporation of the solution, may be obtained as crystals and from which acids precipitate ortho-stannic acid, which is soluble in an excess of the acid. On allowing orthostannic acid to stand under water, it is converted into meta-stannic acid and is then insoluble in acids. Sodium stannate, $Na_2SnO_3 + 3H_2O$, is used in cotton-printing under the name of "preparing-salt."

Metastannic acid is obtained by heating tin with concentrated nitric acid as a white powder which is insoluble in acids. With alkali hydrates it forms salts, metastannates. These are insoluble in the alkali hydrates, but, on the contrary, are soluble in pure water.

Stannic chloride, tin chloride, $SnCl_4$, is obtained on heating tin or stannous chloride in chlorine gas. It is a colorless, fuming liquid which boils at $114°$. With a little water it solidifies to a soft crystalline mass, $SnCl_4 + xH_2O$ (butter of tin); with more water it dissolves completely; on boiling the solution insoluble metastannic acid separates out: $SnCl_4 + 3H_2O = H_2SnO_3 + 4HCl$. The definite crystalline double salt, $SnCl_4 + 2NH_4Cl$, serves in cotton-printing as "pink salt."

Stannic sulphide, tin sulphide, SnS_2, is obtained by treating a stannic salt solution with sulphuretted hydrogen. It forms an amorphous yellow powder which is insoluble in dilute acids. At a red heat it decomposes into $SnS + S$. It is soluble in alkali sulphides, forming sulphostannates which correspond to the stannates. Thus:

$$(NH_4)_2S + SnS_2 = (NH_4)_2SnS_3; \quad K_2S + SnS_2 = K_2SnS_3.$$

Crystalline stannic sulphide may be obtained in golden transparent plates if amorphous stannic sulphide is heated in the presence of ammonium chloride. This latter volatilizes and so regulates the temperature that no decomposition into $SnS + S$ takes place. This modification is known as "mosaic gold" or aurum musivum, and is used as a bronze; it differs from the other sulphides of tin by its insolubility in hydrochloric and nitric acids.

d. Detection of Tin Compounds.

1. When fused with soda on charcoal in the blowpipe flame they yield ductile metallic granules without incrustation.

2. Spongy metallic tin is separated from its solution by metallic zinc in the presence of free hydrochloric acid.

3. When moistened with a cobaltous solution and heated in the oxidizing flame tin compounds turn bluish green.

4. Sulphuretted hydrogen precipitates brown stannous sulphide from stannous salt solutions, and yellow stannic sulphide from stannic salt solutions. Both of these are soluble in yellow ammonium sulphide, forming ammonium sulphostannate: $SnS + (NH_4)_2S_2 = (NH_4)_2SnS_3$. Yellow stannic sulphide may be precipitated from this solution by acids: $(NH_4)_2SnS_3 + 2HCl = 2NH_4Cl + H_2S + SnS_2$.

5. Stannous salts reduce mercuric chloride into insoluble mercurous chloride or into finely divided black mercury; stannic salts do not give this reaction:

$$2HgCl_2 + SnCl_2 = 2HgCl + SnCl_4;$$
$$2HgCl + SnCl_2 = 2Hg + SnCl_4.$$

2. Germanium.

Atomic weight 72.5 = Ge.

Occurs only combined in argyrodite, $Ag_2S + Ag_2GeS_3$, as traces in samarskite, and in euxenite (p. 248). It is obtained, by the reduction of its oxides in a current of hydrogen, as a brittle grayish-white metal which melts at about 900° and has a specific gravity of 5.5. It imparts no coloration to the flame, and only in the induction-spark gives a spectrum which consists of one blue and one violet line. The white germanium sulphide, GeS_2, which is insoluble in dilute acids but soluble in water, and which, like tin sulphide, forms with alkali sulphides sulpho salts, e.g., K_2GeS_3, is characteristic. Argyrodite contains silver sulphogermanate.

3. Lead (Plumbum).

Atomic weight 206.9 = Pb.

Occurrence. Very seldom native, but widely distributed as galena, PbS, seldom as cerussite, $PbCO_3$, wulfenite, $PbMoO_4$, crocoite, $PbCrO_4$, pyromorphite, $Pb_3(PO_4)_2 \cdot PbCl_2$, anglesite, $PbSO_4$, bournonite, $(Pb \cdot Cu)SbS_3$ (p. 181).

Preparation. Nearly entirely from galena by the following methods:
1. By the so-called precipitation method, where the lead sulphide is heated with scrap-iron: PbS + Fe = Pb + FeS.
2. By the roasting process. Lead sulphide is roasted, whereby lead oxide and lead sulphate are produced: $PbS + 3O = PbO + SO_2$; $PbS + 4O = PbSO_4$. The supply of air is then shut off, when the oxidized compounds are transformed into lead by the unchanged lead sulphide present: $PbS + 2PbO = 3Pb + SO_2$; $PbS + PbSO_4 = 2Pb + 2SO_2$.

This lead still contains foreign metals. The silver contained therein is separated according to the method described (p. 238). If this is done by the cupellation process, then the lead is transformed into lead oxide, which is reduced again by means of carbon.

Properties. Bluish-gray shining metal, very soft and ductile, specific gravity 11.4. It leaves a mark on paper; melts at 335°, and is then covered with a gray scum called lead-ash ($Pb_2O + PbO$); at 1700° it vaporizes, and burns into lead oxide on the supply of air. It dissolves readily in nitric acid. When compact it is not attacked by sulphuric or hydrochloric acid, as the lead sulphate or lead chloride formed, on account of its insolubility, protects the lead from further action. In the presence of air it is even attacked by weak organic acids, for instance acetic acid; hence it must not be used for cooking utensils. In dry air it remains unchanged, while in moist air it is covered with a thin layer of PbO. It decomposes water only at white heat; with water containing air it forms in the cold, on the contrary, lead hydroxide, which is somewhat soluble in water.

The action of air and water on lead is of importance, as lead pipes are used in the conveyance of water, and lead salts are poisonous. Ordinary water varies in behavior towards the salts contained therein. The lead is more readily dissolved when chlorides and nitrates are present; but if, on the contrary, the water is hard, containing carbonates and sulphates, then the insoluble coating of lead sulphate or lead carbonate which forms in the lead pipes protects the metal from further action, so that the water that flows through is unaffected. Chemically pure water does not form any protective coating.

Lead occurs divalent as plumbous compounds and tetravalent as plumbic compounds. Most plumbous compounds are isomorphous with the corresponding compounds of the alkaline-earth group, especially with the barium compounds.

a. Alloys of Lead.

Solder (p. 256), type-metal (p. 178), Rose's metal and Wood's metal (p. 264).

b. Plumbous Compounds.

Plumbous oxide, lead oxide, PbO, is prepared by burning lead in air, in the separation of silver from lead (p. 238), or by heating lead carbonate or nitrate. If fusion is prevented, it forms an amorphous yellow powder (Massicot). When quickly cooled it forms a pale-yellow powder, and on slowly cooling a reddish-yellow crystalline powder, which are given different names (litharge, etc.).

Plumbous hydroxide, lead hydrate, $Pb(OH)_2$, is produced as a white precipitate, somewhat soluble in water, on treating a lead salt solution with alkali hydroxide or ammonia. It decomposes into lead oxide and water on heating.

Plumbous oxide and plumbous hydroxide are converted by acid into the corresponding salts and dissolve in an excess of caustic alkali, forming metaplumbates (p. 262). They absorb carbon dioxide from the air and form basic lead carbonate (p. 261).

Plumbous chloride, lead chloride, $PbCl_2$, is produced on treating a lead salt solution with hydrochloric acid or soluble chlorides, which forms a white precipitate slightly soluble in cold water but readily soluble in boiling water.

Plumbous iodide, lead iodide, PbI_2, is obtained by precipitating a plumbous salt solution with potassium iodide. It forms a heavy yellow powder which dissolves in 200 parts boiling water, from which crystalline golden-yellow plates deposit on cooling.

Plumbous sulphate, lead sulphate, $PbSO_4$, occurs naturally as anglesite, and is isomorphous with barium sulphate; it is formed at both electrodes on the discharge of accumulators (lead storage-batteries), according to the following process:

$$Pb + PbO_2 + 2H_2SO_4 \rightleftarrows PbSO_4 + PbSO_4 + 2H_2O,$$
Cathode. Anode.   Cathode. Anode.

while in charging the reverse process takes place and the lead sulphate is reduced to spongy lead on one side and oxidized to PbO_2 on the other (p. 261). It is obtained as a white crystalline precipitate on treating a plumbous salt solution with sulphuric acid or soluble sulphates. Lead sulphate is insoluble in water and acids.

Plumbous nitrate, lead nitrate, $Pb(NO_3)_2$, forms colorless octahedral crystals which are readily soluble in water and which are stable in the air.

Plumbous silicate forms the chief constituent of flint and crystal glass and the glaze of ordinary earthenware (p. 253).

Plumbous sulphide, PbS, occurs as galena in regular crystals having a bluish-gray color (see also p. 262, 3).

Plumbous carbonate, lead carbonate, $PbCO_3$, occurs naturally as cerussite, is isomorphous with aragonite, etc. (p. 223). It is obtained as a white powder on precipitating a plumbous salt solution with ammonium carbonate.

Basic Plumbous Carbonate. If a plumbous salt solution is treated with normal alkali carbonates, white basic plumbous carbonate, whose composition changes with the temperature and concentration of the solutions, precipitates out. This may be represented by the formula $nPbCO_3 + Pb(OH)_2$, where $n = 2, 3, 4$, etc. A carbonate having the formula $2PbCO_3 + Pb(OH)_2$ is the pigment called *white lead*, and is prepared by the action of carbon dioxide upon basic lead acetate according to various methods.

1. French Method. Carbon dioxide is passed into a solution of basic lead acetate (solution of lead oxide in lead acetate), whereby the neutral lead acetate produced remains in solution, while the white lead precipitates. The neutral lead acetate is again transformed into the basic salt by dissolving lead oxide therein.

2. English Method. Lead oxide and lead acetate are rubbed with water and carbon dioxide passed through. The basic lead acetate here formed is decomposed as described in method 1.

3. Dutch Method. Rolled sheets of lead are placed in pots containing some vinegar, and these surrounded by manure. By the fermentation a rise in temperature takes place. The vinegar evaporates and forms basic lead acetate with the aid of the oxygen of the air. This is converted into white lead and lead acetate by the carbon dioxide produced in the fermentation of the manure.

c. Plumbic Compounds.

Plumbic oxide, lead peroxide, PbO_2, may be considered as the auhydride of ortho- and metaplumbic acids; e.g., $PbO_2 + 2H_2O = H_4PbO_4$; $PbO_2 + H_2O = H_2PbO_3$ (see below). It is formed at the positive pole in charging lead accumulators from the lead sulphate covering the lead plate (p. 260). It is also produced on treating minium, $2PbO + PbO_2$ or Pb_3O_4, with nitric acid, when the PbO dissolves, forming lead nitrate, while the PbO_2 remains as an insoluble dark-brown amorphous powder: $Pb_3O_4 + 4HNO_3 = PbO_2 + 2Pb(NO_3)_2 + 2H_2O$. If reducing bodies, such as sugar, oxalic acid, etc., be added at the same time, then the PbO_2 is reduced to PbO and then dissolves in nitric acid.

On heating lead peroxide decomposes into PbO and O; warmed with hydrochloric acid it forms lead chloride with the development of chlorine, and with sulphuric acid it forms lead sulphate with the generation of oxygen:

$$PbO_2 + H_2SO_4 = PbSO_4 + H_2O + O;$$
$$PbO_2 + 4HCl = PbCl_2 + 2H_2O + 2Cl.$$

Orthoplumbic acid, plumbic hydroxide, $Pb(OH)_4$, is known only in the form of salts, the orthoplumbates.

Calcium orthoplumbate, Ca_2PbO_4, is formed on heating CaO with lead oxide or lead peroxide in the air. It is a yellowish mass which is used in Kassner's method for preparing oxygen.

Lead orthoplumbate, $Pb_2(PbO_4)$ or Pb_3O_4, minium, red lead, is obtained by heating PbO in the air to 300°–400°, and forms a scarlet red crystalline powder that on heating above 400° gives off oxygen and is transformed again into lead oxide. With acids its behavior is like that of a mixture of $2PbO + PbO_2$ (see this latter).

Metaplumbic acid, H_2PbO_3, forms on the positive pole, on the electrolysis of lead salts, as a bluish-black body.

Sodium metaplumbate, $Na_2PbO_3 + 3H_2O$, is obtained as colorless crystals on carefully evaporating a solution of PbO_2 in caustic potash solution: $PbO_2 + 2NaOH = Na_2PbO_3 + H_2O$. The watery solution gives with many metallic salt solutions a precipitate of the corresponding metaplumbate.

Lead metaplumbate, $Pb(PbO_3)$, is precipitated from the solution of PbO in caustic alkali by NaClO as a reddish-yellow powder.

Plumbic chloride, lead tetrachloride, $PbCl_4$, is obtained on dissolving PbO_2 in ice-cold hydrochloric acid and treating the solution with NH_4Cl, when yellow crystals of $PbCl_4 + 2NH_4Cl$ separate out. If these crystals are introduced into ice-cold sulphuric acid, the $PbCl_4$ is set free as a yellow liquid which crystallizes at $-15°$.

Plumbic sulphate, $Pb(SO_4)_2$, is produced on the electrolysis of sulphuric acid, making use of lead electrodes, and is a yellowish-white crystalline powder which decomposes with water:

$$Pb(SO_4)_2 + 2H_2O = PbO_2 + 2H_2SO_4.$$

d. Detection of Lead Compounds.

1. They color the non-luminous flame pale blue. The spectrum shows characteristic lines in the green part.

2. Heated with soda upon charcoal they yield soft metallic granules of lead and a yellow coating of lead oxide.

3. Sulphuretted hydrogen precipitates black-lead sulphide from solutions of lead salts. This is insoluble in alkali sulphides and dilute acids.

4. Caustic alkali precipitates white lead hydroxide soluble in an excess of the precipitant.

5. Sulphuric acid precipitates white lead sulphate soluble in caustic alkali and in basic ammonium tartrate.

6. Zinc and iron precipitate lead from its solutions in metallic shining plates (lead-tree).

4. Titanium.

Atomic weight $48.1 = Ti$.

Occurs only in combination, chiefly as TiO_2 in the minerals anatase, rutile, and brookite, which differ from each other only in crystalline form. It is also found in euxenite and oerstedite, and to a less extent in many rocks and in many iron ores. The metal is obtained by heating potassium-titanium fluoride, K_2TiF_6, with potassium. It forms an iron-gray crystalline powder having a specific gravity of 3.55.

5. Zirconium.

Atomic weight 90.6 = Zr.

Zirconium occurs only in combination, especially as zircon or hyacinth, $ZrSiO_4$, also in wöhlerite, oerstedite, and is obtained in the same manner as titanium. It is known in three allotropic modifications, namely, as gray crystals having a specific gravity of 4.15, as graphitic zirconium, and as a black powder.

Zirconium oxide, zirconia, ZrO_2, on heating glows with an intense light and is used as the illuminating body for the oxyhydrogen flame (zircon pencils) (p. 113).

6. Thorium.

Atomic weight 232.5 = Th.

Occurs only combined in thorite, $ThSiO_4$, and with the rare earthy metals in many of the minerals mentioned on p. 248. It is prepared in an analogous manner to titanium and forms a gray powder having a specific gravity of 11.0. Thorium oxide, thoria, ThO_2, glows at a much lower temperature than zirconium oxide and with a more intense light, especially when it contains one per cent. of cerium dioxide, CeO_2. This mixture serves in the preparation of the mantle for the Welsbach light, which is produced when a cotton tissue impregnated with thorium nitrate, $Th(NO_3)_4$, and cerium nitrate, $Ce(NO_3)_3$, is heated. Both of these salts are prepared on a large scale from monazite sand or cerite.

BISMUTH GROUP.

Bismuth. Vanadium. Niobium. Tantalum.

These elements appear trivalent and pentavalent and belong to the nitrogen group (p. 145). Vanadium, niobium, and tantalum are not precipitated by H_2S from their acid nor from their neutral solutions.

1. Bismuth.

Atomic weight 208.5 = Bi.

Occurrence. Chiefly native, as well as bismite or bismuth ocher, Bi_2O_3, bismuthinite, BiS_3, eulytite, $Bi_4(SiO_4)_3$, seldom as tetradymite, $Bi_2Te_3 + Bi_2S_3$.

Preparation. It is separated from the rocks by fusion; from the ores by roasting and reduction of the obtained bismuth oxide by carbon.

Properties. Reddish-white brittle metal having a specific grav‑
ity of 9.8, melting at 268°, and which on slowly cooling crystal‑
lizes in rhombohedra (isomorphous with As and Sb) very similar
to cubes. It is stable in the air, vaporizes at about 1300°, and on
heating it burns into bismuth trioxide. It decomposes water at a
red heat. It is the poorest conductor for heat of the metals, and
expands on cooling. It is insoluble in hydrochloric acid and dilute
sulphuric acid, but dissolves in cold nitric acid and in hot concen‑
trated sulphuric acid, with the development of nitric oxide and sul‑
phur dioxide respectively, forming the corresponding salts:

$$2Bi + 8HNO_3 = 2Bi(NO_3)_3 + 4H_2O + 2NO;$$
$$2Bi + 6H_2SO_4 = Bi_2(SO_4)_3 + 6H_2O + 3SO_2.$$

a. Alloys of Bismuth

are characterized by their ready fusibility and are used in the making
of stereotypes, etc.; Rose's metal (Sn.Pb.Bi) melts at 94°, Wood's
metal (Sn.Pb.Bi.Cd) melts at 63°.

b. Compounds of Bismuth.

Bismuth trioxide, Bi_2O_3, is obtained by burning bismuth or heating
bismuth nitrate, hydroxide, or pentoxide as a yellow powder insoluble in
water and in caustic alkalies.

Bismuth hydroxide, $Bi(OH)_3$, is precipitated from bismuth salt solu‑
tions by caustic alkali, and forms an amorphous powder which is insoluble
in water and caustic alkalies, and at 100° is transformed into

Metabismuth hydroxide, $HBiO_2$ or $OBi(OH)$, which when dry forms
a white amorphous mass: $H_3BiO_3 = H_2O + HBiO_2$.

Bismuth tetroxide, Bi_2O_4, is obtained as a yellowish-brown powder
by the prolonged action of nitric acid upon bismuth pentoxide.

Bismuth pentoxide, bismuthic anhydride, Bi_2O_5, is obtained by heat‑
ing metabismuthic acid, and forms a brown powder that behaves like a
superoxide towards acids.

Bismuthic acid, H_3BiO_4, which is analogous to phosphoric acid, is not
known, but

Metabismuthic acid, $HBiO_3$, which is analogous to metaphosphoric
acid, is known. This is formed on passing chlorine gas into caustic alkali
in which Bi_2O_3 is suspended, when red potassium bismuthate separates out;
this can be decomposed by boiling nitric acid into metabismuthic acid,
which forms a scarlet-red powder and which behaves like Bi_2O_5 towards
acids.

Bismuth salts, Bi_2O_3 and $HBiO_2$, differ from the other analogous com‑
pounds of the nitrogen group in having basic properties and being insol‑
uble in alkalies. By dissolving them in acids and evaporating to crystal‑
lization the normal bismuth salts are obtained; e.g.,

$$BiCl_3; \quad Bi(NO_3)_3 + 5H_2O; \quad Bi_2(SO_4)_3; \quad BiPO_4.$$

Basic Bismuth Salts. The salts of bismuth are soluble in a small quan.

tity of water, but, like those of antimony, are decomposed by considerable water and basic bismuth salts separate out, depending upon the temperature and concentration of the solution. Thus:

$$BiCl_3 \quad + \quad H_2O = BiOCl \quad + 2HCl;$$
$$Bi(NO_3)_3 + 2H_2O = Bi(OH)_2NO_3 + 2HNO_3;$$
$$Bi_2(SO_4)_3 + 4H_2O = Bi_2(OH)_4SO_4 + 2H_2SO_4.$$

Bismuth subnitrate is obtained on mixing a concentrated solution of bismuth nitrate with twenty-five times its volume of boiling water, which forms a white microcrystalline powder:

$$2Bi(NO_3)_3 + 3H_2O = [Bi(OH)_2NO_3 + BiO(NO_3)] + 4HNO_3.$$

c. Detection of Bismuth Compounds.

1. Considerable water precipitates white basic bismuth salts from bismuth salt solution. This basic bismuth salt is insoluble in tartaric acid, differing in this respect from the corresponding antimony compounds.

2. Sulphuretted hydrogen precipitates brownish-black bismuth sulphide, Bi_2S_3, which is insoluble in dilute acids and alkali sulphides.

3. Mixed with soda and heated upon charcoal brittle granules of metallic bismuth and a yellowish-brown incrustation of bismuth trioxide are obtained.

2. Vanadium.

Atomic weight 51.2 = V.

It occurs only in combination, as dechenite, $Pb(VO_3)_2$, as vanadinite, $Pb_3(VO_4)_2$, as carnotite, and in the Thomas slag from Creuzot. The metal is obtained by heating VCl_3 in a current of hydrogen which gives a gray powder having a specific gravity of 5.5 and melting at about 3000°

3. Niobium (Columbium).   4. Tantalum.

Atomic weight 94 = Nb.   Atomic weight 183 = Ta.

Are found only in combination and in fact always together as columbite and tantalite, both $O_3Ta^-Fe^-NbO_3$, also in euxenite, yttrotantalite, pyrochlore, wöhlerite (P. 248). They are prepared in the same way as vanadium and form gray metallic powders whose properties in a pure state are not well known.

CHROMIUM GROUP.

Chromium. Molybdenum. Tungsten. Uranium.

Just as the metallic elements Sn, Zr, Ti, Th are related to the carbon group and the elements Bi, Va, Nb, Ta to the nitrogen group, so the elements Cr, Mo, W, and more remotely U, are related to the sulphur group. Chromium forms the connecting member of this group with that of iron and aluminium, as its compounds are closely related to those of iron on the one side and to those of aluminium on the other. They are stable in the air and decompose water only at a red heat; chromium and tungsten compounds are not precipitated from their acid or neutral solutions by sulphuretted hydrogen.

1. Hexavalent they form, like the elements of the sulphur group, trioxides which are acid anhydrides. They form salts having a constitution analogous to that of the manganates and ferrates, although they are more stable, many of them being similar and isomorphous with the sulphates; e.g., K_2CrO_4, K_2MoO_4.

2. Tetravalent they form (with the exception of chromium) compounds corresponding to the elements of the sulphur group; e.g., UO_2, MoO_2, WO_2, UCl_4.

3. Chromium and molybdenum may form trivalent compounds which are very similar to the trivalent compounds of the iron and the aluminium group; e.g., Cr_2O_3, $CrCl_3$.

4. With the exception of uranium, they form divalent compounds, of which those of chromium are very similar to the magnesium group and the ferrous compounds, although they are less stable; e.g., $CrCl_2$. The corresponding oxygen salts of tungsten and molybdenum are not known.

5. Molybdenum, tungsten, and uranium also occur pentavalent, molybdenum and uranium also octavalent, chromium also nonovalent.

1. Chromium.

Atomic weight 52.1 = Cr.

Occurrence. Never native, generally as chromite or chrome-iron ore, $Fe(CrO_2)_2$, seldom as crocoite, $PbCrO_4$.

Preparation. By heating chromic oxide with aluminium (p. 248).

Properties. Pale-gray shining metal having a specific gravity of 6.8 whose fracture shows large crystals. It is one of the hardest and most refractory (at about 3000°) of the metals; it is non-magnetic, oxidizes only slowly on heating, but burns into chromic oxide on heating to a white heat in oxygen. It dissolves in hydrochloric acid with the evolution of hydrogen, forming $CrCl_2$, and in sulphuric acid, producing $CrSO_4$, but is insoluble in nitric acid.

After lying in the air chromium does not dissolve in dilute acids (inactive chromium). But on heating it with the acid it begins to dissolve, and this continues after washing it off and placing it in the cold acid (active chromium).

Divalent chromium forms chromous compounds, trivalent the chromic compounds, hexavalent forms chromium trioxide and the chromates. All the compounds of chromium are characterized by their beautiful color (χρῶμα, color).

a. Alloys of Chromium

with iron are used as chrome-steel.

b. Chromous Compounds.

These are little known, as they have great power of absorbing oxgyen and being converted into chromic salts.

Chromous oxide, CrO, is not known.

Chromous hydroxide, $Cr(OH)_2$, is obtained by treating chromous chloride with caustic alkali, which gives a yellow precipitate that is quickly converted into chromic oxide with the evolution of hydrogen:

$$2Cr(OH)_2 = Cr_2O_3 + H_2O + 2H.$$

Chromous chloride, $CrCl_2$, is obtained by passing hydrogen over heated chromic chloride which yields a white crystalline powder forming a blue solution with water. This solution takes up oxygen with activity and becomes green with the probable formation of $Cr_2OCl_4(?)$.

c. Chromic Compounds.

These are prepared analogously to the corresponding aluminium compounds.

Chromic oxide, chromium sesquioxide, Cr_2O_3, an amorphous green powder nearly insoluble in acids, is produced on heating $Cr(OH)_3$ or CrO_3. It may be obtained in black crystals isomorphous with Al_2O_3 and Fe_2O_3, by passing the vapors of chromyl chloride through a red-hot tube:

$$2CrO_2Cl_2 = Cr_2O + 4Cl + O.$$

Chromic hydroxide, $Cr(OH)_3$, is precipitated from chromic salt solutions by caustic alkalies or ammonia as a bluish-gray precipitate which, like $Al(OH)_3$, behaves like a weak acid and is soluble in an excess of caustic alkali (p. 270), and precipitating again on boiling. On the other hand it acts like a weak base, like $Al(OH)_3$ and $Fe(OH)_3$, but not combining with the weak acids, such as carbonic acid, sulphurous acid, sulphuretted hydrogen. On heating to 200° in a current of hydrogen it is converted into metachromic hydroxide, $HCrO_2$ or $CrO(OH)$, which is bluish gray and insoluble in dilute hydrochloric acid. The hydrate $Cr_2O(OH)_4$ is used as the beautiful pigment called Guignet s green.

Chromites are the compounds corresponding to the aluminates (p. 251) and are derived from $HCrO_2$; e.g., $Mg(CrO_2)_2$.

Chromic salts are obtained by dissolving chromic hydroxide in the respective acids and evaporating at as low a temperature as possible. They form violet crystals, producing a violet solution in cold water; on heating this solution they become green and on evaporation yield amorphous green masses which consist of a mixture of basic and acid chromic salts. If the green salts are dissolved, then the solution gradually becomes violet, and on careful evaporation violet crystals of the neutral chromic salts separate out.

Potassium-chromium sulphate, chrome alum, $KCr(SO_4)_2 + 12H_2O$ (p. 252), forms deep-violet octahedra and is obtained by the action of sulphur dioxide upon a solution of potassium dichromate treated with sulphuric acid and evaporating the same: $K_2Cr_2O_7 + H_2SO_4 + 3SO_2 = 2KCr(SO_4)_2 + H_2O$.

d. Higher Chromium Compounds.

Chromium trioxide, chromic anhydride, CrO_3, is formed on the electrolysis of chromic sulphate, as well as on treating a concentrated solution of potassium dichromate with an excess of concentrated sulphuric acid: $K_2Cr_2O_7 + H_2SO_4 = K_2SO_4 + 2CrO_3 + H_2O$. On cooling the chromium trioxide crystallizes out in long scarlet-red rhombic crystals; it is deliquescent, readily soluble in water, melts on heating, forming a deep-red liquid, and decomposes at 250°: $2CrO_3 = Cr_2O_3 + 3O$. It has an energetic oxidizing action, so that it destroys many organic bodies (hence its solution cannot be filtered through paper); when poured upon alcohol a faint explosion occurs, reducing it into chromic oxide. It is not attacked by nitric acid; with hydrochloric acid it yields chromic chloride with the generation of chlorine, and with sulphuric acid it forms chromic sulphate with the development of oxygen:

$$CrO_3 + 6HCl = CrCl_3 + 3H_2O + 3Cl;$$
$$2CrO_3 + 3H_2SO_4 = Cr_2(SO_4)_3 + 3H_2O + 3O.$$

Chromyl chloride, chromium oxychloride, CrO_2Cl_2 (chromium trioxide in which one atom of oxygen is replaced by two atoms of chlorine), is prepared by distilling alkali dichromates with common salt and an excess of sulphuric acid (to combine with the water). It forms a fuming, deep-red liquid which decomposes with water into chromium trioxide and hydrochloric acid:

$$K_2Cr_2O_7 + 4NaCl + 3H_2SO_4 = 2Na_2SO_4 + K_2SO_4 + 2CrO_2Cl_2 + 3H_2O;$$
$$CrO_2Cl_2 + H_2O = CrO_3 + 2HCl.$$

(The formation of CrO_2Cl_2 serves to detect the presence of chlorine in the presence of bromine and iodine compounds, these last two not forming an analogous compound.)

Chromic acid, H_2CrO_4 or $HO-CrO_2-OH$, separates on cooling the watery solution of chromium trioxide to 0° in red needles. On warming the solution, it decomposes immediately again into $H_2O + CrO_3$, which latter remains behind on the evaporation of the solution.

Chromates form yellow crystals and are prepared by fusing the chromium compound with bases and an oxidizing agent (see Sodium Dichromate), or by the action of bases upon CrO_3; e.g., $2NaOH + CrO_3 = Na_2CrO_4 + H_2O$. The chromates of the heavy metals and of barium are insoluble in water.

Polychromic acids, $H_2CrO_4 + xCrO_3$, may be considered as obtained by the removal of H_2O from several chromic acid molecules;

e.g., $2H_2CrO_4 = H_2O + H_2Cr_2O_7$ (dichromic acid), $3H_2CrO_4 = 2H_2O + H_2Cr_3O_{10}$ (trichromic acid). These acids are not known free, as they decompose immediately on being formed: $H_2Cr_2O_7 = 2CrO_3 + H_2O$; $H_2Cr_3O_{10} = 3CrO_3 + H_2O$. Their red crystalline salts, the polychromates, are prepared from the chromates by the action of cold acids, and are transformed into chromates by the action of bases. On heating to a red heat the polychromates give off oxygen, and on heating with acids they behave like peroxides (see Potassium and Sodium Dichromate).

Perchromic acid, $HCrO_5$ or $HO-CrO_4$, is known only in the form of deep violet salts and as an unstable deep-blue anhydride, Cr_2O_9 (see Hydrogen Peroxide. p. 119), both of these being readily decomposable, the first producing dichromates and the second chromium trioxide, and at the same time an evolution of oxygen.

Sodium chromate, $Na_2CrO_4 + 10H_2O$, isomorphous with sodium sulphate, crystallizing from its aqueous solution below 18°, and crystallizing at 18°-20° as $Na_2CrO_4 + 6H_2O$, and as anhydrous crystals at 30°. It is prepared analogously to potassium chromate, forming yellow deliquescent crystals. For the preparation on a large scale see Sodium Dichromate.

Potassium chromate, yellow chromate, K_2CrO_4, is obtained by treating a potassium dichromate solution with caustic potash: $K_2Cr_2O_7 + 2KOH = 2K_2CrO_4 + H_2O$, and is produced on fusing every chromium compound with potassium carbonate and saltpeter (p. 270). It forms yellow masses which dissolve in water, and on evaporation yields yellow rhombic crystals isomorphous with potassium sulphate and potassium manganate; on heating with concentrated acids the chromates act like the dichromates.

Lead chromate. Chromate yellow, $PbCrO_4$, occurs as crocoite. It is obtained as a yellow precipitate by precipitating a lead salt solution with potassium chromate. (Used as a paint under the name chrome yellow, Parisian yellow, Leipzig yellow, Hamburg yellow.) At a red heat it decomposes with the generation of oxygen, which oxidizes all organic bodies; hence it is used like copper oxide in the combustion of organic bodies in their elementary analysis. It dissolves in an excess of caustic alkali; if a little caustic alkali is added, and it is then warmed, it becomes red with the formation of basic lead chromate, $PbO + PbCrO_4$, which is used as a pigment (chrome-red, orange-red, cinnabar-red, carmine-red):

$$2PbCrO_4 + 2KOH = (PbO + PbCrO_4) + K_2CrO_4 + H_2O.$$

Potassium dichromate, Red Chromate, Double Potassium Chromate, Acid Potassium Chromate, $K_2Cr_2O_7$. If a saturated solution of potassium chromate is treated with enough sulphuric acid to combine with one-half of the potassium (p. 268), then on cooling potassium dichromate separates out. On a large scale it is obtained by treating sodium dichromate with potassium chloride, when on evaporation the sodium chloride first crystallizes out:

$$Na_2Cr_2O_7 + 2KCl = K_2Cr_2O_7 + 2NaCl.$$

It forms large red triclinic crystals soluble in ten parts of water. On heating it melts without decomposition; on heating to a red heat it decomposes into potassium chromate, chromic oxide, and oxygen:

$$2K_2Cr_2O_7 = 2K_2CrO_4 + Cr_2O_3 + 3O.$$

When heated with concentrated suphuric acid it yields chrome alum and pure oxygen (method of preparing oxygen):

$$K_2Cr_2O_7 + 4H_2SO_4 = 2KCr(SO_4)_2 + 4H_2O + 3O.$$

When heated with concentrated hydrochloric acid it forms chromic chloride, and chlorine is evolved:

$$K_2Cr_2O_7 + 14HCl = 2KCl + 2CrCl_3 + 7H_2O + 6Cl.$$

Sodium dichromate, $Na_2Cr_2O_7 + 2H_2O$, is the substance from which all chromium compounds are prepared. Chrome-iron ore is heated with calcium oxide and an abundant supply of air; the calcium chromate formed is converted into sodium chromate by heating it with a soda solution. This sodium chromate is treated with the proper quantity of sulphuric acid, when from the hot solution anhydrous sodium sulphate separates, and on evaporation sodium dichromate, in the form of red triclinic crystals, is obtained, which is only slightly soluble in water:

$$2Fe(CrO_2)_2 + 4CaO + 7O = 4CaCrO_4 + Fe_2O_3;$$
$$CaCrO_4 + Na_2CO_3 = Na_2CrO_4 + CaCO_3;$$
$$2Na_2CrO_4 + H_2SO_4 = Na_2Cr_2O_7 + Na_2SO_4 + H_2O.$$

e. Detection of Chromium Compounds.

Chromium Salts. 1. They color the borax bead in the blowpipe flame emerald-green.

2. Caustic alkalies precipitate green chromic hydroxide, which dissolves in an excess of the alkali with a green color, and on boiling the solution separates completely. (Separation from aluminium hydroxide, which remains in solution on boiling its alkaline solution.)

3. Ammonia or ammonium sulphide precipitates green chromic hydroxide, which is only very slightly soluble in an excess.

4. Sulphuretted hydrogen does not precipitate chromium salts.

5. On fusion with soda and saltpeter they all give a yellow mass (p. 269) whose solution acidified with acetic acid gives the reactions for the chromates.

Chromates. 1. They are transformed into chromium salts by reducing bodies, e.g., H_2S, SO_2, oxalic acid, alcohol; hence their yellow or red solution turns green (see Potassium-chromium Alum, p. 267).

2. Caustic alkalies, ammonia, ammonium sulphide produce no precipitation in chromate solutions.

3. Lead salts precipitate yellow lead chromate from their neutral

solution, $PbCrO_4$ (p. 269); barium salts, yellow barium chromate, $BaCrO_4$ (p. 226); silver salts, red silver chromate, Ag_2CrO_4.

4. By transforming them into perchromic acid (p. 119).

2. Molybdenum.

Atomic weight 96=Mo.

Occurs only combined in molybdenite, MoS_2, wulfenite, $PbMoO_4$. It is obtained by heating its oxides with aluminium, which forms a silver-white metal, malleable, relatively soft, having a specific gravity of 9, fusing at about 1900°, and being soluble only in concentrated sulphuric acid, nitric acid, and aqua regia.

Compounds of Molybdenum.

Molybdenum trioxide, molybdic anhydride, MoO_3, is obtained on roasting molybdenum or molybdenite which yields white crystals that are readily soluble in ammonia or caustic alkalies, producing molybdates, and insoluble in water and acids.

Molybdic acid, H_2MoO_4, is precipitated as yellow crystals from the molybdates by means of nitric acid.

Phosphomolybdic acid (Sonnenschein's reagent), $H_3PO_4 + 11MoO_3$, serves as a precipitant for alkali salts and alkaloids (which see).

Ammonium molybdate, $(NH_4)_2MoO_4$, is a reagent and precipitant for phosphoric acid and arsenic acid (see these).

Molybdenum trisulphide, MoS_3, is obtained as a black precipitate by treating acidified molybdenum salt solutions with H_2S. It is soluble in alkali sulphides, forming alkali sulphomolybdates; e.g., K_2MoS_4.

3. Tungsten (Wolfram).

Atomic weight 184=W.

Occurs only combined as wolframite, $FeWO_4$, as tungsten or scheelite, $BaWO_4$, or more seldom stolzite. $PbWO_4$. By the reduction of its oxides with aluminium (p. 248) it is obtained as a white, very hard, brittle metal of a specific gravity of 16.6, and fusible at about 2000°. On heating, as well as by warming with nitric acid or aqua regia, it is converted into insoluble WO_3. Tungsten is insoluble in hydrochloric or sulphuric acids. It is prepared on a large scale in order to produce an especially hard form of steel (tungsten steel).

Compounds of Tungsten.

Tungsten trioxide, WO_3, is produced by treating finely powdered tungsten ore or tungsten with nitric acid. It forms a yellow powder which readily dissolves in alkalies or ammonia, forming the corresponding salts, the tungstates or wolframates, but is insoluble in acids.

Tungstic acid, wolframic acid, H_4WO_5, which is readily converted into H_4WO_4, is precipitated as yellow crystals from solutions of tungstates by nitric acid. Peculiarly constituted alkali tungstates, $K_2W_4O_{12}$, have a metallic luster and various colors, and are used as tungsten bronzes.

Sodium tungstate, $Na_2WO_4 + 2H_2O$, is used as a mordant in calico-printing, as well as to make tissues non-inflammable.

Calcium tungstate, $CaWO_4$, serves in the detection of the Röntgen rays, which give a bluish-violet fluorescence therewith.

Phosphotungstic acid, $H_3PO_4 + 11WO_3$ (Scheibler's reagent), is used for the same purposes as phosphomolybdic acid.

4. Uranium.

Atomic weight 238.5= U.

Occurs only combined and chiefly as uraninite, UO_2+2UO_3 (called pitchblende). It is obtained by heating uranium oxides with aluminium, or by the reduction of UCl_4 with sodium, as a silver-white, hard, brittle metal of a specific gravity of 18.7, which melts at about 1500°, is soluble in dilute acids, and which burns when heated into uranic-uranous oxide, UO_2+2UO_3.

Compounds of Uranium.

It occurs tetravalent in the unstable uranous compounds, and hexa-valent in the uranic compounds. These latter all contain the divalent radical UO_2, called uranyl:

$(UO_2)O$,	Uranyl oxide.	$(UO_2)Cl_2$,	Uranyl chloride.
$(UO_2)(NO_3)_2$,	Uranyl nitrate.	$(UO_2)SO_4$,	Uranyl sulphate.

Many uranium salts have a beautiful fluorescence. Uranium and all its compounds have the property of acting upon photographic plates (p. 241) through black paper, and also of causing the illumination of certain phosphorescent substances and of making air a conductor of electricity. This property seems to depend upon the presence of bodies, probably elements, which always accompany uranium. These elements, because of their similarity to the known elements, have been called radiolead, radiothorium (=actinium), radiobismuth (=polonium), radiobarium (=radium).

Uranium oxide, UO_2, is obtained as a black powder on heating the higher uranium oxides in a current of H. It is soluble in hydrochloric and sulphuric acids, forming green uranous salts and also soluble in nitric acid, producing uranyl nitrate. On heating in the air it is transformed into uranous-uranic oxide, UO_2+2UO_3, and colors glass or porcelain black.

Uranyl oxide, $(UO_2)O$, is prepared by dissolving uranium or uranium oxides in nitric acid, evaporating, and heating the uranyl nitrate, $UO_2(NO_3)_2$, obtained to 250°. On warming uranyl oxide with HNO_3 it is changed into uranylic acid, $UO_2(OH)_2$. Uranyl oxide gives a greenish-yellow fluorescence when fused with glass.

Uranyl phosphate, $(UO_2)_2HPO_4+3H_2O$, is precipitated from uranyl salts by phosphates as a yellowish-white powder (quantitative estimation of phosphoric acid).

Uranyl sulphide, UO_2S, is obtained as a black precipitate by treating uranyl salts with ammonium sulphide.

Uranates. Those derived from uranylic acid are not known. If a solution of uranyl salts is treated with caustic alkali, a yellow precipitate of alkali uranates is obtained. These are analogous in composition to the dichromates; e.g., $Na_2U_2O_7$. Sodium uranate is the uranium yellow of commerce.

IRON GROUP.

Manganese. Iron. Cobalt. Nickel.

These decompose water only at higher temperatures and are precipitated from their solutions by H_2S only in the presence of bases (p. 123). The cyanides of manganese, iron, and cobalt form peculiarly complicated compounds (see Cyanogen and p. 102).

1. Divalent they form compounds which are similar to those of the magnesium group, as well as to the cupric compounds, especially by the behavior of their sulphates and their carbonates.

2. Trivalent they form compounds which are constituted analogously to those of the aluminium and chromium groups.

3. Manganese and iron also occur hexavalent, producing acids; manganese also tetravalent and heptavalent.

The six elements of the platinum group as mentioned on p. 55 are allied to the metals of the iron group with regard to chemical properties and their position in the periodic system.

1. Manganese.

Atomic weight 55=Mn.

Occurrence. Only native in meteorites; combined as pyrolusite or brownstone (MnO_2), braunite (Mn_2O_3), manganite, ($MnO.OH$), hausmannite (Mn_3O_4), rhodochrosite · ($MnCO_3$), alabandite (MnS). Traces of manganese are found in many plants and animals.

Preparation. By fusing manganese oxides with carbon or with aluminium (p. 248).

Properties. Grayish-white, very hard, brittle metal of specific gravity 7.5, melting at about 1900°. It oxidizes in moist air, decomposes boiling water with the generation of hydrogen, dissolves in all acids, forming manganous salts, is not, unlike the other metals of this group, attracted to a magnet, nor does it become magnetic.

Divalent manganese forms the manganous compounds, trivalent manganese the manganic compounds, the hexavalent manganese the manganic acid compounds, and heptavalent manganese the permanganic compounds.

a. Alloys of Manganese.

Cupro-manganese, manganese copper (30 per cent. Mn), is a constituent of various alloys of copper having great hardness and toughness (manganese bronzes); spiegel-eisen (10 to 20 per cent. Mn) and ferro-manganese (20 to 70 per cent. Mn) are of importance, as they make varieties of iron denser and tougher. Manganin, see Nickel.

b. Manganous Compounds.

Manganese monoxide, manganous oxide, MnO, is obtained by heating manganous carbonate in the absence of air, or by heating all oxides of manganese in a current of hydrogen. It is a green powder readily soluble in acids, and quickly oxidizes into brown manganous-manganic oxide, Mn_3O_4, in the air.

Manganous hydroxide, $Mn(OH)_2$, is obtained by treating a manganous salt solution with caustic alkali, when a white precipitate is formed which quickly oxidizes in the air into brown manganic hydroxide.

Manganous sulphide, MnS, occurs as manganese blende (alabandite) in black cubes, and is obtained by treating a manganous salt solution with ammonium sulphide, which produces a flesh-colored precipitate that quickly turns brown on oxidizing in the air.

Manganous chloride, $MnCl_2 + 4H_2O$, forms light-red crystalline masses which deliquesce in the air and decompose on heating. The solution of $MnCl_2$ obtained in the preparation of chlorine (p. 133) is mixed with lime and air blown through, when the so-called calcium manganite, $CaMnO_3$, separates (p. 275). This compound can be used like MnO_2 in the preparation of chlorine: $MnCl_2 + 2CaO + O = CaMnO_3 + CaCl_2$ (Weldon's process).

Manganous sulphate, $MnSO_4$, crystallizes below 6° with 7 mol. of water in pink monoclinic prisms (isomorphous with ferrous and cobaltous sulphate, etc.); at ordinary temperatures it forms triclinic prisms with 5 mol. of water (isomorphous with cupric sulphate); 1 mol. of water of crystallization is first given off at higher temperatures than the others. With alkali sulphates it forms double salts which have an analogous constitution and are isomorphous with the corresponding magnesium salts, etc. (p. 229).

Manganous carbonate, $MnCO_3$, occurs as manganese spath (rhodrocrosite) in rose-red crystals which are isomorphous with calc spar and iron spar, etc.

c. Manganic Compounds.

The manganic salts are very unstable and are decomposed by water or on heating.

Manganic oxide, Mn_2O_3, occurs in braunite as brown crystals, and is obtained as a black powder by gently heating the oxides and hydroxides of manganese in the air.

Mangano-manganic oxide, Mn_3O_4 (mangano-manganite, see below), occurs as hausmannite and is obtained as a reddish-brown po e (manganese brown, bister) on strongly heating all oxides and hydroxides of manganese in the air. Both of the above oxides are decomposed by hot nitric acid into manganous nitrate and manganese dioxide:

$$Mn_3O_4 + 4HNO_3 = 2Mn(NO_3)_2 + MnO_2 + 2H_2O;$$
$$Mn_2O_3 + 2HNO_3 = Mn(NO_3)_2 + MnO_2 + H_2O.$$

Cold sulphuric acid dissolves them, forming red solutions which contain a mixture of manganous and manganic sulphates. Cold hydrochloric acid acts in a similar manner. With hot sulphuric acid and hydrochloric acid, on the contrary, they behave like manganese dioxide (see below).

Manganic hydroxide, $Mn(OH)_3$, is obtained by allowing manganous hydroxide to stand in the air, producing a brownish-black powder which behaves towards acids like manganic oxide. It decomposes readily into metamanganic hydroxide, $MnO(OH)$, which occurs as manganite. Mixtures of manganic hydroxide with ferric and aluminium hydroxides form the colors umber and ocher.

Manganic chloride, $MnCl_3$, has not been isolated. If manganic hydroxide or manganic oxide is dissolved in hydrochloric acid, manganic chloride is formed, which on warming immediately begins to decompose with the generation of chlorine and the formation of manganous chloride: $MnCl_3 = MnCl_2 + Cl$.

Manganic sulphate, $Mn_2(SO_4)_3$, is obtained on carefully heating manganese dioxide with concentrated sulphuric acid to 140°, which gives an amorphous dark-green powder that at 160° decomposes into $2MnSO_4 + SO_2 + O_2$. With alkali sulphates, etc., it forms stable manganese alums (p. 252). In the air it quickly decomposes into manganic hydroxide, but with water much quicker. Manganese alum is decomposed in the same manner by water.

Manganic carbonate, $Mn_2(CO_3)_3$, is not known.

d. Higher Manganese Compounds.

Manganese dioxide, MnO_2, occurs in hard gray crystals, as well as in soft fibrous masses as pyrolusite or brownstone. On strongly heating it gives off part of its oxygen (p. 107), and on warming with concentrated sulphuric acid it dissolves with the formation of manganous sulphate and oxygen is set free: $MnO_2 + H_2SO_4 = MnSO_4 + H_2O + O$ (method of preparing oxygen). It is insoluble in nitric acid or dilute sulphuric acid. It dissolves in warm hydrochloric acid with the generation of chlorine and the formation of manganous chloride (p. 132).

Manganous Acids. MnO_2 may be considered as the anhydride of the acids $H_2MnO_3 (MnO_2 + H_2O)$, $H_4MnO_4 (MnO_2 + 2H_2O)$, $H_2Mn_2O_5 (2MnO_2 + H_2O)$, which are not known free, but whose salts are precipitated by oxygen from the manganous salts in the presence of bases. These salts are called manganites: calcium manganite, $CaMnO_3$ (p. 274), manganous manganite, $(Mn_2)MnO_4$ (see above), potassium manganite, $K_2Mn_2O_5$.

Manganic acid, H_2MnO_4, and manganic anhydride, MnO_3, are not known free.

Manganates are formed as dark-green masses whenever a manganese compound is fused with bases, oxides, or carbonates in the air (or in the presence of salts, giving up oxygen):

$$3MnO_2 + 6KOH + KClO_3 = 3K_2MnO_4 + KCl + 3H_2O.$$

Many manganates, especially of the alkali metals, are soluble in water; if alkali hydroxides are present, they dissolve unchanged with a green color; others also become red in water or acids (even on standing in the air by the CO_2 contained therein), or by chlorine, due to the formation of permanganates and the separation of brown MnO_2; e.g., $3K_2MnO_4 + 2H_2O = 2KMnO_4 + MnO_2 + 4KOH$ (on account of this change in color potassium manganate was formerly called the "mineral chameleon"). If the green solution is carefully evaporated, we obtain the manganate in dark-green crystals, of which the alkali manganates are isomorphous with the alkali sulphates and chromates. The manganates act either directly as oxidizing agent or indirectly by their conversion into permanganates.

Manganese heptoxide, Mn_2O_7, permanganic anhydride, is obtained on slowly introducing potassium permanganate into cold concentrated sulphuric acid: $2KMnO_4 + H_2SO_4 = K_2SO_4 + Mn_2O_7 + H_2O$. It is a dark-green heavy liquid which forms readily decomposable violet vapors and which on heating decomposes into $2MnO_2 + 3O$ with explosive violence, and gradually on standing. It has energetic action on paper, alcohol, etc., inflaming them on contact therewith.

Permanganic Acid, $HMnO_4$. If a barium permanganate solution is treated with the necessary quantity of dilute sulphuric acid, a deep-red solution of permanganic acid is obtained. This latter decomposes even in the light or on warming, with the generation of oxygen: $2HMnO_4 = H_2O + 3O + 2MnO_2$.

Permanganates are produced from the manganates by the action of chlorine or acids (see above), and are obtained as dark-violet crystal on the evaporation of the respective solution. These crystals are mostly isomorphous with the corresponding perchlorates. The permanganates are soluble in water or dilute acids without decomposition, and in the presence of bases they are converted into manganates: $2KMnO_4 + 2KOH = 2K_2MnO_4 + H_2O + O$. On heating they give off oxygen, $10KMnO_4 = 3K_2MnO_4 + 7MnO_2 + 2K_2O + 12O$, and in solution they also readily give off a part of their oxygen to oxidizable bodies, and hence they are powerful oxidizing and disinfecting agents. They set chlorine free from hydrochloric acid, they oxidize sulphur dioxide into sulphuric acid, ferrous salts into ferric salts, oxalic acid to carbon dioxide, most organic compounds into carbon dioxide and water, and with H_2O_2 they give off oxygen (p. 118). If the oxidation takes place in the presence of an acid, then colorless manganous salts are produced: $2KMnO_4 + 3H_2SO_4 = K_2SO_4 + 2MnSO_4 + 3H_2O + 5O$; while if the oxidation takes place in neutral or alkaline solution the oxides of manganese separate out: $2KMnO_4 + H_2O = 2MnO_2 + 2KOH + 3O$. When dry they explode with many oxidizable bodies, and with sulphuric acid they generate the readily explosible Mn_2O_7.

Potassium permanganate, $KMnO_4$, is obtained by the action of carbon dioxide upon potassium manganate solutions (p. 275) until they have attained a red color: $3K_2MnO_4 + 2CO_2 = 2KMnO_4 + MnO_2 + 2K_2CO_3$. On evaporating this liquid dark-violet rhombic prisms are obtained which are soluble in 16 parts water.

Calcium permanganate, $Ca(MnO_4)_2$ is soluble in 4 parts water and hence its aqueous solution has a greater oxidizing and disinfecting power than $KMnO_4$.

e. Detection of Manganese Compounds.

1. When fused with borax in the outer blowpipe flame they give an amethyst-red bead.

2. When heated with soda and saltpeter they give a bluish-green mass of sodium manganate.

3. Ammonium sulphide precipitates flesh-colored manganous sulphide (also from manganates and permanganates):

$$K_2MnO_4 + 4(NH_4)_2S = K_2S + MnS + 4H_2O + 2S + 8NH_3.$$

2. Iron (Ferrum).
Atomic weight 55.9 = Fe.

Occurrence. Native only in meteoric iron, but combined in small quantities in river-, sea-, and spring-waters, and in large quantities in many minerals which often form extensive deposits:

Ferro-ferric oxide, Fe_3O_4, as magnetic iron ore.

Ferric oxide, Fe_2O_3, as red hæmatite, specular iron ore; bloodstone; with clay as red clay-ironstone and called itabiryte.

Ferric oxide and hydroxide, $Fe_2O_3 + 2Fe(OH)_3$, as brown hæmatite or limonite.

Iron bisulphide, FeS_2, when regular as iron pyrites and when rhombic called marcasite.

Ferro-ferric sulphide, $5FeS + Fe_2O_3$, as magnetic pyrites.

Ferrous carbonate, $FeCO_3$, as spathic iron ore.

Ferrous and ferric silicates occur in many minerals and in most rocks, and reach the soil after weathering. Iron nitride is found in lava.

Preparation. Pure iron is not obtained by the reduction of iron oxide with carbon, as this latter forms in part a carbide with the iron. It is obtained by heating ferric oxide or ferrous chloride in a current of hydrogen which forms a gray powder that inflames spontaneously in the air and burns into ferric oxide (iron pyrophorous). If the reduction takes place at higher temperatures, it is no longer inflammable (reduced iron, ferrum reductum). It is obtained in compact masses by fusing powdered iron in the oxyhydrogen blowpipe or by fusing the purest wrought iron (piano-wire) with iron oxide, when this iron oxide takes up all the impurities.

Properties. Crystalline silver-white masses which melt at about 1800° and have a specific gravity of 7.8. Iron is rather soft, weldable and malleable, is attracted by magnets, and is itself magnetic, but loses its magnetism as soon as the magnet is removed. It is unchangeable in dry air; in moist air it is covered with ferric hydroxide (rust); on heating in the air it is covered with black ferrous-ferric oxide (forge-scales). When finely powdered it decomposes water with the evolution of hydrogen at ordinary temperatures, but when compact this takes place only at a red heat. It dissolves in dilute acids with the generation of hydrogen and the formation of ferrous salts; in concentrated sulphuric acid it forms ferric sulphate and generates SO_2; in hot concentrated nitric acid it forms ferric nitrate with the generation of NO.

If iron is dipped in concentrated nitric acid and then washed, it is not further acted upon by nitric acid and does not precipitate copper from the solutions of its salts (passivity of iron). In the molten state it can take up 5 per cent. of carbon, which exists either mechanically mixed in the form of graphite or alloyed with the iron as iron carbide (FeC_4, FeC_3, Fe_3C_2, p. 186). When finely powdered it may also combine with carbon monoxide (p. 189).

Iron forms three series of compounds:

Bivalent iron forms the ferrous compounds whose salts are white or green and have great similarity to those of the magnesium group.

Trivalent iron forms the ferric compounds, which, in contradistinction to the corresponding compounds of manganese, cobalt, and nickel, are very stable and behave like the aluminium and chromium salts. The salts of this series are brown or yellow.

Hexavalent iron forms ferric acid, which, like manganic acid, is known only in combination.

a. Alloys of Iron.

Pure iron has no technical uses, but only as alloyed with carbon, manganese, chromium, tungsten, and silicon. Iron is classified according to the amount of carbon it contains and the properties imparted thereby, as follows:

1. Malleable iron contains 0.1 to 1.6 per cent. carbon; is fused with difficulty, is forgeable and malleable.

a. Steel contains 0.6 to 1.6 per cent. carbon, is the only form of iron that can be tempered, is light gray in color, granular and not fibrous, not as tough as wrought iron, more easily melted (at about 1400°), and does not readily rust.

If melted steel is allowed to cool slowly, it becomes more pliable and softer than crude iron. But if it is quickly plunged in water, it becomes brittle and hard so that it can scratch glass, and in fact the higher the temperature to which it is heated and the colder the liquid used in cooling it the harder does it become.

As the degree of hardness cannot be well controlled by simply cooling, it is best to heat the steel to be hardened to a certain temperature and then allow it to cool slowly. Polished steel on heating first becomes pale yellow, then brown, violet, pale blue, dark blue, depending upon the heat applied.

b. Wrought iron contains less than 0.5 per cent. carbon and melts at about 1500°. It is the softest of all varieties of iron. When it contains 0.3 per cent. carbon or less it has a fibrous fracture, and when it contains more the fracture is granular. When fibrous it is more resistant to fracture.

As forms of iron which can be tempered can now be prepared by the addition of larger amounts of manganese, chromium, tungsten, and silicon to iron containing minimum amounts of carbon, it is not possible to differentiate between steel and wrought iron by the amount of carbon contained in them. Many intermediary products exist between steel and wrought iron, so that it is often impossible to tell which is which. For this reason malleable iron is classified according to its condition after its preparation.

2. Pig or cast iron (crude iron) contains 2.3 to 5 per cent. carbon, is hard and brittle, melting at about 1100° and suddenly passing into the liquid state without previously softening. For this reason it is not forgeable nor malleable when heated.

a. Gray pig iron is produced by slowly cooling pig iron, when a part of the carbon contained therein separates out as black plates of graphite and gives the iron a dark-gray color. It is not very hard, but brittle, and contracts uniformly on cooling; hence it is used as cast iron in the preparation of cast-iron ware.

b. White pig iron is produced by quickly cooling the pig iron, when the carbon remains in combination with the iron. From crude iron containing manganese we obtain on slowly cooling a variety of iron called "spiegel-eisen." It is very hard and brittle and contracts unevenly on cooling. It is used in the manufacture of steel and wrought iron.

Iron containing 1.6 to 2.3 per cent. carbon is not used for technical purposes. Pig iron is the material from which all other varieties of iron are prepared. In its obtainment ores containing chiefly iron oxide and iron carbonate are made use of, and these are reduced by heating with carbon. Ores containing sulphur are seldom used, as the sulphur must first be removed by roasting and converted into iron oxide.

1. *Preparation of Pig or Crude Iron.* The ores are roasted, in order to make them more porous and to form oxides, then mixed with coke (or coal) and slag-forming substance (flux), and then introduced into the blast-furnace, which is previously filled with red-hot coals (charcoal was formerly used; anthracite cannot be used directly, but must first be converted into coke). The iron ores nearly always contain clay and sand, which are both infusible and prevent the coalescence of the melted particles of iron; hence substances (flux) are added to the ore before the iron melts which form a readily liquefiable silicate (slag) with the above bodies.

The flux consists of sand or clay when the ores are poor in silicates, and of limestone when the ores are rich in silicates. The slag causes the molten particles of iron to conglomerate, dissolves the foreign constituents of the iron, and prevents the oxidation of the crude iron by the oxygen of the air-blast. As the iron ores are reduced only at very high temperatures, the active combustion is maintained by a blast opening in the lower part of the furnace.

As fast as the coke burns and the ore with the flux melts, the material sinks and is replaced from above continuously, so that a blast-furnace may be in operation for several years. The molten iron collects at the base of the furnace, and the molten slag floats upon the iron and is drawn off through openings on the side. As soon as the melted iron reaches the slag apertures it is drawn off through openings at the bottom and run off into sand moulds.

In the upper part of the furnace (heating zone) the material is warmed and dried. In the next part (the reduction zone) the iron oxide is reduced into spongy iron by the carbon monoxide produced in the lower part of the furnace by the combustion: $Fe_2O_3 + 3CO = 2Fe + 3CO_2$. In the lower part of the furnace, where the combustion is very energetic, due to the air introduced by the blast, the coke burns into carbon dioxide, which is reduced to carbon monoxide by passing through the upper layers of red-hot coal: $CO_2 + C = 2CO$.

The temperature of the reduction zone is not sufficient to melt the iron; but it sinks with the flux to the lower and hotter part of the furnace and here recombines with the carbon, forming readily fusible crude iron (carbonization zone). The soft crude iron now sinks to a still hotter part, where it melts (fusion zone), and where the slag forms from the flux and the other bodies admixed, and which prevents the oxidation of the iron as it passes the openings of the air-blast (combustion or oxidation zone). The iron collects below the air-blast, and the slag which floats upon it prevents its further oxidation.

2. *Preparation of Wrought Iron. a.* In the Bessemer process fused crude iron is introduced into a large pear-shaped vessel (converter) and compressed air blown through the bottom, and the carbon, silicon, etc., is completely burnt out. In this manner 10 tons of pig iron can be converted into wrought iron in 20 minutes. As soon as all the carbon has been burnt out (as shown by the disappearance of the green lines of carbon from the spectrum of the flame) the molten, slag-free wrought iron is poured out by tipping the converter.

In the Bessemer process the phosphorus contained in many iron ores is not burnt out, and this makes the iron brittle. In order to prevent this the converter is lined with limestone containing some magnesia, when all the phosphorus is converted into calcium phosphate (Thomas-Gilchrist process, basic process). This is collected as Thomas slag, which when ground becomes a valuable fertilizer, as it contains 50 per cent. calcium phosphate (Thomas phosphate, p. 223).

b. Pig iron is melted on an open hearth or on a reverberatory hearth and continuously stirred (hence the name "puddle process"), and exposed to a current of air. By this means nearly all the carbon is burnt into CO_2, while the phosphorus, sulphur, and silicon which always exist to a slight extent in pig iron are also burnt into oxides, and the surface of the molten iron is also oxidized; the ferric oxide produced forms a slag with the silicon dioxide. The mass, which consists of pasty wrought iron and slag (the blooms), is hammered or rolled while white-hot and the slag pressed out.

3. *Preparation of Steel. a.* By the Bessemer process (2*a*), where the air-blast is stopped when the flame is highest (Swedish method), or by adding the proper amount of pig iron to the wrought iron produced in the converter and allowing the air-blast to blow through this mixture for a moment in order to melt the mass (English method of preparing Bessemer steel).

b. Pieces of wrought iron are embedded in carbon powder and heated in closed clay boxes, when the carbon unites with the iron. This cementation steel contains more carbon on the outside than on the inside and hence is reforged or remelted (cast steel).

c. By the Siemens-Martin process. By melting pig and wrought iron together in especially constructed furnaces at a very high temperature.

d. By puddling in a similar manner as described under the preparation of wrought iron, but the decarbonization is not allowed to progress so far as in wrought iron.

b. Ferrous Compounds.

Ferrous oxide, FeO, is obtained as a very unstable black powder when hydrogen or CO is passed over ferric oxide heated to 300°.

Ferrous hydroxide, $Fe(OH)_2$, obtained by precipitating ferrous salt solutions with caustic alkali, is a light-green precipitate which quickly oxidizes into brown ferric hydroxide in the air.

Ferrous sulphide, FeS, is obtained by heating iron with the proper amount of sulphur as a bronze-colored crystalline mass, insoluble in water, which readily fuses and which dissolves in acids with the generation of H_2S: $FeS + 2HCl = FeCl_2 + H_2S$. If an intimate mixture of iron and sulphur powder is moistened with water, a combination takes place even at ordinary temperatures. Ammonium sulphide precipitates amorphous black ferrous sulphide from all iron salts; ferric salts are first reduced to ferrous salts by this agent, and at the same time sulphur separates:

$$2FeCl_3 + (NH_4)_2S = 2FeCl_2 + 2NH_4Cl + S;$$
$$FeCl_2 + (NH_4)_2S = FeS + 2NH_4Cl.$$

In the air at a gentle heat it is oxidized in part into ferrous sulphate; with higher heat sulphur dioxide and ferric oxide are produced: $2FeS + 7O = Fe_2O_3 + 2SO_2$.

Ferrous chloride, $FeCl_2 + 4H_2O$, is obtained by dissolving iron in hydrochloric acid and evaporating in the absence of air. It forms pale-green prisms which deliquesce and which cannot be obtained anhydrous without decomposition. Anhydrous ferrous chloride can be obtained as white, fusible, volatile plates on heating iron in hydrochloric acid gas.

Ferrous iodide, FeI_2. Powdered iron is placed in water and the corresponding quantity of iodine added, when a greenish solution is obtained from which on evaporation $FeI_2 + 4H_2O$ separates out as bluish-green monoclinic crystals. It oxidizes in the air into iron oxide and iodine separates; this decomposition is greatly retarded by the addition of sugar, hence the druggist keeps this body in sugar or adds sugar to its solution.

Ferrous sulphate, green vitriol, $FeSO_4 + 7H_2O$, is obtained pure by dissolving iron in dilute sulphuric acid and evaporating the solution. It forms pale-green monoclinic crystals which effloresce in dry air (p. 115), and in moist are covered with brown basic ferric sulphate. It may be obtained as a crystalline powder by precipitating the solution with alcohol. Ferrous sulphate is prepared on a large scale by allowing roasted iron pyrites, FeS_2, to lie in the air after moistening, when it is oxidized into ferrous sulphate. This is extracted with water and allowed to crystallize. On roasting the iron pyrites one-half of the sulphur is eliminated as sulphur dioxide and is used in the manufacture of sulphuric acid. Ferrous sulphate is also prepared in the obtainment of copper by the wet method (p. 234). It is soluble in 1.8 parts cold water, insoluble in alcohol, and •on heating in the air it is transformed into ferric oxide (p. 130).

With alkali sulphates it forms isomorphous double salts corresponding to the magnesium compounds, of which iron ammonium sulphate, $FeSO_4 + (NH_4)_2SO_4 + 6H_2O$, so-called Mohr's salt, is characterized by its stability in the air. Like magnesium sulphate, ferrous sulphate loses its seventh molecule of water first at 260°.

Ferrous sulphate (siccum), $2FeSO_4 + 3H_2O$, is obtained as a white powder by heating ferrous sulphate until it has lost 35 per cent. of its water and still contains 15 per cent.

Ferrous phosphate, $Fe_3(PO_4)_2$, occurs with $8H_2O$ as vivianite, is obtained as a white precipitate, which is insoluble in acetic acid, but soluble in other acids, by treating ferrous salt solutions with sodium phosphate. It is oxidized in the air and becomes grayish blue.

Ferrous carbonate, $FeCO_3$, occurs as spathic iron ore in yellowish crystalline masses or in rhombohedra (which are isomorphous with carbonates of calcium, magnesium, zinc, manganese, as well as with nickelous and cobaltous carbonates). It is obtained by treating a solution of ferrous salt with alkali carbonate with the exclusion of air. It is a white precipitate which in the air is quickly changed into brown ferric hydroxide; as the addition of sugar retards the oxidation for a long time, a mixture of ferrous carbonate with sugar forms the official Ferri carbonas saccharatus. Ferrous carbonate is somewhat soluble in water containing carbon dioxide and occurs in what are called chalybeate waters.

c. Ferric Compounds.

Ferric oxide, Fe_2O_3, occurs extensively in nature (p. 277). It is obtained as an amorphous reddish-brown powder, difficultly soluble in acids, by heating green vitriol (in the manufacture of fuming sulphuric acid, p. 130), also from ferrous or ferric hydroxides.

Ferric hydroxide, $Fe(OH)_3$, occurs as brown hæmatite (p. 277), and is produced as rust on exposing iron to moist air. It is obtained by mixing a ferric salt solution with an excess of caustic alkali or ammonia carbonate (p. 284) which gives a reddish-brown precipitate insoluble in water and readily soluble in acids. It may be dried by careful heating, forming an amorphous granular mass. When freshly precipitated it serves as an antidote in arsenical poisoning (p. 174).

With cane-sugar, $C_{12}H_{22}O_{11}$, it forms a so-called iron saccharate, $2Fe(OH)_3 + C_{12}H_{22}O_{11} + 7H_2O$, which is readily soluble in water.

Sirupus ferri oxydati is a solution of iron saccharate in dilute sugar sirup.

Freshly precipitated ferric hydroxide dissolves in ferric chloride with the formation of iron oxychloride, $4Fe(OH)_3 + FeCl_3$. A similar iron oxychloride solution may also be obtained by the action of HCl upon freshly precipitated ferric hydroxide in certain proportions (Liquor ferri oxychlorati). In these solutions the chlorine does not exist as chlorine ions and hence does not give any precipitate with silver nitrate. If this solution is placed in a dialyzer (p. 47), a colloidal solution of ferric hydroxide free from $FeCl_3$ is obtained as a brown liquid (dialyzed iron oxide solution). Small quantities of alkalies, alkali salts, or sulphuric acid, as well as boiling, precipitate the ferric hydroxide as a red gelatinous mass.

Ferric sulphide, Fe_2S_3, may be obtained by fusing iron together with the proper quantity of sulphur, like the other sulphides of iron.

Ferric chloride, $FeCl_3$, is obtained in an anhydrous form as dark-green plates by heating iron in chlorine gas. It is obtained in solution by the action of hydrochloric acid upon ferric hydroxide or upon iron. In the latter case ferrous chloride solution is obtained which is oxidized by chlorine or nitric acid: $FeCl_2 + Cl = FeCl_3$; $3FeCl_2 + 3HCl + HNO_3 = 3FeCl_3 + 2H_2O + NO$.

On the evaporation of this solution until it has a specific gravity of 1.57 it solidifies, forming a yellow crystalline mass, $FeCl_3 + 6H_2O$. On further evaporation to a thick sirup reddish-brown crystals having the formula $FeCl_3 + 3H_2O$ separate out on cooling. These salts are very deliquescent, and readily soluble in water, alcohol, and ether.

Ferric sulphate, $Fe_2(SO_4)_3 + 9H_2O$, is obtained by the addition of nitric acid to a solution of ferrous sulphate which has previously been treated with sulphuric acid. On evaporation to sirupy consistency colorless crystals are obtained:

$$6FeSO_4 + 3H_2SO_4 + 2HNO_3 = 3Fe_2(SO_4)_3 + 2NO + 4H_2O.$$

Ferric-ammonium sulphate, $Fe(NH_4)(SO_4)_2 + 12H_2O$, ferric alum, ammonium iron alum, forms amethyst-colored octahedra which are soluble in water.

Ferric carbonate, $Fe_2(CO_3)_3$, is not known; if a ferric salt solution is treated with alkali carbonate, ferric hydroxide is produced:

$$2FeCl_3 + 3Na_2CO_3 + 3H_2O = 2Fe(OH)_3 + 6NaCl + 3CO_2.$$

d. Higher Iron Compounds.

Ferric acid, H_2FeO_4, and its anhydride, FeO_3, like the corresponding compounds of manganese, are not found in the free state.

Ferrates, with the exception of potassium ferrate, K_2FeO_4, are unknown. Potassium ferrate is obtained by fusing iron and potassium nitrate together: $Fe + 2KNO_3 = K_2FeO_4 + 2NO$. It forms red prisms isomorphous with potassium sulphate and chromate which are soluble in water and whose solution is decolorized by an excess of water, by acids, or by standing in the air:

$$2K_2FeO_4 + 5H_2O = 4KOH + 2Fe(OH)_3 + 3O.$$

e. Detection of Iron Compounds.

1. Ammonium sulphide precipitates black ferrous sulphide from all solutions of iron salts. This precipitate is readily soluble in acids (p. 282).

2. Potassium ferrocyanide produces with ferrous salt solutions a white precipitate which quickly changes to light blue. In ferric salt solutions it immediately produces a deep-blue precipitate of Prussian blue (See Cyanogen Compounds).

3. Potassium ferricyanide produces immediately a deep-blue precipitate called Turnbull's blue with ferrous salts. With ferric salt it only forms a reddish-brown coloration.

4. Tannic acid (which see) produces no change in ferrous salt solutions, but with ferric salts it immediately gives a bluish-black precipitate (pigment of common ink).

5. Potassium sulphocyanide (which see) produces no change with ferrous salts, but with ferric salts it gives a blood-red coloration of soluble iron sulphocyanide, $Fe(CNS)_3$.

3. Cobalt.

Atomic weight $59 = Co$.

Occurrence. Native only in meteoric iron; combined as safflorite, $CoAs_2$, cobaltite, $CoS.As$, cobalt pyrites or linnæite, $CoS + Co_2S_3$, cobalt bloom or erythrite, $Co_3(AsO_4)_2$.

Preparation. In all cobalt ores cobalt is in part replaced by isomorphous Ni, Fe, Mn. The separation of these metals takes place in a similar manner as in quantitative analysis. We obtain finally by this method cobaltous oxide, which is mixed with flour into a dough, this pressed into small cubes, dried, and heated to a white heat in large crucibles surrounded by charcoal powder (cube cobalt of commerce).

Properties. Reddish-white, ductile, tough, shining metal having a specific gravity of 8.9, melting at about 1800°; it is attracted to a magnet and, like iron, is temporarily magnetic and passive (p. 278). It does not change in moist air; decomposes water at a red heat; dissolves slowly in hydrochloric or sulphuric acid, but rapidly in nitric acid, forming the corresponding cobaltous salts and H or NO respectively.

As divalent it forms the cobaltous compounds; these behave similar to the magnesium compounds (p. 229) and are very stable. The salts of this series are mostly red when they contain water, and blue when anhydrous, and are isomorphous with the corresponding ferrous compounds.

As trivalent, cobalt forms cobaltic compounds; only a few simple salts of this series are known, but more complex salts are known (see below).

a. Cobaltous Compounds.

Cobaltous oxide, CoO, is obtained by heating cobaltous hydroxide or cobaltous carbonate in the absence of air. It forms a greenish powder which, on heating in the air, forms cobaltous-cobaltic oxide.

Cobaltous hydroxide, $Co(OH)_2$. Caustic alkalies precipitate from cobaltous salt solutions blue basic cobaltous salts; these, on boiling, are converted into rose-red cobaltous hydroxide, which, on standing in the air, turns brown by oxidation. It also dissolves in an excess of ammonia, forming cobaltic-amine salts (see below).

Cobaltous sulphide, CoS, is obtained as a black precipitate, insoluble in dilute hydrochloric acid, when a cobaltous salt solution is precipitated with ammonium sulphide.

Cobaltous sulphate, $CoSO_4 + 7H_2O$, crystallizes in monoclinic brownish-red prisms and behaves like the sulphates of the magnesium group.

Cobaltous chloride, $CoCl_2 + 6H_2O$, forms red monoclinic prisms. Writing with a pale-red solution of this salt can be read only after warming, as the previously imperceptible rose-color becomes blue in this operation, due to the formation of anhydrous cobaltous chloride (sympathetic ink, hygrometers).

Cobaltous nitrate, $Co(NO_3)_2$, has the color and the properties of the chloride; it is used in blowpipe analysis.

Cobaltous silicate forms a constituent of blue glass. The blue color called smalt is potassium-cobaltous silicate.

b. Cobaltic Compounds.

Cobaltic oxide, cobalt sesquioxide, Co_2O_3, is obtained as a black powder by gently heating cobaltous nitrate. On stronger heating it is, like cobaltous oxide, converted into black

Cobaltous-cobaltic oxide, $Co_3O_4 = (CoO + Co_2O_3)$.

Cobaltic hydroxide, $Co(OH)_3$, is produced when chlorine is passed into cobaltous hydroxide suspended in caustic alkali. It forms a brownish-black powder: $Co(OH)_2 + H_2O + Cl = Co(OH)_3 + HCl$.

Cobaltic salts. Cobaltic oxide and cobaltic hydroxide, on being well cooled, dissolve in acids with a brownish-yellow color. The cobaltic salts thus obtained cannot be isolated because on evaporation, or on gently warming, they decompose with the evolution of oxygen or chlorine and are transformed into cobaltous salts, hence only a few are known; e.g., $Co_2(SO_4)_3 + 18H_2O$, as well as the corresponding alum. We also know of the *cobaltic-amine salts*, especially the deep-red *purpureo-cobaltic salts*, $CoCl_3(NH_3)_5$, the pale-red *roseo-cobaltic salts*, $CoCl_3(NH_3)_5(H_2O)$, the brownish yellow *luteo-cobaltic salts*, $CoCl_3(NH_3)_6$, which are precipitated by HCl from solutions of $Co(OH)_2$ in NH_3 when they have stood in the air. Salts of *cobaltic-hydrocyanic acid*, $H_3Co(CN)_6$ (known free), and of *cobaltic-nitrous acid*, $H_3Co(NO_2)_6$ (not known free), are also known. The potassium salt of the latter, potassium cobaltic nitrite, $K_3Co(NO_2)_6$, which separates as a yellow precipitate when a cobaltous salt solution is treated with acetic acid and potassium nitrite, is used in the separation of cobalt from nickel and in the detection of potassium, and forms the color called cobalt-yellow.

These last compounds have complex ions instead of one cobalt ion, and hence do not give the following reactions for cobalt.

ʽc. Detection of Cobalt Compounds.

1. When fused with borax they yield a beautiful blue vitreous mass.

2.. Ammonium sulphide precipitates black cobaltous sulphide, which is insoluble in dilute hydrochloric acid (all other metals, with the exception of nickel, which are precipitable by ammonium sulphide form sulphides which are soluble in dilute hydrochloric acid).

3. Potassium nitrite precipitates yellow potassium cobaltic nitrite from cobaltous salt solutions acidified with acetic acid (see above).

4. Caustic alkali precipitates blue basic cobaltous salts (p. 286).

5. Ammonia precipitates blue basic cobaltous salts, which are soluble in excess of ammonia with brown coloration, gradually becoming red; cobaltic-amine salts are hereby produced.

4. Nickel.

Atomic weight $53.7 = Ni$.

Occurrence. Native only in meteoric iron; combined as coppernickel or niccolite, NiAs, nickel glance or gersdorffite, NiAsS, nickel pyrites, NiS, and garnierite, $Ni_5Si_4O_{13} + Mg_5Si_4O_{13} + 3H_2O$ and also in the cobalt ores.

Preparation. Since the discovery of garnierite, nickel is nearly entirely obtained from this mineral by means of the blast-furnace, in the same way as iron. In all other nickel ores the nickel is partly replaced by isomorphous Co, Fe, Cu; the separation of these metals takes place in several roundabout ways; the nickel oxide finally obtained is reduced similarly to cobaltous oxide (see Cobalt).

Properties. Silver-white, ductile, tough metal, of specific gravity 8.8, unchangeable in the air, decomposing water at a red heat, melting at about 1600°, slowly dissolved by hydrochloric and sulphuric acids, and rapidly by nitric acid with the evolution of H or NO. It is attracted by a magnet and may, like iron, be temporarily magnetic as well as passive. On account of its silver-like color and stability, it is extensively used to cover other metals (nickel plate). Like iron, it combines with carbon monoxide (p. 189). Nickel compounds correspond in their constitution and properties to those of cobalt, and are obtained in the same manner.

a. Alloys of Nickel.

Alloys of nickel, copper, and zinc are called German silver, argentan. German and United States nickel coins consist of 25 per cent. nickel and 75 per cent. copper. Nickel steel for armor-plates contains 4–5 per cent. nickel.

Alloys of nickel with copper (called nickelin, constantan), also with copper and manganese (manganin), have low electrical conductivity and hence are used for electrical resistances.

b. Nickelous Compounds.

Nickelous oxide, NiO, forms a gray powder and is produced by heating
Nickelous hydroxide, $Ni(OH)_2$. This is obtained from nickelous salts by alkali hydroxide as a green precipitate, soluble in ammonia with blue color and, contrary to $Co(OH)_2$, stable in the air.

Nickelous sulphide, NiS, is black and behaves like CoS.

Nickelous salts when anhydrous are yellow and green when they contain water.

c. Nickelic Compounds.

Nickelic hydroxide, $Ni(OH)_3$, a black powder, and
Nickelic oxide, Ni_2O_3, forming black masses, are prepared like and have similar properties to the analogous cobalt compounds.

Nickelic salts are not known even in solution, as nickelic oxide as well as nickelic hydroxide is soluble in acids even in the cold, forming nickelous salts and evolving oxygen or chlorine, and hence acts like a superoxide (see Cobaltic Salts).

d. Detection of Nickel Compounds.

1. When fused with borax they give a dark-red mass which on cooling turns pale yellow.

2. Caustic alkali precipitates green nickelous hydroxide $Ni(OH)_2$.

3. Ammonia partly precipitates green nickelic hydroxide, which is soluble in an excess of ammonia with blue color.

4. Nickel salts behave like the cobalt salts towards ammonium sulphide; they are not precipitated by potassium nitrite.

GOLD AND PLATINUM GROUP.

Gold. Platinum. Iridium.

Osmium. Palladium. Rhodium. Ruthenium.

These metals, with the exception of palladium and osmium, are not acted upon by nitric acid and are only dissolved by aqua regia or other liquids containing chlorine. They do not decompose water even at higher temperatures, and are precipitated as sulphides from their acid solutions by sulphuretted hydrogen. These sulphides are soluble in alkali sulphide solutions, forming alkali sulphosalts (see p. 176).

Their oxides decompose on heating into the metal and oxygen (with the exception of ruthenium and osmium); hence they are called noble metals (p. 200).

Their lower oxides are weak bases; the higher oxides, with the exception of those of palladium, are acidic oxides. Gold occurs mono- and trivalent.

The six metals related to gold form the platinum group. They are all found native and in fact alloyed with each other in the platinum ores. They have similarities to the iron group, and those arranged in columns in the following table show great chemical similarity:

Manganese.	Iron.	Cobalt.	Nickel.
Ruthenium.		Rhodium.	Palladium.
Osmium.		Iridium.	Platinum.

The metals of the first group occur di-, tri-, tetra-, and octavalent, oxidize readily in the air, and form when hexavalent salts corresponding to the unknown acid H_2FeO_4. Manganese and ruthenium also occur heptavalent in the acids $HMnO_4$ and $HRuO_4$. Osmium and ruthenium form volatile tetroxides.

The metals of the second group occur di-, tri-, and tetravalent, and are stable in the air and form when trivalent stable salts corresponding to the unknown acid $H_3Co(NO_2)_6$, etc. They also form complex amine salts with NH_3 (p. 286).

The metals of the third group occur di- and tetravalent, nickel also occurring as trivalent, and likewise form amine salts with NH_3.

The metals of all three groups, as well as gold, form, as cyanides, complex compounds with alkali cyanides, and also as chlorides with alkali chlorides.

Ru, Rh, Pd, having a specific gravity of 11.8 to 12.1, are called the light platinum metals, while Os, Ir, Pt, having a specific gravity of 21.1 to 22.4, are called the heavy platinum metals.

1. Gold (Aurum).

Atomic weight 197.2=Au.

Occurrence. Native or mixed with silver in the crystalline rocks or in the sands obtained on the weathering of the same. Nearly all river sand contains small amounts of gold.

Preparation. The sand or the rocks containing gold are stamped, and then washed with water, when all lighter particles are removed, while the gold remains (gold washing). The sands may also be mixed with mercury, with which it forms an amalgam, and the mercury separated from the dissolved gold by distillation.

Ores poor in gold, or the ores after the amalgamation process, are ground with a watery solution of potassium cyanide with an excess of air, when all the gold goes into solution as potassium aurocyanide, $KAu(CN)_2$ (p. 290), and the gold is separated from the solution by the galvanic current or by metallic zinc. Or the roasted ore can also be treated with chlorine, and the dissolved gold chloride precipitated by ferrous sulphate (process, p. 290).

The gold thus obtained is separated from the silver, which is nearly always present, either by boiling with nitric acid (p. 160), or with concentrated sulphuric acid, which leaves the gold undissolved.

Pure gold is obtained by dissolving the commercial metal in aqua regia and treating the solution with ferrous sulphate, which reduces the gold chloride and precipitates the gold as a brown powder. This is fused with borax and saltpeter:

$$AuCl_3 + 3FeSO_4 = Au + Fe_2(SO_4)_3 + FeCl_3.$$

Properties. Shining, yellow metal, nearly as soft as lead, having a specific gravity of 19.3, melting at 1064°. It is the most ductile of all metals and can be beaten into very thin sheets (gold-leaf) which have a green color by transmitted light. When finely divided, as when precipitated from solution by organic substances (on the skin) or by fine silver (toning of photographic prints), it appears red or bluish violet. Neither oxygen nor sulphur nor acids combine with gold directly, but liquids containing chlorine (aqua regia, chlorine-water) dissolve it readily with the formation of auric chloride, $AuCl_3$; potassium cyanide in the presence of oxygen also dissolves gold readily, forming potassium aurocyanide:

$$2Au + 4KCN + H_2O + O = 2KAu(CN)_2 + 2KOH.$$

Colloidal gold, which with water forms a blue or purplish-red solution, is also known (pp. 199 and 292). Gold may be precipitated as a dark-brown powder from solutions of its salts by other metals and many reducing agents.

Gold forms compounds having a composition analogous to those of thallium; thus as monovalent it forms aurous compounds, as trivalent it forms auric compounds. Besides the haloid salts no simple salts are known, but, on the contrary, a few complex salts.

a. Alloys of Gold.

Gold is alloyed with silver or copper, making it harder and more readily fusible without losing its color. The amount of gold in an alloy was formerly expressed in carats (pure gold = 24 carats); good gold ware is 14 carats, i.e., it consists of 14 parts gold and 10 parts copper. At the present time we generally designate pure gold as 1000, and denote the amount of alloys in parts per 1000. United States and German gold coins have a fineness of $\frac{900}{1000}$, i.e., contain one-tenth copper.

b. Aurous Compounds.

Aurous oxide, Au_2O, is obtained by the action of alkali hydroxides upon aurous chloride, giving a dark-violet powder which decomposes at 250° into $Au_2 + O$, and by HCl into $AuCl_3 + Au$.

Aurous chloride, AuCl, is produced on heating auric chloride to 180°, and forms a white powder decomposing into $Au + Cl$ upon further heating, and into $AuCl_3 + 2Au$ on boiling with water.

c. Auric Compounds.

Auric oxide, auric anhydride, gold oxide, Au_2O_3, is produced from magnesium aurate by treatment with concentrated nitric acid: $Mg(AuO_2)_2 + 2HNO_3 = Mg(NO_3)_2 + H_2O + Au_2O_3$. It is a brown powder which behaves like auric hydroxide and decomposes at 250° into $Au_2 + O$.

Auric hydroxide, auric acid, $Au(OH)_3$, is prepared from magnesium aurate (see below) by treatment with dilute nitric acid, $Mg(AuO_2)_2 + 2HNO_3 + 2H_2O = Mg(NO_3)_2 + 2Au(OH)_3$, giving a reddish-brown powder which decomposes in the light, with the generation of oxygen. It dissolves in hydrochloric acid, forming auric chloride; it is not attacked by oxyacids, and dissolves in alkali hydroxides, forming aurates.

The *aurates* are derived from metauric acid, $HAuO_2$, which is not known free. If an auric chloride solution is gently warmed with magnesium oxide, we obtain a yellow precipitate of magnesium aurate, $Mg(AuO_2)_2$; if auric chloride solution is treated with caustic potash, auric hydroxide is precipitated, and this dissolves in an excess of the caustic potash, forming potassium aurate, $KAuO_2$, which on careful evaporation may be obtained as yellow needles: $Au(OH)_3 + KOH = KAuO_2 + 2H_2O$.

Auric chloride, $AuCl_3$, so-called brown gold chloride, is obtained as a brown crystalline, deliquescent mass on heating powdered gold in chlorine gas. If heated to 180°, it decomposes into $AuCl + Cl_2$, and on slowly evaporating its watery solution it separates in yellow needles of $AuCl_3 + 4H_2O$.

Hydrochlorauric acid, $HAuCl_4$, so-called yellow gold chloride, separates with four molecules of water in yellow crystals on the evaporation of the solution obtained by dissolving gold in aqua regia.

Chloroaurates are prepared by evaporating the mixed solutions of auric chloride and the respective metallic chloride. They form yellow crystals, e.g., sodium chloroaurate, $NaAuCl_4$ (gold salt of the photographer).

Gold oxide ammonia, $Au_2O_3 + 4NH_3$, is obtained as a yellowish-brown, readily explosive precipitate when a gold chloride solution is treated with ammonia.

Auric sulphide, gold sulphide, Au_2S_3, is obtained as a black precipitate from gold salt solutions by sulphuretted hydrogen; it is only soluble in aqua regia, and in alkali sulphides it dissolves, forming sulphoauric salts (p. 176), which are derived from sulphoauric acid, H_3AuS_3, which is not known free.

d. Detection of Gold Compounds.

1. When fused with soda upon charcoal they yield golden ductile granules.

2. Zinc, iron, copper, and many other metals, also reducing agents, such as ferrous sulphate, arsenic trioxide, sulphur dioxide, oxalic acid, etc., precipitate metallic gold as a brown powder from gold solutions. On rubbing this powder with a hard object it takes a metallic luster.

3. Sulphuretted hydrogen precipitates black auric sulphide, readily soluble in yellow ammonium sulphide, forming $(NH_4)_3AuS_3$.

4. Stannous chloride solution which contains some stannic chloride causes purple-red precipitations in gold solution. This precipitate consists of tin hydroxide and colloidal gold, called purple of Cassius.

2. Platinum.
Atomic weight 194.8 = Pt.

Occurrence. Always native and mixed with the other platinum metals (p. 289) in platinum ores.

Preparation. The platinum ores are treated with aqua regia. The osmium and iridium remaining undissolved, the solution obtained is evaporated partly and treated with ammonium chloride, when the platinum precipitates as ammonium platinic chloride, $(NH_4)_2PtCl_6$, mixed with some ammonium iridium chloride $(NH_4)_2IrCl_6$. This precipitate is heated, when the platinum containing iridium remains as a gray porous mass (spongy platinum); this is fused in the oxyhydrogen blowpipe flame and cast in moulds. The solution obtained on the filtration of the ammonium-platinic chloride contains the chlorides of palladium, rhodium, ruthenium, and a part of the iridium, which can be precipitated by means of iron and separated according to different methods.

Properties. White soft metal having a specific gravity of 21.4, fusible only at 1770°, malleable at white heat, soluble only in aqua regia, and not directly oxidizable by any oxidizing agent; hence platinum vessels are extensively used in chemical manipulations. As platinum has the same expansion coefficient as glass, platinum wire is used to conduct electricity into vacuous glass vessels (incandescent electric lamps).

Platinum is attacked on fusing with hydroxides, sulphides, nitrates, and cyanides of the alkalies; it forms readily fusible alloys with phosphorus, arsenic, antimony, boron, silicon, and most metals, so that these metals and very readily reducible compounds must not be heated in platinum vessels. On heating with carbon or silicates, platinum takes up carbon and silicon, becomes brittle, and hence platinum vessels must not be heated on a coal fire or over a smoky flame.

Platinum may be obtained in a finely divided state as a black powder (platinum black) by precipitating it from its solution by zinc or iron, or by heating its solution with caustic alkali and organic reducing substances, such as glucose, alcohol, and glycerine.

Colloidal platinum, which gives a deep-black solution in water, is obtained when the electric arc plays between platinum wires under water (p. 53), and may be precipitated from this solution by salts.

Platinum when finely divided has strong catalytic properties (p. 66), especially accelerating the union of many gases (p. 105). This seems to be brought about by the property of platinum, especially when finely divided, of condensing (absorption, p. 49) oxygen and more particularly hydrogen, so that these gases, even at ordinary temperatures, bring about chemical changes which otherwise require higher temperatures.

Platinum occurs divalent in the platinous compounds and tetravalent in the platinic compounds.

a. Alloys of Platinum.

Platinum alloyed with 2 to 3 per cent. iridium is ordinarily used, as it is harder and more resistant to chemicals.

b. Platinous Compounds.

Platinous **oxide**, PtO, is obtained as a gray powder on carefully heating $Pt(OH)_2$.

Platinous **hydroxide**, $Pt(OH)_2$, is obtained as a black powder on warming $PtCl_2$ with caustic alkali. Both of these compounds dissolve in bases and acids with the formation of salts.

Platinous **chloride**, $PtCl_2$, produced on heating hydrochlorplatinic acid to 280°, is a green powder, insoluble in water, which readily forms o   e double salts with the alkali chlorides, Na_2PtCl_4, which are derived from bl

Hydrochlorplatinous acid, H_2PtCl_4, which is obtained on evaporating $PtCl_2$ with HCl. It forms yellow crystals.

Platosamines is the name given to the compounds of platinous salts with ammonia; e.g., $PtCl_2(NH_3)$, $PtCl_2(NH_3)_x$, $Pt_2Cl_2(NH_3)$, etc., in which, in place of one platinum ion, a complex ion with one or two divalent platinum atoms exists (p. 84).

c. Platinic Compounds.

Platinic **oxide**, PtO_2, produced as a black powder on heating $Pt(OH)_4$, and behaves like this towards acids and bases.

Platinic **hydroxide**, $Pt(OH)_4$, platinic acid, precipitates as a reddish-brown powder by treating a solution of platinic chloride with caustic alkali. It dissolves in an excess of caustic alkali, forming potassium platinate; e.g., $Pt(OH)_4 + 4KOH = K_4PtO_4 + 4H_2O$. It dissolves in acids, forming the corresponding platinic salts; e.g., $Pt(SO_4)_2$.

Platinates are also produced on fusing platinum with alkali hydroxides.

Platinic **chloride**, $PtCl_4$, is produced by the action of chlorine upon platinum at very high temperatures and hence is better obtained by heating H_2PtCl_6 in a current of chlorine. It forms reddish-brown crystalline masses which readily dissolve by water. On the evaporation of its watery solution $PtCl_4 + 5H_2O$ separates in yellow prisms.

Hydrochlorplatinic acid, $H_2PtCl_6 + 6H_2O$, platinum chloride of commerce, is obtained as red, deliquescent crystals by evaporating a solution of platinum in aqua regia.

Chloroplatinates. Those of potassium, ammonium, calcium, rubidium, as well as numerous organic bases, are difficultly soluble in water, hence H_2PtCl_6 is used in chemical analysis.

Platinic **sulphide**, PtS_2, is obtained as a black precipitate, insoluble in all acids, with the exception of aqua regia, by passing H_2S into platinic salt solutions; alkali sulphides dissolve it slowly with the formation of sulphoplatinic salts (p. 176), which are derived from

Sulphoplatinic acids, $H_4Pt_3S_8$ and H_2PtS_6, which can be separated from the corresponding sulpho salts.

Platinamines are the compounds of platinic salts with ammonia, **e.g.,**

$PtCl_4(NH_3)$, $PtCl_4(NH_3)_x$, etc., and which have a constitution analogous to that of the cobaltic-amine salts (p. 286).

d. Detection of Platinum Compounds.

1. When fused with soda on charcoal they yield gray porous platinum (spongy platinum) without incrustation.

2. Sulphuretted hydrogen precipitates black platinic or platinous sulphide from its solution. On the addition of ammonium chloride or potassium chloride to its solution in aqua regia, the corresponding hydrochlorplatinic acid salt is precipitated (p. 208), or finely divided platinum is precipitated from its solutions by metallic zinc.

3. Palladium.

Atomic weight 106.5 = Pd.

This metal occurs only native and in the platinum ores (preparation, p. 292). It is white, has a specific gravity of 11.9, and melts at about 1500°; when finely divided as so-called palladium black it dissolves on boiling with concentrated hydrochloric acid, sulphuric acid, or nitric acid. It has the property, especially when finely divided, of absorbing hydrogen, even 980 times its volume, with the formation of a solid solution (p. 65) which appears like palladium and has an energetic reducing power like nascent hydrogen. Palladium occurs divalent in the stable palladous compounds and tetravalent in the less stable palladic compounds; e.g., $PdCl_4$.

Compounds of Palladium.

Palladous iodide, PdI_2, is precipitated from solutions of iodides by palladium salts as a black powder, and is used in the quantitative estimation of iodine in the presence of chlorine and bromine.

Palladous chloride, $PdCl_2$, forms brown masses.

Palladic chloride, $PdCl_4$, is known only in solution; from both of these alkali chlorides precipitate the salts of hydrochlorpalladous acid, H_2PdCl_4, and hydrochlorpalladic acid, H_2PdCl_6, which are not known free.

### 4. Iridium.	### 5. Rhodium.	### 6. Ruthenium.
Atomic weight 193 = Ir.	Atomic weight 103 = Rh.	Atomic weight 101.7 = Ru.

These metals occur only native in the platinum ores (preparation, p. 292), are insoluble when pure in acids and aqua regia, but when alloyed with platinum are partly soluble. $IrCl_3$, $RhCl_3$, $RuCl_3$, are formed by heating the respective metal in a current of chlorine; these give crystalline, rather soluble complex salts with 2 mols. of alkali chlorides: K_2RuCl_5, Na_2IrCl_5, $(NH_4)_2RhCl_5$. On dissolving Ru and Ir in aqua regia $RuCl_4$ and $IrCl_4$ are formed, which with 2 mols. of alkali chlorides yield rather soluble compounds which are isomorphous with the corresponding platinum salts; e.g., K_2RuCl_6, $(NH_4)_2IrCl_6$. $RhCl_2$ and $RhCl_4$ are not known; on the other hand $RhCl_3$ is found.

Iridium is light gray, ductile, has a specific gravity of 22.4, and melts at 1950°. Its name is derived from the various colors of its compounds.

Rhodium is light gray, ductile, has a specific gravity of 12.1 and melts at about 1860°. Its name is derived from its red double chlorides (ῥοδόεις, rose-red).

Ruthenium is steel-gray, hard, and brittle, has a specific gravity of 12.3, and melts at about 2000°.

7. Osmium.

Atomic weight 191 = Os.

This metal occurs only native in platinum ores (preparation, p. 292), is steel-gray, hard, and heaviest of all bodies, specific gravity 22.5, and melts at about 2500°. On account of its infusibility osmium is used instead of the carbon filament in incandescent electric light (Auer's osmium lamp). When finely divided it is oxidized into osmium tetroxide on heating to a red heat, as well as by nitric acid or aqua regia. Compact osmium is insoluble in acids and aqua regia; alloys of osmium and iridium as found in platinum ores are likewise insoluble therein.

Compounds of Osmium.

Osmium tetroxide, perosmic anhydride, OsO_4 (preparation, see above), forms colorless ˙ prisms melting at about 100° and subliming at a somewhat higher temperature and is readily soluble in ˙ water. Its odor is disagreeably pungent ($\acute{o} s\mu\acute{\eta}$, odor), its vapors attack the eyes and respiratory organs violently; it is used in histology as osmic acid or perosmic acid in order to harden as well as to stain various tissues, as organic and other reducing bodies separate finely divided metallic osmium from its solution.

Osmic acid, H_2OsO_4, is precipitated from potassium osmate by dilute inorganic acids as a black po o powder.

Potassium osmate, $K_2OsO_4 + 2H_2O$, forms red octahedra, and is produced by fusing osmium with $KOH + KNO_3$.

Osmium chlorides, $OsCl_2$, $OsCl_3$, $OsCl_4$, are produced on heating osmium in a current of chlorine, and form double salts with alkali chlorides which are analogous to the ruthenium compounds; e.g., Na_2OsCl_6.

PART THIRD.

ORGANIC CHEMISTRY, OR CHEMISTRY OF THE CARBON COMPOUNDS.

CONSTITUTION.

The number of known carbon compounds is much greater than the compounds of all the other elements together, and new organic compounds are being continually prepared artificially.

The number of atoms which form the molecule of an organic compound is as a rule very much larger than in inorganic compounds; thus, for example, cane-sugar, $C_{12}H_{22}O_{11}$, contains 45 atoms and stearin, $C_3H_5(C_{18}H_{35}O_2)_3$, 173 atoms.

This multiplicity and complexity of organic compounds can be explained (see Isomerism, p. 300) by the fact that the C atoms are tetravalent and have the power of uniting with each other by means of one of the bonds to a much higher degree than the atoms of any other element. This bondage can also take place in a great many different ways.

Carbon is a tetravalent element and exists in its simplest compound (with the exception of CO) with all four valences satisfied,

$$\overset{\text{I}}{} \quad \overset{\text{II}}{} \quad \overset{\text{II}}{}$$

e.g., CH_4, CO_2, CS_2; these four bonds (affinities, valences, p. 28) of carbon are alike in behavior.

When tetravalent carbon atoms combine together, then in the simplest case one bond of one atom saturates one bond of the other atom. Thus if two carbon atoms combine in this way, then two of the eight bonds are necessary for the union of the C atoms; if three carbon atoms combine, then four of the twelve valences are necessary for the mutual union; if four carbon atoms combine, then six bonds are necessary for the mutual union, etc. Thus:

Each new carbon atom introduced therefore contains only two free bonds to which other atoms or groups of atoms, depending upon their valence, can be united. If we consider these valences satisfied, for instance, with hydrogen, then we obtain the compounds CH_4, C_2H_6, C_3H_8, C_4H_{10}, etc., or CH_4, H_3C-CH_3, $H_3C-CH_2-CH_3$, $CH_3-CH_2-CH_2-CH_3$, etc. Each of these hydrocarbons (as well as their derivatives) differs from the previous one by CH_2 and hence we can represent this series by the general formula C_NH_{2N+2}, where N can represent each full number; e.g., $C_{30}H_{62}$.

None of these combinations can take up more atoms or groups of atoms than are represented by the formula C_NH_{2N+2}, because the remaining valences of the C atoms are necessary for the mutual union; hence all combinations which contain C atoms which are united together with one bond each, are called saturated compounds.

Besides saturated compounds we know of others which contain C atoms united together by more than one valence. The individual hydrocarbons of these series (as well as their derivatives) differ also from one another by CH_2; e.g.,

CH_2-CH_2. $CH_2-CH-CH_3$. $CH_2-CH-CH_2-CH_3$.

$CH\equiv CH$. $CH\equiv C-CH_3$. $CH\equiv C-CH_2-CH_3$.

The compounds C_2H_4, C_3H_6, C_4H_8, etc., correspond to the general formula C_NH_{2N}, the compounds C_2H_2, C_3H_4, C_4H_6, etc., to the general formual C_NH_{2N-2}, where N represents a full number from 2 on.

The union of the C atoms by several bonds does not denote a firmer mutual union of these atoms, as such combinations are in fact much more readily split than those where the C atoms are simply united.

Compounds with multiple-united C atoms can be converted into compounds having the C atoms in simple union; thus the compounds having the formula C_NH_{2N} and C_NH_{2N-2} can be converted into compounds having the formula C_NH_{2N+2} by the action of hydrogen. In these cases the C atoms with double and treble bonds are changed to simple bonds.

Bésides these unsaturated compounds a still larger group of C compounds (p. 327) are known which contain fewer hydrogen atoms, etc., in the molecule than the saturated compounds, but which still behave like saturated ones in that they are not converted into saturated compounds having the formula C_NH_{2N+2} by the addition of atoms or groups of atoms, as, for instance, by H atoms. This deviation is to be explained by the fact that in these compounds the C atoms do not form an open chain, but rather a ring-formed closed chain (a carbon ring or nucleus), i.e., the beginning and end member of the chain are united together (see below, Figs. 3, 4).

In the formation of carbon chains and carbon rings other multivalent atoms may take part besides the C atoms (see below, Figs. 1, 2, 5, and 6).

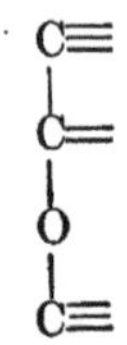

Fig. 1. 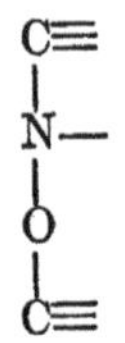   Fig. 2. 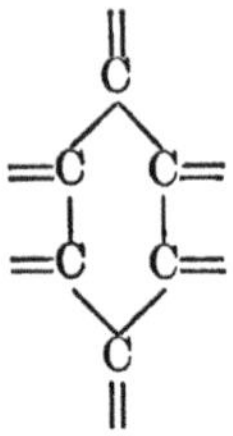   Fig 3. 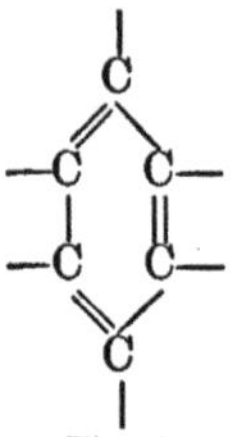   Fig. 4. 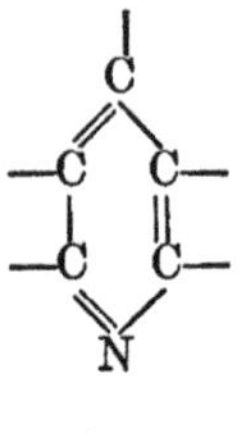   Fig. 5. 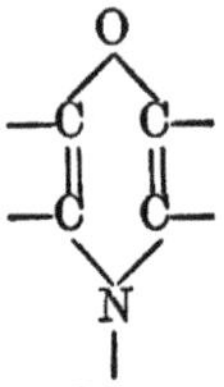   Fig. 6.

Open chains which contain only C atoms are called *homocatenic*, while those containing other atoms besides C atoms are called *heterocatenic*. Closed chains which contain only C atoms are called *homocyclic*, while those containing other atoms are called *heterocyclic*, and the compounds of such chains and rings have the corresponding name.

We also know of C compounds having several atomic rings which are united together directly or by means of C or other atoms. They contain condensed atomic rings, that is, atomic rings which belong to several atoms:

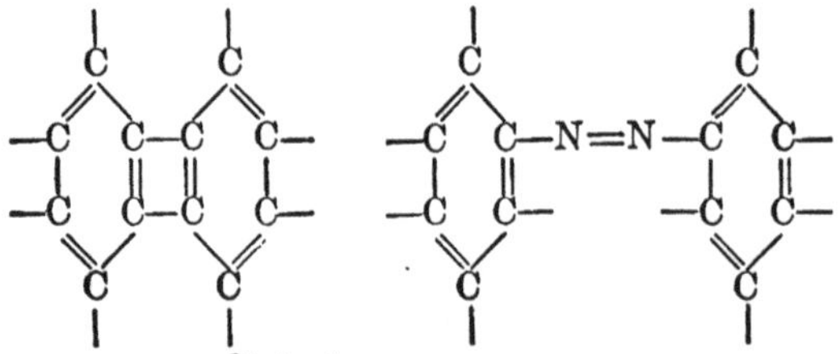

Chain ring. 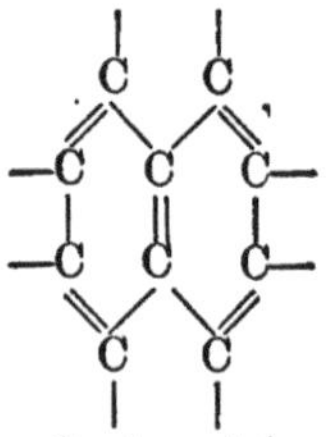   Condensed ring.

All series whose individual members always increase by CH_2 are called *homologous* and *isologous series:*

$$CH_4, C_2H_6, C_3H_8, C_4H_{10}, C_5H_{12}, \text{etc.}$$
$$\underline{\hspace{1cm}} C_2H_4, C_3H_6, C_4H_8, C_5H_{10}, \text{etc.}$$
$$\underline{\hspace{1cm}} C_2H_2, C_3H_4, C_4H_6, C_5H_8, \text{etc.}$$

Combinations belonging to a homologous series nearly always have analogous properties, so that the study of a single member is generally sufficient for the determination of the properties of the entire series.

SUBSTITUTION.

If one or more atoms in an organic compound be replaced (substituted) by one or more atoms of corresponding valence, we obtain derivatives of the compound; thus one or more and even in some cases all the hydrogen atoms in a hydrocarbon can be replaced by other elementary atoms or by groups of atoms. Every hydrocarbon thus forms the starting-point for a series of compounds which all contain the same number of carbon atoms. The following substitutions are of general interest:

1. One hydrogen atom can be replaced by another univalent atom or by a univalent group of atoms; thus from CH_4 we obtain

$$CH_3\text{-}OH, \quad CH_3\text{-}Cl, \quad CH_3\text{-}NH_2, \quad CH_3\text{-}CH_3.$$

2. Two hydrogen atoms can be replaced by two univalent atoms or one bivalent atom or group of atoms; thus from CH_4 we obtain

$$CH_2\text{=}Cl_2, \quad CH_2\text{=}(OH)_2, \quad CH_2\text{=}O, \quad CH_2\text{=}NH, \quad CH_2\text{=}CH_2.$$

3. Three hydrogen atoms can be replaced by three univalent, one trivalent, or one bivalent and one univalent atom or group of atoms; thus from CH_4 we obtain

$$CH\text{≡}Cl_3, \quad CH\text{≡}N, \quad CH\text{≡}CH, \quad CH\begin{smallmatrix}\diagup Cl. \\ \diagdown O. \end{smallmatrix}$$

4. Finally, all four hydrogen atoms can be replaced by uni-, bi-, etc.; atoms or groups of atoms; thus from CH_4 we obtain

$$C\text{≡}Cl_4, \quad C\text{≡}O_2, \quad N\text{≡}C\text{-}Cl, \quad C\text{≡}(CH_3)_4.$$

When uni-, bi- and trivalent hydrocarbon radicals such as $\text{-}CH_3$, $\text{=}CH_2$, $\text{≡}CH$ are substituted in place of the hydrogen atoms in CH_4;

we obtain saturated and unsaturated hydrocarbons which are richer in carbon and in which substitutioñs can take place as in CH_4; e.g.,

$$\begin{array}{ccccc}
CH_3 & CH_3 & CH_3 & CH_3 & CH_3 \\
| & |\;H_2 & |\;H_2 & |\;H_2 & |\;H_2 \\
CH_3 & C & C & C & C \\
 & \backslash Cl & \backslash OH & \backslash CN & \backslash CH_3
\end{array}$$

$$\begin{array}{ccccc}
CH_3 & CH_3 & CH_2 & CH_2 & CH_2 \\
| & \| & \| & \| & \| \\
CH_2 & CH_2 & CH & CH & CH \\
| & |\;H_2 & | & |\;I & |\;OH \\
CH_3 & C & CH_3 & C & C \\
 & \backslash OH & & \backslash H_2 & \backslash H_2.
\end{array}$$

Organic chemistry can therefore be defined as the chemistry of the hydrocarbons and their derivatives. The international nomenclature given on p. 337 depends upon the derivation of all organic compounds from the hydrocarbons having an equal number of C atoms.

ISOMERISM.

The number of C compounds is considerably increased by the existence of·numerous isomeric compounds (pp. 30, 80), i.e., those which have the same qualitative and quantitative composition but different properties.

Isomerism (ἴσος, equal, μέρος, part) of C compounds is to be explained either by the fact that the molecular weights of the respective compounds are multiples of each other (see under 1), or by the fact that the atoms constituting the molecule have a different arrangement in the molecule (see under 2 and 3).

On the other hand many chemical compounds behave in such a manner that two different structural formulæ may be ascribed to them; thus hydrocyanic acid reacts according to the formula $N \equiv C-H$ and also according to the formula $C \equiv N-H$. That property of a compound which causes it so to behave that two different structural formulæ can be ascribed to it is called *tautomerism* (ταὐτό, same), *pseudomerism*, or *desmotropism*. It is due to the presence of a mixture of two structural isomers which are undergoing continuous transformations from one into the other.

1. General Isomerism,

generally called *polymerism*, is conditioned upon differing molecular weights of the compounds. We call all compounds having unlike chemical and physical properties, the same elementary and percentage composition, but different molecular weights, *polymers;* thus,

$$CH_2O, \qquad C_2H_4O_2, \qquad C_3H_6O_3, \qquad C_6H_{12}O_6,$$
Formaldehyde. Acetic acid. Lactic acid. Dextrose.

$$C_2H_4, \qquad C_3H_6, \qquad C_4H_8, \qquad C_5H_{10},$$
Ethylene. Propylene. Butylene. Pentylene.

2. Specific Isomerism,

generally called isomerism simply, and less often *metamerism*, is produced by a different intramolecular arrangement of the atoms although the molecule has the same size. Compounds having unlike chemical and physical properties but the same percentage and elementary composition and the same molecular weight are called *isomers*.

The atoms form in a certain sense the building-stones from which the structure of the molecules is erected. As it is possible to build two entirely different structures from the same number of stones, so it follows that the varying arrangement of the atoms in the molecule may account for the existence of compounds which consist of the same number of atoms of the same elements but having unlike · chemical and physical properties.

This isomerism of the C compounds (with the exception of stereo-isomerism, p. 302) can only be explained by the fact that the C atoms are arranged in the molecule in different relative positions; thus in the following compounds, which are derived from CH_4 by the introduction of the univalent radical $^-CH_3$, no different arrangement of the C atoms is possible: $CH_3^-CH_3$, $CH_3^-CH_2^-CH_3$.

If the univalent radical CH_3 is again introduced, we obtain the hydrocarbon C_4H_{10}. The substitution may take place in several ways, either at the C atom which lies at the ends of the chain, or at the middle C atom; thus we obtain two compounds C_4H_{10}:

$$CH_3^-CH_2^-CH_2^-CH_3 \qquad\qquad \begin{matrix} CH_3 \\ CH_3 \end{matrix}\!\!\Big\rangle CH^-CH_3.$$

In the next member, C_5H_{12}, three cases are possible:

$$CH_3^-CH_2^-CH_2^-CH_2^-CH_3 \qquad\qquad CH_3 \qquad CH_3$$
$$C$$
$$\begin{matrix} CH_3 \\ CH_3 \end{matrix}\!\!\Big\rangle CH^-CH_2^-CH_3 \qquad\qquad CH_3 \qquad CH$$

With the higher members of this hydrocarbon series (C_NH_{2N+2}) the number of possible isomers increases very rapidly according to the law of permutations:

Number of C atoms	6	7	8	9	10	11	12	13
Possible number of hydrocarbons	5	9	18	35	75	159	355	802

H drocarbons having continuous carbon chains without branches are called *normal hydrocarbons*.

Isomers which depend upon differing arrangement of the C chain are also called *chain isomers*.

As we can obtain isomeric combinations by the introduction of the univalent CH_3 group in different positions, so we can obtain the same by the introduction of other atoms or groups of atoms. In compounds like CH_3Cl, CH_2Cl_2, $CHCl_3$, only one formation is possible (see Stereoisomerism), also in CH_2Cl-CH_3. In compounds like $C_2H_4Cl_2$, $C_2H_4(OH)_2$, etc., two arrangements of the atoms are possible, as follows:

$$CH_3-CHCl_2 \qquad CH_2Cl-CH_2Cl.$$
$$CH_3CH(OH)_2 \qquad CH_2(OH)-CH_2(OH).$$

In the next hydrocarbon, $CH_3-CH_2-CH_3$, two different compounds are possible by the substitution of a chlorine atom, namely,

$$CH_3-CH_2-CH_2Cl \qquad CH_3-CHCl-CH_3.$$

As above stated, two compounds having the formula C_4H_{10} are possible. If only one H atom is substituted in these compounds, four different combinations are possible; thus, on the introduction of the OH group we obtain four compounds C_4H_9OH, called butyl alcohols:

$$
\begin{array}{llll}
CH_3 & CH_3 & CH_3 \quad CH_3 & CH_3 \quad CH_3 \\
| & | & \diagdown \diagup & \diagdown \diagup \\
CH_2 & CH-OH & CH & C-OH \\
| & | & | & | \\
CH_2 & CH_2 & CH_2OH & CH_3 \\
| & | & & \\
CH_2OH & CH_3 & &
\end{array}
$$

Isomers which are due to the various positions of the introduced atoms, etc., while the order of the members of the C chain remains the same, are also called *position isomers*.

With the unsaturated compounds the number of possible isomers is still greater than with the saturated because substitution takes place at different points and also because the bonds of the C atoms are multiple at certain points.

In regard to the special form of isomerism which occurs with compounds with ring-formed C atoms, we refer to isocarbocyclic and heterocyclic compounds.

3. Stereoisomerism.

A number of compounds having similar qualitative and percentage composition are known which do not differ, or which differ in only a slight degree, in their chemical properties, but which exhibit certain differences in their physical properties (for example, variation in behavior towards polarized light) and which are therefore called *physically isomeric* (also *optically isomeric*) compounds.

One kind of physical isomerism is due to a varied arrangement or grouping of the molecules themselves, since it appears only in solid compounds; thus, for example, the crystallization of one and the same substance in two or more crystal forms (dimorphism and polymorphism, p. 35), and further the property possessed by certain substances of diverting the ray of polarized light only when in the solid state, which property is not retained when the substances are in the liquid or dissolved state, namely, when the molecules are free to move about one another (p. 38).

The second kind of physical isomerism can be caused only by the arrangement of the atoms in the molecules, and is exhibited in the case of certain isomeric compounds which have the power of diverting the polarized ray of light equally strongly to the right or to the left when they are in the melted or dissolved condition, namely, when the molecules are free to move about one another (optical isomerism). This kind of physical isomerism is also exhibited in the case of certain unsaturated isomeric compounds which differ from each other in all their physical and in certain of their chemical properties.

This second sort of physical isomerism cannot be explained by the theories of structure or constitution (p. 30), since it is not apparent from these how a difference in the structure of the given isomeric compounds can be possible, but it can be explained if the spatial arrangement of the atoms in the molecule is taken into consideration.

Those compounds whose isomerism can only be explained on the assumption of a different spatial arrangement of the atoms in the molecule are called *stereoisomeric* ($\sigma\tau\epsilon\rho\epsilon\delta\nu$, solid body) and, this isomerism is called *stereo-* or *space-isomerism*, also *allo-* or *geometric isomerism*. The formulas representing these compounds are called *stereochemical formulas*, and instead of speaking of the structure or constitution of such bodies the expression "configuration of the molecules" is employed.

The structural or constitutional formulas represent the bonding together of the carbon atoms and the distribution of the atoms and radicals combined with them in such a manner that the formula always lies in a single plane; but since all bodies, including molecules, must extend in three dimensions, this is not expressed in the original formulas; indeed these are actually contradictory to many facts.

For example, by assuming that the four valences of carbon all lie in one plane, of all compounds of the formula Ca_2b_2 (where a and b represent different monovalent groups *), there would be two isomers possible, namely, one where a and a and b and b were adjoining, and one where they were separated from each other:

$$a\diagdown \mkern-4mu \diagup b \quad\quad a\diagdown \mkern-4mu \diagup b$$

while in fact only one such compound (for example, only one methylene chloride, CH_2Cl_2) is known and nothing points to the existence of two such isomers. Of all compounds having the formula $Cabcd$ there must exist for each three isomers, namely,

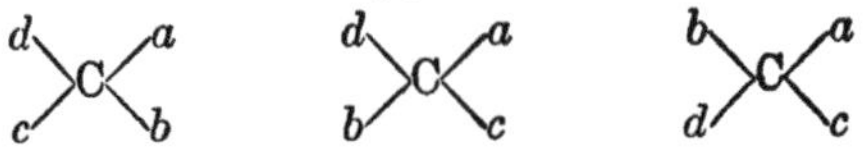

$$d\diagdown C\diagup a \quad d\diagdown C\diagup a \quad b\diagdown C\diagup a$$

while actually only two isomers are known, for example, two chlorbromiodomethanes, $CHClBrI$.

When a spatial configuration is assumed for the carbon compounds, in accordance with the theory proposed by Le Bel and van't Hoff all of the previously mentioned difficulties and contradictions disappear.

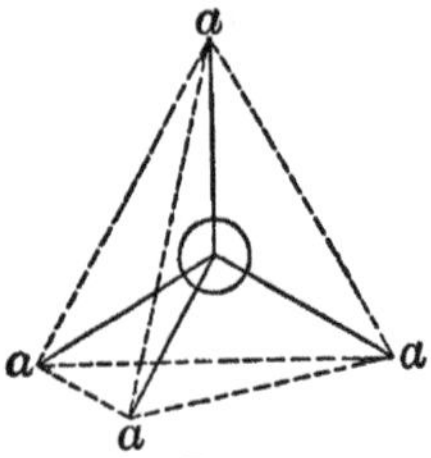

This theory is based on the assumption that the four valences of every carbon atom are equally distributed symmetrically about the space surrounding the carbon atom, and that the direction of every valence forms the same angle with the direction of every other valence, which agrees with the equal value of each of the four valences. This is best represented by placing the carbon atom at the center of a tetrahedron, the four valences being directed towards the four solid angles of the tetrahedron.

FIG. 1.

If we imagine that in the molecule of the compound $Caaaa$ three of the similar groups a are replaced by dissimilar groups, then the following isomers are possible:

a. Isomers in the case of simple saturated compounds.

α. If a group b or still another group c takes the place of one or two of the groups a, then only one configuration can be conceived, since both figures can be made to correspond by turning, so that it is immaterial on which angle of the tetrahedron the substitution takes place

* The expression "group" as used in this chapter denotes "an atom or an atomic group."

(Figs. 2 and 3). Therefore mono- and disubstitution products of methane can exist only in one modification, a fact justified by experience.

β. If a group *d*, different from *a*, *b*, *c*, is introduced in place of one of the remaining *a* groups, then two different configurations are obtained,

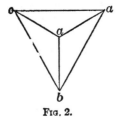 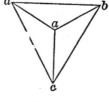

Fɪɢ. 2.   Fɪɢ. 3.

according to whether the group *a* standing to the right or that to the left in the figure is replaced by *d*. These two configurations cannot be made to correspond to one another by turning.

If one pictures one's self as standing at the point of attachment of the group *a*, then in order to pass from *b* over *c* to *d* along the circumference of the circle joining these three points it would be necessary to move in the direction of the hands of a clock in one case (Fig. 4), and in the opposite direction in the other (Fig. 5).

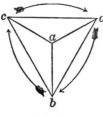 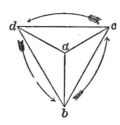

Fɪɢ. 4.   Fɪɢ. 5.

The two systems are not identical, but stand to each other in the same relation as an object stands to its image as reflected in a mirror, and many compounds which correspond to these systems crystallize in two different forms which bear a similar relation to one another. These are called *enantiomorphic crystals* (ἐναντίαος, opposite, enantiomorphism). Compounds of this kind which contain an asymmetric carbon atom can be optically active, and this property, when the arrangement is that shown in Fig. 4, is of an opposite quality from that when the arrangement is as shown in Fig. 5.

As a matter of fact the isomeric compounds with one asymmetric carbon atom are identical in their chemical and mostly also in their physical properties with the exception of their optical behavior, since there are known two modifications of every compound with an asymmetric carbon atom (p. 39), one of which rotates the plane of polarization to the right and the other to the left (optical isomerism). When such compounds crystallize they also exhibit enantiomorphism.

b. Isomerism in complicated saturated compounds.

Here the tetrahedrons corresponding to the asymmetric carbon atoms have a common point of contact. These compounds may differ not only in optical properties, but also in other physical properties; indeed they may show a slight chemical difference. In order to determine the number of isomers in every conceivable case, it must be assumed that both tetrahedrons continually rotate about their common axis in the same or in opposite directions, an assumption which corresponds to the view that a movement of the atoms in the molecule is present.

In the case of the rigid combination of two carbon tetrahedrons even for compounds which do not contain an asymmetric carbon atom (p. 39), as, for example, for $C_2a_2b_2$, there would be three isomers possible; i.e.,

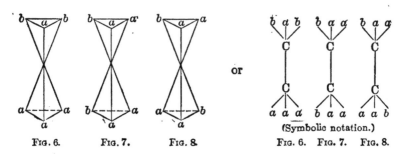

| Fig. 6. | Fig. 7. | Fig. 8. | Fig. 6. | Fig. 7. | Fig. 8. |

(Symbolic notation.)

If the two tetrahedrons can rotate about their common axis, then the arrangement shown in Figs. 7 and 8 would not represent isomerism, but only different phases of rotation, and by turning could all be brought into the same position. There are therefore only two compounds $C_2a_2b_2$ possible; and since, further, all chemical facts go to show that an influence is exerted by the atoms within the molecule, therefore the one isomeric compound $C_2a_2b_2$ would more probably correspond to Fig. 8 than to Fig. 7, since there different groups are opposed. Under such conditions, then, on account of the opposite attraction of *a* and *b*, the rotation can be completely prevented. With reference to these relations the following isomers are possible:

α. If two asymmetric carbon atoms are present in the molecule of a compound, and if one of these carbon atoms is combined with groups which are similar to those combined with the other carbon atom (for example, when they form a so-called symmetrical molecule, as in the case of tartaric acid, HOOC·OH·HC⁻CH·OH·COOH), then, in addition to the two optically opposed forms (i.e., dextrotartaric and lævotartaric acid), and the inactive form which is produced by their combination (e.g., racemic acid), there exists a second inactive form (inactive tartaric acid) which cannot be split up into the two active modifications. In order to explain the occurrence of the latter form, it is assumed that in the molecule of this substance the position of the atoms attached to one of the asymmetric carbon atoms is such as to cause rotation to the right, while the position of the atoms attached to the other asymmetric carbon atoms, namely, the other half of the molecule, is such as to cause rotation to the left, and as a result the rotation of the molecule as a whole is prevented and the compound is inactive optically. This modification cannot be split up into two optically active modifications, because on decomposition the molecule itself

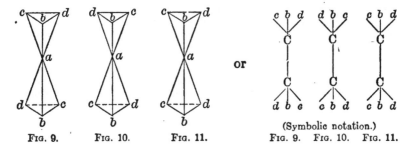

FIG. 9.   FIG. 10.   FIG. 11.   or   (Symbolic notation.)
FIG. 9.   FIG. 10.   FIG. 11.

is destroyed. Thus, for example, in Fig. 11, if the arrangement of the groups *b, c, d*, in the upper tetrahedron corresponds to a rotation to the left, then in the lower half of the molecule the arrangement corresponds to a rotation to the right, while in Fig. 9 the arrangement *b, c, d* in both tetrahedrons corresponds to a rotation to the left, and in Fig. 10 it corresponds to a rotation to the right, representing the two optically opposed modifications.

β. If there are present in the molecule of a compound two asymmetric carbon atoms to which are attached altogether six different groups (forming a so-called asymmetric molecule), for example, *cba*–C–C–*def*, then there are four isomers possible, namely,

$$(1)\ \frac{+A}{+B}, \qquad (2)\ \frac{+A}{-B}, \qquad (3)\ \frac{-A}{+B}, \qquad (4)\ \frac{-A}{-B},$$

where *A* and *B* represent the two tetrahedrons and + represents the arrangement of the atoms in a clockwise direction (i.e., dextrorotatory position), and − represents the arrangement of the atoms in a counterclockwise or lævorotatory position.

γ. If three asymmetric carbon atoms are present in a single asymmetric molecule, for example *hgf*C–C*ab*–C*cde* (as in the pentoses and their corresponding acids), then there are eight isomers possible; if four asymmetric carbon atoms are present in an asymmetric molecule (as in the case of the hexoses), there are sixteen isomers possible; so that in general for every compound having *n* asymmetric carbon atoms there will exist 2^n isomeric modifications (e.g., for four asymmetric carbon atoms $2^4 = 16$ isomers).

All of these isomers are optically active, because in the presence of different atoms or atomic groups attached to the two asymmetric carbon atoms no compensation of the rotatory power can take place.

In addition to these active modifications it is still possible to have inactive forms which are produced by the combination of two active modifications for example, of 1 and 4, 2 and 3, but not of 1 and 2, or 2 and 4 (see β above).

δ. In the case of compounds having three or more asymmetric carbon atoms in a symmetric molecule, the number of possible isomers (as was shown for two asymmetric carbon atoms on p. 306, α) is small and these are in part optically active, but in part optically inactive, and cannot be split up into active modifications (because of intramolecular compensation, as in *a*, Fig. 11). In addition to these other inactive, resolvable mixtures (racemic modifications, p. 39) are possible.

c. Isomerism in unsaturated compounds.

α. Compounds containing carbon atoms connected by double bonds.

In such cases the tetrahedrons corresponding to the asymmetric carbon atoms are connected together at two apices, namely, at one edge, and therefore no rotation of the two tetrahedrons can take place about a common axis. The atoms or atomic groups attached to each of the asymmetric carbon atoms must therefore remain fixed in their original positions (see Figs. 12 and 13).

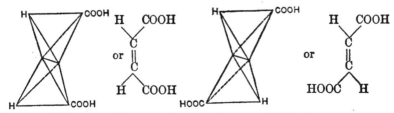

FIG. 12.—Maleïc acid, $C_4H_4O_4$.   FIG. 13.—Fumaric acid, $C_4H_4O_4$.

The conditions here are different on the right and left, although they resemble the conditions which exist when the asymmetric carbon atoms are attached by a single valence, and when two, three, or four different groups are attached to the tetrahedrons it is possible to recognize two isomers of each which show a complete difference in almost all physical and also in certain chemical properties.

Since the atoms or atomic groups which are attached to the asymmetric carbon atoms all lie in one plane, these compounds cannot be optically active.

According to the theory of structure both maleïc and fumaric acid must have the same formula: $HOOC-CH=CH-COOH$. In a spatial representation of these two substances a different structure is evident if for one acid we employ the so-called plane-symmetric or *cis* form (Fig. 12) for one acid and the so-called axial-symmetric or *trans* form (Fig. 13) for the other.

β. Compounds containing carbon atoms connected by triple bonds.

In this case the tetrahedrons corresponding to these carbon atoms are connected by three apices, namely, by one of their faces, and therefore

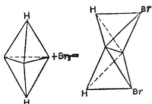

form a double triangular pyramid, so that the two free valences, just as in the above case *a*, lie in one plane. Isomers are here impossible.

If one of the bonds is severed, then the two tetrahedrons swing apart. The two attached atoms or atomic groups will then lie one above the other and the molecule becomes plane-symmetric.

d. Isomerism in nitrogen compounds.

Isomers appear in the case of certain oximes, diazo- and azo-compounds, in which a trivalent nitrogen atom is attached by two of its valences to a carbon or nitrogen atom, which can be explained by the assumption that the three nitrogen valences do not lie in a single plane but are distributed in space like the four free valences of the carbon atom in Figs. 6 to 11. The three valences of nitrogen are imagined as directed towards three corners of a tetrahedron, the nitrogen atom itself being situated at the fourth corner. In this way is represented, for example, the structure of the two isomeric benzaldoximes, $HO—N=CH—C_6H_5$, as shown below. Isomers of this sort are known as *syn* and *anti* forms.

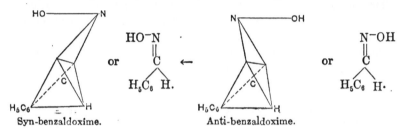

Syn-benzaldoxime.   Anti-benzaldoxime.

DETERMINATION OF THE COMPOSITION, MOLECULAR AND CONSTITUTIONAL FORMULA.

In order to ascribe an empirical and rational molecular formula (p. 26) to an organic compound the following essentials are necessary:

1. The elements, of which the compounds are composed, must be determined (qualitative analysis).

2. The proportion in which the respective elements exist in the compound must be estimated (quantitative analysis).

3. The molecular weight must be determined with reference to the qualitative and quantitative composition.

4. The intermolecular arrangement of the atoms or atomic groups in the molecule (the constitutional formula) must be studied.

1. Qualitative Elementary Analysis.

The elements in an organic compound can generally only be detected after the destruction of the molecule of the compound, which may be done as follows:

The *carbon* in most organic bodies can be detected by their yielding black carbon on heating them in the absence of air. The carbon can be more positively detected by heating the substance with copper oxide (p. 236) or if the substance is volatile, by passing the vapors slowly over heated copper oxide contained in a glass tube. The carbon in this case suffers oxidation at the expense of the oxygen of the copper oxide and is converted into carbon dioxide, which can be detected by passing the gas into lime-water, which becomes cloudy (p. 191).

The *hydrogen* is detected by heating the perfectly dry substance with copper oxide in a glass tube whereby it is burnt into water, which collects on the cold part of the tube.

The *nitrogen* is detected by heating the substance with potassium in a glass tube until the excess of potassium is driven off. The nitrogen present is converted into potassium cyanide, which can be detected by the methods given in connection with the cyanogen compounds. Many nitrogenous compounds also yield ammonia on heating them with soda-lime.

The oxygen of these hydroxides unites with the carbon, forming carbon dioxide, which combines with the alkali, forming carbonates, while the hydrogen at the moment it is set free unites with the nitrogen, forming ammonia. The excess of hydrogen escapes as such or unites with the carbon, forming volatile hydrocarbons; thus

$$NCOH + Ca(OH)_2 = CaCO_3 + NH_3,$$
$$(CN)_2 + 4Ca(OH)_2 = 2CaCO_3 + 2NH_3 + 2CaO + 2H.$$

Many nitrogen compounds such as diazo bodies do not give these reactions because they decompose very readily with the formation of gaseous nitrogen. In these cases we are obliged to proceed according to the Dumas method (p. 312) for the quantitative estimation of nitrogen.

Oxygen is determined in the quantitative analysis of the substance and indeed generally indirectly.

Phosphorus and *sulphur* are detected by fusing the substance with a mixture of sodium carbonate and potassium nitrate, which converts them into phosphoric and sulphuric acids. These are detected in the watery solution of the fused mass after acidification with nitric acid as ammonium phosphomolybdate (p. 169) and barium sulphate (p. 130) respectively. With volatile substances the oxidation into sulphuric and phosphoric acids is brought about by heating the substance with fuming nitric acid in sealed glass tubes.

The *halogens* are converted into calcium halogen salts by heating the substance with calcium oxide (with volatile substances in sealed glass tubes) and then detected in the watery solution, after acidification with nitric acid by silver nitrate (pp. 136, 141, 143).

On heating an organic halogen compound on a platinum wire with copper oxide in a non-luminous flame a blue or green color is given to the flame. On heating organic halogen compounds with fuming nitric acid and silver nitrate in sealed glass tubes an insoluble silver halogen precipitate is obtained.

The remaining elements are to be sought for, if they are not volatile, in the residue obtained on the incineration of the substance, the ash, and if volatile in the colorless liquid finally obtained on boiling the substance with concentrated sulphuric acid.

2. Quantitative Elementary Analysis.

a. Determination of the Elements.

Determination of the Carbon and Hydrogen.

A weighed amount of the substance is mixed with copper oxide and introduced into a tube of infusible glass to the open end of which is attached a weighed calcium chloride tube and to this a weighed potash bulb. The contents of the tube is heated to a red heat whereby the carbon is burnt into carbon dioxide and the hydrogen into water; this water is vaporized and is retained in the calcium chloride tube while the carbon dioxide is absorbed by the caustic potash in the bulb. After complete combustion both tubes are reweighed in order to determine the weight of .carbon dioxide and water formed, from which the amount of C and H in the weighed amount of substance can be calculated (p. 312).

Determination of Nitrogen. Estimation as Ammonia. α. Will-Varrentrap method. A weighed amount of the substance is mixed with soda-lime and introduced into a hard-glass tube the open end of which is connected with an apparatus filled with hydrochloric acid. The contents of the tube is heated and the gases produced allowed to bubble through the hydrochloric acid, which combines with all the ammonia produced (see p. 310). The ammonium chloride formed is precipitated as insoluble ammonium platino-chloride, $(NH_4)_2PtCl_6$ (p. 293), which is collected, dried, weighed, and the nitrogen calculated therefrom.

β. Kjeldahl Method. A weighed amount of the substance is heated with concentrated sulphuric acid and certain metallic oxides (CuO, HgO) which have an oxidizing action, until a complete solution has taken place and the fluid is clear. The ammonia which is produced from the nitrogen in this method and which exists in the liquid as ammonium sulphate is driven off by distillation with caustic alkali and determined as described in method α.

Estimation as Gaseous Nitrogen. Dumas Method. Many artificially prepared organic compounds contain nitrogen in the form of NO, NO_2, NO_3, etc.; this nitrogen is not completely converted into ammonia on heating with soda-lime or on boiling with H_2SO_4; hence their nitrogen must be determined by the following method, which is applicable for all nitrogenous compounds. A weighed amount of the substance is mixed with copper oxide and heated in a glass tube free from air and which contains copper turnings at one end. The gases produced, which consist of water, carbon dioxide, nitrogen, are collected in a graduated glass tube filled with caustic alkali. As the gases pass through the red-hot copper turnings any oxides of nitrogen which may have been produced are decomposed, the oxygen uniting with the Cu and the nitrogen being set free. The carbon dioxide and water are not decomposed, but are absorbed by the caustic alkali, so that the gas collected in the graduated tube consists of pure nitrogen and its volume is readily determined.

From the cubic centimeters of nitrogen obtained (V) the weight of nitrogen (N) can be calculated according to the formula (p. 42), where T represents the temperature, B the barometric pressure, and W the tension of the caustic alkali solution.

$$N = \frac{V(B-W)}{760(1 + 0.003665(T)} \cdot 0.001256.$$

(1 c.c. N weighs 0.00125 gram at $0°$ C. and 760 mm. pressure.)

The estimation of phosphorus, sulphur, and the halogens is performed as described in the qualitative analysis, but here the precipitates obtained are weighed and the amount calculated therefrom.

Estimation of Oxygen. This is generally done indirectly in that all the other constituents are quantitatively determined and their weight subtracted from the weight of the substance analyzed. The difference represents the weight of the oxygen in the substance.

b. Calculation of the Analysis.

In order the better to compare the results of the different analyses, the results obtained are calculated on 100 parts by weight of the substance, and the difference between the sum of the elements estimated and 100 represents the oxygen which was not directly estimated. If these percentages are divided by the atomic weights of the respective elements, we obtain figures which represent the relationship existing between the elements of the compound.

Thus on the qualitative examination of pure acetic acid it was found apparently to contain carbon and hydrogen; 0.395 gram acetic acid yielded on combustion 0.5793 gram CO_2 and 0.2349 gram H_2O. From these results we calculate the amounts of carbon and hydrogen as follows:

$$CO_2 :C \ = CO_2 \text{ found} :C.$$
$$44 :12 \ = 0.5793 \qquad :x. \quad x = 0.158 \text{ C.}$$
$$H_2O :H_2 \ = H_2O \text{ found} :H_2.$$
$$18.02:2.02 = 0.2349 \qquad :x. \quad x = 0.0261 \text{ H.}$$

Calculated in percentage we find the following:

0.395 gram acetic acid : 0.1580 gram C = 100 : x. $x = 40$.
0.395   "   "   " : 0.0261 " H = 100 : x. $x = 6.6$.

From this we see there is a difference of 53.4 grams $(100 - [40 + 6.6] = 53.4)$, which must be the weight of oxygen in the substance, as no other element was found on qualitative testing.

One hundred parts acetic acid therefore consist of:

Carbon.................................... 40.0 parts
Hydrogen. .... 6.6   "
Oxygen. .... 53.4   "

If these figures are divided by the atomic weights of the respective elements, we obtain the relative number of atoms which the compound contains:

$$\frac{40}{12} = 3.3 \text{ C.} \qquad \frac{6.6}{1} = 6.6 \text{ H.} \qquad \frac{53.4}{16} = 3.3 \text{ O.}$$

The atoms in acetic acid have the following relationship to each other: 3.3 : 6.6 : 3.3 (as the atoms are indivisible, hence no fractions are possible, we call the lowest number = 1); then the relationship is 1 : 2 : 1, and hence $CH_2O = 30$ represents the simplest expression for the molecule.

3. Determination of the Molecular Formula.

The formula calculated from the elementary analysis of a compound may indeed represent its true molecular weight, but it may also be a multiple of the formula, hence the determination of the molecular weight must follow the analysis.

The above formula, $CH_2O = 30$, represents the simplest formula for acetic acid, but we know of many compounds having entirely different chemical and physical properites which on elementary analysis also give the same formula CH_2O; hence it is probable that their molecular weights are different.

Before the molecular formula is decided upon it should be noted whether the simplest formula found by analysis coincides with the law of even numbers, which is that the sum of the elements with uneven valences (i.e., univalent and trivalent, such as H, Cl, Br, I and N, P, As) of each C compound must be an even number.

This law can be explained by the fact that the carbon is tetravalent and that the elements combine according to their atomicity. Thus in cyanuric acid, $C_3H_3N_3O_3$, the sum of the N and H atoms$=6$; in ammonium trichloracetate, $C_2Cl_3(NH_4)O_2$, the sum of the Cl, N, and H atoms$=8$. If in the analysis of cyanuric acid the formula $C_3H_2N_3O_3$ was derived there must be some mistake in the analysis, as the sum of the elements with uneven valences is 5, an uneven number.

a. Methods depending upon the determination of the specific gravity of the gaseous compound (gas or vapor density of the compound, p. 18).

If the compound is a gas or can be vaporized without decomposition all that is necessary is to determine the gaseous volume obtained from a known weight of the substance and to calculate from this the weight of the volume of the gaseous compound as compared to an equal volume of oxygen taken as 32.

For example, 0.134 gram acetic acid yield 50 cc. gas (reduced to 0° and 760 mm. pressure, p. 42), and as 50 c.c. oxygen (at 0° and 760 mm.) weigh 50×0.01429 grams$=0.0715$ grams (p. 107), then the molecular weight of acetic acid is 60; e.g.,

Weight of 50 c.c. oxygen		Weight of 50 c.c. acetic acid
0.0715	:	0.134
Molecular weight of O		Molecular weight of acetic acid
$=32$	:	x $(x=60)$.

The vapor density of a gaseous compound may be determined by first weighing a glass vessel free from air and then filled with the gas in question.

The following rapid and easy methods may be used for liquid and solid compounds.

Vapor Density Determination according to A. W. Hofmann. The method ordinarily used (see page 315), is only applicable for those substances which are not decomposed at temperatures above their boiling-point. In other cases a weighed amount of substances is introduced into the vacuum in a barometer tube and vaporized and the volume of the vapor determined. In this manner the vapor density can be determined at temperatures 50 to 100° below the boiling-point of the substance at ordinary pressures, so that the substance is not decomposed.

Determination of the Vapor Density by the Expulsion of Air according to Victor Meyer. The vessel *b*, which is constructed of glass or porcelain and which contains air, is heated in the large tube *c* by the vapors of the liquid contained therein (water, boiling-point 100°; aniline, boiling-point 183°; di henylamine, boiling-point 310°), or by molten lead (at about 1000°) until the temperature is constant or until no more air-bubbles rise from the tube *a* at *f* The graduated tube, which is full of water, is now adjusted over the opening of tube *a* and the glass tube *b* opened at *d* and a weighed amount of the substance quickly introduced and the cork at *d* quickly replaced. The substance rapidly vaporizes and expels from *b* a quantity of air corresponding to its vapor volume, which collects in the graduated tube *g*. The volume of air read off in the tube corresponds to an equal volume of the vapor of the substance in question just as if it were possible to obtain the vapor of the substance at ordinary temperatures.

B. *Methods depending upon the determination of the elevation of the boiling-point and depression of the freezing-point* (p. 19).

The depression of the freezing-point or the elevation of the boiling-point for the molecular weight of a body in dilute solution is constant for various substances when equal quantities of the same solvent are used. In order to find the constant, (K), the freezing-point depression or the boiling-point elevation (T) produced by a few grams (P) of a substance of known molecular weight when dissolved in 100 grams of the solvent is determined and then calculated upon the molecular weight of the respective substance as follows:

$$P : T = M : K; \text{ hence } T \times M = P \times K;$$

hence
$$K = \frac{T \times M}{P}.$$

A few grams (P) of the substance of unknown molecular weight are dissolved in 100 grams of the solvent whose constant (K) has been determined and the depression of the freezing-point or elevation of the boiling-point (T) determined and the molecular weight (M) calculated according to the equation $M = \dfrac{K \times P}{T}.$

For example, 2.721 grams acetic acid (P) dissolved in 100 parts benzol ($K = 49$) depressed the freezing-point 2.222° (T), hence the molecular weight of acetic acid $M = \dfrac{49 \times 2.721}{2.222} = 60$, which corresponds to the figure obtained by the vapor density and shows that the simplest formula obtained for acetic acid by chemical analysis, $CH_2O = 30$, must be double; that is, $C_2H_4O_2 = 60$.

C. Chemical methods.

The physical methods, on account of the ease with which they are performed, have replaced the chemical methods. The latter methods must be different, depending upon the chemical properties of the substance investigated.

If the compound is an acid the molecular weight may be determined after the basicity has been learned by the analysis of its salts, when generally the easily purified silver salt is used, which on incineration leaves metallic silver, which can be directly weighed and the molecular weight of the acid calculated from this result. With a monobasic acid the quantity by weight which unites with 1 atom of silver represents its molecular weight minus 1 atom hydrogen.

For example, acetic acid, which by analysis (p. 313) may have the formula CH_2O or a multiple thereof, is a monobasic acid, and hence in its molecule 1 atom H is replaceable by 1 atom of a univalent metal. If now by analysis we determine how much acetic acid is combined with 1 atom of silver, then we obtain the molecular weight of acetic acid minus 1 atom H.

One hundred parts silver acetate yield 64.68 parts silver on incineration, while 35.32 parts are lost. The acetic acid united with 1 atom of silver (108 parts) in silver acetate will be 59 parts by weight:

$$64.68 : 35.32 :: 108 : x; \quad x = 59.$$
Silver.   At. wt.
           Silver.

Now, as in acetic acid, 1 atom H is replaced by 1 atom Ag, hence its molecular weight is $59 + 1 = 60$, which corresponds to the formula $C_2H_4O_2$. The formula cannot be CH_2O, otherwise the molecular weight would be 30 and cannot be $C_3H_6O_3$, because then the silver salt must contain $1\frac{1}{2}$ atoms silver, which is impossible. The formulæ $C_4H_8O_4$, $C_6H_{12}O_6$ are only possible if the acetic acid was bi-, tri-, etc., basic.

If the compound is a base the molecular weight may be determined by the analysis of its salts. For instance, all organic bases form salts, by direct addition with acids, analogous to ammonia. I HCl is used in the formation of the salt, then the quantity by weight of a monoacid (= monovalent) base which unites with one molecule HCl or of a biacid base which unites with 2 molecules HCl represents the molecular weight of the base.

We generally make use of the hydrochlorplatinic acid salt of the base instead of the HCl salt, as this forms compounds corresponding to the ammonium compound $(NH_4)_2PtCl_6$, having an analogous composition. These can be prepared readily in a pure form generally with water of crystallization and leave platinum, which can be directly weighed on incineration and the molecular weight calculated from its weight.

If the compound is indifferent, then the molecular weight may often be determined by the preparation of simple substitution products, or the quantitative relationship of the individual constituents having known molecular weight, of which the compound is constructed, may be investigated after cleavage.

With compounds containing hydrogen we can generally replace one H atom by a halogen atom and the substitution product obtained can be analyzed. Naphthalin, according to analysis, has the formula C_5H_4 or a multiple thereof. The analysis of bromnaphthalin shows that one atom H is replaced in $C_{10}H_8$ by one bromine atom, while the formulæ $C_{20}H_{16}$, $C_{30}H_{24}$, etc., are excluded, as we know of no compound in which $\frac{1}{16}$, $\frac{1}{24}$, etc., of the hydrogen is replaced by halogens.

If the molecular weight cannot be determined either by physical or chemical methods, the simplest formula found by analysis must be sufficient, and at the same time the law of even numbers must be considered (p. 313).

4. Determination of the Constitutional Formulæ.

If in a compound we have determined the composition as well as molecular weight (empirical molecular formula), still it is impossible to characterize the compound by the formula thus obtained and to prevent a confusion with other compounds. We have seen that a great number of isomeric compounds are possible, i.e., compounds having the same composition and the same molecular weight; thus we learned on page 302 that four compounds having the formula $C_4H_{10}O$ are known, and it is therefore necessary, as soon as it does not follow from circumstances which of these four is meant, that the chemical formulæ simultaneously express the constitution or structure, i.e., the inner construction, of the compound in question.

The construction of a compound as represented by the constitutional formula (rational molecular formula) is not the result of theoretical speculation, but these formulæ must be derived from the study of the decompositions, cleavages, and methods of formation of the respective compound and serve the purpose of giving us an exact representation of the chemical nature of the compound, of calling our attention from what other compounds it is derived and into which it can be converted, etc.

The investigation of the constitutional formula is of special importance for organic compounds (carbon compounds), as by these means important natural products for man have been artificially prepared. This is easier for organic compounds than for inorganic ones, as the carbon, although the multiplicity of its compounds, shows in many ways a more regular behavior than the other elements. We are therefore able to give well-based constitutional formulæ for most organic compounds, while this is not possible for many of the much simpler constructed inorganic compounds.

In the construction of the constitutional formulæ for organic compounds we ordinarily try to decompose the complicated compounds into simpler ones of known constitution (analytical method), or we prepare complicated compounds step by step from simple compounds of known constitution (synthetical method); still the various physical properties of the compound also help us in the elucidation of its constitution.

1. *Chemical Methods.*

By oxidation we differentiate between the primary alcohols which first yield aldehydes and secondary alcohols which, under the same conditions, yield ketones; and tertiary alcohols, which yield acids, having fewer C atoms. Aldehydes differ from the isomeric ketones by yielding, on oxidation, acids having the same number of C atoms, while the ketones yield acids with fewer C atoms. Aromatic compounds which contain the aliphatic side-chain attached to the benzol nucleus are readily oxidized, as of the entire side-chain only that C atom which is directly united to the benzol nucleus is converted into the COOH group, while the other C atoms of the side-chain are split off and oxidized; thus

$$C_6H_4 = (CH_3)_2 + 6O = C_6H_4(COOH)_2 + 2H_2O;$$
$$C_6H_4 = (CH_2-CH_3)_2 + 12O = C_6H_4(COOH)_2 + 2CO_2 + 4H_2O.$$

By reduction we can differentiate between nitro-compounds and their isomers, the nitrite esters. The first yield amines with nascent hydrogen, while the latter yield alcohols:

$$C_2H_5NO_2 + 6H = C_2H_5NH_2 + 2H_2O;$$
Nitroethane.　　　Ethylamine.

$$C_2H_5-O-NO + 6H = C_2H_5OH + NH_3 + H_2O.$$
Ethyl nitrite.　　　Ethyl alcohol.

Primary sulphurous acid esters are converted into alcohols by nascent hydrogen, while the isomeric sulphonic acids give mercaptans:

$$C_2H_5-O-SO-OH + 6H = C_2H_5-OH + 2H_2O + H_2S;$$
Ethyl sulphite.   Ethyl alcohol.

$$C_2H_5-SO_2-OH + 6H = C_2H_5SH + 3H_2O.$$
Ethyl sulphonic acid.   Ethyl mercaptan.

The splitting off of carbon dioxide often gives important information as to the constitution of a compound. Acetic acid heated with lime yields marsh-gas: $C_2H_4O_2 + CaO = CaCO_3 + CH_4$. Benzoic acid heated with lime yields benzol: $C_7H_6O_2 + CaO = CaCO_3 + C_6H_6$. From this it follows that benzoic acid is related to benzol in the same manner as acetic acid to marsh-gas. Those organic bodies which are obtained by splitting off of CO_2 generally have the prefix *pyro*.

The methods of preparation of compounds richer in carbon from those poorer in carbon are of importance in the study of the constitution.

From methane, by the action of chlorine, we obtain methyl chloride, CH_3Cl, from which all the derivatives of methane can be prepared. Carbon and hydrogen unite with the formation of acetylene, C_2H_2, which can be converted into ethylene, C_2H_4, and ethane, C_2H_6, by nascent hydrogen, and from these a large number of other compounds can be obtained. If acetylene is passed through a tube heated red-hot, it is converted into benzol, C_6H_6, from which a great number of cyclic compounds may be obtained.

Sodium combinations of organic substances are decomposed by organic halogen compounds, so that the sodium haloid salt is formed and the two organic residues unite; thus $CH_3-Na + CH_4Cl = CH_3-CH_3$ (ethane) + NaCl.

If carbon dioxide is passed into a sodium compound of a hydrocarbon, then the salt of an acid richer by one atom C is obtained: $CH_3-Na + CO_2$ $= CH_3COONa$ (sodium acetate).

The halogen compounds of the hydrocarbons of the methane series or the sulphonic acids of the benzol series yield cyanides of the hydrocarbons on heating with potassium cyanide (KCN). In these compounds the cyanogen group (CN) is converted into the carboxyl group (COOH) on heating with water:

$$CH_3-Cl + KCN = KCl + CH_3-CN \text{ (methyl cyanide)};$$
$$CH_3-CN + 2HOH = NH_3 + CH_3-COOH \text{ (acetic acid)}.$$

Condensation consists in the union of two or more molecules of the same or different organic substances forming one molecule by union through the C atoms, generally by the elimination of H_2O, HCl, NH_3, CO_2. The components cannot be split off from the new molecule by simple means.

Condensation takes place readily amongst the aldehydes and ketones and can indeed be brought about by the direct action of two substances upon each other. This condensation is generally brought about by the presence of certain bodies such as aluminium chloride, potassium bisulphate, anhydrous sodium acetate, sodium hydroxide, hydrochloric acid, sulphuric acid, zinc chloride, etc.

Aluminium chloride brings about the union of chlorine derivatives of the methane series with hydrocarbons of the benzol series: $C_6H_6+C_2H_5Cl$ $=C_6H_5-C_2H_5+HCl.$ $C_6H_6+2C_2H_5Cl=C_6H_4=(C_2H_5)_2+2HCl.$ C_6H_6+ $3C_2H_5Cl=C_6H_3\equiv(C_2H_5)_3+3HCl.$

Concentrated sulphuric acid causes the union of combinations of the benzol series with aldehydes:

$$2C_6H_6+CH_3-CHO=CH_3-CH=(C_6H_5)_2+H_2O.$$

Polymerization depends upon the union of several similar molecules of a simpler constituted organic substance, without the elimination of atomic groups, forming a complicated molecule. The original substances can be readily split from this new molecule. Polymerization takes place with unsaturated hydrocarbons, aldehydes, and cyanogen compounds. If a compound contains a C atom united with a polyvalent element or radical by more than one valency, then at this position a rupture may take place forming a simple union, whereby the molecules having these free affinities unite with one another (see under example 2). In this case a union of the molecules does not entirely take place through the C atoms as we find in condensation (see under example 1).

2. *Physical Methods.*

Specific Gravity. This is different for isomeric compounds. The quotient of the specific gravity divided into the molecular weight, called the molecular volume (p. 36) of liquid organic compounds may give us information as to the constitution of the body if they are compared at certain temperatures such as, for instance, at their boiling-points. Thus many homologous series show an increase of the molecular volume of approximately 22 for every additional CH_3 (ethyl alcohol molecular volume $= 62.5$, butyl alcohol molecular volume $= 84.8$, etc.) when they have the same atomic grouping, and in many cases a deviation from this rule shows another atomic grouping.

Chlorine or bromine atoms substituted for H atoms in organic compounds occupy a greater volume when they are united to the same C atom as compared with different C atoms, etc.

Melting- and Boiling-points. Isomeric bodies have different melting- and boiling-points. In general an organic body is more readily fusible and more volatile the simpler the molecule is constituted, and the more complicated the molecule is, the higher are the melting- and boiling-points, and the molecules suffer decomposition readily by heat.

The boiling-point determination is an important aid in the investigation of the constitution of organic bodies. Homologous compounds (i.e., those which differ from each other by CH_2) with the same atomic grouping show approximately the same variation in boiling-point, while isomeric compounds with different atomic grouping differ markedly in boiling-point, thus:

Boiling-point.

Ethyl alcohol,	C_2H_6O	78.4°	Difference 19
Normal propyl alcohol,	C_3H_8O	97.4°	
Normal butyl alcohol,	$C_4H_{10}O$	115.0°	" 19.6
Acetic acid,	$C_2H_4O_2$	119.0°	" 22
Propionic acid,	$C_3H_6O_2$	141.0°	
Normal butyric alcohol,	$C_4H_8O_2$	162.0°	" 21

Amongst the isomeric aliphatic compounds the normal compounds always have the highest boiling-point. The boiling-point becomes lower the more branches the carbon chain has.

Of the disubstitution derivatives of benzol the ortho compounds nearly always have a lower boiling-point than the meta and para compounds.

The melting-points also show certain relationships to the constitution of the compounds.

Among the disubstitution products of benzol the para compounds nearly always have a higher melting-point than the ortho or meta compounds. With the normal acids of the formic and oxalic acid series the introduction of one CH_2 group causes a rise in the melting-point, while the next CH_2 group causes a lowering, so that the members of the series with uneven C atoms have a lower melting-point than both neighboring members with even C atoms; thus,

Formic acid, CH_2O_2, melts at $+8.4°$; Acetic acid, $C_2H_4O_2$, at $+17°$

Propionic acid, $C_3H_6O_2$, melts at $-22°$; Butyric acid, $C_4H_8O_2$, at $-8°$

With isomeric bodies the melting-point is higher the more side chains are present.

Refraction of Light. The molecular refraction of liquid organic compounds is equal to the sum of the atomic refraction (p. 38).

The univalent elements in organic compounds have a constant atomic refraction, while with the polyvalent elements the refraction is raised according to the double or treble bonds (also for the C atoms).

In the homologous series the molecular refraction increases at a constant rate, namely, 4.5 for each CH_2 group.

Rotation of the Plane of Polarization. This also gives us information as to the constitution of an organic compound, as an optically active organic compound must contain one or more asymmetric carbon atoms (p. 39).

TRANSFORMATIONS AND DECOMPOSITIONS.

1. Action of Chemical Agents.

Oxygen at ordinary temperatures acts upon only a few combinations, while at a red heat they all suffer combustion.

Active or nascent oxygen (from manganese dioxide, or potassium bichromate and sulphuric acid) unites either directly, or removes H in the form of water, or brings about both changes simultaneously.

Halogens have a substituting action.   Unsaturated compounds are first converted in saturated ones by addition:

$$C_2H_4 + 2Cl = C_2H_4Cl_2; \quad C_2H_4Cl_2 + 4Cl = C_2H_2Cl_4 + 2HCl.$$

Iodine is substituted only in the presence of an oxidizing substance (HgO, HIO_3, etc.) which destroys the HI produced.   This HI would otherwise reduce the iodide formed; e.g., $CH_4 + I_2 = CH_3I + HI$; $CH_3I + HI = CH_4 + I_2$.

In the presence of water the halogens have an oxidizing action in that they decompose water, setting free oxygen, which acts upon the organic compounds: $H_2O + 2Cl = 2HCl + O$.

Ferric chloride has a weak oxidizing action in that it is converted into $FeCl_2$: $C_6H_4(OH)_2 + 2FeCl_3 = C_6H_4O_2 + 2HCl + 2FeCl_2$.

Hydrochloric, hydrobromic, and hydroiodic acids replace the alcoholic hydroxyl groups (p. 333) by chlorine, bromine, and iodine: $C_2H_5OH + HCl = C_2H_5Cl + H_2O$.   In the presence of an excess of HI the iodide formed is reduced (see above).

Sulphuric acid acts upon the alcoholic hydroxyl groups in the same way as upon the hydroxyl groups of the metals:

$$C_2H_5OH + \frac{HO}{HO}{>}SO_2 = \frac{C_2H_5O}{HO}{>}SO_2 + H_2O.$$

With cyclic compounds (P. 327) sulphonic acids are produced with the splitting off of water:

$$C_6H_6 + H_2SO_4 = C_6H_5 - SO_3H + H_2O.$$

Many organic bodies are decomposed by concentrated sulphuric acid, which removes water from them.   On boiling with concentrated sulphuric acid nearly all organic bodies are completely destroyed (p. 311).

Nitric acid acts upon the alcoholic hydroxyl groups in the same manner as it does the hydroxyl groups of the metals: $C_2H_5 - OH + HNO_3 = C_2H_5 - NO_3 + H_2O$.   On aromatic bodies it acts in replacing one or more H atoms by NO_2: $C_6H_6 + HNO_3 = C_6H_5NO_2 + H_2O$.   In many cases nitric acid has an oxidizing action when often a part of the carbon is converted into oxalic acid or carbon dioxide.

Nascent hydrogen (sodium amalgam or aluminium amalgam, p. 243, +water) has a reducing action whereby a simple addition, a removal of oxygen, or both simultaneously, may take place: $C_6H_5NO_2 + 6H = C_6H_5NH_2 + 2H_2O$.   Hydrogen removes chlorine, bromine, and iodine from chlorine, bromine, and iodine substitution products and replaces them so that the original compound is regenerated in this way.

Sulphuretted hydrogen, ammonium sulphide, tin, or tin chloride and hydrochloric acid, zinc powder and caustic alkali, zinc, or iron with acids, also HI, have a similar reducing action as hydrogen.

Alkali hydroxides decompose the esters or compound ethers in watery or alcoholic solutions:

$$\underset{\text{Ethyl acetate.}}{C_2H_3 - C_2H_5O_2} + KOH = \underset{\text{Ethyl alcohol.}}{C_2H_5OH} + \underset{\text{Potassium acetate}}{C_2H_3KO_2}.$$

Alkyl halogens, etc., give hydroxyl derivatives with watery solutions of caustic alkalies: $C_2H_5Cl + KOH = C_2H_5OH + KCl$.

Alcoholic caustic alkali splits off all halogen atoms as halogen hydrogen compounds: $CH_2Cl-CH_3 + KOH = CH_2=CH_2 + KCl + H_2O$.

Solid caustic alkali when fused with organic bodies oxidizes them, in that O is substituted for H_2, and this last set free: $C_2H_5OH + KOH = C_2H_3KO_2 + 4H$. Often complicated molecules are thereby split into several simpler ones:

$$C_6H_{12}O_6 + 6NaOH = 3C_2Na_2O_4 + 18H.$$
$$\text{Dextrose.} \qquad\qquad \text{Sodium oxalate.}$$

Phosphorus trichloride and phosphorus tribromide replace the hydroxyl groups by chlorine or bromine respectively: $3C_2H_5OH + PCl_3 = 3C_2H_5Cl + H_3PO_3$.

Phosphorus pentasulphide replaces the oxygen of the hydroxyl by sulphur:

$$5C_2H_5OH + P_2S_5 = 5C_2H_5SH + P_2O_5.$$

Water combines directly, or, as is generally the case, on heating under pressure it causes a splitting (saponification or hydrolysis, which see).

2. Action of Heat.

If non-volatile organic substances are heated in the absence of air (dry distillation) their elements regroup themselves so that besides carbon monoxide, carbon dioxide, water, etc., a large number of organic compounds of simpler composition are produced and carbon remains behind. The large number of volatile compounds thus produced can be in part condensed by cooling apparatus, the other part remaining in the gaseous state. That portion which is condensed consists of two layers, a watery layer which contains various bodies in solution and another generally dark layer, which is called the tar. Other compounds which are volatile without decomposing at certain temperatures are decomposed with the deposition of carbon when they are passed in the gaseous state through red-hot tubes.

In the dry distillation of brown and anthracite coal, etc., bituminous shale and peat (as in the fabrication of illuminating-gas), street gas, ammoniacal liquor (p. 148), tar and coke are obtained as decomposition products. The street gas contains volatile hydrocarbons of the methane, ethylene, acetylene, and aromatic series as well as some carbon dioxide, carbon monoxide, ammonia, air, and vapor of water, and when imperfectly purified, also sulphuretted hydrogen and sulphur dioxide (see Potassium Ferrocyanide).

Coal-tar is a mixture of many constituents which may be separated by fractional distillation and may be divided into the following four groups, according to their chemical behavior:

a. The hydrocarbons. These form the chief constituent, are indifferent, that is, are not soluble in acids or bases, and belong chiefly to the cyclic compounds (p. 327).

b. The phenols form the second greatest portion of coal-tar, are soluble in alkalies but insoluble in dilute acids.

c. Basic nitrogenous bodies.   These occur to a small extent; hence their preparation from coal-tar is not practical.   They are soluble in acids but insoluble in alkalies, and belong to the cyclic compounds (p. 327).

d. In the fractional distillation of coal-tar a black viscous mass remains called tar asphalt, which is used in making tar-paper, etc.

In the dry distillation of wood the same gases are obtained as in the distillation of coal, but a watery fluid which contains chiefly acetic acid, wood-alcohol (methyl alcohol), acetone, and wood-tar, which consists chiefly of phenols, creosote, paraffines, while charcoal remains. Pix liquida, officinal wood-tar, is obtained on the dry distillation of coniferous trees, while aqua picis is obtained on shaking this with water.

The black tough product remaining after the fractional distillation of wood-tar is called pitch.

By the dry distillation of bituminous shale containing petrified fishes, occurring in Tyrol, an oily fluid is obtained which contains about 11 per cent. combined sulphur. On treating this oil with sulphuric acid ichthyol sulphonic acid, $C_{28}H_{36}S(SO_3H)_2$, is formed, whose ammonium salt, ichthyol, has found use in medicine.   Both bodies are brown liquids.

3. Action of Ferments.

Ferments are those nitrogenous organic bodies which under certain circumstances have the power of causing by their simple presence chemical transformation in certain organic compounds without themselves suffering any decomposition (catalysators, p. 66).   They are divided into organized and unorganized ferments.

The *unorganized ferments or enzymes* are compounds closely related to the proteid bodies and will be discussed with these.   They are secreted from the organized ferments and are therefore the cause of their action, or they are a product of living cells; for example, the cells of the salivary glands, the pancreas, etc.   They can only transform a certain quantity of the decomposable body, and their action consists of a hydrolytic cleavage, i.e., it takes place with the taking up of the elements of water (p. 87).

The *formed or organized ferments* are low fungi of which a small number are sufficient to transform a large amount of the decomposable body, as they rapidly increase in number at the same time that they cause the cleavage processes.   In order that they shall multiply and be active the presence of organic nitrogen and inorganic salts is necessary.   Their action depends presumably upon the production of enzymes; still only a few have been isolated up to the present time.

The cleavage of organic substances into simpler ones by the aid of organized ferments is called *fermentation* in the broad sense and includes the processes of fermentation of varieties of sugar, putrefaction, decay, and acetic acid fermentation, etc. In these processes we are also producing hydrolytic cleavages (p. 87), and generally also oxidations, if oxygen is present.

Of the fungi causing these processes the Saccharomyces bring about fermentation, the Schizomycetes putrefaction, and the molds decay. For the practice of fermentation it is important to know that the Schizomycetes grow only in acid-free solutions, the Saccharomyces in 0.5 per cent. acid, and the molds only in 5 per cent. acid.

The general conditions for fementation are as follows:

Presence of air as carrier of the germs. If the decomposition has begun, then the further supply of air is not necessary.

Presence of water.

Temperatures above 0° and below 100°, as at 0° the organisms cannot develop and at 100° they are nearly all destroyed. Each kind of fermentation takes place best between certain temperature limits, which differ for each one.

Absence of anti-fermentive and anti-putrefactive agents, such as arsenious acid, chlorine, metallic salts, salicylic acid, xanthogenates, sulphur dioxide, carbon disulphide, tannic acid, carbolic acid, creosote, thymol, etc., which have an antiseptic or anti-zymotic action. Alcohol, common salt, sugar, also belong to this group in that they remove the water necessary for the decomposition.

Putrefaction is the fermentation of certain organic substances (especially protein bodies and substances related thereto) with the development of volatile, disagreeable-smelling bodies. Putrefactive bodies are in general the same as the decomposition products obtained by the action of acids or alkalies on the substances in question.

Fermentation in a narrow sense or real fermentation is the destruction of sugars by the action of ferments. We have various kinds, depending upon the products, e.g., alcoholic, butyric acid, lactic acid, and mucilaginous fermentation.

The sugars when pure are unfermentable, but they are split into simpler compounds by the action of organized ferments when the general conditions given above are observed; also certain dilution of solution, and when protein substances and inorganic salts are present. This is the reason why fruit-juices containing sugar undergo fermentation when exposed to the air, from which they take up the fungi spores. Pure sugar solutions and concentrated (evaporated) fruit-juices containing sugar do not undergo this fermentation.

Decay is the gradual oxidation (p. 119) by means of the atmospheric oxygen of the intermediary products produced in putrefaction and their conversion into final products, carbon dioxide, water, and ammonia, and eventually nitric acid (p. 161). If putrefaction takes place in the absence of air, then masses rich in carbon always remain, while if the process takes place in the presence of air (although often after a very long period), the substance disappears entirely and it decomposes completely into carbon dioxide, water, and ammonia.

Mouldering is the very slow decomposition of the products produced in putrefaction which takes place in the presence of a very small supply of air, when, on account of the lack of oxygen the hydrogen unites in part with the carbon forming marsh-gas.

Humus substances are those brown or black uncrystallizable bodies which are formed before the final products, water, carbon dioxide, and ammonia, in the decay and destruction of plant and animal bodies as well as by the action of strong acids or alkalies upon carbodydrates (sugar, starch, etc.). They are odorless and tasteless and actively absorb moisture and ammonia from the air, and hence are important food for the plants. The carbohydrates seem to be closely related to many humus substances.

Humus substances are found in the upper layers of arable soil, in peat, in brown coal, in decomposing wood, in many spring-waters, and the yellowish-brown sediment from many waters, etc. Some of these substances contain nitrogen and are weak acids, and can be extracted by dilute caustic alkalies and in part reprecipitated from the brown solution thus obtained. Little is known in regard to the chemical properties of these substances. They are called humin, $C_{40}H_{30}O_{15}$, ulmin, $C_{20}H_{16}O_7$, huminic acid, $C_{20}H_{12}O_6$, ulminic acid, $C_{20}H_{13}O_6$, geinic acid, $C_{20}H_{12}O_7$, quellic acid, $C_{12}H_{12}O_8$.

CLASSIFICATION.

The carbon compounds may be divided into the following classes according to their constitution (p. 297):

1. *Aliphatic compounds* are those whose molecules contain only open atomic chains of C atoms, or of C atoms and other atoms (p. 298, Figs. 1 and 2).

The aliphatic compounds or fatty bodies are so called because the animal and plant fats (ἄλειφας, fat) belong to this group. They are also called *acyclic* or *catenic compounds* as they contain open atomic chains, and *methane derivatives* because they may be considered as derived from methane, CH_4, by substitution.

2. *Isocarbocyclic compounds* are those whose molecules contain ring-shaped closed atomic chains (atomic ring) which consist only of C atoms. They are also called *homocarbocyclic* or in short *carbocyclic compounds*.

3. *Heterocarbocyclic compounds* are those whose molecules contain ring-shaped closed atomic chains, but which contain one or more polyvalent atoms besides the C atoms (p. 298, Figs. 5 and 6).

We also know of aliphatic compounds which contain closed chains composed of various elements; thus ethers, ureids, lactones, lactides contain these, although they have only slight similarity with the hetero-carbocyclic compounds. The iso- and heterocarbocyclic compounds (= cyclic compounds) formerly bore the general name *aromatic compounds or benzene derivatives*, as many of the compounds belonging to this class have an aromatic odor, or because they can be considered as derived from benzene, C_6H_6, by substitution.

4. *Alicyclic compounds* are those compounds which in chemical properties stand between the aliphatic and the cyclic compounds and contain also ring-shaped closed atomic chains which are composed only of C atoms, or of C atoms and other atoms. But as in the cyclic compounds the C atoms of the ring occur with only one free valence, in the alicyclic compounds one or all of the C atoms of the ring may exist with two free valences (p. 298, Fig. 3). In the following pages the alicyclic compounds will not be treated as a special class, but be discussed with the aliphatic, iso- or heterocarbocyclic compounds from which they are derived.

5. *Compounds of unknown constitution.* These diminish more and more as the science of chemistry advances, and are included in the following pages with the compounds of known constitution, as their constitution has been sufficiently elucidated so that they can be classified in one of the four above-mentioned classes.

I. ALIPHATIC COMPOUNDS.

CONSTITUTION.

All carbon compounds whose molecule contains open chains of C atoms (or of C atoms and other atoms) belong to the aliphatic series:

$$CH_3-CH_2-CH_3, \quad CH_2=CH-CH_3, \quad CH_3-CO-CH_3, \text{ etc.},$$

still other compounds with ring-shaped grouping of the atoms will also be discussed, these latter being directly derived from the aliphatic compounds and are called alicyclic compounds (see p. 327).

NOMENCLATURE.

Hydrocarbons. The saturated hydrocarbons have the termination *ane* and from C_5H_{12} are designated by the Greek numerals. The unsaturated hydrocarbons correspond to the radicals with even valences and have the same names (see below).

Radicals. The unsaturated atomic groups which occur unchanged in a large number of compounds are called radicals. The most important radicals are those derived from the saturated hydrocarbons CH_4, C_2H_6, etc. (p. 297). If one or more hydrogen atoms are abstracted from these, we obtain radicals with different valences which unite either with atoms or groups of atoms until the limit compound C_NH_{2N+2} is reached. Thus if one H is abstracted from ethane, C_2H_6, then the monovalent radical C_2H_5 remains, while if two H are removed we obtain the divalent radical C_2H_4, and if three H atoms are taken away the trivalent radical C_2H_3 remains, etc.

The radicals are designated as follows:

Saturated hydrocarbons......	CH_4	C_2H_6	C_3H_8	C_4H_{10}
	Methane.	Ethane.	Propane.	Butane.
Monovalent............	CH_3	C_2H_5	C_3H_7	C_4H_9
(Ending in -yl).	Methyl.	Ethyl.	Propyl.	Butyl.
Divalent..............	CH_2	C_2H_4	C_3H_6	C_4H_8
(Ending in -ylene or -ene).	Methylene.	Ethylene.	Propylene.	Butylene.
Trivalent..............	CH	C_2H_3	C_3H_5	C_4H_7
(Ending in -enyl or -ine).	Methenyl.	Ethenyl (Vinyl).	Propenyl (Glyceryl).	Butenyl (Crotonyl).

Radicals.

328

In accordance with the law of even numbers (p. 314) the radicals with uneven valence cannot exist free. If they are released from their compounds, they generally double up or they decompose:

$$CH_3I + CH_3I + 2Na = H_3C-CH_3 + 2NaI.$$
2 mol. methyl iodide.   1 mol. dimethyl.

The radicals with even valences can exist in the free state, when they are released from compounds in which the affinities set free belong to two neighboring carbon atoms, so that they can unite together:

$$ClH_2C-CH_2Cl + 2Na = H_2C = CH_2 + 2NaCl.$$
Ethylene chloride.   Ethylene.

From this it follows that the radical $CH_3-CH=$ cannot be obtained from the compound CH_3-CHCl_2 because both free valences cannot be united with the molecule itself.

These radicals are also called mono-, di-, etc., valent alcohol radicals, after their most important compounds, the alcohols. The monovalent radicals are also called *alkyl*, the divalent *alkylene*, and the trivalent *alkenyl* radicals.

A special nomenclature for organic bodies can be based upon the radicals. Thus methane, CH_4, can be considered as a combination of the methyl radical with hydrogen and hence called methyl hydride, CH_3-H, methyl alcohol, CH_3-OH, as methyl hydroxide; dichlormethane, $CH_2=Cl_2$, as methylene chloride; trichlormethane, $CH≡Cl_3$, as methenyl chloride, etc.

Among the monovalent radical we differentiate between the primary, secondary, and tertiary, according to whether the unsaturated C atom is united to one, two, or three other C atoms; thus,

$$CH_3-CH_2-CH_2-CH_2- \qquad CH_3-CH_2{\Large\diagdown}CH- \qquad (CH_3)_3≡C-$$
$$CH_3{\Large\diagup}$$

Primary butyl.   Secondary butyl.   Tertiary butyl.

On satisfying the free valence of these radicles with $-OH$, $-I$, etc., we obtain the primary, secondary, and tertiary compounds (PP. 333, 334).

The radical $-COOH$ is called *carboxyl*, $-O(CH_3)$ ·*methoxyl*, $-O(C_2H_5)$ *ethoxyl*, $-S(CH_3)$ *sulfmethyl*, $-COH$ *carbinol*, etc.

Letters or figures are added to the names of certain isomeric compounds in order to designate the position of certain atoms or radicals in the molecule. Thus we designate the terminal C atom of acids as well as the side-chain of cyclic compounds by ω or 1, the one combined therewith by α or 2, etc., and the terminal C atom of the remaining compounds by α or 1, etc.

Correspondingly we differentiate between

$CH_3\text{-}CHI\text{-}COOH$
α-iodopropionic acid.

$CH_2I\text{-}CH_2\text{-}COOH$
β-iodopropionic acid.

$CH_3\text{-}CH(OH)CH(CH_3)\text{-}COOH$
α-methyl-β-oxybutyric acid.

$CH_2(OH)\text{-}CH_2\text{-}CH_2\text{-}COOH$
γ-oxybutyric acid.

$CH_3\text{-}CH_2\text{-}CCl_3$
α-trichlorpropane.

$CH_3\text{-}CCl_2\text{-}CH_2Cl$
α β β-trich.orpropane.

$CH_2Cl\text{-}CHCl\text{-}CH_2Cl.$
α β γ-trichlorpropane

Optically active bodies (p. 39) are designated by *d*- (dextro) when they turn the ray to the right, by *l*- (lævo) when they turn it to the left, and by *i*- when they represent the inactive modification.

Amines, amin bases, monamines, are to be considered as ammonia whose H atoms are entirely or in part replaced by alkyl radicals (see p. 329). According as one, two, or three alkyls are introduced we obtain primary amines or amid bases, secondary amines or imid bases, tertiary amines or nitril bases (but not nitriles):

$$N{\raisebox{0.5ex}{$\diagup$}}^H_H\raisebox{-0.5ex}{$\diagdown$}_H \qquad N{\raisebox{0.5ex}{$\diagup$}}^H_H\raisebox{-0.5ex}{$\diagdown$}_{CH_3} \qquad N{\raisebox{0.5ex}{$\diagup$}}^H_{CH_3}\raisebox{-0.5ex}{$\diagdown$}_{CH_3} \qquad N{\raisebox{0.5ex}{$\diagup$}}^{CH_3}_{CH_3}\raisebox{-0.5ex}{$\diagdown$}_{CH_3}$$

Ammonia.   Methylamine   Dimethylamine.   Trimethylamine

Diamines are derived from 2 molecules NH_3, in which each NH_3 molecule has one H atom replaced by one valence of one or more divalent alcohol radicals (alkylenes); thus,

$H_2N\text{-}C_2H_4\text{-}NH_2$
Ethylendiamine.

$NH=(C_2H_4)_2=NH$
Diethylendiamine.

$N\equiv(C_2H_4)_3\equiv N.$
Triethylendiamine.

The triamines, tetramines, etc., have an analogous derivation where each valence of one or more multivalent alcohol radicals are introduced into three, four, etc., molecules of NH_3; thus, $CH_3\text{-}C\equiv(NH_2)_3$, ethenyltriamine, $(CH_2)_6N_4$, hexamethylene tetramine.

Nitrosamines are amines in which an H atom is replaced by the nitroso group $-NO$; thus, $(CH_3)_2=N-(NO)$, dimethyl nitrosamine.

Nitramines are amines in which one H atom is replaced by the nitro group $-NO_2$; thus, $CH_3\text{-}NH\text{-}NO_2$, methylnitramine.

Imines, imin bases, are obtained when two H atoms in one molecule of ammonia are replaced by one alkylene radical: $C_3H_6=NH$, propylenimine.

Nitriles (not nitril bases, see above) may be considered as ammonia in which all three H atoms are replaced by an alkenyl radical thus: $(CH_3\text{-}C)\equiv N$, or also as a combination of alkyl with the radical cyanogen $-C\equiv N$, e.g., $CH_3\text{-}C\equiv N$, methyl cyanide.

Oximes, oximido compounds, are to be considered as hydroxylamin, $H_2N(OH)$, in which two H atoms are replaced by an alkylene. They contain the divalent oxime group $=N(OH)$, and are isomeric with the nitroso compounds, i.e., the combination of the alkyls with the NO group; hence they are also called isonitroso compounds. $CH_3-CH=NOH$, acetoxime, is isomeric with CH_3-CH_2-NO, nitrosoethane.

Amidoximes are oximes where the C atom connected with the oxime group $=NOH$ has a NH_2 group attached; thus, $CH_3-C(NH_2)=NOH$, ethenylamidoxime.

Hydroximic acids and *hydroxamic acids* are oximes where the C atom connected with the oxime group $=NOH$ has a $-OH$ group attached; thus, $CH_3-C(OH)=NOH$.

Hydrazines are produced when the hydrogen of hydrazin, H_2N-NH_2 (p. 151), is replaced by alkyls; thus, $(CH_3)HN-NH_2$, methyl hydrazin, $(CH_3)_2N-NH_2$, dimethyl hydrazin.

Hydrazones may be considered as hydrazines in which two H atoms are replaced by an alkylene; thus, $(CH_3)HN-N=(CH-CH_3)$.

Ammonium bases are to be considered (although not proven) as ammonium hydroxide NH_4-OH, in which four H atoms are substituted by alkyls; thus, $(CH_3)_4N-OH =$ tetramethyl ammonium hydroxide.

Metallic organic compounds are those compounds obtained by the combination of metals with the alcohol radicals:

I	II	IV	V
$NaCH_3$	$Hg(CH_3)_2$	$Pb(CH_3)_4$	$Sb(CH_3)_5$

Alcohols are hydrocarbons in which one or more H atoms are replaced by the hydroxyl group $-OH$. According to the number of hydroxyls they are called monatomic, diatomic, triatomic alcohols, etc.:

$C_3H_7(OH)$ propyl alcohol $=$ Monatomic alcohol.
$C_3H_6(OH)_2$ propylene alcohol $=$ Diatomic "
$C_3H_5(OH)_3$ propenyl alcohol $=$ Triatomic "

As bodies containing more than one $-OH$ united to one C atom are unstable, we know only of diatomic alcohols with at least 2 carbon atoms, etc.:

The alcohols may also be considered as compounds of the alcohol radicals with hydroxyls—each according to their atomicity—mono-, di-, tri-, etc., or as metallic hydroxides in which the metal is replaced by the alcohol radicals:

$K(OH)$, $C_3H_7(OH)$, $Ca(OH)_2$, $C_3H_6(OH)_2$, $Al(OH)_3$, $C_3H_5(OH)_3$.

As the metallic hydroxides combine with acids with the elimination of water, forming neutral or acid salts, so also do the alcohols (see Esters and Ester Acids, p. 332).

Ethers are the anhydrides of the alcohols and are formed by the removal of H_2O from their hydroxyl groups; thus, $2C_2H_5{}^-OH$ (ethyl alcohol) $= H_2O + C_2H_5{}^-O{}^-C_2H_5$ (ethyl ether).

As the metallic oxides are obtained from the metallic hydroxides by the abstraction of water, so also may ethers be obtained from the alcohols by abstraction of water. As in the anhydride formation two molecules of hydroxide of a monovalent metal are necessary, so also with the hydroxides of monovalent radicals two molecules are necessary, while with the hydroxides of divalent metals or radicals water can be abstracted from one molecule:

$$2KOH = K_2O + HOH, \qquad 2CH_3OH = (CH_3)_2O + HOH,$$
$$Ca(OH)_2 = CaO + HOH, \qquad C_2H_4(OH)_2 = C_2H_4O + HOH.$$

The ethers may also be considered as the oxides of the alcohol radicals; thus, $C_2H_5{}^-O{}^-C_2H_5$, ethyloxide, $C_2H_4{}^=O$, ethylene oxide.

They may be considered as metallic oxides whose metals are replaced by equivalent alcohol radicals:

$C_2H_5{}^-OH$,	ethyl alcohol,	corresponds to $K{}^-OH$.
$C_2H_5{}^-O{}^-C_2H_5$,	ethyl ether,	" " $K{}^-O{}^-K$.
$C_2H_4(OH)_2$,	ethylene alcohol,	" " $Ca(OH)_2$.
$C_2H_4{}^=O$,	ethylene ether,	" " $Ca{}^=O$.

The ethers of monatomic alcohols may be considered as alcohols in which the H atom of the hydroxyl group is replaced by an alcohol radical.

Ethers with two similar alcohol radicals are called simple, while those with two different radicals are called mixed ethers. These latter are the anhydrides of two different alcohols; thus, $CH_3{}^-OH$ (methyl alcohol) $+ C_3H_7OH$ (propyl alcohol) $= HOH + CH_3{}^-O{}^-C_3H_7$ (methyl propyl ether).

Mercaptans, thioalcohols, thiols, are those organic compounds which correspond to the hydrosulphides of the metals; thus, $CH_3{}^-SH$, methyl mercaptan, corresponds to $K{}^-SH$. They are the sulphur alcohols, and their metallic derivatives are called *mercaptides;* thus, $(CH_3S)_2Hg$, mercuric mercaptide.

Sulpethers, thioethers, are related to the mercaptans in the same way as the ethers to the alcohols or the hydrosulphides of the metals to the sulphides:

$$2K{}^-HS = K_2S + H_2S; \qquad 2CH_3{}^-SH = (CH_3)_2S + H_2S.$$

Esters, compound ethers, correspond to the neutral salts of the metals.

Ester acids, acid esters, ether acids, correspond to the acid salts of the metals.

Both are formed by the replacement of the replaceable hydrogen of inorganic or organic acids (p. 335) by alcohol radicals according as all or a part of the replaceable hydrogen of the acid is replaced, just as in the formation of salts. Thus monobasic acids yield only esters, while multibasic acids yield both esters and ester acids:

$$KOH + HCl = KCl + H_2O.$$
$$C_2H_5OH + HCl = C_2H_5Cl + H_2O.$$
Ethyl alcohol.   Ethyl chloride.
$$KOH + H_2SO_4 = KHSO_4 + H_2O.$$
$$C_2H_5OH + H_2SO_4 = C_2H_5-HSO_4 + H_2O.$$
Ethyl alcohol.   Ethyl sulphuric acid.
$$K_2O + H_2SO_4 = K_2SO_4 + H_2O$$
$$(C_2H_5)_2O + H_2SO_4 = (C_2H_5)_2SO_4 + H_2O.$$
Ethyl ether.   Ethyl sulphate.

Isomeric Alcohols, Aldehydes, Ketones. The isomeric alcohols are divided into *primary*, *secondary*, and *tertiary* according to their chemical behavior, due to the position of the HO groups.

We have seen (p. 302) that in the substitution of one H atom by a univalent element or radical in propane, C_3H_8, two isomers are possible, namely, $CH_3-CH_2-CH_2(OH)$ and $CH_3-CH(OH)CH_3$; also that from the two butanes, C_4H_{10}, four isomeric substitution products are possible; in other words, that by the substitution of one OH group in place of one H atom the existence of two or of four alcohols, respectively, are made possible. These four alcohols are known, and the structure given on page 302 corresponds in fact to their chemical behavior.

Primary alcohols by oxidation are converted into aldehydes (from alcohol dehydrogenatus) by the removal of two intraradical hydrogen atoms (belonging to the alcohol radicals). On further action the aldehydes take up one O atom very readily and are converted into acids (P. 335). Each primary alcohol has a corresponding aldehyde and an acid with the same number of C atoms; thus,

Methyl alcohol, $CH_4O + O = H_2O + CH_2O$, Methyl aldehyde.
Methyl aldehyde, $CH_2O + O = CH_2O_2$,   Formic acid.

This behavior can be explained if we admit that in the primary alcohols one hydroxyl replaces one hydrogen atom of a CH_3 group and hence we have still two oxidizable H atoms in the group; thus,

CH₃	CH₃	CH₃	CH₃
CH₃	H₂=COH	H-C=O	HO-C=O
Ethane.	Ethyl alcohol.	Ethyl aldehyde.	Acetic acid.

Primary alcohols contain the group $^-CH_2(OH)$.

Aldehydes contain the group $^-CH=O$.

Secondary alcohols are isomeric with the primary alcohols, and by oxidation they at first also lose two hydrogen atoms and form ketones which are isomeric with the aldehydes. On further oxidation the ketones yield acids which contain a less number of carbon atoms; thus,

$$CH_3-CH(OH)-CH_3+O = CH_3-CO-CH_3+H_2O.$$
Secondary propyl alcohol.   Propyl ketone.

$$CH_3-CO-CH_3+4O = CH_3-COOH+CO_2+H_2O.$$
Propyl ketone.   Acetic acid.

This behavior is to be explained by the fact that in the secondary alcohols the hydroxyl replaces one H atom of the $^=CH_2$ group, and the group $^=CH(OH)$ thus produced cannot be converted into the $CO(OH)$ group without the splitting off of a neighboring C atom. As the $^=CH(OH)$ group can exist only in such saturated compounds as contain at least three C atoms, then the lowest secondary alcohol must contain three C atoms.

Secondary alcohols contain the group $^=CH(OH)$.

Ketones contain the group $^-CO^-$, which is united to carbon atoms on both sides.

Tertiary alcohols are compounds, isomeric with the primary and secondary alcohols, which on oxidation are immediately transformed, without the formation of intermediate products, into various acids, or ketones and acids, containing less amount of carbon:

$$(CH_3)_2(C_2H_5) \ C-OH+3O = CH_3-COOH+CH_3-CO-CH_3+H_2O.$$
Tertiary amyl alcohol.   Acetic acid.   Propyl ketone.

They contain the hydroxyl in the place of the H atom of the $CH^{\equiv}$ group, and as the $CH^{\equiv}$ group can exist only in saturated compounds, which contain at least four C atoms, the lowest tertiary alcohol must contain four C atoms:

$$\begin{matrix} CH_3 \\ CH_3 \end{matrix} > C < \begin{matrix} OH \\ CH_3 \end{matrix} \text{ or } (CH_3)_3 \equiv C-OH, \text{ tertiary butyl alcohol.}$$

These alcohols contain no more oxidizable hydrogen attached to the hydroxylated carbon, hence they change immediately in the early stages of oxidation

The tertiary alcohols contain the group $\equiv C^{..}OH$, which is combined to the C atoms by its three valences.

Sometimes the isomeric alcohols are considered as derivatives of methyl alcohol or carbinol, $CH_3 \cdot OH$, whose H atoms are replaced by alkyls and are called:

$C_3H_7-CH_2 \cdot OH$, propyl carbinol, primary butyl alcohol;

$(CH_3)_3 \equiv C \cdot OH$, trimethyl carbinol, tertiary butyl alcohol;

$(C_2H_5)(CH_3)=CH \cdot OH$, methyl ethyl carbinol, secondary butyl alcohol.

Acids are formed on the oxidation of primary alcohols, when the two hydrogen atoms which are united with the hydroxyl group to the same C atom are replaced by an O atom, or when an oxygen atom is added to the aldehydes (p. 333). The carboxyl group ^-COOH is characteristic of the acids. All acids contain two hydrogen atoms less and one oxygen atom more than the alcohols from which they are derived. While in inorganic acids nearly all the H atoms present are replaceable with the formation of salts, with the organic acids only so many replaceable H atoms are available in the formation of salts or esters (p. 332), as there are ^-COOH groups present.

The basicity of organic acids is dependent upon the number of carboxyl groups contained, while the atomicity is dependent upon the number of hydroxyl groups present:

CH_3^-COOH,	$CH_2(OH)COOH$,	$HOOC^-COOH$,
Acetic acid.	Glycolic acid.	Oxalic acid.
Monatomic and monobasic.	Diatomic, but monobasic.	Diatomic and bibasic.

CH_3^-COOK,	$CH_2(OH)^-COOK$,	$KOOC^-COOK$,
Potassium acetate.	Potassium glycolate.	Potassium oxalate.

$\begin{matrix} CH_3^-COO \\ CH_3^-COO \end{matrix} \!\!>\!\overset{II}{Ca}$,	$\begin{matrix} CH_2(OH)COO \\ CH_2(OH)COO \end{matrix} \!\!>\!\overset{II}{Ca}$,	$\begin{matrix} COO \\ COO \end{matrix} \!\!>\!\overset{II}{Ca}$,
Calcium acetate.	Calcium glycolate.	Calcium oxalate.

The hydrogen present in organic acids but not available for the formation of salts or esters (intraradical) may also be replaced by atoms or groups of atoms whereby the ability of the acid to form salts, etc., is in no wise changed; thus,

$CH_2(NO_2)^-COOH$,	$CH_2(NH_2)^-COOH$,	$CH_2(OH)^-COOH$,
Nitroacetic acid.	Amidoacetic acid.	Oxyacetic acid.

$CH_2(NO_2)^-COONa$,	$CH_2(NH_2)^-COOK$,	$CH_2(OH)^-COONH_4$,
Sodium nitroacetate.	Potassium amidoacetate.	Ammonium oxyacetate.

Imido-acids are derived from the acids by the replacement of the $O^=$ atom of the ^-COOH group by $^=NH$. They are only known in the form of their esters, the imido ethers; thus, $CH_3^-C(NH)O(CH_3)$.

Thio-acids. The oxygen of the COOH groups of acids can be replaced by sulphur and from these acids many compounds are derived:

CH_3^-COSH,	$CH_2(NH_2)^-COSH$,	CH_3^-CSSH,
Thioacetic acid.	Amidothioacetic acid.	Dithioacetic acid.

CH_3^-COSK,	$CH_3COS(C_2H_5)$,	$(CH_3^-CO)_2S$,
Potassium thioacetate.	Ethyl thioacetate.	Thioacetic anhydride.

Imidothio-acids are derived from the thio acids by replacing the $^=O$ atoms of the $^-CO\!\cdot\!SH$ groups by $^=NH$. They are only known in the form of esters, the imidothio-ethers.

Acid radicals, *acyls,* are the organic acid residues united with the hydroxyl groups, just as with the inorganic acids, but not known free. Their name is formed by adding *yl* to the Latin name of the acid; thus,

$$H-CO, \qquad CH_3C-O, \qquad OC-CH_2-CO,$$

Formyl. Acetyl. Malonyl.

Acid anhydrides are formed from the acids by splitting off of H_2O from the carboxyl groups:

$$2CH_3-COOH = CH_3-COO-CO-CH_3 + H_2O.$$

Acetic acid. Acetic anhydride.

As in the monobasic inorganic acids two molecules must join together in order that water is split off, so the same is true for the organic monobasic acids:

$$2HNO_2 = N_2O_3 + H_2O; \quad 2CH_3-COOH = (CH_3CO)_2O + H_2O.$$

On the contrary, one molecule of water can be split off from multibasic inorganic acids; thus, $H_2SO_4 = SO_3 + H_2O$; so also many organic acids yield the same:

$$C_2H_4 <^{COOH}_{COOH} = C_2H_4 <^{CO}_{CO} > O + H_2O.$$

Succinic acid. Succinic anhydride.

The anhydrides of the monobasic acids are comparable with the anhydrides of the monatomic alcohols (the ethers) in that the ethers contain two alcohol radicals united to an oxygen atom, while the acid anhydrides contain two acid radicals under the same conditions. They also may be considered as acids whose carboxyl hydrogen is replaced by the same acid radical which is contained in the acid.

Amido-acids, amino-acids, glycocolls, are produced when the NH_2 group (amido group) replaces the intraradical H atoms of the acid; thus, $CH_2(NH_2)-COOH$, amidoacetic acid.

Acid amides, amides, are to be considered as ammonia in which the hydrogen is in part or wholly replaced by monovalent acid radicals (amines, p. 330):

$$\begin{matrix} H \\ H \\ H \end{matrix} \Big\rangle N \qquad \begin{matrix} H \\ H \\ CH_3CO \end{matrix} \Big\rangle N \qquad \begin{matrix} H \\ CH_3CO \\ CH_3CO \end{matrix} \Big\rangle N \qquad \begin{matrix} CH_3CO \\ CH_3CO \\ CH_3CO \end{matrix} \Big\rangle N .$$

Ammonia. Acetamide. Diacetamide. Triacetamide.

We differentiate between primary, secondary, or tertiary amides according to whether one, two, or three H atoms are replaced.

Acid imides, imides, are produced when two H atoms of ammonia are replaced by a bivalent acid radical; thus, $HN(-OC-CH_2-CO-)$, malonimide.

Thiamides are derived from the amides by replacement of S for O; thus, $CH_3-CS-NH_2$, thiacetamide.

Amidines, amimides, are derived from the amides by the replacement of $=NH$ for $=O$; thus, $CH_3-C(NH)NH_2$, ethenylamidine.

Amine Acids, Acid Amides. Among the polybasic acids there exist neutral amides and also acid amides or amine acids according to whether two or more carboxyl groups be present; thus,

$$C_2H_4 \Big\langle \begin{array}{l} CO(NH_2) \\ COOH \end{array} \qquad\qquad C_2H_4 \Big\langle \begin{array}{l} CO(NH_2) \\ CO(NH_2) \end{array}$$

Succinaminic acid.   Succinamide.

In the amides all the hydroxyls of the carboxyl groups are replaced by the amido group, NH_2. In the amine acids the hydroxyls of the carboxyl groups are only in part replaced by NH_2.

Hydrazides are obtained when acid radicals replace the hydrogen in hydrazin, H_2N-NH_2; thus, $H_2N-NH(CH_3-CO)$ = acetyl hydrazid.

Mixed Compounds. The different atomic groups which we have seen are characteristic for the various compounds may exist simultaneously in the same molecule, producing bodies having mixed functions; e.g., aldehyde alcohols or aldoses with the groups CHO and CH_2OH (or CHOH or COH), aldehyde acids with the groups CHO and COOH, ketone alcohols or ketoses with the groups CO and CH_2OH, ketone-acids with the groups CO and COOH, amido-alcohols with the groups NH_2 and CH_2OH, etc., amido-ketones with the groups NH_2 and CO, amido-aldehydes with the groups NH_2 and CHO.

2. International Nomenclature.

This is used for the present only in the publication of chemical investigations and large chemical works. It is based upon substitution; each compound is considered as a derivative of a hydrocarbon, and this is obtained if we replace all the other atoms or atomic groups present in the respective C compound by the corresponding number of hydrogen atoms. Accordingly $CH_2=NOH$ is not called formoxime, but methanoxime; $CH_3-CO-NH_2$ not acetamide, but ethanamide; CH_3-CH_2-CN not propionitrile, but propane-nitrile.

Saturated hydrocarbons with branchless chains terminate in *ane;* therefore we have only one butane and one pentane, etc. Hydrocarbons with branched chains are to be considered as derivatives of normal hydrocarbons, and their name is formed from the longest normal chain which the formula contains to which the name of the side chain is added; thus,

$\begin{array}{l} CH_3 \\ CH_4 \end{array}\rangle CH-CH_3$ is not called isobutane, but methyl propane;

$\begin{array}{l} CH_3-CH_2 \\ CH_3-CH_2 \end{array}\rangle CH-CH_2-CH_2$ is not called heptane or triethyl methane, but rather ethyl pentane.

When a carbon radical is introduced into a side chain, then we designate this by the prefix *metho-, etho-,* etc., in place of methyl, ethyl, which are only used in substitution in the main chain. Thus the following hydrocarbon is not called isopropyl heptane, but methoethyl heptane:

$$CH_3-CH_2-CH_2 \diagdown$$
$$\begin{matrix} CH_3 \\ CH_3 \end{matrix} >CH \diagup CH-CH_2-CH_2-CH_3.$$

The Arabic numerals are used to designate which carbon atom of the chief chain has the side chain attached; the numbers beginning at that end of the chief chain closest to the side chain; thus,

| 1 | 2 | 3 | 4 | 5 | | 5 | 4 | 3 | 2 | 1 |

$$\begin{matrix} CH_3 \\ CH_3 \end{matrix} >CH-CH_2-CH_2-CH_3 \quad \text{and} \quad CH_3-CH_2-CH \diagdown \begin{matrix} CH_2-CH_3 \\ CH_3 \end{matrix}.$$

Methyl pentane 2   and   Methyl pentane 3.

This numbering of the members of the hydrocarbons is also retained for all the substitution products:

$$CH_2Cl-CH_2Cl, \qquad \begin{matrix} CH_3 \\ CH_3 \end{matrix} >CH-CH_2-CH_2-CH_2Br.$$

1, 2-dichlorethane.   2-methyl-5-brompentane.

Unsaturated hydrocarbons with one double bond have the final syllable *-ene,* with two *-diene,* with three *-triene,* etc.; for example, ethene instead of ethylene, $H_2C=CH_2$, hexadiene instead of diallyl, $CH_2=CH-CH_2-CH_2-CH=CH_2$, etc. Hydrocarbons with treble bonds end in *-ine, -diine, -triine;* thus, hexadiine, $H_3C-C - C-C≡CH_3$. If double and treble bonds occur simultaneously, we make use of the termination *-enine* and *-dienine,* etc. If necessary, the position of the multiple bonds can be designated by numbering the C atom at which it occurs:

$$CH_3-CH=CH-CH_3, \text{ butene 2.}$$

Hydrocarbons with closed chains of 3, 4, 5, 6 CH_2 groups, for instance tri-, tetra-, penta-, hexamethylene are called *cyclanes* and have the name of the corresponding saturated hydrocarbons with the prefix cyclo; e.g., cyclohexane instead of hexamethylene.

Radicals. Monovalent radicals end in *-yl;* e.g., CH_3-CH_2-, ethyl, $CH_2=CH-$, ethenyl, $CH-C-$, ethinyl.

Radicals with alcoholic functions end in *-ol;* e.g., $-CH_2-CH_2OH$, ethylol, $-CH=CHOH$, ethenylol.

Aldehyde radicals end in *-al;* e.g., $-CH_2-CHO$, ethylal.

Acid radicals have the termination *-oyl* added to the original hydrocarbon; e.g., CH_3-CO-, acetyl = ethanoyl.

Alcohols have the final syllable *-ol* attached to the hydrocarbon from which it is derived. The polyatomic alcohols have di-, tri-, tetra-, etc., added to the -ol; e.g., C_2H_5-OH, ethyl alcohol = ethanol, $C_2H_4=(OH)_2$, glycol = ethandiol, $C_3H_5≡(OH)_3$, glycerin = propantriol. Unsaturated alcohols end, according to the bonds, in *-enol* or *-inol;* e.g., $CH_2=CH-OH$, ethenyl alcohol = ethenol, $CH_2=CH-CH_2OH$, propenyl alcohol = propenol, $HC≡C-CH_2OH$, propargylic alcohol = propinol, etc.

The designation *enol structure* comes from ethenol, and all compounds which like ethenol contain the $=C=COH-$ group are called *enols.*

Ethers are designated by placing *-oxy* between the hydrocarbons from which they are produced: ethyl ether, $(C_2H_5)O$ = ethanoxyethane, ethylamyl ether = pentaoxyethane.

Sulphur ethers have *-thio*, *-dithio*, and *-sulfon* introduced; e g., $C_6H_5-S-C_2H_5$, benzenthioethane, $C_6H_5-S-S-C_6H_5$, benzendithiobenzene, $C_6H_5-SO_2-C_6H_5$ = benzensulfonbenzene.

Aldehydes have the termination *-al* attached to the hydrocarbon, sulphaldehydes *-thial*; for example, HCOH, methanal, CH_3 CSH, ethanthial.

Ketones have the final syllable *-on*, the diketones, triketones, thioketones, the termination *-dion*, *trion*, *-thion;* e g., $CH_3-CO-CH_3$, propanon, $CH_3-CO-CH_2-CO-CH_3$, pentadion 2, 4.

Acids are designated by adding -acid, -diacid, -triacid to the original hydrocarbon with the ending *ic;* e g., CH_3-COOH, ethanic acid instead of acetic acid, $C_2H_4(COOH)_2$, butanic diacid instead of succinic acid or with mixed acids to the name of the respective alcohol, ketone, etc.; e.g., 2-propanolic acid instead of ethylidene lactic acid, $CH_3-CHOH-COOH$, 3-propanolic acid instead of ethylene lactic acid, $CH_2-OH-CH_2-COOH$, etc. Thioacids with a S atom simply united to the C atom are called *thiolic acids;* with doubly united S atom, *thionic acids;* e.g., $CH_3-CO(SH)$, ethanthiolic acid, $CH_3-CS(OH)$, ethanthionic acid, $CH_3CS(SH)$, ethanthiolthionic acid.

Salts and **esters** end in *-ate;* thus, $(CHO_2)_2Ca$, calcium methanate instead of calcium formate, $(COO)_2K_2$, potassium ethandiate instead of potassium oxalate.

CLASSIFICATION.

All aliphatic compounds are formed by replacing all or a part of the H atoms of the limit hydrocarbons (see p. 340) by other atoms or groups of atoms. Consequently the aliphatic compounds will be treated in the following pages, according to the number of side groups they contain, as compounds of mono-, di-, tri-, etc., valent alcohol radicals (p. 328), or in short as mono-, di-, tri-, etc., valent compounds. In connection with these compounds we shall discuss those compounds which have a close genetic relationship, while the carbohydrates will be specially discussed in order to make the subject easier.

COMPOUNDS OF MONOVALENT ALCOHOL RADICALS.

1. Monovalent Alcohol Radicals.

General formula C_NH_{2N+1}.

They cannot exist in the free state (p. 328). If an atom of H is substituted in the saturated hydrocarbons C_NH_{2N+2}, then the hydrocarbon residue C_NH_{2N+1} behaves like a monovalent radical. These are called alkyls, specially methyl, ethyl, propyl, etc. (p. 328), which names are often used as roots in the nomenclature of compounds derived therefrom.

2. Saturated Hydrocarbons.

Limit Hydrocarbons or Paraffins.

General form $Clau_N H_{2N+2}$.

	Boiling-point.			Boiling-point.
Methane, CH_4	$-164°$	Octane,	C_8H_{18}	$124°$
Ethane, C_2H_6	$- 93°$	Nonane,	C_9H_{20}	$149°$
Propane, C_3H_8	$- 45°$	Decane,	$C_{10}H_{22}$	$173°$
Butane, C_4H_{10}	$+ 1°$	Undecane,	$C_{11}H_{24}$	$195°$
Pentane, C_5H_{12}	$38°$	Dodecane,	$C_{12}H_{26}$	$219°$
Hexane, C_6H_{14}	$71°$	Hexadecane,	$C_{16}H_{34}$	solid
Heptane, C_7H_{16}	$99°$	etc.		

Properties. These compounds are called *limit hydrocarbons* because they cannot unite with any more atoms or groups of atoms (p. 297). Their names end in -*ane*. They are not attacked in the cold by such energetic oxidizing agents as concentrated sulphuric acid or nitric acid, hence they are also called paraffins (*parum*, without, *affinis*, affinity).

The lower members are colorless gases, the middle members colorless liquids, and the higher ones ($C_{14}H_{30}$ and above) are colorless solids, all being almost or altogether insoluble in water; the gaseous members are somewhat soluble in alcohol, while the liquids are readily soluble therein. The solubility of the higher members decreases as the amount of C increases. From butane on we find isomeric paraffins which differ from each other in density and boiling-point. When ignited they burn with a luminous and smoky flame (p. 92).

Chlorine has a substituting action forming HCl and mono-, di-, trichloride, etc., depending upon the extent of action; thus, $CH_4 + 2Cl = CH_3Cl + HCl$. Bromine has a less energetic action. In regard to the action of iodine see p. 322. Other derivatives can be readily prepared from the halogen derivatives (p. 302).

According to the chemical structure two isomers of butane must exist, and the number of isomers increases rapidly with the increase in the amount of C in the compound (p. 301). The boiling-point of the normally constituted member is always higher than that of its isomers.

Preparation. 1. All may be constructed from methane; thus if we heat its chloride, bromide, or iodide with sodium (or zinc, p. 348, 8), we obtain ethane: $2CH_3I + 2Na = 2NaI + C_2H_6$. If from this we prepare C_2H_5I and heat it with CH_3I and sodium, we obtain the next hydrocarbon, etc. (Fittig-Würtz method): $C_2H_5I + CH_3I + 2Na = 2NaI + C_3H_8$; $C_3H_7I + CH_3I + 2Na = 2NaI = C_4H_{10}$.

2. The fatty acids (p. 345) decompose into carbon dioxide and paraffins by electrolysis or by heating with caustic alkalies:

$$CH_3-COOH + 2NaOH = CH_4 + Na_2CO_3 + H_2O.$$

3. The unsaturated hydrocarbons form paraffins with nascent hydrogen: $H_2C=CH_2 + 2H = H_3C-CH_3$.

4. By the action of nascent hydrogen on the halogen derivatives of the aliphatic hydrocarbons:

$$CHCl_3 + 6H = CH_4 + 3HCl.$$

5. By heating the zinc alkyls (p. 382) with water:

$$Zn(C_2H_5)_2 + 2HOH = Zn(OH)_2 + 2C_2H_6.$$

6. By heating the alcohols with hydriodic acid:

$$C_2H_5-OH + 2HI = C_2H_6 + H_2O + I_2 \text{ (p. 322)}.$$

Formation. They are obtained in large quantities in the dry distillation of peat, brown coal, bituminous shale, Boghead and Cannel coal (as in the manufacture of illuminating-gas, p. 323). The gases obtained thereby consist of methane, ethane, propane, butane, besides other hydrocarbons. From the tar which is produced at the same time a mixture of paraffins is obtained by fractional distillation.

Wood- and coal-tar contain, on the contrary, chiefly aromatic hydrocarbons, especially the solid constituents naphthalene and anthracene (which see).

Occurrence. American petroleum consists of a mixture of different paraffins; Caucasian petroleum contains chiefly aromatic hydrocarbons and the hydrogen addition products of the same, such as the naphthenes, C_NH_{2N}. Galician and German petroleums stand between these two kinds according to their composition. In the gases which are evolved at the petroleum wells and also dissolved in the crude petroleum we find methane (60–90 per cent.) as well as some ethane, propane, and butane.

Crude petroleum is an oily brown fluid which loses its volatile constituents on standing in the air, becomes thicker, and finally forms the natural asphalt. From American crude petroleum by fractional distillation various hydrocarbons are separated into the following groups:

Petroleum ether, ligroin, consists of pentane and hexane and distils at 40°–70° as a colorless liquid.

Gasoline (low and high boiling-point) is used for making illuminating-gas, for heating purposes, and when vaporized and mixed with air explodes with force; hence it is used as a motive force.

Benzine consists of hexane, heptane, and octane, distils at 50°–120°, and is soluble in 5–6 parts alcohol.

Refined petroleum, kerosene, contains the paraffins distilling between 150° and 300°. The vapors should not ignite when heated below 38° (100° F.), otherwise the oil contains paraffins having a lower boiling-point and may explode if burned in a lamp.

Vulcan oil. Lubricating oil is the thick, impure constituent distilling above 300°, and is used as a lubricant.

Paraffin oil, vaseline oil, obtained from the vulcan oil by decolorization with animal charcoal, is a colorless oily liquid, contains the paraffins distilling above 360°, and is used in the preparation of paraffin salves.

Vasogene, vasol, is a mixture of paraffin oil and ammonium oleate which serves as a basis for salves.

Vaseline, petrolatum, mineral fat, is the name given to the product obtained by purifying the residues from the petroleum distillation. It forms a yellow or colorless soft mass which melts at 32°–35°, and is used as a substitute for fats, as it does not become rancid.

Solid paraffins. Ordinarily only the solid limit hydrocarbons are called paraffins. They occur comparatively pure in the mineral kingdom as mineral wax, ozocerite, and can also be obtained from that portion of American and Indian petroleum which has the highest boiling by cooling strongly, as well as from the various kinds of tar. These crude paraffins are freed from foreign constituents by means of sulphuric acid and decolorized by animal charcoal, and occur in commerce as white opalescent or crystalline masses which melt between 40° and 80° and then called paraffin, ceresin, belmontin, and are used chiefly in the manufacture of candles.

3. Monohydric Alcohols.

General formula $C_NH_{2N+1}OH$.

Methyl alcohol,	CH_3OH.	Nonyl alcohols,	$C_9H_{19}OH$.
Ethyl alcohol,	C_2H_5OH.	Decyl alcohols,	$C_{10}H_{21}OH$.
Propyl alcohols,	C_3H_7OH.	Dodecyl alcohols,	$C_{12}H_{25}OH$.
Butyl alcohols,	C_4H_9OH.	Tetradecyl alcohols,	$C_{14}H_{29}OH$.
Amyl alcohols,	$C_5H_{11}OH$.	Cetyl alcohol,	$C_{16}H_{33}OH$.
Hexyl alcohols,	$C_6H_{13}OH$.	Ceryl alcohol,	$C_{27}H_{55}OH$.
Heptyl alcohols,	$C_7H_{15}OH$.	Melissyl alcohol,	$C_{30}H_6OH$
Octyl alcohols,	$C_8H_{17}OH$.		etc.

Properties. The lower members are colorless fluids soluble in water, while the intermediate members are less soluble in water and have an alcoholic or fusel-oil odor and a burning taste. The members from $C_{12}H_{26}O$ and above are solids, insoluble in water and mostly without taste or odor.

Above propyl alcohol we have isomeric alcohols which are differ-entiated by their behavior on oxidation and by their varying boiling-points and are called primary, secondary, and tertiary alcohols (p. 333).

All alcohols are neutral in reaction, but behave like bases in that they combine with acids forming salt-like combinations which are called esters or ester-acids (p. 332).

Halogens do not have a substituting action upon the alcohols, but an oxidizing action instead:

$$C_2H_5OH + 2Cl = C_2H_4O + 2HCl.$$

On treatment with HCl, HBr, and HI or, better, with the halogen derivatives of phosphorus, the alcohols yield simple chlorine, bromine, or iodine derivatives of the hydrocarbons (p. 332).

The alcohols form crystalline compounds with $CaCl_2$, which decompose again with water (p. 349).

The alcohols, like water, are acted upon by the alkali metals with the evolution of hydrogen, and crystalline metallic *alcoholates* are obtained which, decompose with water into alcohol and alkali hydroxide, and readily give up the metal for other monovalent elements or radicals:

$$HO^-H + Na = Na^-OH + H;$$
$$CH_3^-OH + Na = CH_3^-ONa + H;$$
$$CH_3^-O^-Na + CH_3I = CH_3^-O^-CH_3 + NaI.$$

As only one H atom is replaceable, this one must have a different position from the other H atoms.

On heating primary and secondary alcohols with concentrated sulphuric acid they are converted into their anhydrides, the ethers (p. 332).

This process must not be considered as simply an abstraction of water by means of the sulphuric acid (see Ethyl Ether).

On heating with an excess of concentrated sulphuric acid or other dehydrating bodies the divalent hydrocarbons are obtained: $C_2H_5OH = C_2H_4 + H_2O$.

The primary, secondary, and tertiary alcohols behave on oxidation as described on p. 334.

Occurrence. Various alcohols are found in nature in the ethereal oils and waxes, seldom free, but generally as esters.

Formation. Ethyl alcohol and certain of its homologues are produced in the fermentation of grape-sugar.

Preparation of Primary Alcohols. **1.** The primary alcohols are obtained by the action of nascent hydrogen upon the aldehydes:

$$CH_3-CH_2-COH + 2H = CH_3-CH_2-CH_2(OH).$$
Propyl aldehyde.   Propyl alcohol.

2. By the action of moist silver oxide (also by KOH) on the halogen mono-substitution derivatives of the hydrocarbons:

$$C_2H_5I + AgOH = C_2H_5-OH + AgI.$$
Ethyl iodide.   Ethyl alcohol.

3. By the heating of their esters (better the ester acids) with water, alkaline hydroxides, or acids:

$$C_2H_5(C_2H_3O_2) + KOH = C_2H_5-OH + K(C_2H_3O_2).$$
Ethyl acetate.   Ethyl alcohol.

4. By treating primary amines with nitrous acid:

$$C_2H_5NH_2 + HNO_2 = C_2H_5OH + N_2 + H_2O.$$

Preparation of Secondary Alcohols. **1.** The secondary alcohols are obtained by the action of nascent hydrogen upon the ketones:

$$CH_3-CO-CH_3 + H_2 = CH_3-CH(OH)-CH_3.$$
Acetone.   Isopropyl alcohol.

2. From the secondary iodides (p. 349) by boiling with KOH:

$$CH_3-CHI-CH_3 + KOH = CH_3-CH(OH)-CH_3 + KI.$$

Preparation of Tertiary Alcohols. **1.** The tertiary alcohols are obtained from the tertiary iodides (p. 349) by boiling them with water.

2. From zinc alkyls by means of acid chlorides (ketones, *q.v.*, are formed by short action):

$$Zn(CH_3)_2 + CH_3-COCl + HOH = ZnCl(OH) + (CH_3)_3 \equiv COH.$$

4. Fatty Acid Series.

General formula $C_NH_{2N}O_2$.

Formic acid,	CH_2O_2.	Lauric acid,	$C_{12}H_{24}O_2$.
Acetic acid,	$C_2H_4O_2$.	Myristic acid,	$C_{14}H_{28}O_2$.
Propionic acid,	$C_3H_6O_2$.	Palmitic acid,	$C_{16}H_{32}O_2$.
Butyric acid,	$C_4H_8O_2$.	Stearic acid,	$C_{18}H_{36}O_2$.
Valeric acid,	$C_5H_{10}O_2$.	Arachidic acid,	$C_{20}H_{40}O_2$.
Caproic acid,	$C_6H_{12}O_2$.	Lignoceric acid,	$C_{24}H_{48}O_2$.
Caprylic acid,	$C_8H_{16}O_2$.	Cerotic acid,	$C_{26}H_{52}O_2$.
Capric acid,	$C_{10}H_{20}O_2$.	Melissic acid,	$C_{30}H_{60}O_2$.

Properties. The monatomic, monobasic acids of the monohydric alcohols are also called fatty acids because the higher members are found combined in the fats. From butyric acid upward we have isomeric fatty acids, which differ from each other by having different boiling- or melting-points. ·

The lower members at ordinary temperatures are caustic, pungent liquids, soluble in water with a strong acid reaction, while the intermediate members have an odor similar to perspiration and are less soluble in water. From $C_{10}H_{20}O_2$ upward the acids are solids, insoluble in water, generally without taste or odor and can only be distilled without decomposition in a vacuum.

All the members of this series are readily soluble in alcohol and ether and are, with the exception of formic acid, acted upon only with difficulty by oxidizing agents. Carbon dioxide expels them from their componnds.

Besides the replacement of the hydrogen of the COOH group by metals (salt formation), by alcohol radicals (ester formation) and acid radicals (anhydride formation), we can also replace the hydroxyl of the COOH group by atoms or groups of atoms, also the non-replaceable (intraradical) hydrogen.

In this last case bodies having all the properties of the acids are obtained, as the COOH is still intact (p. 335). These acids are designated α, β, γ, see p. 330.

Every fatty acid can be reduced to its aldehyde if a salt of the acid is heated with a formic acid salt (a formate):

$$CH_3-COONa + H-COONa = CH_3-CHO + Na_2CO_3.$$

Sodium   Sodium   Ethyl-   Sodium
acetate.   formate.   aldehyde.   carbonate.

On the distillation of a salt of the fatty acids with caustic alkalies the carboxyl group is split and replaced by hydrogen, yielding hydrocarbons which contain one C atom less than the acid used: e.g.,

$$CH_3-COONa + NaOH = CH_3-H + Na_2CO_3.$$

Sodium acetate.   Methane.

On the dry distillation of the calcium salts of the fatty acids ketones are obtained:

$$(CH_3COO)_2Ca = CH_3-CO-CH_3 + CaCO_3.$$

If a galvanic current is passed through a concentrated solution of an alkali salt of the fatty acids, the alkali metal is set free at the negative pole, with a decomposition of the water and the liberation of hydrogen, while at the positive pole carbon dioxide and a saturated hydrocarbon are set free:

$$2CH_3-COONa = C_2H_6 + 2CO_2 + 2Na.$$

From this behavior, as well as from the means of formation, it is evident that the fatty acids contain alkyl as well as a carboxyl groups.

Occurrence. Certain of the fatty acids occur free in nature, while a great number occur as esters, especially as waxes and fats (which see).

Preparation. 1. By the oxidation of primary alcohols and aldehydes:
$$CH_3-OH + O_2 = CH_2O_2 + H_2O.$$

2. By boiling the alkyl cyanides (nitriles) with acids or alkalies when the carbon of the cyanide group is converted into the carboxyl group, while the nitrogen is split off as ammonia:

$$CH_3-CN + 2H-O-H + HCl = CH_3-COOH + NH_4Cl.$$
$$CH_3-CN + H-O-H + KOH = CH_3-COOK + NH_3.$$
Methyl cyanide.   Potassium acetate.

3. By the action of carbon dioxide upon sodium alkyle:
$$CH_3Na + CO_2 = CH_3-COONa \text{ (sodium acetate).}$$

4. From the alkyl derivatives of acetoacetic ester (which see) by alcoholic potash.

5. From the oxy-fatty acids (which see) by heating with HI:
$$CH_2(OH)-COOH + 2HI = CH_3-COOH + H_2O + I_2.$$

6. By the oxidation of secondary and tertiary alcohols and ketones by chromic acid we obtain fatty acids which contain less carbon than the original substance.

5. Methane and its Derivatives.

Methane, Methyl Hydride, Marsh-gas, Fire-damp, CH_4.

Occurrence. It is formed in the slow decomposition of organic substances with lack of air, and is therefore found in coal-mines, as well as in the gases which are evolved from marshy ground and stagnant water. In certain localities it streams up from fissures in the earth (holy fire of Baku), and it is also evolved from the earth with petroleum. It is also found in the salt-mines of Wieliczka and in the hot springs of Aachen and Weilbach. Methane is produced in the dry distillation of many organic bodies (wood and coal), hence it occurs to a considerable extent in illuminating-gas (40–50 per cent.). Methane exists in the intestinal gases often in large quantities (even to 38 per cent.), and originates chiefly from the decomposition of the cellulose. It is formed when hydrogen acts upon carbon at 1200°, and in connection with ethane and acetylene when the electric current passes between carbon poles in hydrogen gas.

Preparation. 1. A mixture of carbon disulphide and sulphuretted hydrogen is passed over red-hot copper:

$$CS_2 + 2H_2S + 8Cu = CH_4 + 4Cu_2S.$$

2. Ordinarily it is prepared by heating sodium acetate with sodium hydroxide : $CH_3-COONa + NaOH = CH_4 + Na_2CO_3.$

Properties. Colorless and odorless, readily inflammable gas, nearly insoluble in water and liquefiable at $-82°$ under a pressure of 55 atmospheres. With air it forms an explosive mixture (fire-damp of the coal-mines): $CH_4 + 4O = CO_2 + 2H_2O$; mixed with a corresponding quantity of chlorine we obtain in diffused daylight chlorine substitution products, CH_3Cl, CH_2Cl_2, $CHCl_3$, CCl_4; this action takes place with explosive violence in the sunlight.

Methyl chloride, monochlormethane, CH_3Cl (preparation see page 348), is a colorless gas having a pleasant odor and liquefiable at $-25°$, inflammable, and burning with a beautiful green flame. It occurs in commerce liquefied in sealed glass tubes, and is used in medicine.

Trichlormethane, Chloroform, $CHCl_3$. *Preparation.* 1. In large quantities by distilling ethyl alcohol or acetone with chloride of lime, whereby trichloraldehyde (chloral) is first formed, which is further decomposed into chloroform and calcium formate by the action of the caustic lime always present in the chloride of lime.

2. So-called chloral-chloroform, or English chloroform, is obtained by distilling trichloraldehyde (chloral) with caustic alkali, which decomposes into alkali formate and chloroform (see Trichloraldehyde).

Properties. Colorless liquid having a sweetish odor, boiling at 62°, and solidifying at $-70°$. It is only slightly soluble in water, but soluble in ether, alcohol, and fatty oils; it burns with difficulty and is used as an anæsthetic and as a solvent for bromine, iodine, alkaloids, phosphorus, resins, etc.

By the action of air and light chloroform is in part decomposed into carbon oxychloride: $CHCl_3 + O = COCl_2 + HCl$. The addition of some alcohol and keeping it in the dark prevents this decomposition; hence commercial chloroform contains about 1 per cent alcohol and has a specific gravity of 1.489. On warming with alcoholic potash it is decomposed into potassium chloride and formate (p. 352) while on warming with alcoholic potash and ammonia it yields potassium cyanide: $CHCl_3 + NH_3 + 4KOH = CNK + 3KCl + 4H_2O$. On heating chloroform with aniline and alcoholic potash the characteristic odor of phenyl carbylamine is produced (detection of traces of chloroform; see Carbylamines). Recently pure chloroform has been prepared by cooling ordinary chloroform to crystallization (Pictet's chloroform).

Tri-iodomethane, Iodoform, CHI_3. *Preparation.* 1. By the action at ordinary temperatures of iodine and caustic alkali upon ethyl alcohol, acetone, and many other organic substances containing a methyl group (but not from methyl alcohol): $C_2H_5OH + 8I + 6KOH = CHI_3 + H\text{-}COOK + 5KI + 5H_2O$.

2. By the electrolysis of an aqueous KI solution containing a little alcohol, whereby $I+KOH$ are produced, and these acting upon the alcohol form CHI_3.

Properties. Yellow hexagonal crystalline plates which melt at 120°, have a persistent odor, are volatile, and soluble in ether, alcohol, glycerine, fatty oils, but not soluble in water.

Tribrommethane, bromoform, $CHBr_3$, prepared after the manner of iodoform, is a colorless liquid boiling at 148° to 150°, and having an odor similar to chloroform and a sweetish taste.

Tetrachlormethane, carbon tetrachloride, CCl_4, is obtained as the final product in the action of chlorine upon methane, methyl chloride, chloroform, and by the action of chlorine upon boiling carbon disulphide. It is a colorless liquid boiling at 78°, and serves as a solvent and an extraction medium in place of the inflammable carbon disulphide, ether, etc.

Halogen Derivatives of the Limit Hydrocarbons in General.

1. All the halogens can be replaced atom for atom by nascent hydrogen and the corresponding hydrocarbon is obtained.

2. As the halogens can be readily exchanged for other monovalent atoms or groups of atoms, these derivatives serve in the preparation of many other compounds. The iodides are the best suited for this purpose, the bromides less so, and the chlorides still less so.

3. On heating with alkali hydroxides (better with moist silver oxide, Ag_2O+H_2O) the halogens are replaced by hydroxyl: $CH_3I+KOH=CH_3-OH+KI$.

4. By the action of caustic alkalies in alcoholic solution upon the halogens unsaturated hydrocarbons are produced:

$$CH_3-CH_2-CH_2Cl+KOH=CH_3-CH=CH_2+KCl+H_2O.$$
$$\text{Propyl chloride.} \qquad\qquad \text{Propylene.}$$

5. On heating with potassium hydrosulphide or sulphide the halogens are replaced by HS or S:

$$CH_3I+KSH=KI+CH_3SH \text{ (methyl mercaptan)};$$
$$2CH_3I+K_2S=2KI+(CH_3)_2S \text{ (methyl sulphide).}$$

6. Ammonia replaces the halogens by the amido group NH_2:

$$CH_3I+NH_3=CH_3-NH_2+HI.$$

7. On heating the halogens with zinc we obtain the zinc compounds of the alkyls: $2CH_3I+2Zn=Zn(CH_3)_2+ZnI_2$.

8. On heating an excess of halogen alkyls with zinc (or sodium, p. 341) we obtain the higher hydrocarbons: $2CH_3I+Zn=C_2H_6+ZnI_2$.

9. The halogen derivatives burn with a greenish flame, and the halogens can only be detected after the destruction of the organic substance (see p. 310).

10. *Preparation.* The monoderivatives are prepared by the action of the nascent halogen acids upon the monohydric alcohols by distilling the latter with sodium chloride and sulphuric acid, or by distilling with bromine or iodine and phosphorus. Phosphorus trihalogenide is first

formed, which is decomposed by the alcohols in the same manner as with water:

$$PBr_3 + 3HOH = H_3PO_3 + 3HBr.$$
$$PBr_3 + 3CH_3(OH) = H_3PO_3 + 3CH_3Br.$$

They are also produced by the action of the halogens upon the limit hydrocarbons. If iodine derivatives are to be prepared according to this last method, oxidizing bodies must be added at the same time in order to destroy the HI formed (see p. 322).

Diderivatives are prepared by the action of the halogens upon the hydrocarbons, $C_NH^4_N$ (see Olefines). By the further action of the halogens upon mono- and diderivatives all of the hydrogen can be replaced atom for atom.

Secondary derivatives (P. 329) are obtained by the action of halogen acids upon the secondary alcohols, also upon the alkylenes from C_3H_6 upward, when the halogens are attached to the carbon atom poorest in hydrogen and not to the terminal carbon atoms:

$$CH_3-CH=CH_2 + HI = CH_3-CHI-CH_3.$$

On treating polyhydric alcohols with HI we generally obtain their secondary iodine derivatives.

Tertiary derivatives (p. 329) are prepared by the action of halogen acids upon tertiary alcohols (see also Olefines).

Nitrochloroform, $CCl_3(NO_2)$, chlorpicrin, is obtained from many hydrocarbons by the simultaneous action of chlorine and nitric acid as a colorless irritating fluid.

Methyl Hydroxide, Methyl Alcohol, Wood Spirits, Wood Alcohol, Carbinol, CH_3^-OH. *Occurrence.* Methyl alcohol exists with acetic acid and acetone, etc., in the watery distillation products of wood (crude wood vinegar), also as salicylic acid ester in the Galtheria procumbens (oil of wintergreen). It is formed to a slight extent in the alcoholic fermentation of many fruits.

Preparation. The crude wood vinegar is neutralized with lime in order to neutralize the acetic acid and the product distilled. The crude wood spirit thus obtained is treated with calcium chloride, with which it forms the compound $CaCl_2 + 4CH_3(OH)$, and this can be heated above 100° without decomposing, while the volatile impurities are driven off. The residue is treated with water, when the compound decomposes into its constituents, and distilled, when the methyl alcohol passes over.

Properties. Colorless fluid, boiling at 67° and having a peculiar odor. It is inflammable and soluble in water and yields formaldehyde and then formic acid on oxidation.

Methyl Aldehyde, Formaldehyde, Methanal, CH_2O or $H-CH=O$. *Preparation.* 1. Methyl alcohol vapors and air are passed over incandescent spirals of platinum or copper and the vapors condensed, when we obtain a solution of methyl aldehyde in methyl alcohol.

2. On heating paraformaldehyde (see below) we obtain pure gaseous formaldehyde.

Properties. A colorless, irritating gas which is strongly antiseptic, condenses at $-20°$ into a colorless liquid, solidifies at $-90°$, and is soluble in water.

On heating formaldehyde with dilute caustic alkali it splits into methyl alcohol and formic acid: $2CH_2O + KOH = CH_4O + H-COOK$, and on standing with ammonia it yields hexamethylentetramine: $6CH_2O + 4NH_3 = (CH_2)_6N_4 + 6H_2O$. On standing with calcium hydrate formaldehyde yields a mixture of sugars, $C_6H_{12}O_6$, (Loew's formose), and with protein bodies (not with peptones) it forms insoluble elastic masses, and with gelatine insoluble brittle masses called glutol.

Formalin, formol, is a 35 per cent. watery solution; *formalithe* is infusorial earth impregnated with 40 per cent. formaldehyde solution; *ichthoform* is a combination with ichthyol, and *tannoform*, a compound with tannin.

Paraformaldehyde, Trioxymethylene, $(CH_2O)_3$, separates on the evaporation of the watery or alcoholic solution of formaldehyde as colorless crystals insoluble in water, melting at 154°, and which on further heating decompose into gaseous formaldehyde (generation of formaldehyde for disinfecting purposes). When pressed with carbon it forms *carboformal* cakes which when ignited generate formaldehyde and are used for disinfecting purposes.

The Aldehydes in General. 1. They are colorless, neutral, irritating bodies; the lower members are colorless fluids (formaldehyde being a gas) having an irritating odor, soluble in water, and volatile without decomposition. As the C atoms increase they lose their odor, solubility, etc., and the higher members are solids which are odorless and only volatile without decomposition in vacuo. They have a strong reducing action because they are readily oxidized to the corresponding acid; thus the aldehydes decompose an ammoniacal silver solution with the separation of metallic silver. As no evolution of gas takes place in this reduction, the silver deposits as a mirror upon the surface of the glass; hence aldehydes produce a silver mirror.

2. The alkali bisulphites give crystalline insoluble compounds with the aldehydes: $CH_3CH=O + NaHSO_3 = CH_3CH(OH)(NaSO_3)$.

3. The aldehydes are readily transformed into polymers. especially in the presence of small quantities of acids or salts (particularly $ZnCl_2$) at ordinary temperatures. At higher temperatures condensation takes place (see p. 319 and Ethyl Aldehyde).

4. Heated with caustic alkalies the aldehydes yield resinous masses of high molecular weight (aldehyde resins).

5. On oxidation they yield the corresponding acid.

6. By nascent hydrogen they are converted into their alcohols.

7. With hydroxylamin the aldehydes give aldoximes (p. 331):

$$CH_3CH=O + H_2N-OH = CH_3-CH=N(OH) + H_2O.$$

8. By addition with alcohols they give so-called *alcoholates:*

$$CH_3C-H=O + C_2H_5OH = CH_3-CH\big\langle{}^{OH}_{OC_2H_5};$$

while with alcohols with the elimination of water the aldehydes form so-called *acetals:*

$$CH_3CH=O + 2(C_2H_5OH) = CH_3-CH(OC_2H_5)_2 + H_2O \text{ (ethylal)}.$$

With mercaptans (p. 332) they form *mercaptals*, corresponding to the acetals:

$$CH_3-CH=O + 2(C_2H_5-SH) = CH_3-CH(SC_2H_5)_2 + H_2O.$$

9. The aldehydes give cyanides of the next highest oxyacid (see Lactic Acid) when treated with hydrocyanic acid:

$$CH_3CH=O + HCN = CH_3-CH(OH)(CN).$$

10. With ammonia they form aldehyde ammonias, $CH_3-CH(OH)(NH_2)$, which (with the exception of that with formaldehyde, p. 350) yield pyridin bases on heating.

11. With 1 molecule of phenylhydrazin, $C_6H_5-HN-NH_2$, the aldehydes and ketones unite with the elimination of water, generally forming insoluble compounds, the hydrazons:

$$CH_3CH=O + C_6H_5-HN-NH_2 = C_6H_5-HN-N=CH-CH_3 + H_2O.$$

Phenylhydrazin therefore serves in the detection of the aldehyde and ketone groups and in the synthesis of sugars (see Carbohydrates).

12. *Preparation.* *a.* With the exception of formaldehyde they are prepared by the careful oxidation of the corresponding primary alcohols (p. 333).

b. By distilling the corresponding acids or their salts with a formate:

$$CH_3COONa + HCOONa = Na_2CO_3 + CH_3CH=O.$$

Sod. acetate.   Sod. formate.   Acetaldehyde.

c. From the dihalogen derivatives of the hydrocarbons by boiling with water: $CH_3-CHCl_2 + H_2O = CH_3-CH=O + 2HCl.$

Methylmercaptan, CH_3-SH, occurs to a slight extent in human intestinal gases, as well as in the urine after eating asparagus, and is produced in the dry distillation and putrefaction of many foodstuffs. It is a colorless liquid boiling at 60° and has a nauseating odor.

Methylsulphide, $(CH_3)_2S$, methylsulphur ether, is a liquid having an unpleasant odor and boiling at 116°.

Mercaptans and Sulphur Ethers in General (p. 332). 1. Both are colorless volatile liquids nearly insoluble in water and having a nauseating odor similar to leeks or onions.

2. The mercaptans are faintly acid and readily form salts with bases and metallic oxides, the so-called mercaptides, $(CH_3-S)_2Hg$. In the air they oxidize with the formation of disulphides: $2CH_3-HS + O = (CH_3)_2S_2 + H_2O.$ When oxidized by nitric acid they yield sulphonic acids (which see):

$$CH_3SH + 3O = CH_3-SO_3H.$$

3. The sulph-ethers give with alkyl iodides the *sulfin or sulfonium iodides*, in which the sulphur exists as tetravalent sulphur: $(CH_3)_2S + CH_3I = (CH_3)_3SI$. These bodies give with moist silver oxide bases which are exactly similar to the alkali hydroxides, the *sulfin or sulfonium bases:* $(CH_3)_3S(OH)$. On oxidation the sulph-ethers first yield *sulfoxides*, $(CH_3)_2S + O = (CH_3)_2SO$, then *sulfones*, $(CH_3)_2S + O_2 = (CH_3)_2SO_2$.

4. Formation of mercaptals, see p 351, 8, and of mercaptols, p. 371, 2.

5. Preparation, see p. 348, 5.

Formic Acid, CH_2O_2 or $H-COOH$. *Occurrence.* In ants and caterpillars, from which it can be obtained by distillation with water; in certain mineral waters, in peat, in pine needles, in nettles, in the bee, and hence in honey, in human urine, leucamic blood, in various animal secretions, in the soap-tree, and in the fruits of the tamarind.

Formation. 1. In the oxidation of methyl alcohol, of sugar, and of starch by chromic acid or by manganese dioxide and sulphuric acid.

2. By the action of hydrogen or water upon CO_2, or of hydrogen upon CO under the influence of the dark electric discharge.

3. Formic acid is also produced by the decomposition of chloroform, bromoform, and iodoform by the aid of alcoholic solution of caustic potash:

$$CHCl_3 + 4KOH = H-COOK + 3KCl + 2H_2O;$$

or from hydrocyanic acid by keeping the same in watery solution:

$$CNH + 2H_2O = HCOO(NH_4).$$
Hydrocyanic acid.   Ammonium formate.

Preparation. 1. Formic acid is prepared in larger quantities by heating oxalic acid:

$$HOOC-COOH = H-COOH + CO_2.$$
Oxalic acid.   Formic acid.

This decomposition takes place best in the presence of glycerine at 100°–110°, otherwise the free oxalic acid in part sublimes undecomposed. If the temperature is raised above 110°, then glycerin monoformate is formed, and this on further heating yields allyl alcohol, C_3H_5-OH (which see).

2. By passing carbon monoxide over the alkali hydroxides at a pressure of several atmospheres or at 200°, $NaOH + CO = H-COONa$, and distilling the formate produced with dilute mineral acids.

Properties. Colorless irritating liquid, causing blisters when applied to the skin. It solidifies at 1° and melts at 9°, while it boils at 99°. It mixes with water and alcohol and its vapors are inflammable. Formic acid has a reducing action, as it is readily oxidized to carbon dioxide, $H-COOH + O = CO_2 + H_2O$, and therefore precipitates silver as a black powder from its solutions, and white mercurous chloride from a solution of mercuric chloride (detection). When warmed with concentrated sulphuric acid it decomposes into carbon monoxide and water: $CH_2O_2 = CO + H_2O$; and when heated to redness with caustic alkali it generates hydrogen:

$$H-COOK + KOH = K_2CO_3 + 2H.$$

Officinal formic acid contains 75 per cent. water and 25 per cent. formic acid.

Formic acid spirits, formerly obtained by distilling ants with alcohol, is now prepared by dissolving 1 part formic acid in 24 parts dilute alcohol.

Formates are all soluble in water and crystalline. They are prepared by dissolving metallic oxides in the diluted acid and evaporating the solution, etc.

Ammonium formate, $H-COO(NH_4)$, is prepared by neutralizing formic acid with ammonia and evaporating, when colorless prisms are obtained. If these are heated to 180°, they decompose into liquid formamide (p. 336) and water: $H-COO(NH_4) = H-CO-NH_2 + H_2O$, which boils at 200° and when quickly heated decomposes into $CO + NH_3$. Ammonium formate and formamide on heating with phosphorus pentoxide yield hydrocyanic acid (HCN = formonitrile, p. 384): $H-COO(NH_4) = HCN + 2H_2O$. Thus from a harmless body there is a transition into a violent poison by the splitting off of water. On the other hand, dilute hydrocyanic acid can readily take up water and be transformed into ammonium formate(p. 352).

6. Ethane and its Derivatives.

Ethane, ethyl hydride, C_2H_6 or H_3C-CH_3 (p. 340), is obtained on heating methyl iodide with zinc, $2CH_3I + Zn = ZnI_2 + CH_3-CH_3$, and also, mixed with methane and acetylene, when the electric arc plays between carbon poles in hydrogen gas. It forms a colorless gas which liquefies at 4° and a pressure of 46 atmospheres.

Ethyl chloride, C_2H_5Cl, is obtained in the same manner as methyl chloride, and is a colorless, pleasant-smelling liquid which produces insensibility, boils at 12°, and therefore is used in the production of local anæsthesia.

A mixture of tri-, tetra-, and pentachlorethane was formerly used as an anæsthetic in place of chloroform.

Ethyl bromide, C_2H_5Br, is prepared by distilling ethyl alcohol with sulphuric acid and potassium bromide, when ethyl sulphuric acid is first produced and then decomposes by the KBr into potassium bisulphate and ethyl bromide. It is a colorless liquid boiling at 39°, and is insoluble in water but soluble in alcohol. It decomposes when exposed to air and light, and serves as an anæsthetic. It must not be mistaken for the poisonous ethylendibromide, $C_2H_4Br_2$.

Ethyl Hydroxide, Ethyl alcohol, Alcohol, Spirits of Wine, Ethanol, C_2H_5OH. *Occurrence.* In small quantities free, or as esters in certain plants, and also in animal organs; with acetone in the urine of many diabetics. It is formed to a slight extent in the dry distillation of many organic substances; hence it is found in coal-tar and bone-oil, etc.

Formation. 1. From aldehyde by reduction (p. 350). 2. From ethyl chloride by means of alkali hydroxide (p. 348). 3. Direct from C and H as follows: If the electric spark is passed between carbon poles surrounded by hydrogen, acetylene, C_2H_2, is produced, which with nascent hydrogen forms ethylene, C_2H_4. This unites with sulphuric acid, forming ethylsulphuric acid, $C_2H_5HSO_4$, which when distilled with caustic alkalies yields alcohol:

$$C_2H_5HSO_4 + 2KOH = C_2H_5-OH + K_2SO_4 + H_2O.$$

Preparation. On a large scale alcohol is prepared by fermentation (p. 325) of various kinds of sugars, for instance grape-sugar, invert-sugar, maltose, which is brought about by yeast or so-called Saccharomyces. The sugars decompose into alcohol and carbon dioxide:

$$C_6H_{12}O_6 = 2C_2H_6O + 2CO_2.$$
Grape-sugar.   Alcohol.   Carbon dioxide.

In the manufacture of alcohol the purer forms of sugar are not used, but rather molasses or bodies rich in starch, such as potatoes, where the starch is first converted into fermentable maltose by the action of diastase (see Ferments).

The fermented liquid is distilled, when the alcohol, which boils at 79°, passes over with some water, while the non-volatile bodies and the greater part of the water remain (which when obtained from potatoes is a valuable food for cattle on account of the proteids it contains).

If the alcohol containing water is distilled several times (or only once in specially constructed, so-called column apparatus) and the first portion collected, then we obtain an alcohol which contains only 8-10 per cent. water and is sold in commerce as spirits. The water cannot be entirely removed by further distillation, hence we are obliged to treat the alcohol with substances which have a greater attraction for water than alcohol (for this purpose we allow the spirits to stand over CaO or $CaCl_2$ for 24 hours) and then distilling. The product is anhydrous alcohol or absolute alcohol.

In alcoholic fermentation small quantities of glycerine, succinic acid, as well as mixtures of propyl, isobutyl, amyl alcohols, and esters of fatty acids, are always formed besides the ethyl alcohol. This mixture is called *fusel-oil;* it has a higher boiling-point than the alcohol, and hence distils over only to a slight extent with the alcohol. These products give the peculiar odor and taste to the crude forms of spirits, but they can be removed by passing the vapors of the crude spirits over charcoal. The fusel-oil is separated by distillation from the liquid which remains (phlegma) after the repeated distillation of the crude spirits (rectification).

Properties. Colorless, nearly odorless liquid having a burning taste; it boils at 78.5°, has a specific gravity of 0.79 at 15° C., and solidifies at −131°, forming a white crystalline mass. It burns with a bluish hardly perceptible flame. Alcohol attracts water with avidity and mixes therewith in all proportions, whereby heat is evolved and a diminution in volume occurs. It dissolves resins, fats, volatile oils, bromine, iodine, etc., also many salts and gases. By oxidizing agents or by certain ferments it is converted into aldehyde and acetic acid, and with sodium it forms crystalline sodium ethylate, $C_2H_5^-ONa$ (p. 343).

The amount of alcohol in watery liquids is determined from the specific gravity with proper regard for the temperature. As a contraction takes place on mixing alcohol and water, the specific gravity of such mixtures has been determined and tables are given which give the composition of such mixtures. If the liquid contains other bodies in solution, the alcohol must first be distilled off and the amount of alcohol determined from the specific gravity of the distillate. Small amounts of alcohol can be detected by converting it into iodoform, which is recognized by its odor and crystalline form.

Solid alcohol is obtained by dissolving 8 to 10 per cent. sodium soap or stearic acid in commercial alcohol.

Methylated spirit, which is used for technical purposes, contains 3 vol. per cent. of a mixture of methyl alcohol and pyridin bases.

Rectified spirit is an alcohol containing 9 to 10 vol. per cent. water and has a specific gravity of 0.830 to 0.834.

The word spirits is also used to denote the colorless alcoholic solutions of many medicinal substances, such as spirits of camphor, juniper, lavender, peppermint, etc.

Tinctures are alcoholic extracts of plants having medicinal properties; still the word is used for certain alcoholic and alcohol-ether solutions of salts, etc., used in medicine.

Liqueurs and cordials consist of dilute alcohol and sugar flavored with ethereal oils, tinctures, etc., and generally colored.

Alcohol containing 50 to 65 per cent. water and having a different taste depending upon admixture with fusel-oil, etc., derived from the materials used in the fermentation, forms certain beverages. We differentiate between

Whiskey, which obtained from crude starchy products, such as fermented malted grains (corn, rye, barley, oats), from potatoes, and from rice (arrac).

When obtained from fermented molasses it is called *rum*, from the cherry-kernel, *kirchwasser*, from the plum-kernel, *plum-brandy* or Slibowitz, from malted barley and rye meal with hops and rectified from juniper-berries it is called *gin*.

Brandy from alcoholic liquids. True brandy is obtained on the distillation of wine. The better kinds are called cognac.

Wine is an alcoholic liquid, completely fermented, which is obtained by the spontaneous fermentation of grape-juice (p. 325), without the addition of yeast and without distillation. Wines contain 8–10 to 20 per cent. (southern wines) alcohol, besides the constituents of the grape-juice. Malt wines are produced by the fermentation of a barley wort (see below) with a southern wine-yeast.

Beer is an alcoholic (3–4 per cent.) liquid still undergoing fermentation and prepared from starchy bodies (generally barley) without distillation. The chief operations are the following:

Malting consists in the preparation of diastase (see Ferments) by means of the artificial germination of barley grains and the interruption of this germination by heating the malt.

The *mash* consists in extracting the malt with water, when the diastase converts the starch into soluble sugar (maltose) and dextrin.

Hops and *boiling*. The solution (*wort*) obtained in the mash is treated with hops and boiled, when it becomes concentrated and the proteids coagulate and are precipitated by the tannic acid contained in the hops. The wort becomes clear and the constituents of the hops give the beer an agreeable taste and aids in preserving the same.

The *fermentation* is introduced by adding yeast to the wort. After completion of the chief fermentation the "new beer" is separated from the yeast and allowed to undergo "after-fermentation." Upon completion of the fermentation the beer is put in kegs, where a further decomposition of the sugars takes place, although the yeast-fungus is present only to a slight extent. The bungs are inserted loosely, and not driven home until a fortnight before delivery, so that the carbon dioxide is in part absorbed and produces the foam.

Ethyl nitrite, $C_2H_5{}^-O^-N^-O$, is obtained pure by distilling nitrous acid ($KNO_2 + H_2SO_4$) with alcohol. It is a colorless liquid boiling at 18° and with an odor similar to apples. If alcohol is distilled with nitric acid, a part of the acid is reduced to nitrous acid by the alcohol; at the same time aldehyde is formed and the nitrous acid then acts upon the undecomposed alcohol, producing ethyl nitrite:

$$C_2H_6O + HNO_3 = CH_3{}^-CHO + H_2O + HNO_2;$$
$$C_2H_6O + HNO_2 = C_2H_5{}^-O^-N^-O + H_2O.$$

The distillate thus obtained, a solution of ethyl nitrite in alcohol, is called sweet spirits of nitre.

Nitroethane, $C_2H_5{}^-NO_2$, isomeric with ethyl nitrite, contains, like all nitro bodies, the alkyl directly united to the nitrogen. Nascent hydrogen converts it into ethylamine, while it changes ethyl nitrite into ethyl alcohol (p. 318). The nitro compounds are obtained by treating the alkyl iodides with silver nitrate.

Ethyl nitrate, $C_2H_5{}^-NO_3$. Nitric acid has an oxidizing action upon alcohol, depending upon the quantity of nitrous acid it contains (see Ethyl Nitrite). If the distillation takes place in the presence of urea, then the nitrous acid is destroyed, $CO(NH_2)_2 + 2HNO_2 = 3H_2O + CO_2 + 4N$, and we obtain pure ethyl nitrate as a colorless liquid boiling at 86°, with a pleasant odor, differing from ethyl nitrite.

Ethyl sulphate, sulphuric acid ethyl ester $(C_2H_5)_2SO_4$ (see below), is obtained by passing sulphuric anhydride vapors into ethyl ether. It is a colorless aromatic-smelling liquid boiling at 220°:

$$(C_2H_5)_2O + SO_3 = (C_2H_5)_2SO_4.$$

The Neutral Esters in General. 1. They are mostly liquids having a low boiling-point, often of a pleasant odor, neutral, volatile without decomposition, and insoluble or nearly so in water. The esters of the lower fatty acids have odors like fruits and serve as artificial fruit flavors. Nearly all fruit odors can be obtained by admixture of these various esters. The esters of the acids rich in C, and also those of the aromatic acids, are mostly crystalline.

2. They are decomposed, like the analogous inorganic salts, by boiling with alkali hydroxides. The alkali metal combines with the acid, and the alcohol radical with the OH is set free as alcohol. This process is called *saponification* (see Glycerin):

$$FeSO_4 + 2KOH = Fe(OH)_2 + K_2SO_4.$$
$$C_2H_5(CH_3COO) + KOH = C_2H_5OH + K(CH_3COO).$$
Ethyl acetate.   Alcohol.   Potassium acetate.

They are also decomposed by superheated steam or boiling with acids into alcohol and acid with the taking up of water. This process is called hydrolysis.

3. Heated with ammonia the esters yield acid amides and alcohols:

$$CH_3-COO-C_2H_5 + NH_3 = CH_3-CO-NH_2 + C_2H_5OH.$$
Ethyl acetate.   Acetamide.

With hydrazin they yield hydrazids (p. 331), and with hydroxylamin the hydroximic acids (p. 331).

4. *Preparation.* a. If alcohols are treated with acids, ester formation takes place: $CH_3COOH + C_2H_5OH = CH_3-COO-C_2H_5 + H_2O$. The water formed takes part in the reaction and causes a corresponding equilibrium (see Ethyl Sulphuric Acid). If the acids or their salts are distilled with the alcohol in the presence of sulphuric acid, or if HCl vapors are passed through the mixture during the distillation, then the water combines with the acid and the reaction is complete.

b. By the action of halogen alkyls upon the silver salt of the acid in question:

$$3C_2H_5Cl + Ag_3PO_4 = (C_2H_5)_3PO_4 + 3AgCl.$$

c. In regard to the preparation of the esters of the halogen acids see p. 353.

Ethyl sulphuric acid, $(C_2H_5)HSO_4$, is obtained by mixing ethyl alcohol with sulphuric acid. The mixture always contains free sulphuric acid and unchanged alcohol even when equal molecular weights of the substances are used, as well as when an excess of sulphuric acid or alcohol is used, as a certain state of equilibrium always exists pp. 61, 62): $C_2H_5\bar{\ }OH + H_2SO_4 \rightleftharpoons C_2H_5\bar{\ }HSO_4 + HOH$.

In order to obtain pure ethyl sulphuric acid from this mixture it is neutralized by barium carbonate, which forms insoluble barium sulphate with the excess of sulphuric acid, while the barium ethyl sulphate remains in solution. This solution is decomposed by the necessary amount of sulphuric acid, the barium sulphate filtered off, and the filtrate evaporated in a vacuum over sulphuric acid. The pure ethyl sulphuric acid thus obtained is a colorless oily fluid which is readily decomposable. Heated with water it decomposes into sulphuric acid and alcohol: $(C_2H_5)HSO_4 + H_2O = C_2H_5OH + H_2SO_4$; while when heated alone it yields ethyl sulphate and sulphuric acid: $2(C_2H_5)HSO_4 = (C_2H_5)_2SO_4 + H_2O$.

The Acid Esters or Ester Acids in General. 1. They are acid-reacting, odorless liquids readily soluble in water, not volatile without decomposition (see above), of acid character, forming esters and salts. In general they behave like the neutral esters (which see).

2. *Preparation.* In a manner similar to that of ethyl sulphuric acid.

Ethyl oxide, ethyl ether, ether, $C_2H_5\bar{\ }O\bar{\ }C_2H_5$, incorrectly called sulphuric ether.

Formation. From ethyl iodide and sodium ethylate:

$$C_2H_5\bar{\ }ONa + C_2H_5I = C_2H_5\bar{\ }O\bar{\ }C_2H_5 + NaI.$$

Preparation. On a large scale by heating 1 part alcohol and 2 parts sulphuric acid to 140° (with more sulphuric acid ethylene, C_2H_4, is obtained), whereby two reactions take place, the first being the formation of ethyl sulphuric acid and water:

$$C_2H_5\bar{\ }OH + H_2SO_4 = (C_2H_5)HSO_4 + H_2O.$$

The ethyl sulphuric acid at 140° in the presence of more alcohol (see Ethyl Sulphuric Acid) decomposes into ether and free sulphuric acid:

$$(C_2H_5)HSO_4 + C_2H_5\bar{\ }OH = C_2H_5\bar{\ }O\bar{\ }C_2H_5 + H_2SO_4.$$

The water formed and the ether are distilled off, leaving the sulphuric acid behind, and if more alcohol is allowed to flow into the vessel the reaction proceeds without interruption. In this manner a small quantity of sulphuric acid transforms a large quantity of alcohol into ether, and this is the reason why the action of the sulphuric acid was formerly considered only as a contact substance (p. 9).

Properties. Colorless, peculiar pleasant-smelling liquid with burning taste and having a specific gravity of 0.72. It boils at 35° and solidifies at $-129°$; it inflames readily and burns with a luminous flame. Ether readily dissolves fats, resins, ethereal oils, sulphur, phosphorus, bromine, iodine, etc., and is slightly soluble in water, but soluble in all proportions in alcohol. Ether vapor is very inflammable and with air forms an explosive mixture. Inhaled it causes stupor and then complete unconsciousness, and hence is used in place of chloroform as an anæsthetic. On account of its volatility it has a strong cooling action, and therefore the vaporization of ether is used in the production of low temperatures, also as a local anæsthetic.

Anæsthesia ether is chemically pure ethyl ether. Spiritis ætheris, Hoffmann's anodyne, is a mixture of 1 part ether with 3 parts alcohol.

Ethers in General. 1. They are neutral, very stable bodies which do not, like the alcohols, unite with acids and are not oxidized by the halogens, but are substituted. Methyl ether is a gas, while the next members are volatile liquids having a characteristic "ethereal" odor, and the higher members of the series rich in carbon are solids. They are oxidized by nitric acid, etc.

2. All the hydrogen atoms of ethers show a similar behavior, and metallic sodium does not act upon them (p. 343).

3. On heating with water (in the presence of some sulphuric acid) in sealed tubes they are converted into the alcohols with the taking up of water.

4. *Preparation.* *a.* The simple ethers are obtained by heating the respective alcohol with sulphuric acid, or by the action of halogen alkyls upon the sodium alcoholates (p. 358).

b. The mixed ethers (those with two different alkyls) are obtained by the action of the potassium compound of an alcohol upon the iodide of another alkyl:

$$CH_3-OK + C_2H_5-I = CH_3-O-C_2H_5 + KI;$$
Potassium methylate.   Ethyl iodide.   Methyl-ethyl ether.

or by heating two different monovalent alcohols with sulphuric acid:

$$(CH_3)HSO_4 + C_2H_5OH = CH_3-O-C_2H_5 + H_2SO_4.$$

Ethyl peroxide, $C_2H_5-O \cdot O-C_2H_5$, is obtained by the action of ethyl sulphate upon alkaline H_2O_2 solution. It is a liquid having a faint ethereal odor and boiling at 65°.

Ethyl Aldehyde, Acetaldehyde, Ethyliden Oxide, Ethanal, C_2H_4O or $CH_3-CH=O$. *Preparation.* 1. By the gentle oxidation of alcohol by means of manganese dioxide or potassium bichromate and sulphuric acid, when the aldehyde may be distilled off.

2. By the distillation of an acetate with a formate (p. 351,[12]):

$$CH_3{}^-COONa + H^-COONa = CH_3{}^{\cdot\cdot}CH^=O + 2Na_2CO_3.$$
Sod. acetate.   Sod. formate.   Aldehyde.   Sod. carbonate.

3. On a large scale in the purification of crude alcohol by means of charcoal a part of the alcohol is oxidized by the air condensed in the pores of the charcoal. This part first passes over in the distillation, and the pure aldehyde is obtained from this portion by fractional distillation.

Properties. Colorless irritating liquid having a peculiar odor, and boiling at 21°, and oxidizing in the air into acetic acid.

Paraldehyde, $(C_2H_4O)_3$. Traces of mineral acids, phosgene gas, zinc chloride, etc. (p. 350,3), at ordinary temperature transform ethyl aldehyde into its polymer paraldehyde with the development of heat. Paraldehyde is a colorless liquid boiling at 124° without decomposition, soluble in alcohol, ether, and in 9 parts water, and can be made crystalline by strongly cooling.

Metaldehyde, $(C_2H_4O)_6$, is produced by the action of traces of mineral acids upon aldehyde below 0°. It forms white crystals which sublime on heating with a partial formation of aldehyde.

Both modifications do not show the properties of the simple aldehydes (p. 350). They are retransformed into ordinary aldehyde by distilling them with dilute sulphuric acid. If aldehydes are treated at high temperatures with traces of mineral acids, etc., condensation takes place with the elimination of water (p. 319):

$$2CH_3CHO = H_2O + CH_3{}^-CH^=CH^-CHO \text{ (crotonic aldehyde)}.$$

Aldol, $C_4H_8O_2$, is produced when aldehyde is allowed to stand several days in contact with dilute hydrochloric acid at 15° C.:

$$CH_3{}^-CHO + CH_3{}^-CHO = CH_3{}^-CH(OH)^-CH_2{}^-CHO.$$

In this so-called aldol condensation one H atom of one molecule passes over into the other molecule.

Aldol is a thick odorless liquid which readily undergoes polymerization and which is converted with the elimination of water into crotonic aldehyde, C_4H_6O (see above), on heating. Aldol is an aldehyde alcohol which is derived theoretically from the corresponding dihydric alcohol and is also called β-oxybutyl aldehyde.

Trichloraldehyde, Chloral, C_2HCl_3O or $CCl_3{}^-CHO$. *Preparation.* By passing chlorine into ethyl alcohol and distilling the crystalline chloral alcoholate, $CCl_3{}^-CH(OH)(O \cdot C_2H_5)$, thus obtained with sulphuric acid.

If chlorine is allowed to act upon ethyl aldehyde, no substitution of the intraradical hydrogen takes place, but acetyl chloride, $CH_3{}^-COCl$, is obtained. If chlorine acts upon alcohol, crystalline chloral alcoholate (see p. 351,8) is obtained, and from this the chloral is set free by sulphuric acid:

$$CCl_3-CH(OH)(O \cdot C_2H_5) + H_2SO_4 = CCl_3-CHO + (C_2H_5)HSO_4 + H_2O.$$

Properties. Thick liquid with peculiar odor, boiling at 98°, and which gives all the reactions of the aldehydes. It is oxidized into trichloracetic acid, CCl_3COOH, by means of nitric acid, and by caustic alkalies it is decomposed into chloroform and a formate (p. 347):

$$CCl_3-CHO + KOH = CHCl_3 + HCOOK.$$

This process takes place in the preparation of chloroform, in that the chlorine of the chloride of lime forms chloral with the alcohol, and the calcium hydroxide which is always present in the chloride of lime decomposes this chloral into chloroform and calcium formate:

$$2CCl_3-CHO + Ca(OH)_2 = 2CHCl_3 + \frac{HCOO}{HCOO}\Big\rangle Ca.$$

When chloral is treated with a little water it solidifies to colorless crystals of chloral hydrate, $CCl_3-CH(OH)_2$, which melt at 58° and are readily soluble in water with a bitter taste. This is the form in which chloral is used in medicine. Chloral hydrate is an excellent solvent for many bodies, such as resins, starch, etc.

Chloral formamide, $CCl_3-CH\Big\langle {OH \atop CO(NH_2)}$, is obtained by bringing chloral, CCl_3-CHO, and formamide, $H-CO-NH_2$, together. It forms colorless, odorless, but bitter crystals which are soluble in 20 parts water and which melt at 114°.

Acetic Acid, $C_2H_4O_2$ or CH_3-COOH. *Occurrence.* Acetates are found in the animal kingdom, and free acetic acid is found in the feces, urine, perspiration, and several parenchymatous gland extracts. Pathologically it is found in leucæmic blood and gastric juice. It occurs free or as calcium or potassium acetate in the juice of many plants. Triacetin, $C_3H_5(C_2H_3O_2)_3$, occurs in the oil of the Evonymus europæus and Croton Tiglium; octyl acetate, $(C_8H_{17})C_2H_3O_2$, in the seeds of Heracleum giganteum and Heracleum sphondylium. Besides other products it is found in the putrefaction, fermentation, and dry distillation of many organic bodies.

Preparation. 1. From alcohol. If 8 to 15 per cent. alcohol is allowed to slowly trickle over wood shavings previously moistened with vinegar and pressed into barrels with false bottoms, the alcohol is finely divided and spread over a greater surface, when oxidation takes place rapidly by the action of the air (under the influence of the vinegar fungus). In order that the oxidation is complete the liquid is passed through several times.

Instead of alcohol we make use of wine, beer, fermented grain-mash, fruit-juices. The liquid thus obtained is called *vinegar* (wine-, malt-, or fruit-vinegar) and contains 5–8 per cent. acetic acid. It also contains other organic bodies which give a yellowish-brown or red color to the liquid.

Pure alcohol is not oxidized in the air either alone or when diluted with water; nevertheless all oxidizing agents, and also air or oxygen in the presence of platinum-black (p. 292), converts·the alcohol first into alde-hyde and then into acetic acid. Fermented alcoholic liquors on the con-trary, when exposed to the air soon become sour of themselves, if they do not contain too much alcohol. This is dependent upon the fact that these liquids contain certain salts and nitrogenous compounds which are necessary for the further development in the liquid of the spores of the acetic acid fungus, bacterium aceti, which are always found in the air and which take the part of oxygen-carriers (analogous to platinum-black). In the manufacture of vinegar the wood shavings serve as nutrition for the fungus.

2. From wood. The watery product obtained on the dry distilla-tion of wood and which contains 5–6 per cent. acetic acid, also methyl alcohol, acetone, and tar-oil, is a brown liquid and is sold as crude wood vinegar, pyroligneous acid (crude), and when purified by distil-lation occurs in commerce as a yellow liquid, called purified wood vinegar or rectified pyroligneous acid.

3. Anhydrous acetic acid is obtained by saturating the above-mentioned varieties of vinegar with sodium carbonate, evaporating and heating the residue to 250°. By this means the organic con-taminations are destroyed, while the sodium acetate remains un-changed and anhydrous. The acetic acid is obtained from this purified sodium acetate by distillation with sulphuric acid:

$$2CH_3\text{-}COONa + H_2SO_4 = 2CH_3\text{-}COOH + NaSO_4.$$

Properties. Pure acetic acid, *glacial acetic acid*, forms a colorless crystalline mass which melts at 17° to a colorless corrosive liquid having a specific gravity of 1.05, producing blisters on the skin, and boiling at 118°. It dissolves many organic substances, also sulphur and phosphorus.

Commercial acetic acid contains 30 per cent., vinegar essence, 70 per cent. acetic acid.

As acetic acid with 4 per cent. water has the same specific gravity as that with 30 per cent. water, it is impossible to determine the amount of acetic acid in watery solutions by means of the specific gravity, and this can only be done by chemical means.

Acetates are, with the exception of the silver and mercurous salts and certain basic salts, readily soluble in water. On heating, all acetates are decomposed and leave a residue; the alkali acetates leaving carbonate, and the other acetates metallic oxides or metals.

Detection. On heating an acetate with sulphuric acid the characteristic odor of acetic acid is developed; if alcohol is added to this mixture, ethyl acetate is produced, which is detected by its odor. On heating dry alkali acetates with As_2O_3 the disagreeable-smelling alkarsine (see Arsines) is obtained.

Potassium acetate, $C_2H_3KO_2$, is obtained by dissolving potassium hydroxide or potassium carbonate in acetic acid and evaporating to dryness. It is a white crystalline powder which absorbs moisture from the air.

Sodium acetate, $C_2H_3NaO_2+3H_2O$, forms colorless efflorescent crystals which are soluble in 1 part water.

Ammonium acetate, $C_2H_3(NH_4)O_2$, decomposes on heating into water and acetamide: $CH_3\text{-}COONH_4 = CH_3\text{-}CO\text{-}NH_2 + H_2O$.

Basic aluminium acetate, $HO\text{-}Al = (C_2H_3O_2)_2$, is only known in solution. An 8 per cent. watery solution having a specific gravity of 1.048 is used as a local astringent.

Lead acetate, $Pb(C_2H_3O_2)_2 + 3H_2O$, is obtained by dissolving lead oxide in acetic acid and evaporating. It forms colorless prisms which are soluble in 2.3 parts water and which have a sweetish taste; hence it is also called sugar of lead.

Basic Lead Acetate, Lead Subacetate, $Pb(C_2H_3O_2)_2 + xPbO$. Solutions of lead acetate readily dissolve lead oxide with the formation of basic salts which may be precipitated by alcohol as white crystalline powders having the following structure: ·

$$CH_3\text{-}COO\text{-}Pb\text{-}O\text{-}Pb\text{-}O\text{-}Pb\text{-}OOC\text{-}CH_3;$$
$$CH_3\text{-}COO\text{-}Pb\text{-}O\text{-}Pb\text{-}OOC\text{-}CH_3.$$

Such a solution of lead oxide in lead acetate solution is called vinegar of lead. This solution quickly absorb carbon dioxide, and hence they become cloudy in the air as well as on the addition of ordinary water, by the precipitation of basic lead carbonate

(Goulard's solution). A mixture of 1 part basic lead acetate with 49 parts distilled water forms what is called lead-water, which only becomes slightly cloudy when exposed to the air.

Cupric acetate, $Cu(C_2H_3O_2)_2 + H_2O$, obtained by dissolving verdigris (p. 237) or copper oxide in acetic acid and evaporating the solution, is a dark-green crystalline body and soluble in water.

Basic cupric acetate, $Cu(C_2H_3O_2)_2 + xCuO$, which has a composition analogous to that of the basic lead acetates, occurs in commerce as verdigris. It is a bluish or greenish crystalline powder insoluble in water, which is obtained when copper plates are exposed to vinegar or wine residues, undergoing acetic acid fermentation in air. Verdigris is produced in a similar manner when foods containing vinegar or undergoing acid fermentation are kept in copper vessels (p. 237).

Cupric-aceto-arsenite, $Cu(AsO_2)_2 + Cu(C_2H_3O_2)_2$, Schweinfurter green, is a beautiful green poisonous powder, used as a paint.

Zinc acetate, $Zn(C_2H_3O_2)_2 + 2H_2O$, is obtained by dissolving zinc oxide in acetic acid. It forms white shining plates, which are soluble in water and alcohol.

Ethyl acetate, acetic ether, acetic acid ethyl ester, $CH_3 \! - \! COO \! - \! C_2H_5$, is obtained by distilling dry sodium acetate with ethyl alcohol and sulphuric acid. It is a colorless, readily inflammable liquid with a refreshing odor, having a specific gravity of 0.90 and boiling at 74°.

Acetyl acetic acid, aceto-acetic acid, diacetic acid, β-ketobutyric acid, $C_4H_6O_3$ or $CH_3 \! - \! CO \! - \! CH_2 \! - \! COOH$ (acetic acid in which a hydrogen atom of the methyl group is replaced by the acetyl radical, $CH_3 \! - \! CO \! - \!$). It is obtained from its esters (see below) as a thick acid liquid, miscible with water and readily decomposing into CO_2 and acetone, $CH_3 \! - \! CO \! - \! CH_3$, on warming. Its salts and esters are colored violet-red with ferric chloride. The potassium and sodium salts are sometimes found in the urine of diabetics, etc.

Ketonic Acids in General. 1. Ketonic acids have, as they contain the CO group besides the COOH group, the character of an acid as well as a ketone. We differentiate between α-, β-, γ-ketonic acids (p. 329), depending upon the nearness or remoteness of the CO group to the COOH group. The β-ketonic acids readily decompose into CO_2 and the corresponding ketone.

2. Ketonic acids are converted into the corresponding alcohol acid by nascent hydrogen:

$$CH_3 \! - \! CO \! - \! CH_2 \! - \! COOH + H_2 = CH_3 \! - \! CH(OH) \! - \! CH_2 \! - \! COOH.$$
$$\beta\text{-oxybutyric acid}$$

3. Like the ketones (p. 371) they unite with alkali bisulphites, hydroxylamin, phenylhydrazin.

4. *Preparation.* The α-ketonic acids are prepared from the cyanides of the acid radicals (p. 346) by hydrolysis:

$$CH_3-CO-CN+2HOH=CH_3-CO-COOH+NH_3.$$

β-ketonic acids are obtained from their esters (see below) by saponification with dilute cold caustic alkali. γ-ketonic acids (see Levulinic Acid) are prepared from the β-ketonic acids with α-halogen fatty acids (p. 366, 4c).

Acetyl Acetic Acid Ethyl Ester, acetoacetic ester, β-ketobutyric acid ethyl aster, $CH_3 \cdot CO \cdot CH_2 \cdot COO \cdot C_2H_5$. (In regard to preparation see p. 366, 8.) It forms a neutral liquid with a fruit-like odor, boiling at 181°, and is only very slightly soluble in water.

Acetoacetic ester has, like all β-ketonic acid esters, great importance for chemical syntheses, as the one H atom of the methylene group can be readily replaced by sodium and this then replaced in turn by various radicals by the action of different organic halogen compounds. The second H atom of the methylene group can be made to follow the same procedure, and the compounds obtained can be made to split in two ways by heating with caustic alkalies; thus into alkyl ketones (ketone cleavage) or alkyl acetic acids (acid cleavage).

Acetoacetic ester shows besides this, like all β-ketonic acid esters, the phenomena of tautomerism (p. 300); hence it follows that derivatives with ketone structure as well as with "enol" structure (p. 338) may be formed. For example, if an acid chloride is allowed to act upon the sodium compound of the acid, we obtain a derivative with ketone structure: $CH_3^-CO^-CH(CH_3^-CO)^-COO^-C_2H_5$; while if an acid chloride is allowed to act upon a mixture of acetoacetic ester with pyridin, we obtain a derivative with "enol" structure:

$$CH_3^-C(CH_3^-CO)^-CH^-COO^-C_2H_5.$$

The β-ketonic-acid Esters in General. 1. In the cold the alkali salt of the β-ketonic acids are formed by the action of dilute aqueous caustic alkali, and from these the β-ketonic acid can be set free by treatment with the proper quantity of sulphuric acid and isolated by shaking with ether.

2. On boiling with dilute aqueous caustic alkali or with dilute sulphuric acid a ketone is obtained besides alcohol and carbon dioxide (ketone cleavage):

$$CH_3-CO-CH(CH_3)-COO-C_2H_5+2KOH=$$
$$CH_3-CO-CH_2-CH_3+K_2CO_3+C_2H_5-OH.$$

3. On boiling with concentrated alcoholic caustic alkali two acids (or their alkali salts) are obtained besides alcohol. One of these acids is always acetic acid (or CO_2) (acid cleavage):

$$CH_3-CO-CH(CH_3)-COO-C_2H_5+2KOH=$$
$$CH_3-COOK+CH_3-CH_2-COOK+C_2H_5OH.$$

4. By the aid of their sodium compounds a great many syntheses are possible, of which the following are the most important:

a. If ethyl iodide is allowed to act upon the above, the Na is replaced by ethyl, and in this compound the second hydrogen can be replaced by sodium, and this again replaced by an alcohol radical:

$CH_3-CO-CH(C_2H_5)-COO(C_2H_5)$, ethylacetoacetic ester;
$CH_3-CO-C(C_2H_5)_2-COO(C_2H_5)$, diethylacetoacetic ester.

b. By the action of acid chlorides the corresponding acid combinations are produced:

$CH_3-CO-CH(CH_3CO)-COO(C_2H_5)$, diacetoacetic ester;
$CH_3-CO-C(CH_3CO)_2-COO(C_2H_5)$, triacetoacetic ester.

c. By the introduction of chlorinated esters we obtain di- and tri-basic acids:

$CH_3-CO-CHNa-COO-C_2H_5 + CH_2Cl-COOC_2H_5 =$

$NaCl + CH_3-CO-CH \Big\langle {CH_2-COO-C_2H_5 \atop COO-C_2H_5}$ (acetylsuccinic acid ethyl ester).

If α halogen fatty acid esters are used in the above, we obtain the γ-ketonic acids as the product; thus, from acetylsuccinic acid ethyl ester we get $CH_3-CO-CH_2-CH_2-COOH$ (levulinic acid) $+ C_2H_5-OH + CO_2$.

d. Two molecules of the ester may be united by means of ethylene bromide:

$$CH_3-CO-CH-COO-C_2H_5 \atop {>C_2H_4 \atop CH_3-CO-CH-COO-C_2H_5.}$$

5. The compounds given above as ketone derivatives may also be obtained under certain circumstances as the enol derivatives (p. 338).

6. The H atoms of the methylene group are replaceable by NH_2, $2NH_2$, Cl, 2Cl, $=NOH$, $=NH$, etc.

7. With aldehyde ammonias, anilines, phenols, phenylhydrazins, we obtain compounds of the pyridin, chinolin, cumarin, pyrazol groups with the elimination of water or of alcohol.

8. *Preparation.* Sodium ethylate is heated with the acid ester, the sodium compound decomposed by the addition of the corresponding quantity of acetic acid, and the ester split off purified by distillation:

$C_2H_5-ONa + 2CH_3-COO-C_2H_5 = CH_3-CO-CHNa-COOC_2H_5 + 2C_2H_5-OH$.
Sod ethylate.　　Ethyl acetate.　　　Sodiumacetoacetic ester.

$C_2H_5-ONa + CH_3-COO-C_2H_5 + C_6H_5-COO-C_2H_5 =$
　　　　　　　Ethyl acetate.　　　Benzoic acid ethyl ester.

$C_6H_5-CO-CHNa-COO-C_2H_5 + 2C_2H_5-OH$.
Sodium benzoylacetic ester.

If one of the esters used is a formic acid ester, then aldehyde acid esters are obtained: $C_2H_5-ONa + H-COOC_2H_5 + CH_3-COO-C_2H_5 =$
　　　　　　　　　Ethyl　　　　　Ethyl
　　　　　　　　formate.　　　　acetate.

$H-CO-CHNa-COO-C_2H_5 + 2C_2H_5-OH$.
Sodium formyl acetic ester.

Acetyl chloride, CH_3^-COCl, is prepared by the distillation of phosphorus trichloride with acetic acid: $3CH_3COOH + PCl_3 = 3CH_3^-COCl + H_3PO_3$. It is a colorless, irritating liquid boiling at 55°.

The Acid Halogenides in General. 1. They are irritating, fuming liquids which readily exchange their halogens for other elements or radicals.

2. On boiling with water they yield the original acid:

$$CH_3^-COCl + HOH = CH_3^- COOH + HCl.$$

3. With alcohols they form esters:

$$CH_3^-COCl + C_2H_5^- OH = CH_3^-COO-C_2H_5 + HCl.$$

4. With ammonia, acid amides are produced:

$$CH_3COCl + NH_3 = CH_3^-CO^- NH_2 + HCl.$$

5. When heated with salts of organic acids they form acid anhydrides:

$$CH_3^-COCl + CH_3^-COONa = (CH_3CO)_2O + NaCl.$$

6. Ketones or tertiary alcohols are formed with zinc alkyls.

7. *Preparation.* In the same manner as acetyl chloride.

Acetyl oxide, acetic anhydride, $(CH_3CO)_2O$, prepared from acetyl chloride and sodium acetate (see above, 5), is a colorless liquid with an odor like acetic acid and boiling at 180°. It does not at first mix with water, but gradually it decomposes into acetic acid therewith:

$$CH_3^- CO^-O^-OC^-CH_3 + H_2O = 2CH_3^-COOH.$$

The Acid Anhydrides in General. 1. They are liquids or solids having a neutral reaction and soluble in alcohol and ether.

2. With water they are transformed gradually into the free acid, and with alcohols they form the esters of these acids:

$$(C_2H_3O_2)_2O + 2C_2H_5^-OH = 2C_2H_5(C_2H_3O_2) + H_2O.$$

3. *Preparation.* They cannot be prepared by simply abstracting water from the acids, for instance by P_2O_5, but are obtained from the alkali salts of the acids by the action of acid chlorides (see Acid Chlorides, 5).

Acetamide, $CH_3^-CO^-NH_2$, forms colorless crystals which melt at 78° and boil at 222° and are readily soluble in water and alcohol.

The Amides in General. 1. They are generally crystallizable, volatile bodies. The primary amides are neutral in reaction, but as they contain the basic amido group, they unite, like ammonia, directly with acids, forming salt-like compounds: $(CH_3^-CO^- NH_2)HNO_3$. The amido group has also the power, due to the acid radical, of having one of the H atoms replaced by metals: $(CH_3-CO-NH)_2Hg$, mercuric acetamide.

Secondary and tertiary amides are indifferent bodies.

2. On boiling with acids or alkalies they decompose into their components with the taking up of water:

$$CH_3^-CO-NH_2 + H_2O = CH_3COOH + NH_3.$$

3. On heating with phosphorus pentoxide they lose 1 molecule of water and are converted into the nitriles (p. 330):

$$CH_3-CO-NH_2 = H_2O + CH_3-CN.$$
Acetamide.   Acetonitrile.

4. HNO_2 decomposes the primary amides in the same manner as the amido-acids (p. 369, **3**).

5. Brominated amides yield amines (which see) with caustic alkalies.

6. *Preparation.* *a.* By the action of ammonia upon the esters of the organic acids (p. 357, 3, *c*), or upon the acid halogenides (p. 367, 4).

b. By the dry distillation of the ammonium salts of the fatty acids:

$$CH_3-COO-NH_4 = CH_3-CO-NH_2 + H_2O.$$

Chloracetic Acids. If chlorine is passed into boiling acetic acid, the hydrogen of the methyl group is replaced and we obtain the following, according to the extent of action:

Monochloracetic acid, CH_2Cl^-COOH, colorless crystals which melt at 62° and which readily deliquesce.

Dichloracetic acid, $CHCl_2^-COOH$, a liquid above 0°.

Trichloracetic acid, CCl_3^-COOH, prepared by the oxidation of chloral, forms readily soluble rhombic crystals which melt at 55° and which decompose into chloroform on heating with caustic alkali:

$$CCl_3^-COOH + KOH = CHCl_3 + KHCO_3.$$

These compounds have a strong caustic action and are transformed into acetic acid again by hydrogen.

The Halogen Fatty Acids in General. 1. Some are fluids and others solids and have great similarity to the original acid, but have still more marked acid character than the acid from which they are derived.

2. They readily exchange their halogen for other elements or radicals and hence serve in the preparation of acids containing the $-NO_2$, $-NH_2$, $-OH$, $-CN$, $-HSO_3$, etc., groups: $CH_2Cl-COOH + KCN = CH_2(CN)-COOH + KCl$.

3. In regard to the isomers α-, β-, etc., acids see p. 330.

4. *Preparation.* By the direct action of halogens upon fatty acids or of HCl, HBr, HI, on the oxyfatty acids:

$$CH_2(OH)COOH + HI = CH_2I-COOH + H_2O.$$

Oxyacetic acid, glycollic acid, $CH_2(OH)COOH$, is obtained on warming monochloracetic acid with alkali hydrates:

$$CH_2ClCOOH + KOH = CH_2(OH)COOH + KCl.$$

Oxyacetic acid is the first member of a new series of acids whose members are all obtained in the same manner from chlorinated fatty acids and which will be considered with the divalent compounds.

Thioacetic acid, thiacetic acid, CH_3-COSH, is obtained by the action of P_2S_5 upon acetic acid and is a colorless liquid boiling at 100° and smelling like acetic acid and H_2S, into which it decomposes with water.

Amidoacetic acid, glycocoll, glycin, glue-sugar, $CH_2(NH_2)^-COOH$, is obtained by warming monochloracetic acid with ammonia:

$$CH_2Cl^-COOH + NH_3 = CH_2NH_2^-COOH + HCl.$$

It can also be obtained with other bodies from hippuric acid (which see), toluric acid, phenaceturic acid, bile-acids, silk, spongin, and glue (hence the name glycocoll) by boiling with acids or alkalies. Uric acid decomposes into urea and glycocoll on heating with HI. Amido-acetic acid is a colorless solid crystallizing in rhombic crystals which melt at 170° and which decompose at higher temperatures. It is soluble in water, giving a sweetish taste thereto.

The Amido-acids in General. 1. They are colorless, mostly crystalliza-ble, neutral solids which give salt-like combinations with acids as well as bases Many are produced in the putrefaction of glue and protein bodies.

2. They differ from the amides in that the amido-group is firmly united (like the amines) and cannot be split off by boiling with caustic alkalies.

3. By the action of nitrous acid they (like the amides and amine acids) exchange OH for NH_2; thus,

$$CH_2(NH_2)^-COOH + HNO_2 = CH_2(OH)COOH + 2N + H_2O.$$

4. If nitrous acid is allowed to act upon their esters, then isodiazo fatty acid esters are produced. These may be considered as fatty acid esters in which two H atoms are replaced by the $-N=N-$ group (see Diazo Com-pounds):

$$(CH_3)^-OOC-CH_2(NH_2) + HNO_2 = (CH_3)OOC-CH(N=N) + 2H_2O.$$
Amidoacetic acid methyl ester.   Isodiazoacetic acid methyl ester

5. *Preparation.* By heating the monohalogen fatty acids with ammo-nia or by the reduction of the corresponding nitro fatty acid with hydrogen.

Methylamidoacetic acid, sarcosin, $C_3H_7NO_2$, is obtained by warm-ing monochloracetic acid with methylamine, $NH_2(CH_3)$:

$$\begin{matrix} CH_2Cl \\ | \\ COOH \end{matrix} + NH_2{}^-CH_3 = \begin{matrix} CH_2N(CH_3)H \\ | \\ COOH \end{matrix} + HCl;$$

also by heating creatine, theobromine, caffeine with barium hydroxide. It is a colorless neutral solid crystallizing in rhombic crystals which are soluble in water and melt at 215° and are volatile without decom-position at higher temperatures.

Betaine, lycin, oxyneurine, $C_5H_{11}NO_2$ or $CH_2-N(CH_3)_3$, is the internal an-
$$\overset{|\qquad\quad|}{CO-O}$$
hydride of trimethylhydroxylamidoacetic acid: $HOOC-CH_2-N(CH_3)_3(OH)$. It is a heterocarbocyclic compound (p. 326), and forms the type of a series of similarly constituted compounds which have been called betains (see

Trigonellin). It is prepared by the careful oxidation of choline and is found in the cotton-seed, buckthorn, in the sugar-beet and hence also in the molasses from beet-root sugar, occurring also as a non-poisonous ptomaine. It forms colorless deliquescent crystals which on heating evolve trimethyl-amine (preparation of this last from beet-root molasses).

Glycocholic acid, $C_{26}H_{43}NO_6$, occurs in ox and human bile, and decomposes on boiling with water or alkalies into glycocoll and cholalic acid: $C_{26}H_{43}NO_6 + H_2O = C_2H_5NO_2 + C_{24}H_{40}O_5$. It forms crystalline needles which are insoluble in water but soluble in alcohol.

Hyoglycholic acid, $C_{27}H_{43}NO_5$, occurs in pig bile and decomposes when treated as above into glycocoll and hyocholalic acid (see below).

Cholalic or cholic acids are the monobasic acids of unknown constitution which with glycocoll and taurin (which see) form the bile-acids, which are found combined with alkalies in the bile ($\chi o \lambda \acute{\eta}$). Free cholalic acids are found in the intestine and urine in jaundice.

The cholalic acids of human and ox bile, $C_{24}H_{40}O_5$, of pig bile (*hyocholalic acid*, $C_{25}H_{40}O_4$), of goose bile (*chenocholalic acid*, $C_{27}H_{44}O_4$), as well as the *choleic acid*, $C_{25}H_{42}O_4$, found in ox bile and *fellic acid*, $C_{23}H_{40}O_4$, found in human bile, form monobasic, colorless, bitter crystals which are difficultly soluble in water and ether but readily soluble in alcohol, and this solution having a dextrorotatory power. They differ from each other by their melting-point. On boiling with acid, by putrefaction in the intestine or by heating, they loose water and are converted into amorphous compounds, the dyslysines, for example, $C_{24}H_{36}O_3$, which are insoluble in water and alkalies. The cholalic acids and their compounds, the bile acids, give Pettenkofer's reaction, which consists in treating the solution with two-thirds volume concentrated sulphuric acid, so that the temperature does not rise above 60°, and then adding 3 to 5 drops of a cane-sugar solution (or furfurol), when a beautiful violet coloration is obtained.

7. Propane and its Derivatives.

Propane, propyl hydride, C_3H_8 or $CH_3{-}CH_2{-}CH_3$ (p. 340). Two series of isomeric compounds are derived from the above, depending upon whether the $CH_3{-}$ or the $CH_2{=}$ group is substituted. In the first case we obtain the normal compounds and in the second case the isopropyl compounds.

Normal or **primany propyl alcohol,** C_3H_8O or $CH_3{-}CH_2{-}CH_2OH$, is formed in the fermentation of certain forms of sugars and wine residues and can be separated from the fusel-oil (p. 354) by fractional distillation. It is a colorless, pleasant-smelling liquid which boils at 96° and yields propylaldehyde and propionic acid on oxidation.

Propyl aldehyde, C_3H_6O or $CH_3{-}CH_2{-}CHO$, is obtained on the oxidation of propyl alcohol or by the distillation of formic acid with salts of propionic acid (p. 351, 12). It forms a liquid similar to ethyl aldehyde and boils at 49°.

Propionic acid, $C_3H_6O_2$ or $CH_3{-}CH_2{-}COOH$, is found in perspiration, gastric juice, in the fruit of the Gingko biloba, in the fly-agaric, in crude wood vinegar, and in the mineral water of Weilheim and Brückenau. It is prepared by the oxidation of propyl alcohol or from ethyl-cyanide (p. 346, 2). It is a strong irritating liquid, boiling at 141° and readily soluble in water. It can be separated from its watery solution as a liquid swimming on the surface by the addition of calcium chloride thereto. This property and the fact that its salts have a fatty touch is the origin of its name (πρῶτον, the first; πιον, fat).

α-**Amidopropionic acid,** alanin, $CH_3-CH(NH_2)-COOH$, is a cleavage product of the proteids and forms needles which melt at 250°.

β-**Acetylpropionic acid,** levulinic acid, $CH_3-CO-CH_2-CH_2-COOH$ (p. 366, c), is obtained on boiling most carbohydrates with dilute hydrochloric or sulphuric acids, forming colorless plates which melt at 33°.

Secondary or **isopropyl alcohol,** $CH_3-CH(OH)-CH_3$, is prepared according to the general methods (p. 344) and is a colorless liquid boiling at 83°.

Dimethyl ketone, acetone, C_3H_6O or $CH_3-CO-CH_3$, is found in small quantities in human urine, in the blood, transudations and exudations, and in larger quantities in the urine of diabetics. It is formed in the gentle oxidation of isopropyl alcohol and in the dry distillation of tartaric acid, citric acid, sugar, wood, and hence occurs also in crude wood alcohol. Ordinarily it is obtained by the dry distillation of sodium or calcium acetate (p. 372, 8).

It is a colorless liquid, smelling like peppermint, and boiling at 56°. It is soluble in water, etc., and on oxidation it decomposes into acetic acid and carbon dioxide:

$$CH_3-CO-CH_3 + 4O = CH_3-COOH + H_2O + CO_2.$$

With iodine solution and caustic alkali it yields iodoform (p. 347). Acetone is condensed by hydrochloric acid gas or by concentrated sulphuric acid:

$$2C_3H_6O = H_2O + (CH_3)_2-C=CH-CO-CH_3 \text{ (Mesityl oxide)};$$
$$3C_3H_6O = 2H_2O + (CH_3)_2-C=CH-CO-CH=C-(CH_3)_2 \text{ (Phoron)};$$
$$3C_3H_6O = 3H_2O + C_9H_{12} \text{ (Mesitylene)}.$$

The Ketones in General. 1. They are similar in physical properties to the aldehydes, also in their behavior towards acid alkali sulphites, hydrocyanic acid, and phenylhydrazin, but they do not reduce ammoniacal silver solutions (p. 350, 1).

2. They do not combine with alcohols, but on the contrary they combine with mercaptans with the elimination of water and the formation of mercaptals, which are analogous to the mercaptols (p. 351, 8):

$$CH_3-CO-CH_3 + 2C_2H_5-SH = (CH_3)_2-C-(SC_2H_5)_2 + H_2O.$$

3. With hydroxylamin they form oximide- or isonitroso-compounds, which are called acetoximes:

$$CH_3-CO-CH_3 + NH_2OH = (CH_3)_2-C=N-OH + H_2O \text{ (p. 351, 7).}$$

4. By nascent hydrogen they are transformed into secondary alcohols. In these reactions we have side products produced in that each two molecules of the ketone unite together, forming divalent, ditertiary alcohols (see Glycols), which are called pinacones:

$$2CH_3-CO-CH_3 + H_2 = (CH_3)_2-C(OH)-C(OH)=(CH_3)_2.$$

5. They cannot be polymerized, but on the contrary suffer condensations readily (see Dimethylketone).

6. On oxidation the ketones yield acids which contain less carbon atoms in the molecule (p. 334).

7. With ammonia they form ketonamines, whereby 2 or 3 molecules of the ketone combines with 1 molecule of ammonia, with the elimination of H_2O:$(CH_3)_2-C(NH_2)-CH_2-CO-CH_3$, diacetonamine.

8. *Preparation.* *a.* By the oxidation of secondary alcohols.

b. By the dry distillation of fatty acid salts: $2CH_3-COONa = CH_3-CO-CH_3 + Na_2CO_3$. If a mixture of two salts is used we obtained a mixed ketone (with two different alcohol radicals): $CH_3-COONa + C_2H_5-COONa = CH_3-CO-C_2H_5 + Na_2CO_3$. With formates we always obtain aldehydes instead (p. 351, 12).

c. By the action of acid chlorides upon zinc alkyls (p. 367):

$$2CH_3-COCl + Zn(CH_3)_2 = 2CH_3-CO-CH_3 + ZnCl_2;$$
$$2CH_3-COCl + Zn(C_3H_7)_2 = 2CH_3-CO-C_3H_7 + ZnCl_2.$$

d. From the acetoacetic esters by caustic potash (p. 365, 2).

Disulfonethyldimethylmethane, $(CH_3)_2-C=(SO_2-C_2H_5)_2$, *sulfonal,* melts at 126°; also

Disulfonethylmethylethylmethane, $(CH_3)(C_2H_5)=C-(SO_2-C_2H_5)_2$, methyl sulfonal, *trional,* melts at 76°, and

Disulfonethyldiethylmethane, $(C_2H_5)_2-C=(SO_2-C_2H_5)_2$, *tetronal,* which melts at 85°; all three form colorless and tasteless crystals, difficultly soluble in water, and when heated with carbon powder yield the characteristic odor of the mercaptans. They are obtained by the oxidation of the corresponding mercaptols, which are produced from acetone and mercaptans (p. 371, 2).

8. Butane and its Derivatives.

Butanes, C_4H_{10}. Two are known. On substituting one H atom by monovalent elements or radicals we have 4 isomers possible. Thus on the introduction of HO group we have 4 alcohols, all of which have been prepared:

Butane. Primary. Secondary.   Isobutane.   Primary. Tertiary.
    Butyl alcohol.   Isobutyl alcohol.

Primary butyl alcohol, $C_4H_{10}O$, occurs in fusel-oil, especially with wine-yeast fermentation, and is prepared according to the general methods (p. 344); also from glycerine by schizomycetes fermentation. It is a liquid boiling at 117°, with a pleasant odor.

Butyric acid, ethylacetic acid, fermentation butyric acid, $CH_3^-CH_2^-CH_2^-COOH$ or $C_4H_8O_2$. *Occurrence.* It occurs as glycerine ester to a slight extent in butter, in cod-liver oil, croton-oil, in the fruit of the tamarind, of the soap tree, and of Ginko biloba. Its alkyl esters also occur in the ethereal oils of different Compositæ and Umbelliferæ; it occurs free in rancid butter, in the juice of the caterpillar, perspiration, and cheese, combined with bases in the fluid of the spleen, muscle, and in the contents of the large intestine and pathologically in the gastric juice. It is formed by a special fermentation of sugar, starch, lactic acid; also in the putrefaction and oxidation of proteids. It is therefore found in sauerkraut, in sour pickels, in spent tan, in Limburger cheese, etc.

Preparation. 1. According to the general methods (p. 346).

2. Ordinarily it is prepared by a special fermentation of sugars, starch, dextrins, or glycerine by mixing them with water, chalk, and old cheese and allowing this to stand for a long time at 30–40°. The liquid becomes gradually thicker and finally solidifies, with the formation of calcium lactate, $Ca(C_3H_5O_3)_2$, and if this is allowed to stand longer it becomes again fluid, with the evolution of carbon dioxide and hydrogen, and as soon as the development of gas ceases all the lactic acid, $C_3H_6O_3$, has been transformed into butyric acid:

$$2C_3H_6O_2 = C_4H_8O_2 + 4H + 2CO_2.$$
Lactic acid. Butyric acid.

On distillation with sulphuric acid the butyric acid is separated from its calcium salt.

The fermentation above mentioned is brought about by the lactic acid and butyric acid organisms, the mixed spores of which exist in the cheese. The fermentation is stopped by an excess of free acid, hence calcium carbonate, zinc oxide, etc., are added in order to form neutral salts. If we make use of pure lactic or butyric acid bacilli instead of the cheese we obtain lactic or butyric acids immediately.

Properties. Butyric acid is a colorless liquid boiling at 163° with a peculiar odor, especially unpleasant when dilute. It is soluble in water in all proportions. Calcium butyrate, $Ca(C_4H_7O_2)_2 + H_2O$, is less soluble in hot water than in cold and hence separates out from concentrated watery solutions on boiling.

Isobutyl alcohol, $(CH_3)_2{}^-CH^-CH_2OH$, occurs in fusel-oil, especially from beer-yeast, and is a colorless liquid having an odor similar to fusel-oil and boiling at 107°. On oxidation it yields isobutylaldehyde and then isobutyric acid.

Isobutyric acid, $(CH_3)_2{}^-CH^-COOH$, is found free in carobs (Ceratonia siliqua), in the oil of Pastinaca sativa, in feces, and in the putrefactive products of proteids; it exists as ester in Roman camomile-oil. It is a colorless liquid with an unpleasant odor similar to butyric acid and boiling at 154°. It is soluble in 5 parts water. Calcium isobutyrate, $Ca(C_4H_7O_2)_2 + 5H_2O$, is more soluble in hot water than in cold.

9. Pentane and its Derivatives.

Pentanes, C_5H_{12}. The three possible pentanes are known, from which eight structural isomeric alcohols are derived, all of which are known. The four primary pentyl alcohols have correspondingly four pentyl acids.

Isopentyl alcohol, ordinary amyl alcohol, fermentation amyl alcohol, $C_5H_{12}O$ or $(CH_3)_2{}^-CH^-CH_2{}^-CH_2OH$, occurs as ester in Roman camomile-oil and is the chief constituent of fusel-oil, of potato spirits (p. 354), from which it is obtained by fractional distillation. It is a colorless, poisonous, inactive liquid, boiling at 132° and with a characteristic odor. It is soluble in 40 parts water and the vapors cause coughing. Its esters are used in the making of confectionery, cordials, etc.; thus amyl valerianate as apple-oil, amyl acetate as pear-oil, etc. (see Esters, p. 357).

Amyl alcohols, $(CH_3)(C_2H_5){}^-CH^-CH_2OH$. Three are known; the lævorotatory modification is also found in fusel-oil, the dextrorotatory one prepared artificially, and the inactive modification obtained on mixing the other two (p. 39).

Amyl nitrite, $C_5H_{11}{}^-O^-NO$, is produced on passing nitrous acid into hot amyl alcohol, from which amyl nitrite can be distilled off as a yellowish liquid, having a specific gravity of 0.88, boiling at 98°, and with a fruit-like odor; when inhaled causes an increased flow of blood to the head.

Valeric acid, $C_5H_{10}O_2$ or $(CH_3)_2{}^-CH^-CH_2{}^-COOH$, propylacetic acid. *Occurrence.* Free and as ester in the blubber of the Delphinus globiceps, cheese, perspiration of the feet, in the valerian root (Radix valerianæ), in the angelica root, in Viburnum opulus, and in human feces.

Preparation. It was formerly prepared by the distillation of the valerian root, but now it is obtained by the oxidation of fermentation amyl alcohol, whereby the lævorotatory amyl alcohol (see p. 374), which is always present, is oxidized into lævorotatory valeric acid, which boils at 172°.

Properties. Valeric acid is a liquid which boils at 175°, has an irritating odor similar to old cheese, is optically inactive, and dissolves in 12 parts water.

Diamido-valeric acid, ornithin, $C_4H_7(NH_2)_2-COOH$, is produced on boiling ornithuric acid with concentrated hydrochloric acid as well as from arginin (which see) by boiling with baryta-water. On putrefaction it yields putrescin, $C_4H_8(NH_2)+CO_2$ (p. 398).

Tertiary amyl alcohol, $(CH_3)_2=C(OH)-CH_2-CH_3$, dimethyl ethyl carbinol, is prepared by distilling calcium hydroxide with amyl sulphuric acid obtained from isoamylene, C_4H_{10} (see Olefines), and sulphuric acid: $(C_5H_{11})HSO_4+Ca(OH)_2=CaSO_4+C_5H_{11}OH+H_2O$. It is a colorless liquid boiling at 102.5° and having an odor similar to camphor.

10. Compounds with more than Five Carbon Atoms.

As the number of isomers greatly increases with an increase in the carbon atoms (p. 301) only the most important compounds will be treated from now onward.

a. Alcohols.

From hexyl alcohol on, the alcohols are soluble in water with difficulty or are insoluble. From cetyl alcohol on, only solid alcohols are known and the best known are mentioned below.

Hexyl alcohol, $C_6H_{14}O$, occurs as hexylbutyrate in the essential oil of Heracleum giganteum.

Heptyl alcohol, $C_7H_{16}O$. Thirty-eight isomeric alcohols with this formula are possible and thirteen of these have been already prepared artificially.

Octyl alcohol, $C_8H_{18}O$, occurs as ester in the ethereal oil of Heracleum sphondylium, Pastinaca sativa, Heracleum giganteum.

Cetyl alcohol, ethal, $C_{16}H_{34}O$, forms the chief constituent of spermaceti as cetyl palmitate and is found in the coccygeal glands of the goose and duck. It is a white crystalline solid, melting at 50°.

Cetin, spermaceti, is obtained from the cranial cavity of the pot-whale, where it exists as a liquid fat and crystallizes on cooling.

Ceryl alcohol, cerotin, $C_{27}H_{56}O$, exists as ceryl-cerotinate (cerotic acid ester, p. 377) in Chinese wax (plant wax), and is a white crystalline solid, melting at 70°.

Melissyl alcohol, myricyl alcohol, $C_{30}H_{62}O$, exists as myricyl palmitate as the chief constituent of beeswax (p. 377), and is a white crystalline solid, melting at 85°.

b. Acids.

Corresponding to the preceding alcohols we have to mention the following acids, which are derived therefrom by oxidation.

Caproic acid melts at $-2°$, while all the higher fatty acids are solid at ordinary temperatures. Caproic acid and the following acids of the series are soluble with difficulty in water and from lauric acid and beyond they are insoluble in water. Theoretically 8 hexyl acids are possible, 33 octyl acids, and 507 undecyl acids.

Caproic acid, $C_6H_{12}O_2$, occurs as ester in the fruits of the Gingko biloba, seeds of Heracleum sphondylium, flowers of Satyrium hircinum, and is also formed as traces in the butyric acid fermentation. It is a liquid boiling as 205° and having an odor similar to perspiration.

Caprylic acid, $C_8H_{16}O_2$, is a crystalline solid, melting at 16°, and having an odor similar to perspiration.

Capric acid, $C_{10}H_{20}O_2$, is a crystalline solid melting at 30° with a similar odor.

Caproic, caprylic, and capric acids occur as glycerine esters in butter, in cocoanut-oil, and in many fats. They also occur free or as esters in cheese, perspiration, fusel-oil from wine, and in beet-root molasses.

Ethyl caprinate, with some ethyl caprilate, forms œnanthic ether, ($oivos$, wine), also called wine-oil, which gives odor to the wine (but not the bouquet), and is obtained from wine-yeast by distillation.

Amidocaproic acid, leucine, $C_5H_{10}(NH_2)^-COOH$; according to structure it is amidoisobutyl acetic acid, $(CH_3)_2{=}CH^-CH_2{^-}CH(NH_2)^-COOH$.

Occurrence. In river crabs, spiders, butterfly caterpillars, lupin, squash seeds, beet-root molasses, pancreatic juice, in all parenchymous organs and glands, and also in the blood in urine in certain diseases.

Preparation and Formation. It is formed in the putrefaction of proteids (hence occurring in old cheese and hence the old name cheese oxide) and is obtained, besides glycocoll (p. 369), on boiling proteids or gelatine with sulphuric acid or caustic alkalies (Synthesis, see p. 369, 5).

Properties. Shining crystalline leaves or characteristically formed balls or tufts which are soluble in water and hot alcohol. According as to the origin of the material from which it is prepared we have inactive, dextro or lævorotatory leucin. With nitrous acid it is converted into leucinic acid: $C_6H_{12}O_3$ (p. 369, 3).

$$C_5H_{10}(NH_2)COOH + HNO_2 = C_5H_{10}(OH)COOH + 2N + H_2O.$$

Leucinimide, $C_4H_9{-}CH{\Big\langle}{{CO-NH}\atop{NH-CO}}{\Big\rangle}CH{-}C_4H_9$, is formed with leucine in the decomposition of the proteids.

Diamido caproicacid, lysin, $C_5H_9(NH_2)_2COOH$, is a cleavage product of proteins and on putrefaction yields cadaverin, $C_5H_{10}(NH_2)_2 + CO_2$ (p. 398).

Lauric acid, $C_{12}H_{24}O_2$, occurs in the oil of laurel as glycerine ester(laurin), forms white crystals melting at 44°.

Myristic acid, $C_{14}H_{28}O_2$, as glycerine ester (myristin) in nutmeg fat, wool fat, whale-oil; melts at 53°.

Palmitic acid, $C_{16}H_{32}O_2$, occurs in large quantities part as glycerine ester (palmitin) and part free in palm-oil. It forms crystalline masses melting at 62°.

Stearic acid, $C_{18}H_{36}O_2$, occurs to the greatest extent as glycerine ester (stearin) in the solid animal fats (the tallows). It forms colorless leaves and melts at 69°. Palmitic and stearic acids exist free in decomposed pus, cheesy tuberculous masses, adipocere, etc., and combined with alkalies or calcium in excrements, pus, transudations, etc. Stearin candles consist of stearic and palmitic acids.

The mixed glycerine esters of palmitic, stearic, and oleic acids (which see) form most of the fats and their cholesterin esters, the wool fats.

The alkali salts of palmitic, stearic, and oleic acids are soluble in water and alcohol and are called soaps. All their other salts are insoluble. In regard to the preparation of these fatty acids on a large scale see Glycerine and Oleic Acid.

Arachidic acid, $C_{20}H_{40}O_2$, theobromic acid, occurs as glycerine ester in cacao- and earth-nut, and melts at 75°.

Behenic, $C_{22}H_{44}O_2$, exists in behenic oil (oil from the seeds of Moringa oleifera) as glycerine ester and melts at 76°.

Lignoceric acid, $C_{24}H_{48}O_2$, occurs in the beech-wood tar.

Cerotic acid, $C_{26}H_{52}O_2$, melts at 78°, and is found free in beeswax and as ceryl ester i'l Chinese or plant waxes.

Melissic acid, $C_{30}H_{60}O_2$, is obtained from melissyl alcohol by heating with soda lime. It melts at 88°.

II. Combinations with Metalloids.

All metalloids form volatile compounds with the alkyls; these compounds have an analogous composition to their hydrogen compounds. Besides these, combinations with nitrogen and phosphorus are known which contain hydrogen besides alkyls. The alkyl combinations of oxygen (the ethers) and of sulphur (sulpho-ethers) have already been considered. The alkyl compounds of nitrogen, phosphorus, arsenic, antimony are called *amines, phosphines, arsines, stibines.*

a. Compounds of Nitrogen.

Amines, amine bases. We differentiate between the primary or amine bases, secondary or imide bases, and tertiary or nitrile bases (p. 330).

Occurrence. Methyl, ethyl, and propyl combinations are formed in the putrefaction of many organic bodies, especially fish, gelatine, and peptone.

Properties. The amines have an alkaline reaction, are non-poisonous, volatile without decomposition, and combine, like NH_3, directly with acids, forming salts. Their sulphates combine with aluminum sulphate, forming alums; their chlorides give crystalline double salts with platinum chloride similar in composition to that with ammonium chloride. The lower amines are gaseous, similar to NH_3, but differ therefrom in being inflammable, or liquids, while the higher members are colorless and odorless solids. The volatility and solubility in water decreases with the increase in the amount of carbon. Their salts differ from the ammonium salts by their solubility in alcohol.

Primary amines yield alcohols when treated with nitrous acid: $CH_3-NH_2 + HNO_2 = CH_3-OH + H_2O + 2N$. Secondary amines yield nitrosamines: $(CH_3)_2-NH + HNO_2 = (CH_3)_2-N-NO + H_2O$. Tertiary amines are not changed at ordinary temperatures by nitrous acid; primary aromatic amines give diazo-compounds (which see). Further, in regard to the identification of primary amines see Isonitrile (p. 391).

Preparation. 1. By the action of nascent hydrogen upon the cyan-alkyls (Mendius's reaction):

$$CH_3-CN + 4H = CH_3-CH_2-NH_2.$$
Acetonitrile.   Ethylamine.

2. On distilling the ester of isocyanic acid with caustic alkali:

$$OCN(CH_3) + 2KOH = NH_2CH_3 + K_2CO_3,$$

or by distilling a brominated amide with caustic alkali (Hoffmann's reaction), whereby isocyanic acid ester is first formed, and then this yields the amine: $CH_3-CO-NHBr + KOH = CH_3-NCO + KBr + H_2O$.

3. By the reduction of the nitro-alkyls:

$$CH_3-NO_2 + 6H = CH_3-NH_2 + 2H_2O.$$

4. On heating the halogen alkyls with ammonia:

$$C_2H_5I + NH_3 = C_2H_5-NH_2 + HI.$$

If the primary amine thus obtained is heated again with alkyl iodide we obtain a secondary amine, and this treated further in the same manner yields a tertiary amine:

$$C_2H_5I + NH_2(C_2H_5) = HI + NH(C_2H_5)_2 \text{ diethylamine;}$$
$$C_2H_5I + NH(C_2H_5)_2 = HI + N(C_2H_5)_2 \text{ triethylamine.}$$

If we allow an iodide of another alcohol radical to act upon a primary amine, we obtain a mixed amine: $CH_3I + NH_2(C_2H_5) = HI + NH(CH_3)(C_2H_5)$ (methyl ethylamine).

The HI produced combines directly with the amines, forming salts. If the hydro-iodide of the amine thus obtained be distilled with caustic alkali we obtain the free amine: $(C_2H_5)_3N.HI + KOH = (C_2H_5)_3N + KI + H_2O$.

Methylamine, $NH_2(CH_3)$, is found in the herring-brine, in the products of the dry distillation of animal bodies (in animal oil) of wood (in crude wood alcohol), and is also formed by the action of nascent hydrogen upon formonitrile (hydrocyanic acid). It is a colorless combustible gas with an ammoniacal odor, liquefiable below $-6°$. It is the most soluble of all gases, 1 volume water dissolving 1150 volumes at $12°$; this solution shows all the properties of an ammonia solution; it precipitates metallic salts, dissolves copper and silver salts in excess but not cobalt, nickel, and cadmium salts.

Trimethylamine, $\dot{N}(CH_3)_3$, is found in the flowers of the haw-thorn, pear tree, mountain ash, ergot, bone-oil, coal-tar oil, as well as in the herring-brine, to which it gives its odor. It is formed in the putrefaction of animal tissues and gelatine as well as in the dry distillation of many organic bodies. It is obtained by the distillation of herring-brine with caustic alkali or by the dry distillation of "vinasse" from beet-root molasses. It is a colorless liquid boiling at $3°$ and with a pronounced fishy odor.

Ammonium Bases (p. 331). These bases are derived from the hypo-thetical ammonium hydroxide, NH_4^-OH, in which all the hydrogen atoms of the ammonium group NH_4 (p. 150) are replaced by alkyls.

Properties. They are similar to the alkali hydroxides. They are soluble in water, giving a strong alkaline reaction thereto; they deliquesce in the air, saponify fats, precipitate metallic hydroxides from metallic solutions, and form crystalline salts with acids. They are not volatile without decomposition.

Preparation. On heating the tertiary amines (see p. 378, 4) with the iodides of the alcohol radicals:

$$(C_2H_5)_3N + C_2H_5I = (C_2H_5)_4NI \text{ (corresponding to } H_4NI).$$
Triethylamine.   Tetraethyl ammonium iodide.

The tetra-alkyl ammonium salt thus obtained is not decomposed by caustic alkali, but on treating it with moist silver oxide the alkyl ammonium hydroxide is obtained, $(C_2H_5)_4NI + Ag.OH = (C_2H_5)_4N(OH) + AgI$, which separates out on evaporating the alkaline solution under the air-pump.

Tetramethyl ammonium hydroxide, $(CH_3)_4N(OH)$, forms white crys-talline, caustic masses.

Choline, $C_5H_{15}N_2O_2$ (structure below), bilineurine, sincalin, trimethyl-cxyethyl ammonium hydroxide, occurs in the fly agaric, hops, cotton-seed, herring-brine, fresh cadavers, also as constituent of the lecithins widely distributed in the animal kingdom. It is produced on boiling ox brains, yolk of egg, bile, with $Ba(OH)_2$, also from the alkaloids of the white mustard (sinapin) by boiling with alkalies. Choline can be prepared artificially by heating ethylenoxide with trimethylamine and water: $N(CH_3)_3 + C_2H_4O + H_2O = (HO)(CH_3)_3N(C_2H_4.OH)$. It is a deliquescent crystalline solid which has a strong alkaline reaction, non-poisonous, and gives good crystalline salts.

Muscarine, oxycholine, $C_5H_{15}NO_3$, is the poison of the fly agaric, and is also formed in the oxidation of choline with nitric acid. It forms colorless deliquescent crystals.

Neurine, $C_5H_{13}NO$ or $(HO)(CH_3)_3N(C_2H_3)$, trimethylvinyl ammonium hydroxide (monovalent radical vinyl, $CH_2=CH$), is obtained as a cleavage product of the lecithins, respectively of choline. It is also obtained in the short putrefaction of meat and fish (see Ptomaines) by the splitting off of a molecule of water from the choline, a constituent of the lecithins, and is a difficultly crystallizable, deliquescent poisonous body.

b. Compounds of Phosphorus.

Phosphines and **phosphonium** bases are very similar to the above-mentioned nitrogen compounds and are obtained in the same manner.

Methyl phosphine, $PH_2(CH_3)$, is a neutral, spontaneously inflammable gas having an extremely unpleasant odor.

Tetramethyl phosphonium hydroxide, $P(CH_3)_4OH$, decomposes on heating into trimethyl phosphine oxide, which is very stable, and methane:

$$P(CH_3)_4OH = P(CH_3)_3O + CH_4.$$

c. Compounds of Arsenic and Antimony.

Arsines and **Stibines** have no basic properties on account of the more metallic character of these elements, but on the contrary have the property of forming compounds with oxygen, sulphur, and halogens, having the formula $As(CH_3)_3 x_2$ ($x_2 = O$, S, or Cl_2). The arsines and stibines are therefore to be considered as their pentachlorides, sulphides or oxides, in which these elements are entirely or in part replaced by alkyls.

Arsonium and **stibonium** bases are prepared in the same manner as ammonium and phosphonium bases and are, like these, strong bases.

Monomethyl arsine, $AsH_2(CH_3)$, is a gas which liquefies at $0°$ and which readily oxidizes into

Methyl arsine oxide, CH_3-AsO, or into

Methyl arsenious acid, $CH_3-AsO(OH)H$.

Trimethyl arsine, $As(CH_3)_3$, is a colorless liquid boiling at $220°$, and which is obtained by treating zinc methyl with arsenic trichloride: $2AsCl_3 + 3Zn(CH_3)_2 = 3ZnCl_2 + 2As(CH_3)_3$. It combines with oxygen, forming trimethyl arsine oxide, $As(CH_3)_3O$, with the halogens, forming $As(CH_3)_3Cl_2$, etc. With methyl iodide it combines directly, forming tetramethyl arsonium iodide, $As(CH_3)_4I$, which crystallizes in colorless plates.

Dimethyl diarsine, cacodyl, $(CH_3)_2As-As(CH_3)_2$, is a colorless liquid boiling at $170°$, and with a most disagreeable odor, and which readily undergoes spontaneous inflammability in the air.

Dimethyl diarsine oxide, cacodyl oxide, alkarsin, $(CH_3)_2As-O-As(CH_3)_2$, is a colorless liquid boiling at $150°$, and has a most nauseating odor. Both compounds are produced on distilling dry acetates with arsenious oxide:

$4CH_3COOK + As_2O_3 = (CH_3)_2As-O-As(CH_3)_2 + 3K_2CO_3 + 2CO_2$. (Detection of acetates by means of arsenious oxide, p. 363.)

Dimethyl arsenic acid, cacodylic acid, $(CH_3)_2AsO(OH)$, is produced by the oxidation of dimethyl diarsine oxide. It occurs as odorless prisms.

Tetramethyl arsonium hydroxide, $As(CH_3)_4 \cdot OH$, and

Tetramethyl stibonium hydroxide, $Sb(CH_3)_4 \cdot OH$, are bodies quite similar in chemical properties to potassium hydroxide.

d. Compounds with Boron and Silicon.

Bormethyl, $B(CH_3)_3$, is a colorless gas which spontaneously inflames in the air, burning with a greenish flame, and has an unsupportable sharp odor.

Borethyl, $B(C_2H_5)_3$, is a colorless liquid which acts like bormethyl. (Preparation, see Silicon Ethyl.)

Silicon ethyl, $Si(C_2H_5)_4$, is produced from zinc ethyl (p. 382) by the action of silicon chloride. It is a colorless liquid, boiling at 153°. If this is treated with chlorine we obtain monochlorsiliconethyl, $(C_2H_4Cl)-Si \equiv (C_2H_5)_3$, or $SiC_8H_{19}Cl$, a liquid boiling at 185°, which, like the alkyl chlorides, yields an acetic acid ester when heated with alkali acetate: $SiC_8H_{19}Cl + C_2H_3KO_2 = KCl + (SiC_8H_{19})C_2H_3O_2$; this on treating with alkali hydrate is converted into potassium acetate and the alcohol $SiC_8H_{19}-OH$. According to this we must consider silicon ethyl as nonane or nonyl hydride, in which 1 atom of carbon is replaced by silicon. This and its derivatives correspond completely with the nonyl compounds:

Silicononane,	SiC_8H_{20}.	Nonane,	C_9H_{20}.
Silicononyl chloride,	$SiC_8H_{19}Cl$.	Nonyl chloride,	$C_9H_{19}Cl$.
Silicononyl acetate,	$(SiC_8H_{19})C_2H_3O_2$.	Nonyl acetate,	$(C_9H_{19})C_2H_3O_2$.
Silicononyl alcohol,	$SiC_8H_{19}-OH$.	Nonyl alcohol,	$C_9H_{19}-OH$.

12. Metallic Compounds,

or metallo-organic combinations, are known only with alkyl radicals. Furthermore those metals have the power of forming alkyl compounds which, according to their position in the periodic system (p. 54), are closely related to the metalloids. As the basic nature of the metal increases, so does the stability of the corresponding alkyl compound decrease more and more. The metals often unite with a greater number of monovalent alcohol radicals than with monovalent elementary atoms; these compounds, which correspond to the maximum valence of the metals, are volatile liquids which are generally converted into the vaporous form without decomposition. The determination of their vapor density therefore gives a means of estimating the valence of the metals, as well as their atomic weight (p. 21).

Preparation. 1. By the direct action of the metals or their sodium alloys upon the halogen alkyls:

$$ZnNa_2 + 2C_2H_5I = Zn(C_2H_5)_2 + 2NaI.$$

2. By the action of zinc or mercuric alkyl upon metallic chlorides:

$$SnCl_4 + 2Zn(C_2H_5)_2 = Sn(C_2H_5)_4 + 2ZnCl_2.$$

Properties. Colorless liquids, volatile without decomposition, which in part inflame in the air (magnesium, zinc, aluminium alkyls), while others (mercury, lead, tin alkyls) are stable.

Zinc ethyl, $Zn(C_2H_5)_2$, is produced on heating zinc with ethyl iodide, whereby zinc ethyl iodide is first formed, this decomposing into zinc iodide and zinc ethyl on further heating:

$$Zn + C_2H_5I = Zn{<}^{C_2H_5}_{I} \; ; \quad 2Zn{<}^{C_2H_5}_{I} = Zn(C_2H_5)_2 + ZnI_2.$$

Zinc alkyls are decomposed by water into hydrocarbons: $Zn(C_2H_5)_2 + 2H_2O = 2C_2H_6 + Zn(OH)_2$. By the slow action of oxygen, zinc alcoholates (p. 343) are produced: $Zn(C_2H_5)_2 + 2O = Zn(OC_2H_5)_2$. On account of their decomposability zinc alkyls are used in the preparation of many other compounds.

Sodium ethyl, C_2H_5Na, cannot be directly prepared, but is obtained by the action of sodium upon zinc ethyl when the zinc precipitates. A crystalline compound consisting of sodium ethyl and zinc ethyl separates from the resulting solution on allowing it to cool. Pure sodium ethyl cannot be obtained from this mixture. All alkali alkyls show the same behavior, and their solutions absorb carbon dioxide with the formation of salts of fatty acids (p. 346, 3), and are decomposed by water in the same way as zinc alkyls.

MONOVALENT COMPOUNDS OF POLYVALENT ALCOHOL RADICALS.

Polyvalent alcohol radicals have the power, under certain conditions, of replacing monovalent ones in compounds. The most important of these compounds are derived from the tri- and pentavalent alcohol radicals (which see).

COMPOUNDS OF THE CYANOGEN RADICAL.

The monovalent cyanogen radical ^-CN, is in many regards similar to the halogens; thus it forms an acid with hydrogen and unites with the metals and alcohol radicals, forming compounds which are very similar to those with the halogens. Cyanogen as a monovalent radical cannot exist free, but is doubled, like all other monovalent radicals, forming the molecule dicyanogen, NC^-CN. Most of the compounds of cyanogen also form polymeric modifications, which may be considered as derivatives of triazine, $C_3H_3N_3$ (which see).

Potassium ferrocyanide and potassium cyanide form the starting-point in the preparation of the cyanogen compounds.

According to theory, two cyanogen radicals are possible according as the nitrogen is a trivalent element, the so-called nitrile group, $N\equiv C^-$, or as a pentavalent element, the so-called isonitril group, $C\equiv N^-$, so that the elements or groups combined with the cyanogen radicals are united to nitrogen or to carbon. According to another conception the carbon exists as a tetravalent element in the nitriles and as a divalent element in the isonitriles: $N \quad C^-$ or $C=N^-$.

Cyanogen and its compounds with H, ^-OH, ^-SH, $^-NH_2$, Cl, etc., react according to either one or the other formula; still only one of the two isomerides exists free, as the other contains the atoms in unstable equilibrium, which in the preparation of the respective compound is immediately converted into the stable form of the other compound (tautomerism, p. 300). Nevertheless the two isomers of such derivatives of the above-mentioned compounds are known, which are produced by the introduction of alkyls in place of hydrogen (p. 390).

Dicyanogen, cyanogen, C_2N_2 or $N=C^-C\equiv N$. *Occurrence.* To a slight extent in the gases of the blast-furnace and in illuminating-gas.

Formation. By heating ammonium oxalate: $(NH_4)OOC^-COO(NH_4)$ $= C_2N_2 + 4H_2O$. It is, according to this, the nitrile of oxalic acid, $HOOC^-COOH$, and has the structure $N\equiv C^-C\equiv N$. With water it is gradually converted into ammonium oxalate, taking up H_2O.

Preparation. By heating silver or mercuric cyanide (p. 386): $Hg(CN)_2 = C_2N_2 + Hg$; also by heating a solution of copper sulphate with potassium cyanide: $4KCN + 2CuSO_4 = C_2N_2 + 2CuCN + 2K_2SO_4$. It cannot be obtained by the direct union of carbon and nitrogen (see Potassium Cyanide).

Properties. Colorless, irritating, poisonous gas which liquefies at $-12°$; it is inflammable, burning with a purple-red flame into carbon dioxide and nitrogen. Water dissolves 4 volumes, and alcohol 23 volumes. Potassium burns in C_2N_2 to potassium cyanide, KCN, and potassium hydrate absorbs the gas with the formation of potassium cyanide and potassium cyanate:

$$2KOH + C_2N_2 = KCN + NCOK + H_2O \text{ (analogous to chlorine).}$$

Hydrocyanic Acid, Prussic Acid, HCN. *Occurrence and Formation.* Free in Pangium edule, certain Araceæ and Hydrocarpus varieties of Java, and as traces in tobacco-smoke. It occurs combined in the seeds and often in other parts of the Amygdalacæ, Drupacea, Pomacæ, but especially in the bitter almond and the leaves of the cherry-laurel, where it exists as the glucoside amygdalin (see Glucosides), which on standing with water decomposes through the presence of the ferment emulsin, which also exists in these plants, into hydrocyanic acid, sugar, and benzaldehyde. The hydrocyanic acid thus obtained is very dilute (1 part HCN per 1000), and is called bitter almond water when obtained from bitter almonds by distillation with water and some little alcohol, whereby the latter dissolves the benzaldehyde, which distils over.

The African Lotus arabicus contains the glucoside lotusin, which by the ferment lotase, existing in the same plant, is split into hydrocyanic acid, lotoflavin, and dextrin.

Hydrocyanic acid is readily formed on heating ammonium formate: $H-COO-NH_4 = HCN + 2H_2O$, and accordingly, is the nitrile of formic acid (formonitrile, p. 368, 3). Ammonium formate is produced on allowing a watery solution of HCN to stand, whereby water is taken up. Prussic acid is also produced when the dark electric discharge is passed through a mixture of acetylene and nitrogen, or by heating ammonia with chloroform under pressure: $CHCl_3 + NH_3 = HCN + 3HCl$. According to this reaction hydrocyanic acid has the following structure: $N = CH$.

Preparation. Ordinarily by distilling metallic cyanides with dilute inorganic acids:

$$2KCN + H_2SO_4 = K_2SO_4 + 2HCN.$$

It is prepared more readily by distilling potassium ferrocyanide and dilute sulphuric acid, whereby only one-half of the cyanogen in the potassium ferrocyanide is obtained as hydrocyanic acid (p. 388):

$$2K_4(Fe''C_6N_6) + 3H_2SO_4 = K_2Fe''(FeC_6N_6) + 3K_2SO_4 + 6HCN.$$

In both cases a dilute watery solution of HCN is obtained. In order to prepare anhydrous hydrocyanic acid we pass the vapors containing water over $CaCl_2$ and liquefy the gas by a freezing mixture. Nearly anhydrous hydrocyanic acid is obtained on distilling potassium cyanide with 50 per cent. sulphuric acid (p. 386).

Properties. When anhydrous it is a colorless, extremely poisonous, penetrating liquid having an odor similar to bitter almonds, boiling at 27° and crystallizing at −15°. Its vapors when inhaled cause death.

It is a very weak acid, soluble in water, alcohol and ether, and burns with a violet flame, its watery solution quickly changing into ammonium formate (p. 384). The solution of prussic acid is rather stable in the presence of very small amounts of mineral acids. With nascent hydrogen it yields methylamine: $N \equiv CH + 4H = H_2N^-CH_3$, and under certain conditions it is polymerized into colorless crystalline trihydrocyanic acid $(NCH)_3$ (see Triazines), which yields HCN again on heating. With hydrogen peroxide it yields non-poisonous oxamide: $2NCH + H_2O_2 = H_2N^-OC^-CO^-NH_2$ (antidote for HCN poisoning).

Detection. The liquid to be tested is treated with caustic alkali and a few drops of a ferrous and ferric salt solution, then warmed and acidified. If hydrocyanic acid is present, then potassium ferrocyanide is produced, and this gives a deep-blue precipitate with the ferric salt. In the presence of very small quantities of HCN at first only a blue coloration of Prussian blue is obtained (p. 338). If the liquid to be tested is evaporated to dryness with ammonium sulphide, ammonium thiocyanide is produced, and this gives a blood-red coloration with $FeCl_3$ (p. 393). The HCN can be detected in its insoluble compounds or in mixtures by distilling with dilute sulphuric acid, which sets the HCN free, and then testing as above mentioned.

I. Compounds of Cyanogen with Metals.

a. Simple Metallic Cyanides.

The simple metallic cyanides or salts of hydrocyanic acid or cyanides are obtained by the action of hydrocyanic acid upon metallic oxides or metallic hydroxides: $2HCN + HgO = Hg(CN)_2 + H_2O$. They are also obtained if nitrogen and hydrogen are passed over heated carbides of the alkali or alkaline earth metals, while if iron is present the corresponding ferrocyanide combination is produced (p. 387).

The cyanides of the light metals are soluble in water and are decomposed by dilute acids with the generation of HCN. Even the carbonic acid of the air sets hydrocyanic acid free from these cyanides, and this is the reason why they always smell of hydrocyanic acid; on the contrary, they are very stable even at red heat. Heated with concentrated sulphuric acid they develop carbon monoxide: $2KCN + 2H_2O + 2H_2SO_4 = K_2SO_4 + (NH_4)_2SO_4 + 2CO$. The soluble cyanides are violent poisons.

The cyanides of the heavy metals are, with the exception of mercuric cyanide, insoluble in water, and are only decomposed by strong acids and decompose generally into cyanogen and metal on heating them to redness.

They are best obtained by treating soluble metallic salts with potassium cyanide: $AgNO_3 + KCN = KNO_3 + AgCN$.

Potassium Cyanide, KCN. *Preparation.* If nitrogen is passed over a red-hot mixture of carbon and potassium carbonate, potassium cyanide is formed: $K_2CO_3 + 2N + 4C = 2KCN + 3CO$. It is prepared on a large scale, according to the same principle, by heating nitrogenous organic refuse (blood, leather, hoofs, horns) with potassium carbonate (see Potassium Ferrocyanide). At the present time it is prepared by heating potassium carbonate with carbon in ammonia gas, $K_2CO_3 + C + 2NH_3 = 2KCN + 3H_2O$, or by fusing potassium ferrocyanide: $K_4FeC_6N_6 = 4KCN + FeC_2 + 2N$. The finely divided iron carbide is separated by filtering the fused mass through earthenware. In regard to its formation from chloroform see p. 347.

Properties. Colorless, very poisonous cubes, soluble in water with alkaline reaction (hydrolytic dissociation, p. 86) and also in dilute alcohol. On heating it fuses without decomposition and is an important reducing agent because it unites directly with oxygen and also with sulphur. The watery solution soon turns brown, undergoing decomposition, whereby potassium formate and ammonia are produced: $KCN + 2H_2O = CHKO_2 + NH_3$. Potassium cyanide precipitates the corresponding metallic cyanide from solutions of the heavy metals, these being soluble in an excess of the potassium cyanide (see below). Its aqueous solution dissolves finely divided gold, hence it is used for the extraction of the latter (which see).

Silver cyanide, AgCN (or AgNC, p. 391, 2), is precipitated as a white cheesy precipitate from silver salts by potassium cyanide. It is similar to silver chloride, but does not darken when exposed to the light.

Mercury cyanide, mercuric cyanide, $Hg(CN)_2$, is obtained by dissolving mercuric oxide in hydrocyanic acid and evaporating. It consists of colorless crystals which are soluble in water and alcohol. It can also be obtained from Prussian blue (p. 388).

b. Compound Metallic Cyanide.

The cyanides of the heavy metals which are insoluble in water are soluble in a watery solution of potassium cyanide with the formation of crystallizable compounds soluble in water; for example, $AgCN + KCN = KAg(CN)_2$. As these compounds do not respond to the reactions which the ions of the salts from which they are formed give, they cannot be considered as double salts but rather as complex salts which contain a complex anion; for example, $Ag(CN)_2$ (p. 83). They may be divided into two groups, the salts of one group being poisonous and readily split in the cold by inorganic acids with the separation of simple metallic cyanides and formation of hydrocyanic acid, $KAg(CN)_2 + HCl = KCl + HCN + AgCN$, while the salts of the other group are non-poisonous and split off in the cold with dilute acids, peculiar complex acids, so that the compound must be considered as salts of these acids. To this group belong the compounds of ferrous and ferric, manganous and manganic, cobaltic, chromic, and platino cyanides, with the alkali cyanides; thus $Fe(CN)_2 + 4KCN = K_4Fe(CN)_6$, with the anion FeC_6N_6, which decomposes with acids as follows:

$$K_4FeC_6N_6 \ + \ 4HCl = H_4FeC_6N_6 \ + \ 4KCl.$$
Potassium ferrocyanide. Hydroferrocyanic acid.

$$K_2PtC_4N_4 \ + \ 2HCl = H_2PtC_4N_4 \ + \ 2KCl.$$
Potassium platinocyanide. Hydroplatinocyanic acid.

$$K_3CoC_6N_6 \ + \ 3HCl = H_3CoC_6N_6 \ + \ 3KCl.$$
Potassium cobalticyanide. Hydrocobalticyanic acid.

In these acids the hydrogen is not only replaceable by alkali metals, but also by other metals. The salts of hydroplatinocyanic acid have beautiful colors. · Barium platinocyanide, $BaPtC_4N_4$, is used in the detection of Röntgen rays.

Hydroferrocyanic Acid, $H_4Fe''C_6N_6$. (In six condensed hydrocyanic acid molecules we have two hydrogen atoms replaced by a ferrous iron atom.) If an inorganic acid is added to a cold concentrated watery solution of potassium ferrocyanide, $H_4Fe''C_6N_6$ separates out as a white crystalline powder.

Of the salts of this acid, the ferrocyanides, the potassium salt, the ferric and the cupric salts are of importance, as it is in this form that the ferric and cupric compounds are detected.

Potassium Ferrocyanide, Yellow Prussiate, $K_4FeC_6N_6$. *Preparation.* If a ferrous salt solution is treated with potassium cyanide, a precipitate is obtained of ferrocyanide, $Fe(CN)_2$, which dissolves in an excess of potassium cyanide, producing potassium ferrocyanide:

$$Fe(CN)_2 + 4KCN = K_4FeC_6N_6.$$

Powdered iron or ferrous sulphide also dissolves when treated with a watery solution of potassium cyanide. Oxygen is taken up from the air:

$$Fe + 6KCN + H_2O + O = K_4FeC_6N_6 + 2KOH.$$

Potassium ferrocyanide was formerly prepared on a large scale by heating carbonized nitrogenous refuse with potassium carbonate and iron. The carbon and nitrogen united with the potash, forming potassium cyanide (p. 386), while the sulphur contained in the substances combined with the iron, forming iron sulphide. If the fused mass was treated with water, the following transformation took place:

$$FeS + 6KCN = K_4FeC_6N_6 + K_2S.$$

In Germany at the present time nearly all the potassium ferrocyanide is prepared from the iron used in the purification of illuminating-gas. This contains considerable sulphur besides ferric hydrate, as well as the greater part of the cyanogen produced in the distillation of the coal, in the form of iron cyanogen compounds (Prussian blue) and as sulphocyanides.

In order to purify illuminating-gas from its impurities (H_2S, CS_2, CO_2, NH_3, cyanogen compounds), it is not sufficient to wash the gas with water, but the gas is passed over ferric hydroxide. The mass is freed from ammonium carbonate and sulphocyanate by lixiviation and then heated with $Ca(OH_2)_2$, producing calcium ferrocyanide, which is dissolved in water and treated with K_2CO_3, which yields potassium ferrocyanide. The residue yields SO_2 on burning in the air, and is used in the manufacture of sulphuric acid.

Properties. It forms large yellow prisms containing 3 molecules of water. It is soluble in water, and decomposes on heating into nitrogen, potassium cyanide, and iron carbide (p. 386). If heated with dilute sulphuric acid, it yields hydrocyanic acid (process, p. 384), and with concentrated sulphuric acid carbon monoxide is produced:

$$K_4Fe''C_6N_6 + 6H_2SO_4 + 6H_2O = FeSO_4 + 2K_2SO_4 + 6CO + 3(NH_4)_2SO_4.$$

Nitric acid converts it into potassium nitroprusside (p. 389):

$$K_4Fe''C_5N_6 + 3HNO_3 = K_2Fe''(NO')C_5N_5 + 2KNO_3 + CO_2 + NH_3.$$

Ferri-ferrocyanide, Prussian blue, $Fe_4'''(Fe''C_6N_6)_3$, serves in the detection of ferric salts as their solutions if treated with potassium ferrocyanide, give a dark blue precipitate having the above constitution:

$$3K_4Fe''C_6N_6 + 4Fe'''Cl_3 = Fe_4'''(Fe''C_6N_6)_3 + 12KCl.$$

It is decomposed into potassium ferrocyanide and ferric hydrate by alkali hydroxides. On boiling with freshly precipitated mercuric oxide it decomposes into mercuric cyanide, ferrous and ferric hydroxide:

$$Fe_4'''(Fe''C_6N_6)_3 + 9HgO + 9H_2O = 9Hg(CN)_2 + 4Fe(OH)_3 + 3Fe(OH)_2.$$

Potassium ferri-ferrocyanide, $KFe'''(Fe''C_6N_6)$. Soluble Prussian blue. If a ferric salt solution is added to an excess of potassium ferrocyanide solution, a deep-blue precipitate is obtained which is soluble in water as soon as the potassium salt mixed with it is removed by washing.

Potassium ferro-ferrocyanide, $K_2Fe''(Fe''C_6N_6)$, is obtained in the preparation of hydrocyanic acid from potassium ferrocyanide as a white insoluble powder.

Ferro-ferrocyanide, $Fe_2''(Fe''C_6N_6)$, is obtained as a white precipitate, which quickly turns into Prussian blue when exposed to the air on treating a ferrous salt solution with potassium ferrocyanide.

Cupric ferrocyanide, $Cu_2''(Fe''C_6N_6)$, is obtained on mixing a cupric salt solution with a solution of potassium ferrocyanide as a reddish-brown precipitate which is insoluble in dilute acids (Hatchett's brown).

Hydroferricyanic Acid, $H_3Fe'''C_6N_6$. (Six condensed hydrocyanic acid molecules in which three hydrogen atoms are replaced by one ferric iron atom.) It is precipitated as brownish crystals from a cold concentrated solution of potassium ferricyanide by treating this with an inorganic acid.

Of its salts, the ferricyanides, after the potassium salt, the ferrous salt is of importance, as it is used in the detection of ferrous salts.

Potassium ferricyanide, red prussiate, $K_3FeC_6N_6$, is produced on passing chlorine into a watery potassium ferrocyanide solution:

$$K_4FeC_6N_6 + Cl = K_3FeC_6N_6 + KCl.$$

It crystallizes in deep-red anhydrous prisms, has an oxidizing action in the presence of free alkali, and evolves oxygen with barium peroxide in the presence of water and forms barium-potassium-ferrocyanide:

$$BaO_2 + 2K_3Fe'''C_6N_6 = BaK_6(Fe''C_6N_6)_2 + O_2.$$

Ferro-ferricyanide, Turnbull's blue, $Fe_3''(Fe'''C_6N_6)_2$, serves in detecting ferrous salts, as it is formed on treating a ferrous salt solution with potassium ferricyanide. It is a deep-blue precipitate:

$$2K_3Fe'''(C_6N_6) + 3Fe''Cl_2 = Fe_3''(Fe'''C_6N_6)_2 + 6KCl.$$
Pot. ferricyanide.   Turnbull's blue.

With alkali hydroxides it is decomposed into potassium ferricyanide and ferrous hydroxide. The first quickly changes into potassium ferrocyanide, and the ferrous hydroxide is oxidized to ferric hydroxide.

c. *Nitro-prusside Compounds.*

These are formed by the action of nitric acid upon ferrocyanide combinations (process, p. 388) and are derived from

Hydronitroprussic acid, $H_2Fe''(NO')C_5N_6$, which is obtained as dark-red prisms on treating a nitroprusside salt with hydrochloric acid. All soluble nitroprusside compounds are colored a beautiful violet by even the most dilute solutions of metallic sulphides (detection of soluble metallic sulphides).

2. Compounds of Cyanogen with Alkyls.

The hydrogen of hydrocyanic acid can be replaced by alcohol radicals just as with metals. We have seen, p. 383, that two series of isomeric compounds can be derived from cyanogen, although these are not known in cnnection with hydrocyanic acid and the metallic cyanides. Both series are known with the cyanides of the alcohol radicals, and we differentiate between nitriles and carbylamines, or isonitriles. They are not esters, as they are not, like these, transformed by bases into the corresponding alcohol and acid.

a. Nitriles.

Properties. In the nitriles the nitrogen exists as a trivalent atom, so that the alkyl is united to the still free fourth valence of carbon: $N \equiv C^- CH_3$. Nitriles have an ethereal odor and are colorless neutral liquids or solids and less poisonous than the isonitriles. The nitriles poor in carbon are soluble in water.

On heating above 100° with water or with acids and alkalies (p. 346, 2) the C atom of the cyanogen is converted into the carboxyl group and remains combined with the alkyl, while the nitrogen is split off as ammonia:

$$CH_3 C \equiv N + 2H_2O = CH_3 - COOH + NH_3.$$

With nascent hydrogen they are converted into amines:

$$N \equiv C - CH_3 + 4H = H_2N - CH_2 - CH_3.$$

This transformation shows the union of the alcohol radicals to the carbon of the cyanogen.

Preparation. 1. By the dry distillation of ammonium salts or the amides of the fatty acids with dehydrating agents (phosphoric anhydride):

$$CH_3 - COO - NH_4 = CH_3 - C \equiv N + 2H_2O;$$
$$CH_3 - CO - NH_2 = CH_3 - C \equiv N + H_2O.$$

2. By heating potassium cyanide with alkyl iodides:

$$CNK + CH_3I = CH_3 - C \equiv N + KI.$$

(According to this reaction potassium cyanide contains the ^-CN group.)

Acetonitrile, methyl cyanide, $NC^- CH_3$, found in coal-tar, is a pleasant-smelling colorless liquid.

Fulminic acid, $C_2H_2N_2O_2$, perhaps $NC^- (CH_2)(NO_2)$, nitroacetonitrile, is not known free. Compounds are:

Silver fulminate, $NC^- CAg_2(NO_2)$, which forms white needles, exploding with great violence by heating or concussion, and often spontaneously. It is used in the manufacture of explosives.

Mercury fulminate, mercuric acetonitrile, $NC-CHg(NO_2)$, explodes with less violence and is used in the percussion cap.

Both compounds are obtained as colorless crystals if a solution of the respective metal in nitric acid is gradually treated with an excess of alcohol. They decompose into hydroxylamin and formic acid with HCl:

$$C_2HgN_2O_2 + 2HCl + 4H_2O = HgCl_2 + 2CH_2O_2 + 2NH_2(OH).$$

b. Carbylamine or Isonitrile.

Properties. In the isonitriles the nitrogen exists as a pentavalent element (p. 383), so that the alkyls are united to the still free fifth valence of nitrogen: $C\equiv N-CH_3$. Isonitriles are very poisonous, disagreeable-smelling, colorless liquids, either insoluble or difficultly soluble in water.

On heating with water to 180°, also by dilute acids even in the cold, but not by bases, they are split into formic acid and amines, at the same time taking up water. From this it follows that the alcohol radical is united to the nitrogen of the cyanogen:

$$CH_3-N\equiv C + 2H_2O = CH_3-NH_2 + CH_2O_2.$$

Preparation. 1. By warming chloroform and primary amine bases with alcoholic potash solution: $CH_3-NH_2 + CHCl_3 = CH_3-N\equiv C + 3HCl$. (Hofmann's carbylamine test for primary amines; secondary and tertiary amines do not give isonitrile, and hence not the characteristic disagreeable odor.)

2. By the action of alkyl iodides upon silver cyanide:

$$CH_3I + AgNC = CH_3-N\equiv C + AgI.$$

(According to this reaction silver cyanide contains the $N\equiv C$ group.)

3. Compounds of Cyanogen with Halogens, etc.

The hydrogen of hydrocyanic acid is not only replaced by metals and alcohol radicals, but also by halogens and monovalent atomic groups: $-OH$, $-SH$, $-NH_2$.

a. Chlorides and Amides.

Cyanogen chloride, chlorcyan, $NC-Cl$, is obtained by passing chlorine through a solution of mercuric cyanide or a watery solution of hydrocyanic acid:

$$Hg(CN)_2 + 4Cl = HgCl_2 + 2NC-Cl;$$
$$HCN + 2Cl = HCl + NC-Cl.$$

It forms an oily colorless liquid boiling at 15° and whose vapors are irritating and cause the flow of tears.

Cyanuric chloride, solid cyanogen chloride, $N_3C_3Cl_3$, is a derivative of triazine (which see), and is formed by keeping cyanogen chloride, also by passing chlorine into anhydrous hydrocyanic acid, in direct sunlight. It forms shining poisonous crystals which melt at 145°.

Cyanamide, $NC-NH_2$, is obtained, as colorless crystals melting at 40° C, by the action of ammonia upon cyanogen chloride. If warmed with dilute sulphuric or nitric acid it combines with 1 molecule of water and is transformed into urea: $NC-NH_2 + H_2O = H_2N-CO-NH_2$ (urea). Its watery solution is gradually changed into

Dicyanamide, $(NC-NH_2)_2$, which melts at 205° C.

Cyanuramide, melamine, $(NC-NH_2)_3$, is obtained on the polymerization of cyanamide by heating to 150°.C. as colorless crystals having the character of a monoᴜasic acid.

b. Compounds of Cyanic Acid.

Theoretically two compounds of cyanogen and OH are possible, namely, normal cyanic acid, $N \ C-OH$ (cyanogen hydroxide), and isocyanic acid, $OC=NH$ (carbimide). Still in this case, like with hydrocyanic acid, only one cyanic acid and one series of salts (cyanates) are known which probably have the structure NCOH, NCOK, etc. Esters of isocyanic acid, the carbonimides, are known, $OC=N-CH_3$, methyl carbonimide, as well as isocyanuric acid, $(OCNH)_3$, and cyanuric acid, $(NCOH)_3$.

Cyanic acid, NCOH, cannot be isolated from its salts, as it decomposes on being set free, $NCOH + H_2O = CO_2 + NH_3$, or it forms cyanuric acid. It is obtained on heating cyanuric acid, $(NCOH)_3$, as a volatile, irritating liquid, which is stable only under 0° C. As soon as it is removed from the cooling mixture which is used in liquefying it, it passes into cyanelide, $(NCOH)_4$, which is a white amorphous solid.

Potassium cyanate, NCOK, is the substance from which all the other cyanates are prepared and which are obtained therefrom by double decomposition. It is produced on heating potassium cyanide with readily reducible metallic oxides. It is ordinarily prepared by fusing potassium cyanide with red lead (see Urea). It is a white solid, crystallizing in plates, which are readily soluble and which are only slightly poisonous. Its watery solution quickly decomposes:

$$NCOK + 2H_2O = KHCO_3 + NH_3.$$

Ammonium cyanate, $NCO(NH_4)$, is obtained by the action of cyanic acid vapors upon dry ammonia gas as a white crystalline powder (see Urea). On the evaporation of its watery solution it is converted into its isomer urea: $NCO(NH_4) = CO(NH_2)_2$.

Cyanuric acid, $N_3C_3O_3H_3$, is a derivative of triazine (which see) and hence has the preceding formula. It is produced by the action of water upon cyanuric chloride: $N_3C_3Cl_3 + 3HOH = N_3C_3O_3H_3 + 3HCl$. If acetic acid is added to a solution of potassium cyanate, primary potassium cyanurate, $N_3C_3O_3H_2K$, gradually separates out and from this cyanuric acid can be obtained by mineral acids. Cyanuric acid forms, with the addition of 2 molecules of water, large colorless crystals which decompose into cyanic acid on heating. Only one cyanuric acid and one series of salts are known, although the esters of cyanuric acid and isocyanuric acid are known (see above).

c. Compounds of Thiocyanic Acid.

According to theory, thiocyanic acid, $N\equiv C-SH$, and isothiocyanic acid or sulphcarbimide, $SC=NH$, are possible, although only one acid, $NC-SH$, and one series of salts are known.   Still we know of esters of thiocyanic acid as well as isothiocyanic acid.   These last are also called *"mustard oils,"* from the most important members of the group.

Thiocyanic acid, sulphocyanic acid, $NC-SH$, is found in the gastric juice of the dog.   It is obtained as a colorless irritating liquid by the decomposition of mercuric thiocyanate with sulphuretted hydrogen: $(NCS)_2Hg + H_2S = 2NCSH + HgS$.   When anhydrous it is changed into yellow, amorphous thiocyanuric acid, $(NCSH)_3$, at ordinary temperatures.

On heating the thiocyanates with dilute sulphuric acid, avoiding an excess of the acid, we obtain a watery distillate of thiocyanic acid.   With an excess of concentrated sulphuric acid the free thiocyanic acid is decomposed into carbon oxysulphide and ammonia:

$$NCSH + H_2O = COS + NH_3.$$

Sulphocyanic acid and its soluble salts color even very dilute solutions of ferric salts red with the formation of ferri-thiocyanate, $(NCS)_3Fe$.

Potassium thiocyanate, potassium sulphocyanide, $NCSK$, is obtained by fusing sulphur with potassium cyanide.   It forms colorless prisms which are soluble in water and alcohol.

Sodium thiocyanate, sodium sulphocyanide, $NCSNa$, occurs in the saliva and urine of man and other animals.

Ammonium thiocyanate, ammonium sulphocyanide, $NCS(NH_4)$, is obtained on warming hydrocyanic acid with yellow ammonium sulphide or carbon disulphide with alcoholic ammonia: $CS_2 + 4NH_3 = NCS(NH_4) + (NH_4)_2S$.   It is prepared commercially by lixiviating the iron hydrate used in purifying illuminating-gas (p. 388).   It forms colorless prisms, which are converted into its isomer sulphur urea by heating to 170°: $NCS(NH_4) = CS(NH_2)_2$.

Mercuric thiocyanate, mercury sulphocyanide, $(NCS)_2Hg$, is obtained by precipitating KCNS with a mercuric salt.   It is a white amorphous powder which increases greatly in volume on burning (chief constituent of the so-called Pharaoh's serpent).

COMPOUNDS OF DIVALENT ALCOHOL RADICALS.

1. Divalent Alcohol Radicals.

Alkylenes, Alkenes, or Olefines.

General formula C_NH_{2N}.

		Boiling-point.			Boiling-point.
Ethylene	C_2H_4	Gas	Octylene	C_8H_{16}	125°
Propylene	C_3H_6	Gas	Diamylene	$C_{10}H_{20}$	160°
Butylene	C_4H_8	+ 3°	Cetene	$C_{16}H_{32}$	275°
Pentylene	C_5H_{10}	39°	Cerotene	$C_{27}H_{54}$	Solid
Hexylene	C_6H_{12}	70°	Melene	$C_{30}H_{60}$	Solid

If two atoms of hydrogen of the hydrocarbons of the methane series, C_NH_{2N+2}, are substituted, the hydrocarbon residue C_NH_{2N} acts as a divalent radical. Although the hydrocarbon residues having the formula C_NH_{2N+1} (the alkyl) do not exist in the free state, those with the formula C_NH_{2N} (the akylenes) are known free. They are accordingly called ethylenes, or oil-forming gases (French "gas olefiant"), or olefines.

Methylene, CH_2, does not exist. In all reactions which produce methylene free methylene is not produced, but instead its polymers like ethylene, C_2H_4, propylene, C_3H_6, etc. In the alkylenes two of the C atoms present are always united together by two bonds: $CH_3-CH=CH_2$, propylene.

No cases of isomers are possible with ethylene and propylene, while with butylene, C_4H_8, three isomers are possible, namely:

$$CH_3-CH_2-CH=CH_2; \qquad CH_2=C\big\langle {}^{CH_3}_{CH_3}; \qquad CH_3-CH=CH-CH_3.$$
Butylene.   Isobutylene.   Pseudobutylene.

Five isomers of C_5H_{10} are possible. The number of isomers is even greater than with the paraffins.

Properties. The lower members are gaseous, those intermediate are readily volatile liquids, and the higher members above $C_{27}H_{54}$ are all colorless solids. Being unsaturated compounds, they unite directly with two monovalent atoms (H, Br, Cl, I, etc.) or atomic groups, when the double bonds of the C atoms are transformed into single bonds; e.g.,

$$\begin{matrix} CH_2 \\ \| \\ CH_2 \end{matrix} + 2H = \begin{matrix} CH_3 \\ | \\ CH_3 \end{matrix} \qquad\qquad \begin{matrix} CH_2 \\ \| \\ CH_2 \end{matrix} + 2Br = \begin{matrix} CH_2Br. \\ | \\ CH_2Br. \end{matrix}$$
Ethylene.   Ethane.   Ethylene bromide.

With these addition-products two isomers having the formula $C_2H_4X_2$ are readily possible, namely, $XH_2C^-CH_2X$ and $H_3C^-CHX_2$. The first contain the group $H_2C^-CH_2$ and are called ethylene compounds, while the second contain the group H_3C^-CH (p. 329) and are called the ethylidene compounds. This nomenclature is also used with compounds with more than two C atoms (see Lactic Acid).

Isomers of the dihydric alcohols of the ethylidene series, $H_3C^-CH(OH)_2$, are not known, as bodies with more than one HO group united to one C atom are unstable (p. 331).

The olefines are absorbed by concentrated sulphuric acid with the formation of alkyl sulphuric acid esters, when the acid residue attaches itself to the C atom poorest in hydrogen (see below): $C_4H_8 + H_2SO_4 = C_4H_9HSO_4$ (butyl sulphuric acid). Polymerization may also take place with sulphuric acid (or with zinc chloride or boron fluoride).

The olefines unite directly with HCl, HBr, and HI. In these cases we find also that the halogen atoms also attach themselves to the C atom poorest in hydrogen, producing secondary and tertiary compounds:

$$CH_3^-CH_2^-CH{=}CH_2 + HI = CH_3^-CH_2^-CHI^-CH_3;$$
Butylene.   Secondary butyl iodide.

$$CH_2 {=} C\!\!<^{CH_3}_{CH_3} + HI = CH_3 - CI\!\!<^{CH_3}_{CH_3}.$$
Isobutylene.   Tertiary butyl iodide.

With watery hypochlorous acid they form so-called chlorhydrines:

$$CH_2 {=} CH_2 + ClOH = CH_2Cl^-CH_2OH.$$
Ethylene.   Ethylenchlorhydrine.

The olefines are readily oxidized into acids containing less C by $KMnO_4$ or CrO_3, but not by HNO_3, in the cold. By careful oxidation in the presence of water the corresponding alcohols are obtained: $C_NH_{2N}(OH)_2$ (p. 400, 2).

Occurrence. On the dry distillation of many C compounds olefines are obtained; hence they exist to a slight extent in illuminating-gas and in the tar oils from wood, brown, bituminous and anthracite coals. The hydrocarbons C_NH_{2N}, isomeric with the olefines found in the Caucasian petroleums, belong to the aromatic compounds (see Naphthenes).

Preparation. 1. On distilling monohydric alcohols with dehydrating agents such as sulphuric acid, zinc chloride, phosphorus pentoxide:
$$C_2H_5OH = C_2H_4 + H_2O.$$

2. By warming the halogen alkyls with alcoholic caustic **potash:**

$$C_2H_5Br + KOH = C_2H_4 + KBr + H_2O.$$

3. On the electrolysis of the alkali salts of the oxalic acid **series:**

$$KOOC^-C_2H_4^-COOK = C_2H_4 + 2CO_2 + 2K.$$
Potassium succinate.　Ethylene.

4. By the action of alkali metals upon the halogen compounds $C_NH_{2N}X_2$:

$$C_2H_4Br_2 + 2Na = 2NaBr + C_2H_4.$$

Ethylene, ethene, olefiant gas, C_2H_4 or $H_2C^-CH_2$, is produced on the dry distillation of many organic substances and is therefore found in illuminating-gas (about 6 per cent.). It is obtained on distilling ethyl alcohol with six volumes concentrated sulphuric acid (p. 358):

$$C_2H_5OH + H_2SO_4 = C_2H_4 + H_2O + H_2SO_4.$$

It is a disagreeably smelling gas which burns with a luminous flame and slightly soluble in water and alcohol. It liquefies at $-1.1°$ and a pressure of 43 atmospheres or at $-103°$ C.

2. Halogen Compounds of the Alkylenes.

The hydrogen of the alkylenes cannot be directly substituted by halogens, as addition products are formed. Substitution products are obtained by the action of alcoholic caustic alkali upon the addition products: $C_2H_4Cl_2 + KOH = C_2H_3Cl + KCl + H_2O.$

Halogen addition products with ethylene groups (p. 395)—thus $XH_2C^-CH_2X$—are obtained by the action of the halogens upon the olefines or of halogen acids upon dihydric alcohols. Those with ethylidene groups—thus $H_3C^-CHX_2$,—are obtained by the action of the halogens upon the paraffins or by the action of PCl_5, etc., upon the aldehydes or ketones of the methane series:

$$CH_3^-COH + PCl_5 = CH_3CHCl_2 + POCl_3.$$

The first yield acetylene (C_2H_2) by the energetic action of alcoholic caustic potash, while the second yield acetals, $(CH_3^-CH(OC_2H_5)_2)$. The first yield glycols, while the others do not (p. 399). The other properties of the halogen alkylenes coincide with the halogen alkyls.

Methylene chloride, CH_2Cl_2 (p. 349), is a liquid similar to chloroform which boils at 41° C. and is used as a narcotic.

Methylene iodide, CH_2I_2, is a colorless liquid having a high specific gravity (3.3), and is used in separating those constituents of minerals having a higher specific gravity and which sink to the bottom.

Ethylene chloride, $C_2H_4Cl_2$ or CH_2Cl-CH_2Cl, is obtained by the uinon of equal volumes of ethylene and chlorine, or by heating concentrated hydrochloric acid with glycol, $C_2H_4(OH)_2$, to 200° C. It is a colorless, heavy liquid, having an odor and action similar to chloroform and boiling at 85°. It is also called oil of the Dutch chemists, liquor hollandicus, and as it is obtained from ethylene, this last is also called "oil-forming gas."

If an excess of chlorine is allowed to act upon ethylene, we obtain products which are isomeric with the chlorinated ethanes, and finally hexachlor ethane, C_2Cl_6, is obtained.

If ethylene chloride is treated with alcoholic caustic potash first (see above) one molecule of HCl is split off and we obtain

Chlor ethylene, vinyl chloride, CH_2=$CHCl$, a gas which has an odor similar to garlic:

$$C_2H_4Cl_2 + KOH = CH_2\text{=}CHCl + KCl + H_2O.$$

Ethylene bromide, $C_2H_4Br_2$, is a colorless, poisonous liquid boiling at 130° C. and yields acetylene, C_2H_2, with alcoholic caustic potash:

$$C_2H_4Br + 2KOH = C_2H_2 + 2KBr + H_2O.$$

Ethylidene chloride, $C_2H_4Cl_2$ or CH_3-$CHCl_2$, is produced by the action of chlorine upon ethane or by the distillation of aldehyde with phosphorus pentachloride (process, p. 396). It is a pleasant-smelling liquid boiling at 60°.

3. Amines of the Alkylenes.

If one alkylene is introduced in the place of two H atoms of one molecule of ammonia imines are obtained: C_3H_6=HN, trimethylenimine; C_4H_8=NH, tetramethylenimine; C_5H_{10}=NH, pentamethylenimine or piperidin (which see). These imines are not well known, but most of the alkylenes form primary, secondary, or tertiary diamines by replacing one, two, or three H atoms in two molecules of ammonia. The diamines are obtained as basic, colorless liquids by heating alkylene bromides with ammonia·

$$C_2H_4B_2 + 2NH_3 = H_2N - C_2H_4 - NH_2 + 2HBr;$$
Ethylendiamine.

$$2C_2H_4Br_2 + 2NH_3 = HN\diagdown\diagup{\begin{smallmatrix}C_2H_4\\C_2H_4\end{smallmatrix}}\diagdown\diagup NH + 4HBr;$$
Diethylendiamine.

$$3C_2H_4Br_2 + 2NH_3 = N\diagdown\begin{smallmatrix}C_2H_4\\C_2H_4\\C_2H_4\end{smallmatrix}\diagup N + 6HBr.$$
Triethylendiamine.

Ammonium bases are also known. Choline, one of these, contains both alkyl and alkylenes (p. 379). Some of the diamines are poisonous and are the mother-substance of most ptomaines and toxins.

Ptomaines ($\pi\tau\tilde{\omega}\mu\alpha$, cadaver), septicine, ptomatine, putrefaction bases, is the name given to a number of basic, nitrogenous, organic compounds which occur in putrefying animal and also plant proteid matter. They are precipitated from their solutions by the same reagents as the alkaloids (which see), but do not give the same color reactions.

As certain proteids and many other bodies give the same color reactions as the alkaloids, it is possible that the reactions may be obtained with insufficiently purified ptomaines, and this was the reason why it was formerly stated that the ptomaines gave all the reactions of the alkaloids and why they were incorrectly called "cadaver alkaloids." The ptomaines are mostly amine and diamine bases, while the alkaloids are mostly pyridin and quinoline bases.

Certain of the ptomaines are poisonous and are called toxines, while others are not; some are liquid and volatile, while others are non-volatile liquids or crystallizable solids. Amongst these we must mention *neurine, muscarine* (p. 380), *betaine* (p.380), *putrescine, neuridine, cadaverine, saprine* (see below), *mydatoxine*, $C_6H_{13}NO$, *mydine*, $C_8H_{11}NO$, and *mydaleine*, all prepared from putrefying meat; *gadinine*, from putrefying fishes; *mytilotoxine*, $C_6H_{15}NO_2$, from poisonous mussels, the sausage and cheese poison, etc.; *anthracine*, from anthrax bacilli; *tetanine* and *tetanotoxine*, $C_5H_{11}N$, from the tetanus bacillus; *typhotoxine*, $C_7H_{17}NO_2$, from the typhoid bacillus; *samandrine*, the poison of the salamander; *methylguanidine* (p. 412), etc.

The basic bodies, similar to ptomaines, which are regularly produced as decomposition products of proteids in living organisms are called *leucomaines*, to differentiate them from the ptomaines produced by microorganisms.

Ethylen-ethenyl diamine, lysidin, $N\diagdown\begin{matrix}C_2H_4\\C_2H_3\end{matrix}\diagup NH$, and

Diethylendiamine, $C_4H_{10}N_2$ (structure, p. 397) piperazin, are heterocyclic compounds on account of their ring-shaped constitution, and will be treated in connection with these bodies.

Tetramethylendiamine, putrescin, $H_2N-C_4H_8-NH_2$, or $C_4H_{12}H_2$, is formed in the putrefaction of cadavers, in cultures of the cholera bacillus, in many pathological processes, as well as in the putrefaction of ornithin (p. 375). It is poisonous.

Pentamethylendiamine, $N_2H-C_5H_{10}-NH_2$ or $C_5H_{14}N_2$, cadaverin, is obtained from cadavers and is poisonous. It is also produced in the putrefaction of lysin (p. 376).

Neuridine and *saprine*, $C_5H_{14}N_2$, are isomeric with cadaverin, are both non-poisonous, and are produced in the putrefaction of meat.

Hexamethylentetramine, $(CH_2)_6N_4$, urotropin, formin, aminoform, is obtained by the action of formaldehyde upon ammonia (p. 350), and forms colorless crystals; it is a solvent for uric acid, hence is used in medicine.

4. Dihydric Alcohols.

General formula, $C_N N_{2N}(OH)_2$.

If two HO groups are attached to the alkylenes after the rupture of the double bonds of the C atom, we obtain a series of dihydric alcohols which are called *glycols* on account of their sweet taste: ethylene glycol, $HO^-H_2C^-CH_2^-OH$ or $C_2H_6O_2$, propylene glycol, $HO^-CH_2^-CH_2^-CH_2^-OH$, etc. As we have two replaceable hydrogen atoms (hydroxyl groups) in these bodies, it is possible to form two mixed ethers of the same alcohol radical, two glycolates of the alkali metals (p. 343), also two esters of the same acid, etc.:

$$C_2H_4 \diagdown \begin{matrix} O(C_2H_3O) \\ OH \end{matrix}$$

Glycol monacetate.

$$C_2H_4 \diagdown \begin{matrix} OH \\ Cl \end{matrix}$$

Glycol chlorhydrin.

$$C_2H_4 \diagdown \begin{matrix} O(C_2H_5) \\ OH \end{matrix}$$

Glycol ethyl ether.

$$C_2H_4 \diagdown \begin{matrix} ONa \\ OH \end{matrix}$$

Monosodium glycolate.

$$C_2H_4 \diagdown \begin{matrix} O(C_2H_3O) \\ O(C_2H_3O) \end{matrix}$$

Glycol diacetate.

$$C_2H_4 \diagdown \begin{matrix} Cl \\ Cl \end{matrix}$$

Ethylene chloride.

$$C_2H_4 \diagdown \begin{matrix} O(C_2H_5) \\ O(C_2H_5) \end{matrix}$$

Glycol diethyl ether.

$$C_2H_4 \diagdown \begin{matrix} ONa \\ ONa \end{matrix}$$

Disodium glycolate.

The anhydrides of the glycols or alkylene oxides correspond to the anhydrides (ethers) of the monhydric alcohols. As the glycol molecule contains two hydroxyl groups, therefore water can be split off from one molecule in the same way as from the dihydroxyl compounds of the metals:

$$Ca(OH)_2 = CaO + H_2O; \qquad C_2H_4(OH)_2 = C_2H_4O + H_2O$$

Ethylene glycol.   Ethylene oxide.

The only glycols known are those with two C atoms or more (p. 331).

Occurrence. They are not found in nature.

Preparation. 1. From the ethylene bromides, by heating them with silver acetate or potassium acetate:

$$C_2H_4Br_2 + 2Ag(C_2H_3O_2) = C_2H_4 \diagdown \begin{matrix} C_2H_3O_2 \\ C_2H_3O_2 \end{matrix} + 2AgBr.$$

The glycol acetate thus obtained is decomposed, by boiling with caustic alkali, into glycol and alkali acetate:

$$C_2H_4\!\!<^{C_2H_3O_2}_{C_2H_3O_2} + 2KOH = C_2H_4\!\!<^{OH}_{OH} + 2K(C_2H_3O_2).$$

2. By the oxidation of olefines in the presence of water:

$$C_2H_4 + H_2O + O = C_2H_4(OH)_2.$$

Properties. The solubility of the compounds in water increases as the number of alcoholic hydroxyl groups contained in the compound increases, while the solubility in alcohol and especially in ether diminishes. At the same time a marked rise in the boiling-point takes place and the bodies become sweet. Consequently the glycols have a sweeter taste, are more readily soluble in water and only slightly soluble in ether, and boil at about 100° higher than do the corresponding monohydric alcohols. On oxidation the primary glycols yield, besides the numerous intermediary products (p. 402), also diatomic monobasic acids (oxyfatty or lactic acid series, p. 403), as well as diatomic, bibasic acids (oxalic acid series, p. 420). The number of isomers is even greater than in the monhydric alcohols; thus there are six isomeric butylene glycols possible.

Besides the normal or diprimary glycols there are primary-secondary, primary-tertiary, disecondary, secondary-tertiary, and ditertiary glycols (pinacones, p. 372, 4):

CH$_2$(OH)	CH$_2$(OH)	CH$_2$OH	CH$_3$	CH$_3$	CH$_3$ CH$_3$
CH$_2$	CH(OH)	CH$_2$	CH(OH)	CH(OH)	C(OH)
CH$_2$	CH$_2$	C(OH)	CH(OH)	C(OH)	C(OH)
CH$_2$(OH)	CH$_3$	CH$_3$ CH$_3$	CH$_3$	CH$_3$ CH$_3$	CH$_3$ CH$_3$
Diprimary.	Primary-secondary.	Primary-tertiary.	Di-secondary.	Secondary-tertiary.	Ditertiary.

5. Esters and Ethers.

Ethylene chlorhydrate, ethylene chlorhydrin, $C_2H_4\!\!<^{OH}_{Cl}$, is formed by the direct union of HClO with ethylene or when ethylene glycol is warmed with hydrochloric acid:

$$C_2H_4\!\!<^{OH}_{OH} + HCl = C_2H_4\!\!<^{OH}_{Cl} + H_2O.$$

It is a colorless liquid.

Glycol sulphuric acid, $C_2H_4\!\!<^{OH}_{O\cdot SO_2\cdot OH}$, is produced on warming glycol with sulphuric acid.

Ethylenhydrin-sulphuric acid, isathionic acid, oxyethyl-sulphonic acid, $C_2H_4 < {}^{OH}_{SO_2 \cdot OH}$, is isomeric with ethyl sulphuric acid, $C_2H_5-O \cdot SO_2 \cdot OH$, and forms deliquescent crystals on heating ethylene chlorhydrine with potassium sulphite

Sulpho- or *sulphonic acids* (not to be mistaken for the inorganic sulphonic acids, p. 176) contain the sulphur of the monovalent group $-SO_3H$ united directly with the carbon of the radicals, while in the isomeric compounds the sulphurous acid unites the carbon to the oxygen. Perhaps this can be explained by the existence of two sulphurous acids:

$$OS < {}^{OH}_{OH} \qquad O_2S < {}^{H}_{OH} \qquad OS < {}^{O(C_2H_5)}_{OH} \qquad O_2S < {}^{C_2H_5}_{OH}$$

| Symmetric Sulphurous acid. | Unsymmetric | Ethyl sulphurous acid. | Ethyl sulphonic acid. |

Sulphinic acids containing the monovalent group SO_2H are also known; e.g., ethyl sulphinic acid, $C_2H_5-SO_2H$.

Taurine, amidoethyl-sulphonic acid, $C_2H_4 < {}^{NH_2}_{SO_3H}$, exists, combined with cholic acids, as taurocholic acid and chenotaurocholic acid in the bileacids (p. 370) containing sulphur. It forms colorless neutral prisms which are insoluble in alcohol but readily soluble in water.

Taurocholic acid, $C_{26}H_{45}NO_7S$, occurs in ox and human bile and forms silky needles which are easily soluble and decompose into taurine and cholic acid on boiling with alkalies or water (p. 370):

$$C_2 \cdot H_{45}NO_7S + H_2O = C_2H_7NO_3S + C_{24}H_{40}O_5.$$

Chenotaurocholic acid, $C_{29}H_{49}NO_6S$, occurs in goose bile and decomposes into taurine and chenocholic acid (p. 370).

Ethylenethyl ether, glycolethyl ether, $C_2H_4 < {}^{O(C_2H_5)}_{OH}$.

Ethylenediethyl ether, glycoldiethyl ether, $C_2H_4 < {}^{O(C_2H_5)}_{O(C_2H_5)}$, is a colorless liquid boiling at 127°.

Metallic sodium dissolves in the cold in glycol with the formation of sodium glycol, and on heating it forms disodium glycol, $C_2H_4 < {}^{ONa}_{ONa}$. Both form colorless crystals which form the corresponding ether with the halogen derivatives of the alkyls:

$$C_2H_4 < {}^{ONa}_{OH} + C_2H_5I = C_2H_4 < {}^{O(C_2H_5)}_{OH} + NaI.$$

$$C_2H_4 < {}^{ONa}_{ONa} + 2C_2H_5I = C_2H_4 < {}^{O(C_2H_5)}_{O(C_2H_5)} + 2NaI.$$

Ethylidene diethyl ether, acetal, $CH_3-CH < {}^{O(C_2H_5)}_{O(C_2H_5)}$ (p. 351, 8), occurs as a product in the distillation of brandy. It is a colorless liquid boiling at 104° C., and is produced, with aldehyde, in the oxidation of alcohol and also from sodium ethylate and ethylidene bromide:

$$CH_3-CHBr_2 + 2C_2H_5-ONa = CH_3-CH(OC_2H_5)_2 + 2NaBr.$$

Ethylene oxide, ethylene ether, C_2H_4O (structure below), is obtained on the distillation of ethylenchlorhydrine (p. 395) with caustic alkali.

$$C_2H_4 < ^{Cl}_{OH} + KOH = < ^{CH_2}_{CH_2} > O + KCl + H_2O.$$

It is a liquid boiling at 14° and having an ethereal odor which mixes with water and unites gradually therewith, forming ethylene glycol. It unites directly with acids, forming monoglycol esters: $HO-C_2H_4-HSO_4$.

Ethylidene oxide, CH_3-CHO, is ethyl aldehyde (p. 359).

6. Derivatives with Aldehyde and Ketone Groups.

The dihydric, primary alcohols on oxidation yield two aldehydes and two acids, according to whether hydrogen is removed and oxygen introduced in only one or both CH_2-OH groups:

$HO-H_2C-CH_2OH$	$OHC-CH_2OH$	$OHC-CHO$
Glycol.	Glycol aldehyde.	Glyoxal.
$HOOC-CH_2OH$	$HOOC-CHO$	$HOOC-COOH$
Glycolic acid (Monobasic).	Glyoxylic acid (Gloyxalic acid).	Oxalic acid (Bibasic).

It follows from the constitution that these compounds show in part different functions (p. 337). Thus glycolaldehyde has the properties of an alcohol on account of the CH_2-OH group and an aldehyde because of the CHO group; hence it may be called an aldehyde alcohol (aldose). The compounds with two aldehyde groups, like glyoxal, are called dialdehydes. Glyoxylic acid is an aldehyde acid, while glycolic acid is an alcohol acid, etc.

The *secondary glycols* also yield mixed compounds on oxidation (p. 337) and the primary-secondary alcohols (P. 400) yield aldehyde alcohols (aldoses), ketone alcohols (ketoses), ketone aldehydes, alcohol acids, ketone acids:

$CH_3-CH \cdot OH-CH_2OH$	$CH_3-CO-CH_2OH$	$CH_3-CO-CHO$
a-Propylene glycol.	Acetone alcohol.	Methyl glyoxal.
$CH_3-CH \cdot OH-CHO$	$CH_3-CH \cdot OH-COOH$	$CH_3-CO-COOH$
Lactic acid anhydride.	Lactic acid.	Pyroracemic acid.

The *disecondary glycols* yield ketone alcohols and diketones, but no acids (p. 334). All aldoses and ketoses which have an HO group attached to the C atom neighboring the aldehyde or ketone group belong to the class of bodies called sugars.

Glycoaldehyde, $HO \cdot H_2C-CHO$, only obtained indirectly and only in solution. It is a sugar-like body (which see) and reduces Fehling's solution and combines with phenylhydrazin.

Glyoxal, $OHC-CHO$, is an amorphous, colorless solid.

Glyoxalic acid, $OHC-COOH$, glyoxylic acid, occurs in the leaves and unripe fruit of many plants and as traces in acetic acid, and forms colorless crystals with 1 molcule H_2O, which cannot be removed; hence the formula $(HO)_2=HC-COOH$ is often given to this body (see Alcohols, p. 331).

Pyroracemic acid, acetyl formic acid, $CH_3-CO-COOH$, is a ketonic acid (p. 364) and is obtained on heating tartaric acid (p. 428), or by the oxidation of ethylidene lactic acid (p. 406), or by heating glyceric acid (p. 433). It is a liquid having an odor similar to acetic acid and boils at 170°. With nascent hydrogen it yields ethylidene lactic acid, and on boiling it partly decomposes into pyrotartaric acid:

$$2C_3H_4O_3 = C_5H_8O_4 + CO_2.$$

7. Oxyfatty Acid or Lactic Acid Series.

General formula $C_NH_{2N}O_3$.

Carbonic acid,	CH_2O_3	Oxybutyric acids,	$C_4H_8O_3$
Glycollic acid,	$C_2H_4O_3$	Oxyvaleric acids,	$C_5H_{10}O_3$
Lactic acids,	$C_3H_6O_3$	Oxycaproic acids,	$C_6H_{12}O_3$, etc.

Properties. The dihydric, monobasic acids from the glycols **are** called the acids of the lactic acid series, because of their most important member, or the oxyfatty acids, as they have one atom O more than the fatty acids; thus $C_2H_4O_2$, acetic acid; $C_2H_4O_3$, glycollic acid. Carbonic acid, or oxyformic acid, $HO-CO-OH$, which is not known free, because of its symmetrical structure, shows no difference between the two OH groups and is hence bibasic and forms the connection between the above group of acids and the bibasic acids of the glycols (p. 420). For this reason it will be treated of on p. 408.

The other acids are colorless and crystallizable, readily soluble in water but less soluble in ether than the corresponding fatty acids; also less volatile, and cannot be distilled at ordinary pressures without decomposition. On oxidation they yield acids of the oxalic acid series. They behave in regard to their chemical properties like alcohols and acids on account of their structure. The hydrogen of the carboxyl group can be readily replaced by metals or alcohol radicals, producing normal salts or esters respectively:

$$HO \cdot H_2C-COOK; \quad HO \cdot H_2C-COO(C_2H_5).$$

The CH_2-OH group behaves exactly like as in the alcohols, the hydrogen being replaceable by alkali metals and by alcohol or acid radicals:

CH_2OH	CH_2OH	$CH_2 \cdot O(C_2H_5)$	$CH_2 \cdot O(C_2H_5)$
$\vert$	$\vert$	$\vert$	$\vert$
$COOH$	$COO(C_2H_5)$	$COOH$	$COO(C_2H_5)$
Glycollic acid.	Ethyl glycollate.	Ethyl glycollic acid.	Diethyl glycollate.

The number of isomers is, as with the dihydric alcohols, much greater than with the monohydric acids. There are two structural isomers of lactic acid possible,

$$CH_3-CH(OH)-COOH \quad \text{and} \quad CH_2(OH)-CH_2-COOH,$$
Ethylidene lactic acid.   Ethylene lactic acid.

and five of oxybutyric acid. We differentiate between the isomers, according to whether the HO group exists next to the COOH group or whether it is removed. They are designated α-, β-, γ-, etc., acids (p. 329).

The divalent groups which are united with the two hydroxyls are called the radicals of the oxyfatty acids. They are also designated α, β, γ, etc., according to the position where the hydroxyl is lacking:

$-OC-CH_2-$glycolyl, $-CH_2-CH_2-CO-\beta$-lactyl, $CH_3-\overset{|}{C}H-\overset{|}{C}O$ α-lactyl.

The formation of anhydrides may take place as follows:

a The internal anhydrides are produced by the removal of 1 mol. H_2O from the alcoholic hydroxyl and the carboxyl of one molecule of the acid.

These are called *lactones*, and are designated by adding *-olid* to the hydrocarbon from which they are derived:

$$<\!\!\begin{array}{l} CH_2-CH-CH \\ CO\underline{\quad\quad\quad}_2\!\!> \end{array}\!\! O, \quad \begin{array}{l} \text{Butyrolactone,} \\ \text{Butanolid.} \end{array}$$

Contrary to other acid anhydrides, the lactones are chemically indifferent. They are very readily prepared from the γ-oxyfatty acids, which are mostly known only as lactones.

By the removal of 1 mol. H_2O from the alcoholic hydroxyl and a neighboring alcohol radical of β-oxyfatty acids we obtain unsaturated acids: $CH_3-CH(OH)-CH_2-COOH$ (β-oxybutyric acid)$=CH_3-CH=CH-COOH$ (crotonic acid)$+H_2O$.

b. On the removal of 1 mol. H_2O from the carboxyls of 2 mol. acid we do not obtain true acid anhydrides (p. 336):

$$O\!<\!\!\begin{array}{l} OC-CH_2OH \\ OC-CH_2OH \end{array}\!\!.$$

c. The alcoholic anhydrides are formed by the removal of H_2O from two alcohol hydroxides of two acid molecules: $HOOC-CH_2-O-CH_2-COOH$, diglycollic acid.

d. By the removal of H_2O from one alcohol hydroxyl and one carboxyl group of two acid molecules we obtain acid esters, incorrectly called acid anhydrides, as the true acid anhydrides (see above) are not known:

$$\begin{array}{l} COOH \\ | \\ CH_2OH \end{array} + \begin{array}{l} CH_2OH \\ | \\ COOH \end{array} = \begin{array}{l} COO---CH_2 \\ | \quad\quad\quad | \\ CH_2OH \quad COOH \end{array} +H_2O, \quad \text{glycollic anhydride}$$

e. By the removal of 2 mol. H_2O from both alcoholic hydroxyls and of both carboxyls of two molecules of acid (or by the removal of 1 mol. H_2O from the simple esters) neutral double esters ending in *id* are produced:

$$\begin{array}{l} COOH \\ | \\ CH_2OH \end{array} + \begin{array}{l} CH_2OH \\ | \\ COOH \end{array} = 2H_2O+ \begin{array}{l} COO---CH_2 \\ | \quad\quad\quad | \\ CH_2-OOC \end{array}, \quad \text{glycolid.}$$

f. Diacid anhydrides are obtained on the removal of 2 mol. H_2O in such a manner that 1 mol. H_2O comes from the two alcoholic hydroxyls and 1 mol. H_2O from the two carboxyls:

$$CH_2-OH-COOH \atop CH_2-OH-COOH = 2H_2O + O < {CH_2-CO \atop CH_2-CO} >, \text{ diglycollic anhydride.}$$

Preparation. 1. By the gentle oxidation of the glycols by dilute nitric acid or platinum-black.

2. By the action of hydrocyanic acid and hydrochloric acid upon the aldehydes we obtain the acids containing one ethylidene group. Oxycyanides (p. 351, 9) are first formed and the cyanogen group is transformed into the carboxyl group by the presence of the hydrochloric acid (p. 346, 2):

$$CH_3-COH + CNH = CH_3-CH(OH)(CN)$$
$$CH_3-CH(OH)(CN) + 2H_2O = CH_3-CH(OH)-COOH + NH_3.$$
Ethylidene lactic acid.

The acids containing one ethylene group are also produced from the chlorhydrins (see Ethylene Lactic Acid).

3. On boiling the monobrom- or monchlorfatty acids with alkali hydroxides:

$$HOOC-CH_2Cl + KOH = HOOC-CH_2OH + KCl.$$
Monochloracetic acid.   Oxyacetic acid.

The reverse may take place, i.e., the oxyfatty acids can be converted into the brominated fatty acids if they are heated with hydrobromic acid exactly the same as ethyl alcohol is converted into ethyl bromide by hydrobromic acid:

$$HOOC-CH_2OH + HBr = HOOC-CH_2Br + H_2O.$$
Glycollic acid   Bromacetic acid.

If, on the contrary, we heat with hydriodic acid, we obtain fatty acids (p. 346, 5):

$$HOOC-CH_2OH + 2HI = HOOC-CH_3 + H_2O + 2I.$$

4. By the action of nitrous acid upon the amido-acids (p. 369, 3):

$$HOOC-CH_2(NH_2) + HNO_2 = HOOC-CH_2OH + 2N + H_2O.$$
Amidoacetic acid.   Oxyacetic acid

5. By the action of nascent hydrogen upon ketonic acids (p. 364, 2) or ketonic acid ester (p. 365) we obtain also the oxyfatty acids derived from ethylidene.

Glycollic acid, oxyacetic acid, $C_2H_4O_3$, is a constituent of the juice from unripe grapes as well as the juice from Ampelopsis hederacea. It forms colorless crystals which are soluble in water and which yield oxalic acid on oxidation.

Ethylidene Lactic Acids, $CH_3^-CH(OH)^-COOH$, α-oxypropionic acids. Three are known, one optically inactive, one dextrorotatory, and one lævorotatory; their existence is explained by the stereochemical theory (p. 303).

1. *Fermentation Lactic Acid, Inactive Lactic Acid.*

Occurrence. It is produced in a special fermentation (P. 373, 2) of starch, as well as of certain sugars, and therefore occurs in many substances which contain starch, sugar, etc., and which have become sour, such as sour milk, sauerkraut, sour pickles, spent tan, the fermented juice of the beet-root. It is sometimes found in the contents of the stomach and intestine, in the blood taken from cadavers, and in the gray substance of the brain.

Preparation. 1. Ordinarily by the fermentation of certain sugars, etc., according to the method given on p. 373, 2 accompanied with butyric acid. The calcium lactate formed is transformed by means of zinc chloride into the readily· crystallizable zinc. lactate, and this decomposed by H_2S into zinc sulphide, which can be filtered off and the filtrate evaporated on the water-bath and then repeatedly distilled in a vacuum.

2. By the oxidation of secondary propylene glycol, $CH_3^-CH(OH)^-CH_2OH$, as well as by the action of HCl upon aldehyde and hydrocyanic acid (process, p. 405, 2).

3. From pyroracemic acid, $C_3H_4O_3$, by the action of nascent hydrogen:

$$C_3H_4O_3 + H_2 = C_3H_6O_3.$$

Properties. It forms colorless and odorless crystals which melt at 18° C. and which are soluble in water, alcohol, and ether. It is volatile without decomposition only under very diminished pressure, and is hygroscopic, hence it becomes deliquescent on keeping. On gentle oxidation it yields pyroracemic acid: $CH_3^-CH(OH)^-COOH + O = CH_3^-CO^-COOH + H_2O$, but on stronger oxidation acetic acid is obtained:

$$\underset{CH(OH)COOH}{\overset{CH_3}{\mid}} + 2O = \underset{COOH}{\overset{CH_3}{\mid}} + CO_2 + H_2O.$$

Lactic acid containing 25 per cent. water is a thick liquid. Its salts are called *lactates*. *Zinc lactate*, $Zn(C_3H_5O_3)_2$, crystallizes with 3 mol. H_2O, and *calcium lactate*, $Ca(C_3H_5O_3)_2$, which is soluble in alcohol, crystallizes with 5 mol. H_2O (differing from the optically active lactic acids). *Ferrous lactate*, $Fe(C_3H_5O_3)_2 + 3H_2O$, forms greenish-white crystalline crusts which are soluble in water.

Fermentation lactic acid contains one asymmetric C atom, but is optically inactive, as it consists of a dextro- and a lævorotatory modification (p. 306). It is split into these two on the fractional crystallization of its strychnine salts. Lævolactic acid is obtained by the action of the Bacillus acidi lævolactici, as it consumes the dextrolactic acid (p. 39). The fungus Penicillium glaucum produces dextrolactic acid, as it consumes the lævolactic acid. On mixing the two active modifications the inactive lactic acid is obtained again (p. 39).

2. *Sarco- or Paralactic Acid, Dextrorotatory Lactic Acid.* It is prepared as above described and is found in bile and in muscle plasma as alkali salt. After violent work and after rigor mortis the quantity in the muscles may increase so that they have an acid reaction. It is often found in the urine and blood as a pathological constituent.

3. *Lævorotatory Lactic Acid.* Preparation as above. Both optically active modifications coincide with fermentation lactic acid in all their properties, with the exception that their zinc salts crystallize with 2 mol. H_2O and their calcium salts with 4 mol. H_2O. The latter are soluble in alcohol.

Ethylene lactic acid, β-oxypropionic acid, hydracrylic acid, $CH_2(OH)-CH_2-COOH$, is prepared according to the method given on p. 405, 2, namely, by the action of potassium cyanide and hydrochloric acid upon ethylene chlorhydrin:

$$CH_2(OH)-CH_2Cl + KCN = CH_2(OH)-CH_2(CN + KCl;$$
$$CH_2(OH)-CH_2(CN) + 2H_2O = CH_2(OH)-CH_2-COOH + NH_3.$$

It is optically inactive, and is a sirupy liquid which splits into acrylic acid and water on heating: $C_3H_6O_3 = C_3H_4O_2 + H_2O$ (hence its name). It yields the dihydric, bibasic malonic acid on oxidation:

$$\begin{array}{ccc} CH_2-CH_2OH & CH_2CH_2OH & CH_2-COOH \\ | & | & | \\ CH_2OH & COOH & COOH \\ \text{Propylene glycol.} & \text{Ethylene lactic acid.} & \text{Malonic acid.} \end{array}$$

Its very readily soluble zinc salt crystallizes with 4 mol. H_2O, and its calcium salt, which is insoluble in alcohol, crystallizes with 2 mol. H_2O.

Amidoethylene lactic acid, $CH_2(OH)CH(NH_2)-COOH$, serin, is obtained on boiling silk gelatine (sericin) with dilute sulphuric acid.

Cystin, dithiodiamidodiethylidene lactic acid, $C_6H_{12}N_2O_4S_2$ or
$$\begin{array}{l} CH_2-CS(NH_2)-COOH \\ | \\ CH_2-CS(NH_2)-COOH \end{array}$$
, occurs seldom as vesical and kidney calculi in human beings, oftener in dogs. It also occurs to a slight extent in human urine, and is produced on the cleavage of proteins. It forms colorless crystals which are insoluble in water, alcohol, and ether, but readily soluble in ammonia and caustic alkalies. It is lævorotatory and splits on reduction into two molecules of

Cystein, $CH_3C(SH)(NH_2)COOH$, which is insoluble in water and which also occurs as a cleavage product of proteins.

Oxybutyric acids, $C_4H_8O_3$. Of the five isomers of this acid we have learned of the preparation of β-oxybutyric acid in treating of the ketonic acids, p. 364, 2.   This occurs in a lævorotatory form in the urine and blood in many diseases.

Leucinic acid, $C_6H_{12}O_3$, is formed by the action of nitrous acid upon amidocaproic acid (process, p. 376).

Lanopalmitic acid, $C_{16}H_{32}O_5$, occurs in wool-fat.

8. Carbonic Acid and its Derivatives.

Carbonic acid, oxyformic acid, CH_2O_3 or HO^-CO^-OH (p. 190), is not known free, but only as derivatives.   It is a weak bibasic acid which differs from the other acids of this series on account of its different symmetric structure, and it forms the connection between the lactic acid series and the oxalic acid series (p. 420).   Its acid esters are unstable, while its neutral esters are obtained from alkyl iodide and silver carbonate, e.g., $Ag_2CO_3 + 2C_2H_5I = (C_2H_5)_2CO_3 + 2AgI$, as ethereal liquids which are insoluble in water.

a. Sulphoderivatives.

Trisulphocarbonic acid, CH_2S_3 or HS^-CS^-SH, is not known free (p. 192).   Its primary esters are not known, but its secondary esters are produced on treating potassium sulphocarbonate with alkyl iodides:

$$K_2CS_3 + 2C_2H_5I = (C_2H_5)_2CS_3 + 2KI.$$

Disulphocarbonic acids, H_2COS_2.   Two isomers are possible:

Sulphocarbonyl disulphocarbonic acid, $CS{<}{}^{SH}_{OH}$ Carbonyl disulphocarbonic acid, $CO{<}{}^{SH}_{SH}$.

The free acids are not known, nevertheless the secondary esters of carbonyl-disulphocarbonic acid are known and secondary as well as primary esters of sulphocarbonyl-disulphocarbonic acid:

$$HS^-SC^-O(C_2H_5) \quad \text{and} \quad (H_5C_2)S^-CS^-O(C_2H_5).$$

Its sodium salt is obtained by treating carbon disulphide with caustic soda:

$$CS_2 + 2NaOH = NaS^-CS^-ONa + H_2O.$$

Xanthogenic acids is the name given to the primary esters or ether acids of sulphocarbonyl-disulphocarbonic acid:

Methyl xanthogenic acid, $CS{<}{}^{OCH_3}_{SH}$,   Ethyl xanthogenic acid, $CS{<}{}^{OC_2H_5}_{SH}$.

Potassium ethylxanthogenate, $KS^-CS^-OC_2H_5$, deposits as silky yellow needles by mixing carbon disulp ide and potassium ethylate (alcoholic caustic potash): $CS_2 + C_2H_5^-OK = KS^-CS^-OC_2H_5$.   On heating with dilute sulphuric acid ethyl xanthogenic acid separates as an oily liquid.

Monosulphocarbonic acids, H_2CO_2S.   There are two isomers possible, HO^-CS^-OH and HO^-CO^-SH, just as we had with the disulphocarbonic acids.   The free acids, the primary salts and esters are not known, but only the secondary salts and esters. If COS is passed into potassium ethylate, we obtain the potassium salt of carbonylethylmonosulphocarbonic acid:

$$COS + C_2H_5{}^-OK = KS^-CO^-OC_2H_5.$$

b. *Amido-derivatives.*

Like all other organic bibasic acids, carbonic acid forms the same amides and imides (p. 420); for example,

$$CO{<}^{OH}_{OH} \qquad CO{<}^{NH_2}_{OH}, \qquad CO{<}^{NH_2}_{NH_2} \qquad CO{=}NH$$

Carbonic acid.   Carbamic acid.   Carbamide.   Carbimide.

Carbamic Acid, $HO^-CO^-NH_2$, is not known free, but its calcium salt occurs in horses' urine and in the urine of man and dogs when large amounts of calcium salts are partaken of. Ammonium carbamate occurs in commercial ammonium carbonate (p. 218) and is formed by the union of dry ammonia with carbon dioxide, $2NH_3 + CO_2 = NH_2{}^-CO^-ONH_4$. It forms a white crystalline mass which in the presence of water is transformed into ammonium carbonate:

$$H_2N^-CO^-ONH_4 + H_2O = H_4NO^-CO^-ONH_4.$$

Carbamic acid ethylester, $H_2N^-CO^-OC_2H_5$, urethan, forms colorless crystals which are readily soluble.

Carbamide, urea, CN_2H_4O or $H_2N^-CO^-NH_2$.

Occurrence. It is the chief constituent of human urine (2-3 per cent.) and the urine of mammalia, birds, and amphibians. It occurs to a less extent in the blood, liver, kidneys, lymph, etc., and in uræmia it occurs to a very great extent in all human tissues and fluids.

Formation. 1. Like all amides, by heating the corresponding ammonium salt; e.g., from ammonium carbonate:

$$CO{<}^{ONH_4}_{ONH_4} = CO{<}^{NH_2}_{NH_2} + 2H_2O;$$

also by heating ammonium carbamate:

$$CO{<}^{NH_2}_{ONH_4} = CO{<}^{NH_2}_{NH_2} + H_2O.$$

In both cases this takes place by heating to about 130° in sealed vessels.

2. From carbonyl chloride and ammonia:

$$COCl_2 + 2NH_3 = CO(NH_2)_2 + 2HCl.$$

3. Urea is also formed as a cleavage product of guanine, xanthine, creatine, and of uric acid.

Preparation. 1. On evaporating an aqueous solution of ammonium cyanate, when a change in the atomic arrangement takes place:

$$NCO(NH_4) = NH_2\text{-}CO\text{-}NH_2.$$

If potassium ferrocyanide is fused with potassium carbonate and minium or manganese peroxide gradually added, we obtain potassium cyanate, which is dissolved in water and treated with ammonium sulphate, forming ammonium cyanate. This is evaporated to dryness, and we obtain a mixture of potassium sulphate and urea, which last can be extracted by alcohol:

$$
\begin{aligned}
K_4FeC_6N_6 + K_2CO_3 &= 6KCN + FeCO_3; \\
6KCN + 6Pb_3O_4 &= 6NCOK + 18PbO; \\
2NCOK + (NH_4)_2SO_4 &= K_2SO_4 + 2NCO(NH_4).
\end{aligned}
$$

2. Urine is evaporated to a sirupy consistency and then treated with nitric acid, when difficultly soluble urea nitrate (P. 411) crystallizes out. This is treated with barium carbonate, which forms barium nitrate and free urea, and the latter is then extracted from the evaporated residue by alcohol.

Properties. It forms neutral crystalline needles which have a taste similar to that of saltpeter, are readily soluble in water and alcohol, and melt at 132° C. On heating higher it decomposes into biuret, $C_2H_5N_3O_2$, and ammonia:

$$2NH_2\text{-}CO\text{-}NH_2 = NH_2\text{-}CO\text{-}NH\text{-}CO\text{-}NH_2 + NH_3;$$

and if this fused mass is dissolved in water and treated with caustic alkali and a few drops dilute copper sulphate solution, we obtain a violet coloration (Biuret reaction).

On heating with water above 100° or by boiling with acids or alkalies it decomposes into carbon dioxide and ammonia, at the same time taking up water:

$$CO(NH_2)_2 + H_2O = CO_2 + 2NH_3.$$

The same decomposition takes place quickly in the urine at ordinary temperatures by means of certain micro-organisms.

Like all amides, urea is decomposed by nitrous acid (p. 368, 4 and p. 369, 3):

$$CO(NH_2)_2 + 2HNO_2 = CO_2 + 3H_2O + 4N.$$

Urea suffers the same decomposition by chlorine and bromine in the presence of caustic alkalies (sodium hypobromite or chloride of lime):

$$CO(NH_2)_2 + 3NaBrO = CO_2 + 2H_2O + 2N + 3NaBr.$$

If the alkali is in excess, the carbon dioxide is absorbed and only nitrogen is evolved, and the quantity of urea decomposed can be calculated from the volume of the gas (Knop-Hüfner method).

Urea is a monoacid base which combines directly with acids, bases, and salts. Urea nitrate, $CO(NH_2)_2 \cdot HNO_3$, is readily soluble in water, but nearly insoluble in nitric acid. Urea oxalate, $(CO \cdot N_2H_4)_2 \cdot C_2H_2O_4 + 2H_2O$, is very slightly soluble in cold water. (Urea is qualitatively detected in concentrated aqueous solution by nitric or oxalic acid, the precipitates of both salts exhibiting a characteristic crystalline form under the microscope.)

The most important compound of urea is that with mercuric nitrate, as it serves in the quantitative estimation (Liebig-Pflüger method).

If a solution containing up to 4 per cent. urea be treated with dilute mercuric nitrate solution, we obtain the compound $Hg(NO_3)_2 + 2CO(NH_2)_2 + 3HgO$. If after every addition of the mercuric nitrate we treat a few drops of the solution with a soda solution, we obtain, as soon as all the urea is precipitated and the slightest excess of mercuric nitrate exists, a red precipitate of mercuric oxide, because the white precipitate of mercuric urea does not change its color by soda. If a mercuric nitrate solution containing a known amount of Hg is treated with a urea solution until a drop taken out does not turn yellowish red with soda, we can calculate the quantity of urea from the volume of the mercuric nitrate solution used.

Sulphocarbamide, thio-urea, $CS(NH_2)_2$. As ammonium cyanate is transformed into its isomer carbamide by evaporation, so also is ammonium sulphocyanate (p. 393) converted into its isomer sulphocarbamide on warming to 170° C.:

$$OCN(NH_4) = NH_2-CO-NH_2; \qquad SCN(NH_4) = NH_2-CS-NH_2.$$

It forms colorless needles which are soluble in water and alcohol and which, like urea, combine directly with acids, forming compound sulphoureas, which can be prepared from CS_2 and primary amines:

$$CS_2 + 2NH_2 \cdot CH_3 = CS(NH \cdot CH_3)_2 + H_2S.$$

Allylsulphocarbamide, allyl thio-urea, thiosinamine, $NH_2-CS-NH(C_3H_5)$, forms colorless prisms having a leek-like odor, and is prepared from mustard-oil by the action of ammonia.

Guanidine, CH_5N_3 or $NH_2-C(NH)-NH_2$, imido-carbamide, is derived from urea in which the $=O$ is replaced by the divalent imide group $=NH$.

Guanidine is prepared by the oxidation of guanine (p. 417, hence its name). It may be synthetically prepared by heating cyanamide with ammonium chloride:

$$C \lessgtr^N_{NH_2} + NH_4Cl = C(NH) \lessgtr^{NH_2}_{NH_2} + HCl.$$

It forms colorless crystals which are soluble in water and alcohol and is a strong base, combining directly with acids:

$(CH_5N_3) \cdot HNO_3.$
Guanidine nitrate

$(CH_5N_3)_2 \cdot H_2CO_3.$
Guanidine carbonate.

Like urea (p. 413), guanidine also forms acid and alkyl derivatives:

$$C(NH) <^{NH-CO}_{NH-CH_3}>$$
Glycolyl guanidine.

$$C(NH) <^{NH_2}_{NH(CH_3)}$$
Methyl guanidine.

Methyl guanidine, $C_2H_7N_3$ (structure above), belongs to the ptomaines and occurs in the cultures of the cholera bacillus as well as in putrefying meat, and forms poisonous deliquescent crystals.

Arginin, $C_6H_{14}N_4O_2$ or $C(NH) <^{NH_2}_{NH \cdot C_5H_8(NH_2)O_2}$, guanidine diamido-valerianic acid, is produced on the cleavage of proteïds by acids, and decomposes on boiling with baryta-water into urea and ornithin (diamido-valerianic acid).

Creatine, $C_4H_9N_3O_2$ or $C(NH) <^{NH_2}_{N(CH_3)(CH_2 \cdot COOH)}$, methyl guanidine acetic acid. *Occurrence.* Chiefly in the muscle plasma of vertebrate animals and to a less extent in blood, brain, amniotic fluid, and urine.

Preparation. 1. By extracting meat with cold water, boiling the extract in order to coagulate the proteïd, precipitating the phosphates with baryta-water, and evaporating the filtrate until crystallization begins.

2. Creatine can be prepared artificially by heating sarcosine (methyl-amidoacetic acid) with cyanamide:

$$<^{NH(CH_3)}_{CH_2 \cdot COOH} + C \lessgtr^N_{NH_2} = C(NH) <^{NH_2}_{N(CH_3)(CH_2 \cdot COOH)}.$$

Properties. It crystallizes with 1 mol. H_2O in rhombic, neutral, colorless columns which are soluble in 74 parts water and combine with acids and with salts. On boiling with baryta-water it decomposes into sarcosine, $C_3H_7NO_2$, and urea. $C_4H_9N_3O_2 + H_2O = CO(NH_2)_2 + C_3H_7NO_2$. On heating with dilute acids water is split-off and creatinine is obtained.

Creatinine, $C_4H_7N_3O$ or $C(NH) \lessgtr^{NH——CO}_{N(CH_3)CH_2}>$, glycolylmethyl guanidine. *Occurrence.* It occurs in the urine of man and other mammalia and to a very slight extent in ox blood and in milk.

Preparation. 1. Ordinarily prepared by heating creatine with dilute mineral acids.

2. From the mother-liquor of creatine (p. 412, 7).

Properties. Creatinine occurs as colorless, neutral prisms, which are soluble in 11 parts water; this solution reduces Fehling's solution. It combines with acids, salts, and bases, especially on warming, and is thereby converted into creatine. Its most important compound is $(C_4H_7N_3O)_2 \cdot ZnCl_2$, which is precipitated from creatinine solutions by zinc chloride as a difficultly soluble crystalline powder (estimation in urine). Traces of creatinine can be detected in solution (in urine) by the ruby-red coloration produced by a dilute solution of sodium nitroprusside followed by the addition of caustic alkali.

c. Ureïdes and Diureïdes.

The hydrogen atoms of the amido groups in urea can be replaced by alcohol or acid radicals. The alcohol derivatives are called *compound ureas*, a great number of which are known—for example, $(CH_3)HN-CO-NH(CH_3)$—and behave like urea; the acid derivatives with one molecule urea are called *ureïdes* and those with two molecules urea, *diureïdes* or *purines*. Most of the ureides and diureïdes behave like acids although no carboxyl group is present. The acid character is brought about by the carbonyl groups present, which so modify the basic properties of the imid group, NH, that its hydrogen can be replaced by metals. As the ureïdes correspond to the amides, so also by the introduction of acid residues containing carboxyl groups we may produce *ur-acids* corresponding to the amido acids.

Allophanic acid. carboxyl urea, is only known in the form of an ester and forms difficultly soluble cyrstals:

$$CO < \begin{matrix} NH_2 \\ NH-COOH \end{matrix}$$

Biuret, the amide of allophanic acid, is produced on heating urea to 150°–170° C., and forms white needles which are readily soluble:

$$CO < \begin{matrix} NH_2 \\ NH \cdot CO \cdot NH_2 \end{matrix}$$

Allanturic acid, glyoxyl urea, $(-CH(OH)-CO-$, glyoxyl), is obtained on the oxidation of hydantoin and allantoin (p. 414). It is also prepared from uroxanic acid, $C_5H_3N_4O_6$

$$CO < \begin{matrix} NH-CO \\ | \\ NH-CH(OH). \end{matrix}$$

Hydantoin, glycolyl urea, (–CO–CH$_2$–, glycolyl), is obtained by the action of HI upon allantoin or alloxanic acid:

$$CO < ^{NH-CO}_{NH-CH_2} >.$$

Hydantoic acid, glycoluric acid, is obtained from hydantoin by boiling with Ba(OH)$_2$, and forms colorless prisms:

$$CO < ^{NH_2}_{NH-CH_2COOH.}$$

Parabanic acid, oxalyl urea, (–CO–CO–, oxalyl), is produced in the energetic oxidation of uric acid and alloxan:

$$CO < ^{NH-CO}_{NH-CO} >.$$

Dimethylparabanic acid, cholestrophan, is prepared synthetically from parabanic acid or from theine by treating with HNO$_3$:

$$CO < ^{N(CH_3)-CO}_{N(CH_3)-CO} >.$$

Oxaluric acid is obtained by carefully heating parabanic acid with dilute alkalies and its ammonium salt occurs as traces in urine:

$$CO < ^{NH_2}_{NH-CO \cdot COOH.}$$

Alloxan, mesoxalyl urea, is obtained by the careful oxidation of uric acid (p. 419), and forms colorless prisms which color the skin red:

$$CO < ^{NH \cdot CO}_{NH \cdot CO} > CO.$$

Alloxanic acid, mesoxaluric acid, is produced by the action of dilute alkalies upon alloxan, and forms white crystals which are readily soluble:

$$CO < ^{NH-CO-CO}_{NH_2 \qquad COOH.}$$

Dialuric acid, tatronyl urea, is prepared from alloxan by the action of nascent hydrogen. It is oxidized in the air into alloxanthin:

$$CO < ^{NH \cdot CO}_{NH \cdot CO} > CH(OH).$$

Barbituric acid, malonyl urea, is produced by the reduction of dialuric acid; also by heating urea with malonic acid. It forms colorless prisms:

$$CO < ^{NH-CO}_{NH-CO} > CH_2.$$

Allantoin, C$_4$H$_6$N$_4$O$_3$, the diureide of the hydrate of glyoxalic acid (p. 402), is prepared from uric acid by oxidizing with KMnO$_4$. It occurs in the urine of new-born infants, of pregnant women, of calves, in the allantoic fluid of cows, in the horse-chestnut and maple, and forms shining crystals which decompose into urea and allanturic acid (p. 413) by the action of alkalies·

$$CO\big<{\substack{NH-CH-NH \\ | \\ NH-CO-NH_2}}\big>CO.$$

Alloxanthine, $C_8H_4N_4O_7$, the diureide of tatronylic and mesoxalylic acids. It is produced from alloxan by reduction with $SnCl_2$. Its tetramethyl derivative is called amalic acid, and is produced from caffeine (p. 418) by the action of chlorine water:

$$\substack{CO<{\substack{NH-CO \\ NH-CO}}>C \\ CO<{\substack{NH-CO \\ NH-CO}}>C} O.$$

Purpuric acid, imido alloxanthine, $C_8H_4N_4O_6(NH)$, does not occur free; its ammonium salt is called murexid and is used in the detection of uric acid (see 419).

d. Purin Derivatives.

Hypoxanthine, $C_5H_4N_4O$, **xanthine,** $C_5H_4N_4O_2$, and the diureïde uric acid, $C_5H_4N_4O_3$, contain the tetravalent acid radical trioxyacryl, $-OC-\overset{|}{C}=\overset{|}{C}-$, of the hypothetical trioxyacrylic acid, $HOOC-C(OH)_2=C(OH)_2$; these bodies and their methyl and amido derivatives may be derived better from *purin*, $C_5H_4N_4$, than from urea. Hypoxanthine, xanthine, and their derivatives, adenine and guanin, are also called *xanthine* or *nuclein bases*, as they are cleavage products of the nucleoproteids.

Purin, $C_5H_4N_4$ or $\substack{1\ 6\ \ \ 5\ 7 \\ 2 \diagup N=CH-C-NH \diagdown 8 \\ HC< \qquad\ \| \qquad >CH. \\ \diagdown N----C--N \diagup \\ 3\quad\ \ \ 4\quad 9}$

This is a diazindiazole (see Heterocyclic Compounds). The figures given above in the formula serve to designate the constitution of the derivatives. Purin is obtained from uric acid by transforming it into trichlorpurin (p. 419) and replacing the chlorine of this by nascent hydrogen. It forms colorless crystals which are readily soluble and melt at 216°, forming salts with acids and bases. Chlorinated purins serve in·the syntheses of its derivatives.

Adenine, $C_5H_3N_4(NH_2)$, 6-amidopurin, occurs in the nucleins, in the spleen, pancreas, tea, etc. It forms colorless needles with 3 mol. H_2O:

$$\substack{\diagup N=C(NH_2)-C-NH \diagdown \\ HC< \qquad\ \| \qquad >CH. \\ \diagdown N————C-N \diagup}$$

Hypoxanthine, sarcine, $C_5H_4N_4O$, 6-oxypurin (see p. 417):

$$\substack{\diagup NH-CO-C-NH \diagdown \\ HC< \qquad\ \| \qquad >CH. \\ \diagdown N————C—N \diagup}$$

Guanine, $C_5H_3N_4O(NH_2)$, amido hypoxanthine, 2-amido-6-oxypurin (see p. 417):

$$(H_2N)C\begin{array}{c}NH\!-\!CO\!-\!C\!-\!NH\\ \diagdown N\quad\quad\overset{\|}{\underset{C\quad\ \ N}{}}\end{array}\!\!\!CH.$$

Xanthine, $C_5H_4N_4O_2$, 2,6-dioxypurin (see p. 417):

$$OC\begin{array}{c}NH\!-\!CO\!-\!C\!-\!NH\\ \diagdown NH\quad\quad\overset{\|}{\underset{C\quad\ \ N}{}}\end{array}\!\!\!CH.$$

Heteroxanthine, $C_5H_3(CH_3)N_4O_2$, 7-methyl xanthine, occurs as traces in urine and is a white amorphous or crystalline powder:

$$OC\begin{array}{c}NH\!-\!CO\!-\!C\!-\!N(CH_3)\\ \diagdown NH\quad\quad\overset{\|}{\underset{C\quad\quad N}{}}\end{array}\!\!\!CH.$$

Paraxanthine, 1,7-dimethyl xanthine, $C_5H_2(CH_3)_2N_4O_2$, occurs as traces in urine, and forms six-sided colorless crystals:

$$OC\begin{array}{c}N(CH_3)\!-\!CO\!-\!C\!-\!N(CH_3)\\ \diagdown NH\quad\quad\overset{\|}{\underset{C\quad\quad N}{}}\end{array}\!\!\!CH.$$

Theophylline, 1,3-dimethyl xanthine, $C_5H_2(CH_3)_2N_4O_2$, occurs in tea-leaves and forms colorless crystals:

$$OC\begin{array}{c}N(CH_3)\!-\!CO\!-\!C\!-\!NH\\ \diagdown N(CH_3)\quad\quad\overset{\|}{\underset{C\quad\ N}{}}\end{array}\!\!\!CH.$$

Theobromine, 3,7-dimethyl xanthine, $C_5H_2(CH_3)_2N_4O_2$ (see p. 418)

$$OC\begin{array}{c}NH\!-\!CO\!-\!C\!-\!N(CH_3)\\ \diagdown N(CH_3)\!-\!\overset{\|}{\underset{C\quad\quad N}{}}\end{array}\!\!\!CH.$$

Caffeine, 1,3,7-trimethyl xanthine, $C_5H(CH_3)_3N_4O_2$, methyl theobromine (p. 418):

$$OC\begin{array}{c}N(CH_3)\!-\!CO\!-\!C\!-\!N(CH_3)\\ \diagdown N(CH_3)\quad\quad\overset{\|}{\underset{C\quad\quad N}{}}\end{array}\!\!\!CH.$$

Uric acid, 2,6,8-trioxypurin, $C_5H_4N_4O_3$ (see p. 418):

$$OC\begin{array}{c}NH\!-\!CO\!-\!C\!-\!NH\\ \diagdown NH\quad\quad\overset{\|}{\underset{C\!-\!NH}{}}\end{array}\!\!\!CO.$$

Carnine, 1,3-dimethyl uric acid, $C_5H_2(CH_3)_2N_4O_3$, occurs in muscle plasma (hence also in meat extracts), in yeast, and the juice of the sugar-beet. It forms colorless microscopic crystalline masses.

Hypoxanthine, sarcine, $C_5H_4N_4O$ (structure, p. 415). *Occurrence.* Nearly always associated with xanthine in the nucleins; in flesh, especially that of horses or oxen; in many animal tissues and fluids, as, for instance, in the spleen, liver, kidneys, brain, pancreas, urine; also in many plants, such as the pumpkin, malt, and lupin sprouts, tea, sugar-beets, young potatoes, etc.

Preparation. 1. By the action of nitrous acid upon adenine:

$$C_5H_5N_5 + HNO_2 = C_5H_4N_4O + H_2O + N_2.$$

2. Ordinarily from meat extracts, whose aqueous solution is precipitated by silver nitrate and the precipitate dissolved in dilute hot nitric acid and then allowed to stand, when hypoxanthine, guanine, and adenine silver separate out, while xanthine silver remains in solution.

Properties. White crystalline powder which is soluble with difficulty and which combines with acids and bases.

Guanine, amidohypoxanthine, $C_5H_3N_4O(NH_2)$ (p. 416), is found in the nucleins, in the excrement of snails, cephalopods, and spiders, in the pancreas, spleen, liver, lungs, in the young sprouts of various plants, and to a great extent in guano.

Preparation. 1. From Peruvian guano accompanied with uric acid (see p. 418).

2. From meat extracts (see Hypoxanthine).

Properties. It forms colorless crystals or a white amorphous powder which is insoluble in water, alcohol, or ether, and which combines with acids as well as with bases. On oxidation it decomposes into guanidine (CH_5N_3) and parabanic acid (oxalyl urea, $C_3H_2N_2O_3$):

$$C_5H_5N_5O + 3O + H_2O = CH_5N_3 + C_3H_2N_2O_3 + CO_2.$$

Nitrous acid converts it into xanthine:

$$C_5H_5N_5O + HNO_2 = C_5H_4N_4O_2 + 2N + H_2O.$$

Xanthine, $C_5H_4N_4O_2$ (structure, p. 416). *Occurrence.* Seldom alone, but generally associated with hypoxanthine in urinary calculi.

Preparation. 1. Ordinarily from meat extracts (see Hypoxanthine).

2. By the action of nitrous acid upon guanine, as well as upon uric acid (p. 419).

3. From uric acid by reduction, still only indirectly (p. 419).

Properties. White microcrystalline powder which combines with acids and alkalies and which is readily transformed into theobromine and caffeine (which see). When evaporated with chlorine-water xanthine gives a yellowish-red residue which becomes purple-red with ammonia (Weidel's reaction).

Theobromine, dimethylxanthine, $C_5H_2(CH_3)_2N_4O_2$ (structure, p. 416).
Occurrence. In the cacao-bean and as traces in the urine.
Preparation. 1. From the cacao-bean. 2. From uric acid (p. 119).
3. By precipitating an alkaline xanthine solution by lead acetate and heating the precipitate of xanthine lead with methyl iodide:

$$C_5H_2PbN_4O_2 + 2CH_3I = C_5H_2(CH_3)_2N_4O_2 + PbI_2.$$

Properties. White crystalline powder which dissolves with difficulty in water and alcohol and sublimes at 300° C. Theobromine has weak basic properties and when evaporated with chlorine-water it acts like xanthine. Its salts are decomposed by water into the free acid and theobromine.

Theobromine-sodium-salicylate, $C_6H_4(OH)^-COONa + C_7H_7NaN_4O_2$, diuretin, is a white crystalline powder.

Caffeine, guaranine, theine, trimethyl xanthine, $C_8H_{10}N_4O_2$ or $C_5H(CH_3)_3N_4O_2$. *Occurrence.* In coffee, tea, Paraguay tea, in guarana paste, in the cola-nut, and with theobromine in certain kinds of cacao.

Preparation. 1. From coffee. 2. From uric acid (p. 419). 3. If theobromine is dissolved in ammonia and silver nitrate added, a precipitate of theobromine silver, $C_7H_7AgN_4O_2$, is obtained, which when heated with methyl iodide yields caffeine: $C_7H_7AgN_4O_2 + CH_3I = C_7H_7(CH_3)N_4O_2 + AgI$.

Properties. It forms with 1 mol. H_2O colorless, neutral, bitter, shining needles which dissolve with difficulty in water and alcohol. It begins to sublime at 130°, melts at 230.5°, and behaves like xanthine when evaporated with chlorine-water. It possesses weak basic properties, and its salts, like those of theobromine, are readily decomposed by water into acid and base.

Caffeine-sodium-salicylate, $C_8H_{10}N_4O_2 + 2C_6H_4(OH)COONa$, forms a white crystalline powder which is soluble in water.

Uric Acid, $C_5H_4N_4O_3$ (structure, p. 416). *Occurrence.* Abundantly in the pasty urine of birds (hence also in guano), of reptiles, and of invertebrate animals; also in the gout nodules and many calculi; to a less extent in the urine of carnivora, and only as traces in the urine of herbivora. It also occurs pathologically in the blood.

Preparation. 1. Ordinarily by boiling the excrement of snakes or guano, which consists of ammonium and sodium urate, with caustic alkali until all the ammonia has been driven off. After filtering, this solution of alkali urate is poured into dilute HCl, when the uric acid precipitates. Guanine can be obtained from the filtrate by saturating it with NH_3.

2. If partially concentrated urine is treated with one-tenth volume HCl and allowed to stand in a cold place for about 48 hours, all the uric acid will be precipitated out.

Formation. 1. By heating urea with glycocoll to 200° C.:

$$C_2H_3(NH_2)O_2 + 3CO(NH_2)_2 = C_5H_4N_4O_3 + 3NH_3 + 2H_2O.$$

2. By heating trichlorlactic acid amide with urea:

$$CCl_3-CH(OH)CO(NH_2) + 2CO(NH_2)_2 = C_5H_4N_4O_3 + 3HCl + NH_3 + H_2O.$$

3. Malonyl urea (P. 414) yields the isonitroso compound, violuric acid, with HNO_2, and this on reduction gives amido-malonyl urea, amidobarbituric acid, or uramil:

$$CO<^{NH\ CO}_{NH\cdot CO}>C=NOH+4H=H_2O+CO<^{NH\cdot CO}_{NH\cdot CO}>CH-NH_2.$$

Violuric acid.   Uramil

On fusing uramil with potassium cyanate the potassium salt of pseudo-uric acid is obtained, and this differs from uric acid only in that it contains 1 mol. of H_2O more. This can be split off by heating with dilute mineral acids:

$$CO<^{NH\cdot CO}_{NH\cdot CO}>CH-NH-CO-NH_2=CO<^{NH-CO-C-NH}_{NH———C-NH}>CO+H_2O.$$

Pseudo-uric acid.   Uric acid.

Properties. Uric acid forms small white scales, and when they form slowly whetstone-shaped crystals are obtained. It is without odor or taste, nearly insoluble in water, alcohol, ether, and acids. It is a weak bibasic acid and forms chiefly primary salts called urates, most of which are soluble in water with difficulty. The most soluble of the uric acid compounds are the combinations with piperazine, lithium, and utropin (p. 398).

On heating uric acid it decomposes into carbon dioxide, ammonia, urea, cyanuric acid, and on careful oxidation (with cold nitric acid) it yields urea and mesoxalyl urea (alloxan, p. 414):

$$C_5H_4N_4O_3+O+H_2O=CO<^{NH\cdot CO}_{NH:CO}>CO+CO<^{NH_2}_{NH_2}.$$

On further oxidation (warming with nitric acid) the alloxan decomposes into oxalyl urea (parabanic acid, p. 414), from which we infer that uric acid contains the $C<^{N-C}_{N-C}>C$ group. Uric acid is oxidized into the diureïd allantoin (p. 414) by $KMnO_4$: $C_5H_4N_4O_3+H_2O+O=C_4H_6N_4O_3+CO_2$, which shows that uric acid contains two urea residues.

On the action of halogen alkyls upon a watery solution of alkali urates we obtain the alkyl derivatives of uric acid. If tetramethyl uric acid thus obtained be treated with $POCl_3$, we obtain chlorcaffeine, which yields caffeine with nascent hydrogen:

$$3C_5N_4O_3(CH_3)_4+POCl_3=3C_5N_4O_2(CH_3)_3Cl+PO(OCH_3)_3;$$

and trimethyl uric acid in the same manner yields theobromine.

By the action of $POCl_3$ upon uric acid we obtain 2.6.8-trichlorpurin, which with nascent hydrogen is changed into purin, hence the purin structure of uric acid.

The reduction of uric acid into xanthine can only be performed as follows: Sodium ethylate converts 2.6.8-trichlorpurin into 2.6-dioxyethyl-8-chlorpurin, which on heating with HI replaces the ethyl groups, and the chlorine with hydrogen, so that 2.6-dioxypurin=xanthine is the result.

Detection. On evaporating uric acid with nitric acid a yellowish-red residue is obtained which turns purple-red on moistening with ammonia, and beautiful blue on the subsequent addition of caustic alkali (murexid test, p. 415).

9. Oxalic Acid Series.

General formula $C_NH_{2N-2}O_4$.

Oxalic acid	$C_2H_2O_4$	Suberic acid	$C_8H_{14}O_4$
Malonic acid	$C_3H_4O_4$	Azelaic acid	$C_9H_{16}O_4$
Succinic acid	$C_4H_6O_4$	Sebacic acid	$C_{10}H_{18}O_4$
Pyrotartaric acid	$C_5H_8O_4$	Brassylic acid	$C_{11}H_{20}O_4$
Adipic acid	$C_6H_{10}O_4$	Roccelic acid	$C_{17}H_{32}O_4$
Pimelic acid	$C_7H_{12}O_4$	etc.	

Properties. The dihydric, bibasic acids of the glycols are called the acids of the oxalic acid series. They are crystalline solids, not volatile without decomposition, and generally soluble in water. As they contain two hydroxyl groups they are diatomic, and as they contain two carboxyl groups they are bibasic and form neutral and acid salts as well as esters:

$$HOOC-COO(C_2H_5)$$
Primary or acid ester.

$$(C_2H_5)OOC-COO(C_2H)_5$$
Secondary or neutral ester.

The amide derivatives are similar in properties and method of formation to the amides of the monobasic acids. On account of the presence of the two carboxyl groups we have besides the real amides (the diamides) also acid amides or amine acids:

$$(H_2N)OC-CO(NH_2)$$
Oxamide.

$$(H_2N)OC-COOH$$
Oxamic acid.

On replacing two H atoms in a molecule of ammonia by a divalent acid radical we obtain imides, $C_2H_4{<}^{CO}_{CO}{>}NH$, succinic acid imide.

These derivatives may also be prepared from the acid and neutral ammonium salts by removing water:
Acid ammonium salts minus $H_2O = $ *amic acids:*

$$HOOC-COO-(NH_4) = HO\overset{\bullet}{O}C-CO-NH_2 + H_2O.$$
Acid ammonium oxalate.   Oxamic acid.

Acid ammonium salts minus $2H_2O = $ *imides:*

$$C_2H_4{<}^{COONH_4}_{COOH} = C_2H_4{<}^{CO}_{CO}{>}NH + 2H_2O.$$
Acid ammonium succinate.   Succinimide.

Neutral ammonium salts minus $2H_2O = amides:$

$$H_4N-OOC-COO-NH_4 = NH_2-OC-CO-NH_2 + 2H_2O.$$
Ammonium oxalate.   Oxamide.

Neutral ammonium salts minus $4H_2O = nitriles:$

$$H_4N-OOC-COO-NH_4 = NC-CN + 4H_2O \text{ (p. 423)}.$$

Anhydrides may be formed similarly to the oxyfatty acids even with one molecule (see Succinic Acid), when the carboxyls occur united to different carbon atoms. If the carboxyls are united to one C atom, then on heating fatty acids are obtained and at the same time CO_2 is split off: $HOOC^-CH_2^-COOH = CH_3^-COOH + CO_2$.

With the exception of oxalic acid all the acids of this series may be considered as alkylene dicarbonic acids, which coincides with the methods of formation and with their decomposition.

If the potassium salts are decomposed by the electric current we obtain alkylenes, carbon dioxide, and potassium:

$$C_2H_4 {<}^{COOK}_{COOK} = C_2H_4 + 2CO_2 + 2K.$$

When heated with alkali hydroxides they decompose into ethanes and carbon dioxide; thus suberic acid yields hexane:

$$C_6H_{12} {<}^{COOH}_{COOH} = C_6H_{14} + 2CO_2.$$

The possible isomers of the dibasic acids correspond to the isomers of the alkylenes. The two carboxyl groups may be united to two different atoms or to one carbon atom. No isomers are possible of the first two acids—oxalic acid, $HOOC-COOH$, and malonic acid, $HOOC-CH_2-COOH$—while two are possible of the third acid:

$$HOOC-CH_2-CH_2-COOH \quad \text{and} \quad CH_3-CH {<}^{COOH}_{COOH}.$$
Ethylene succinic acid.   Ethylidene succinic acid.

There are four isomers possible of the fourth member, $C_3H_6(COOH)_2$, etc. The isomers, which have both carboxyl groups united to one C atom, may be said to be derived from malonic acid (p. 423) and their structure is called the malonic acid structure.

We call the divalent acid residues united with the two hydroxyls the acid radicals (p. 336):

$$ {<}^{CO-}_{CO-} \text{ oxalyl}, \quad CH_2 {<}^{CO-}_{CO-} \text{ malonyl}, \quad C_2H_4 {<}^{CO}_{CO} \text{ succinyl}.$$

Preparation. 1. If the halogen alkylenes are heated with potassium cyanide we obtain alkylene cyanides, which yield the corresponding acid on boiling with acids or caustic alkalies (p. 346, 2):

$$CN-CH_2-CH_2-CN + 4H_2O = HOOC-CH_2-CH_2-COOH + 2NH_3;$$
Ethylene cyanide.   Ethylene succinic acid.

$$CN-CH(CN)-CH_3 + 4H_2O = HOOC-CH(COOH)-CH_3 + 2NH_3.$$
Ethylidene cyanide.   Ethylidene succinic acid.

2. The monohalogen fatty acids are transformed into the cyanfatty acids and the cyanogen group is converted into the carboxyl group by boiling with acids or caustic alkalies:

$$HOOC-CH_2-CH_2-CN + 2H_2O = HOOC-CH_2-CH_2-COOH + NH_3;$$
β-Cyanpropionic acid.   Ethylene succinic acid.

$$HOOC-CH(CN)-CH_3 + 2H_2O = HOOC-CH(COOH)-CH_3 + NH_3.$$
a-Cyanpropionic acid.   Ethylidene succinic acid.

3. By the oxidation of the acids of the fatty acid and oleic acid series as well as of neutral fats by means of HNO_3.

4. By the oxidation of the normal glycols and oxyfatty acids (mentioned on pp. 400, 402).

5. The homologues of oxalic acid are produced from malonic acid ethyl-ester in a similar manner to acetic acid from aceto-acetic ester.

Oxalic Acid, $C_2H_2O_4$ **or** $HOOC-COOH$. *Occurrence.* It exists combined as acid potassium salt in many plants, such as in the varieties of Runex and Oxalis, from whose juice oxalic acid used to be obtaind by evaporation. It also occurs in many plants as calcium oxalate, partly in solution and partly as crystals. Oxalic acid also occurs as calcium salt in the animal kingdom, especially in normal and pathological urine, as urinary sediment, in the caterpillar excrement, as vesicle calculi (mulberry calculus), and as potassium oxalate in many organs.

Formation. 1. By passing CO_2 over fused sodium we obtain sodium oxalate: $2CO_2 + 2Na = C_2Na_2O_4$.

2. From fats, carbohydrates (sugar, starch, gums, cellulose), and many other organic bodies by oxidation with nitric acid. On fusing these bodies with caustic alkalies we obtain alkali oxalates.

Preparation. 1. By fusing sawdust (cellulose, $C_6H_{10}O_5$) with alkali hydroxide at 250° to 300° (p. 323):

$$C_6H_{10}O_5 + H_2O + 6NaOH = 3C_2Na_2O_4 + 18H.$$

2. By heating alkali formates: $2H-COONa = NaOOC-COONa + 2H$.

The alkali oxalate thus obtained is dissolved in water and the solution mixed with calcium hydroxide (milk of lime) and the insoluble calcium oxalate which separates out treated with sulphuric acid, when insoluble calcium sulphate precipitates and a solution of oxalic acid is obtained.

Properties. On evaporating an aqueous solution fine mono-clinic crystals, $C_2H_2O_4 + 2H_2O$, separate out. On heating to 100° C. or drying over sulphuric acid we obtain anhydrous oxalic acid, which appears as a white powder, and on carefully heating to 150° C. sublimes and on rapidly heating decomposes into carbon dioxide and formic acid (p. 352): $C_2H_2O_4 = CH_2O_2 + CO_2$. When heated with caustic alkalies it splits into alkali carbonate and hydrogen: $C_2K_2O_4 + 2KOH = 2K_2CO_3 + 2H$, and when heated with concentrated sulphuric acid it decomposes into water, carbon monoxide, and carbon dioxide: $C_2H_2O_4 = CO + CO_2 + H_2O$.

The salts of oxalic acid, the *oxalates*, are, with the exception of the alkali salts, insoluble in water or soluble with difficulty.

Potassium bioxalate, $C_2HKO_4 + H_2O$.

Potassium quadroxalate, $C_2HKO_4 + C_2H_2O_4 + 2H_2O$, salt of sorrel, forms colorless crystals soluble in water.

Ammonium oxalate, $C_2(NH_4)_2O_4$, forms colorless rhombic crystals which are readily soluble in water. It occurs in Peruvian guano, and splits into oxamide and water on being heated:

$$(H_4N)OOC-COO(NH_4) = (H_2N)-OC-\overset{..}{C}O(NH_2) + 2H_2O.$$

Oxamide, $C_2(NH_2)_2O_2$, is a white powder insoluble in water. When heated with phosphorus pentoxide it loses 2 mol. H_2O and oxalonitrile is produced (cyanogen gas):

$$(H_2N)OC-CO(NH_2) = NC-CN + 2H_2O.$$

The reverse is obtained when an aqueous solution of cyanogen is allowed to stand in the presence of mineral acids (p. 383).

Ammonium acid oxalate, $C_2H(NH_4)O_4$, consists of colorless crystals which are difficultly soluble in water and yield oxamic acid and water on being heated:

$$HOOC-COO(NH_4) = HOOC-CO(NH_2) + H_2O.$$

Oxamic acid, $C_2H(NH_2)O_3$, is a white insoluble powder.

Calcium oxalate, $C_2CaO_4 + H_2O$ (p. 225), is precipitated from a calcium salt solution by oxalic acid or an oxalate as a white crystalline powder. It is insoluble in acetic acid and serves in the detection of oxalic acid as well as for the soluble calcium compounds.

Malonic acid, $C_3H_4O_4$ or $HOOC-CH_2-COOH$, is obtained on the oxidation of ethylene lactic acid (p. 407) and malic acid (p. 426). It occurs as a calcium salt in the sugar-beet and melts at 132° C. All compounds which contain two COOH groups attached to the same C atom split off 1 mol. CO_2 on heating. Correspondingly malonic acid decomposes on heating into $CO_2 + CH_3COOH$.

Malonic acid diethylester, $H_5C_2-OOC-CH_2-COO-C_2H_5$, and the other esters of malonic acid are used, like aceto-acetic ester (p. 365), for numerous syntheses as the H atoms of the methylene group are readily replaceable by sodium and the Na then readily replaceable by radicals, so that by this means a large number of higher bibasic acids can be prepared.

As all these acids contain two COOH groups united to the same C atom, therefore on heating, CO_2 is split off (see Malonic Acid); hence the malonic acid ester syntheses are also used in the obtainment of monobasic acids:

$$HOOC-C(CH_3)(C_2H_5)-COOH = HC(CH_3)(C_2H_5)COOH + CO_2.$$

Methyl ethyl malonic acid.   Active valeric acid.

Oxymalonic acid, tartronic acid, $HOOC-CH(OH)-COOH$ or $C_3H_4O_5$, is obtained from brommalonic acid, as well as in the oxidation of glycerine (p. 434) and the tartaric acids (p. 428).

Dioxymalonic acid, $C_3H_4O_6$ or $HOOC-C(OH)_2-COOH$. This acid may be considered as mesoxalic acid hydrate (p. 434).

Normal Succinic Acid, Ethylendicarbonic acid, $C_4H_6O_4$ or $HOOC-C_2H_4-COOH$. *Occurrence.* In amber (succinum, fossil coniferous resin), in resin and turpentine of certain conifers, in many brown coals, in wormwood and poppy. In the animal kingdom it occurs in the thymus gland, spleen, thyroid gland, hydrocele and echinococcus fluids, sometimes also in the blood, saliva, and urine. It is produced to a slight extent in alcoholic fermentation, as well as in the putrefaction of meat.

Preparation. According to the general method described on page 421 as well as by heating malic or tartaric acid with hydriodic acid (p. 427). It is prepared on a large scale by the dry distillation of amber or by the fermentation of calcium malate with putrid cheese at 30–40° C.

Properties. Colorless crystals having a faint acid taste and soluble in water and alcohol. Its salts are called *succinates.* Succinic acid melts at 180° C. and distils at 235°, when it partly splits into water and succinic anhydride, $C_2H_4 {<}^{CO}_{CO}{>}O$. Its vapors cause coughing. Heated with bromine it yields monobromsuccinic acid, $C_4H_5BrO_4$, and dibromsuccinic acid, $C_4H_4Br_2O_4$, used in the synthesis of malic and tartaric acid. All these compounds form colorless needles.

Amidosuccinic acid, or aspartic acid, $HOOC-C_2H_3(NH_2)-COOH$, occurs free in the acid secretion of many sea-snails, and also in beet-root molasses, and is split off from proteins by treatment with dilute sulphuric acid, etc. Nitric acid converts it into malic acid (p. 426). It is obtained from asparagine by boiling with acids or alkalies (see below). It consists of rhombic crystals which are soluble in hot water.

The neutral solution is dextrorotatory at ordinary temperatures, and inactive at 75° C., while at higher temperatures it is lævorotatory. The acid solution is dextrorotatory and the alkaline solution lævorotatory (p. 39).

Asparagine, $HOOC^-C_2H_3(NH_2)^-CO^-(NH_2)$, amido-succinamic acid, the monamide of amidosuccinic acid, occurs in the potato, beets, asparagus, licorice, marsh-mallow, oyster-plant, and in many other plants, especially in the sprouts. It crystallizes with 1 mol. H_2O as colorless prisms from the pressed juice of these plants on evaporation. These crystals are tasteless and lævorotatory. It is readily soluble in water and insoluble in alcohol and ether and forms salts with acids as well as with bases.

On boiling with water, but quicker with acids or alkalies, it splits into aspartic acid:

$$C_2H_3(NH_2) <^{CO(NH_2)}_{COOH} + H_2O = C_2H_3(NH_2) <^{COOH}_{COOH} + NH_3.$$

Nitrous acid converts it into malic acid:

$$C_2H_3(NH_2) <^{CO(NH_2)}_{COOH} + 2HNO_2 = C_2H_3(OH) <^{COOH}_{COOH} + 2H_2O + 4N.$$

Dextro-asparagine, which has a sweet taste, is obtained from the wicken sprouts and inactive asparagine is prepared synthetically.

Isosuccinic acid, $CH_3-CH(COOH)_2$, ethylidene dicarbonic acid, is obtained from the α-cyanpropionic acid (p. 422, 2) and forms colorless crystals, which are more soluble in water than the ordinary succinic acid. It melts at 130° C. and decomposes on further heating into carbon dioxide and propionic acid.

Pyrotartaric acid, $C_5H_8O_4$ or $HOOC-C_3H_6-COOH$. Four isomers are possible and known, according to the structural theory, and of these we must mention:

Common pyrotartaric acid, $CH_3-CH(COOH)-CH_2-COOH$, or methyl succinic acid, is produced on the dry distillation of tartaric acid (p. 428). It melts at 112°, and is inactive and can be split into the two optically active modifications.

Normal pyrotartaric acid, $HOOC-H_2C-CH_2-CH_2-COOH$, or glutaric acid, is prepared synthetically and melts at 97° C.

Amido normal pyrotartaric acid, amido glutaric acid, glutamic acid, $HOOC-CH(NH_2)-CH_2-CH_2-COOH$, occurs with aspartic acid in many plants and in beet-root molasses, and is obtained with other amido acids by treating proteins (which see) with dilute sulphuric acid, etc. It forms rhombic crystals which are dextrorotatory in aqueous solution. The inactive modification is also known and the lævorotatory modification can be obtained from this by the action of molds.

Glutamine, $HOOC-CH(NH_2)-CH_2-CH_2-CO(NH_2)$. Glutamic acid amide occurs with asparagine in many plants and forms colorless needles which are dextrorotatory in acid solution.

10. Malic Acid, Tartaric Acid, and Citric Acid.

These acids, which should be treated of in connection with the trihydric and tetrahydric compounds, stand in very close connection to succinic acid, hence they will be discussed at this place.

Malic acid, $C_4H_6O_5$ or $HOOC-CH(OH)-CH_2-COOH$, oxysuccinic acid.

Occurrence. In the juice of most sour fruits.

Preparation. 1. From unripe apples, grapes, or mountain ash by evaporating the juice from these, filtering and precipitating with lead acetate. The precipitate of lead malate obtained is decomposed by $H_2S : C_4H_4PbO_5 + H_2S = C_4H_6O_5 + PbS$. The PbS is filtered off and the filtrate evaporated to point of crystallization.

Formation. 1. By treating asparagine (p. 425) or aspartic acid (p. 424) with nitrous acid:

$$C_2H_3(NH_2) <^{COOH}_{COOH} + HNO_2 = C_2H_3(OH) <^{COOH}_{COOH} + H_2O + 2N.$$

Aspartic acid.   Malic acid.

2. By boiling bromsuccinic acid with silver hydroxide:

$$C_2H_3Br <^{COOH}_{COOH} + AgOH = C_2H_3(OH) <^{COOH}_{COOH} + AgBr.$$

If malic acid is heated with hydriodic acid (p. 427) succinic acid is obtained; also in the fermentation of calcium malate.
3. On heating tartaric acid with HI (p. 427).

Properties. Malic acid is a trihydric, bibasic (p. 335) acid; forming white needle-shaped, deliquescent crystals having a pleasant acid taste. It is oxidized by chromic acid into malonic acid:

$$C_2H_3(OH) <^{COOH}_{COOH} + 2O = CH_2 <^{COOH}_{COOH} + CO_2 + H_2O.$$

Natural malic acid is lævorotatory, while that obtained from dextro-tartaric acid is dextrorotatory and that from succinic acid derivatives is optically inactive. This inactive modification splits into the dextro and lævo forms (p. 39). The salts of malic acid are called *malates* and are readily soluble in water with the exception of the lead salt.

Ferrous and ferric malate occur to about 30 per cent. in the extract obtained from powdered iron and crushed sour apples.

Fumaric acid and maleïc acid, $HOOC-CH=CH-COOH$ or $C_4H_4O_4$. Malic acid decomposes on heating into these two stereoisomeric, dibasic acids (structure, p. 308).

Fumaric acid occurs free in many plants; thus, in Fumaria officinalis, in Iceland moss, and in certain fungi, especially in champignons. It sublimes at 200° C. without melting and is difficultly soluble in water and is non-poisonous.

Maleïc acid melts at 130° C., is readily soluble in water, and is poisonous. Both these acids are converted into ordinary succinic acid by nascent hydrogen and into malic acid on heating with water; on the electrolysis of their salts acetylene is produced: $C_4H_2K_2O_4 = C_2H_2 + 2CO_2 + 2K$.

Tartaric acid, $C_4H_6O_6$ or $HOOC-CH(OH)-CH(OH)-COOH$, dioxysuccinic acid, is tetrahydric and bibasic. Four tartaric acids of the same structure are known, namely, dextrotartaric acid, lævotartaric acid, inactive racemic acid (which can be split), and inactive mesotartaric acid (which cannot be split) (p. 306), which differ from each other chiefly by optical properties and which can be transformed into one another. If solutions of equal amounts of dextro- and lævotartaric acids are mixed and evaporated we obtain racemic acid, which is readily split into the above acids. If dextro- or lævotartaric acid is heated with water to 170° or boiled for a longer time with an excess of caustic alkali they become inactive, as one-half of the respective acid is changed to the variety having the opposite rotation; hence on evaporation we obtain racemic acid. In these operations some mesotartaric acid is also alway produced, which under the same conditions is in part converted into racemic acid (equilibrium condition, p. 62).

Synthetically a mixture of optically inactive racemic and mesotartaric acids are obtained by boiling dibromsuccinic acid with moist silver oxide:

$$\begin{array}{c} CHBr-COOH \\ | \\ CHBr-COOH \end{array} + 2AgOH = \begin{array}{c} CH(OH)-COOH \\ | \\ CH(OH)-COOH \end{array} + 2AgBr.$$

On heating with HI the tartaric acids first yield malic acid and then succinic acid:

$$C_2H_2(OH)_2 < {}^{COOH}_{COOH} + 2HI = C_2H_3(OH) < {}^{COOH}_{COOH} + H_2O + 2I;$$
Tartaric acid.   Malic acid.

$$C_2H_3(OH) < {}^{COOH}_{COOH} + 2HI = C_2H_4 < {}^{COOH}_{COOH} + H_2O + 2I$$
Malic acid.   Succinic acid.

On oxidation the tartaric acids yield bibasic dioxytartaric acid, $HOOC-CO-CO-COOH$ (P. 443), or tartronic acid, $HOOC-CH(OH)-COOH$ (p. 434), and then CO_2 and formic acid. When fused and then cooled they give the isomer metatartaric acid, $C_4H_6O_6$, an amorphous deliquescent mass; at 180° soluble tartaric anhydride, tartrelic acid, $C_4H_4O_5$, forming deliquescent crystals, is obtained. On heating further until the mass becomes infusible, insoluble tartaric anhydride, $C_4H_4O_5$, a white powder, is produced. These three compounds are reconverted into tartaric acid by boiling with water.

On further heating tartaric acids they give off an odor of burnt sugar, and amongst other bodies we find

Pyrotartaric acid, $C_5H_8O_4$ or $HOOC-C_3H_6-COOH$ (p. 425), and
Pyroracemic acid, $C_3H_4O_3$ or $CH_3-CO-COOH$ (p. 403).

1. Mesotartaric acid, antitartaric acid. *Preparation.* From dribomsuccinic acid (besides racemic acid, see p. 427), also by the oxidation of erythrite, sorbite, maleïc acid, and phenol.

Properties. Colorless plates having the formula $C_4H_6O_6 + H_2O$. It forms a potassium acid salt which is more soluble than the corresponding salts of the other tartaric acids. It is optically inactive and cannot be transformed into active tartaric acids (P. 306, *a*), but it may be converted into racemic acid.

2. Racemic acid. *Occurrence.* Sometimes in the mother-liquor of crude tartar.

Preparation. From dextrotartaric acid (p. 427), also by the oxidation of mannite, dulcite, or mucic acid and fumaric acid. Synthetically, besides mesotartaric acid, from dibromsuccinic acid and also by boiling glyoxal with hydrocyanic acid and hydrochloric acid:

$$<^{COH}_{COH} + 2HCN + 4H_2O = <^{CH(OH)COOH}_{CH(OH)COOH} + 2NH_3.$$

Properties. Colorless, triclinic, optically inactive crystals, crystallizing with 1 mol. H_2O. It is less soluble than dextro- or lævotartaric acids in water, and its calcium salt (calcium racemate) is precipitated from solutions of the free acid by the addition of $CaCl_2$ in contradistinction to the other tartaric acids. In watery solution it completely decomposes into its components dextro- and lævotartaric acids, which can be separated from each other by the following method:

One-half of a racemic acid solution is saturated with ammonia and the other half with caustic soda. These two solutions are mixed and allowed to crystallize, when we obtain two varieties of crystals of $C_4H_4Na(NH_4)O_6$, which differ from each other by some having a certain small face on the right side and others having the same face on the left side (p. 305). If these crystals are separated we find that the first are dextrorotatory and is the salt of dextrotartaric acid, while the other is lævorotatory and is the salt of lævotartaric acid.

3. Lævotartaric acid. This acid behaves exactly like dextrotartaric acid with the exception of its lævorotatory power and the above-mentioned shape of the crystals of its salts.

4. Dextrotartaric Acid, Ordinary Tartaric Acid. *Occurrence.* Free and as acid potassium salt widely distributed in the vegetable kingdom, especially in the juice of the grape. The alcohol produced in the fermentation of the grape-juice precipitates the acid potassium tartrate. This deposit from wine is called *crude tartar* and forms gray or reddish crystalline crusts which also contain calcium tartrate and yeast residues, etc.

Preparation. The crude tartar is boiled with water, and chalk, producing soluble neutral potassium tartrate and insoluble calcium tartrate:

$$2C_4H_5KO_6 + CaCO_3 = C_4H_4K_2O_6 + C_4H_4CaO_6 + H_2O + CO_2.$$

Potassium   Potassium   Calcium
bitartrate.   tartrate.   tartrate.

The dissolved potassium tartrate is precipitated as calcium tartrate by the addition of calcium chloride solution. The combined precipitates of calcium tartrate are decomposed by sulphuric acid, the calcium sulphate filtered off, and the filtrate containing the free tartaric acid evaporated to crystallization:

$$C_4H_4K_2O_6 + CaCl_2 = C_4H_4CaO_6 + 2KCl.$$
$$C_4H_4CaO_6 + H_2SO_4 = C_4H_6O_6 + CaSO_4.$$

Properties. Tartaric acid is tetravalent and bibasic, and forms large monoclinic prisms which melt at 170° and are readily soluble in water and alcohol but difficultly soluble in ether. Its solutions are dextrorotatory. Its salts are called *tartrates.*

Acid potassium tartrate, $C_4H_5KO_6$, cream of tartar, potassium bitartrate, is obtained by the recrystallization of crude tartar (see above). It is a white crystalline powder not readily soluble in water, and is precipitated from concentrated potassium salt solutions by the addition of an excess of tartaric acid solution, hence potassium salts are used in detecting tartaric acid. The must from the tropical tamarind is rich in potassium acid tartrate, malic, tartaric, and citric acids, besides sugar.

Potassium tartrate, $2C_4H_4K_2O_6 + H_2O$, consists of neutral crystals soluble in 0.7 part water.

Sodium-potassium tartrate, $C_4H_4KNaO_6 + 4H_2O$, Rochelle salts, is obtained by saturating a potassium acid tartrate solution with soda. It consists of colorless crystals soluble in 1.4 parts water:

$$2\frac{K}{H} > C_4H_4O_6 + Na_2CO_3 = 2\frac{K}{Na} > C_4H_4O_6 + H_2O + CO_2.$$

Effervescent powder is a mixture of tartaric acid, sodium bicarbonate, and sugar.

Seidlitz powders contain sodium bicarbonate and sodium potassium tartrate mixed together in one paper and tartaric acid in the second paper. When the contents of the two papers are mixed with water sodium tartrate is produced and carbon dioxide is evolved with effervescence.

Potassio-antimonous Tartrate, $2C_4H_4K(SbO)O_6 + H_2O$, Tartaremetic. In this compound the H atom of one carboxyl is replaced by the monovalent group antimonyl, SbO (p. 180). It is obtained by boiling potassium acid tartrate, antimony oxide, and water, filtering and evaporating the solution until white, sweet crystals are obtained which are soluble in 17 parts water and which cause vomiting:

$$2\frac{K}{H} > C_4H_4O_6 + Sb_2O_3 = 2\frac{K}{SbO} > C_4H_4O_6 + H_2O.$$

Calcium tartrate, $C_4H_4CaO_4 + 4H_2O$, precipitates from neutral tartrate solutions by the addition of calcium chloride as a white crystalline powder which is soluble in dilute acids. Cold solutions of alkalies also dissolve it, but on boiling the solution it precipitates again (separation from malic and citric acids).

Aluminium aceto-tartrate, $Al_2(C_2H_3O_2)_2(C_4H_4O_6)(OH)_2$, alsol, forms colorless gummy masses which are soluble in water.

Borax-tartar, $KNaC_4H_4O_6 + 2K(Bo)C_4H_4O_6$, is obtained by dissolving borax and cream of tartar in water and evaporating to dryness.

Citric acid, $C_6H_8O_7$ or $CH_2(COOH)^-C(OH)(COOH)^-CH_2(COOH)$, oxypropantricarbonic acid, on account of its occurrence and behavior on heating, etc., will be treated of here.

Occurrence. Free in the lemon, currant, cranberry, and other sour fruits, in the beet-root, and to a slight extent in milk.

Preparation. 1. From lemon-juice by neutralizing with lime and heating to boiling, when calcium citrate (p. 431) separates out. This is decomposed by dilute sulphuric acid, the calcium sulphate formed filtered off, and the filtrate evaporated to crystallization.

2. By fermenting a grape-sugar solution by means of the Saccharomycetes citromyces.

Formation. Synthetically from dichloracetone, $CH_2Cl-CO-CH_2Cl$ which is converted into $CH_2-Cl-C(OH)(CN)-CH_2Cl$ by HCN (p. 371 1), and this then transformed by HCl into dichloracetonic acid, $CH_2Cl-C(OH)(COOH)-CH_2Cl$ (p. 346, 2). On treating this with KCN the chlorine is replaced by CN and if this product is heated again with HCl citric acid is obtained:

$$CH_2(CN) - C(OH)(COOH) - CH_2(CN) + 4H_2O =$$
$$CH_2(COOH) - C(OH)(COOH) - CH_2(COOH) + 2NH_3.$$

Properties. Large colorless rhombic prisms with 1 mol. H_2O, which melt at 100° and lose their water of crystallization at 150°. On further heating it forms aconitic acid, $C_6H_8O_6$, then itaconic acid and mesaconic acid, $C_5H_6O_4$ (see below). Citric acid is readily soluble in water and alcohol. It is tetravalent and tribasic and hence forms three series of salts (*citrates*) and esters. On oxidation it yields oxalic acid, acetic acid, and acetone.

Calcium citrate, $Ca_3(C_6H_5O_7)_2 + 4H_2O$, is not readily soluble in cold water, but insoluble in hot water; hence a cold saturated solution precipitates completely on boiling (detection of citric acid).

Magnesium citrate, $Mg_3(C_6H_5R_7)_2 + 14H_2O$, mixed with citric acid, sodium bicarbonate and sugar, forms the ordinary effervescent magnesia or magnesii citras effervescens.

Silver citrate, itrol, $Ag_3C_6H_5O_7$, is a colorless powder.

Ferric citrate, $Fe(C_6H_5O_7) + 3H_2O$, is obtained by dissolving ferric hydroxide in citric acid and evaporating, when an amorphous brown mass soluble in water is obtained. Iron-ammonium citrate, $2Fe(C_6H_5O_7) + (NH_4)_2C_6H_6O_7 + 2H_2O$, forms amorphous brownish-red masses soluble in water.

Tricarballylic acid, propantricarbonic acid, $C_6H_8O_6$ or $C_3H_5(COOH)_3$, occurs in unripe beet-root and is prepared from aconitic acid (see below) by the action of nascent hydrogen or by heating citric acid with HI.

Aconitic acid, equisetic acid, $C_6H_6O_6$ or $C_3H_3(COOH)_3$. This tribasic acid occurs in Aconitum napellus, Equisetum fluviatile, in sugar-cane, and in the sugar-beet, and is produced by heating citric acid to 175°:

$$C_3H_4(OH)(COOH)_3 = C_3H_3(COOH)_3 + H_2O.$$

Itaconic acid, methylene succinic acid, $C_5H_6O_4$ or $CH_2 = C_2H_2(COOH)_2$, and its isomer.

Citraconic acid, methylmaleïc acid, $C_5H_6O_4$, are obtained by heating citric or aconitic acid above 175°.

Mesaconic acid, methylfumaric acid, $HOOC - (CH_3)C = CH - COOH$ or $C_5H_6O_4$, a stereoisomer of citraconic acid is prepared by heating itaconic or citraconic acids with water to 200°.

COMPOUNDS OF TRIVALENT ALCOHOL RADICALS.

1. Trivalent Alcohol Radicals.

General formula C_NH_{2N-1}.

These, like all radicals with uneven valence, do not exist free; They are called methenyl, ethenyl, propenyl, etc., or glyceryl, butenyl.

crotonyl, etc. The trivalent compounds are derived from the saturated hydrocarbons in a manner similar to that of the mono- and di- valent compounds in that three atoms are replaced by other atoms or by atomic groups. The relationship of these compounds to each other is simply as follows:

Propyl alcohol,	C_3H_8O.	Propionic acid,	$C_3H_6O_2$.
Propylene alcohol,	$C_3H_8O_2$.	Lactic acid,	$C_3H_6O_3$.
Propenyl alcohol,	$C_3H_8O_3$.	Glyceric acid,	$C_3H_6O_4$.

2. Trihydric Alcohols.

General formula $C_NH_{2N-1}(OH)_3$.

The properties which the glycols show is also found in the trihydric and polyhydric alcohols, and these become more and more pronounced as the number of hydroxyl groups increases. Thus from glycerine on oxidation we obtain a trihydric, monobasic acid, as well as a trihydric, dibasic acid; also aldehyde alcohols, ketone alcohols, etc. (p. 337). We do not know of any trihydric alcohols containing less than three C atoms (p. 331).

Glycerine, $C_3H_5(OH)_3$ or $CH_2(OH)^-CH(OH)^-CH_2(OH)$, Propenyl Alcohol, Glycerol. *Occurrence.* It is the only trihydric alcohol found in nature and occurs as glycerine acetate (triacetin), $C_3H_5(C_2H_3O_2)_3$, in ethereal oils of the spindle-tree, and forms esters with the higher fatty acids and oleic acid, which constitute the animal and vegetable fats. It is produced to a slight extent in alcoholic fermentation and hence exists in beer and wine.

Preparation. 1. As a by-product in the manufacture of soaps' and plasters, where the fats (fatty acid esters of glycerine) are decomposed by alkalies or lead hydroxide in a manner similar to other esters (p. 357, 2). It is formed, for example, when fats are boiled with lead hydroxide, producing an insoluble lead salt of the fatty acids and glycerine:

$$2C_3H_5(C_{18}H_{35}O_2)_3 + 3Pb(OH)_2 = 2C_3H_5(OH)_3 + 3Pb(C_{18}H_{35}O_2)_2.$$
Tristearin.

2. As a by-product in the manufacture of stearin candles (p. 440), where fats are treated with superheated steam or warmed with con-

centrated sulphuric acid to 120° C., when the fat takes up water and splits into glycerine and the fatty acids which separate out:

$$C_3H_5(C_{18}H_{35}O_2)_3 + 3H_2O = C_3H_5(OH)_3 + 3C_{18}H_{36}O_2.$$

Tristearin.   Glycerine.   Stearic acid.

The solution of glycerine thus obtained according to methods 1 or 2 (after neutralization or removal of dissolved lead oxide by H_2S) is carefully concentrated by evaporation, the salts which separate removed by filtration, and the crude glycerine thus obtained purified by distillation with superheated steam.

Synthetically. Propenyl bromide with silver acetate yields glycerine acetic acid ester, which is saponified by bases:

$$C_3H_5Br_3 + 3AgC_2H_3O_2 = C_3H_5(C_2H_3O_2)_3 + 3AgBr;$$
$$C_3H_5(C_2H_3O_2)_3 + 3KOH = C_3H_5(OH)_3 + 3KC_2H_3O_2.$$

Or propenyl chloride is heated with water to 170°:

$$C_3H_5Cl_3 + 3HOH = C_3H_5(OH)_3 + 3HCl.$$

Properties. Thick, colorless, sweet (hence the name) liquid having a specific gravity of 1.27, and when free from water it solidifies to a white crystalline mass at 0° C. It distils without decomposition at 290° C., and is soluble in water and alcohol but insoluble in ether and fatty oils. It dissolves the alkalies and alkaline earths and many metallic oxides, forming compounds similar to the alcoholates. With dehydrating agents it gives acrolein, C_3H_4O (p. 439), and when carefully oxidized it gives, according to the conditions, aldehydes and acids or mixed compounds. As glycerine contains three hydroxyl groups, it forms three series of esters (the *glycerides*) and mixed ethers.

3. Derivatives of Trivalent Alcohols.

Glyceric aldehyde, $C_3H_6O_3$ or $CHO-CH(OH)-CH_2OH$, is an aldehyde alcohol and forms colorless sweet crystals.

Dioxyacetone, $CH_2 \cdot OH-CO-CH_2OH$, the isomeric ketone alcohol, forms sweet crystals.

Both compounds are obtained as a mixture (called *glycerose*) by the careful oxidation of glycerine. They readily unite, forming inactive lævulose, $C_6H_{12}O_6$, also called α-acrose. Glyceric aldehyde and dioxyacetone are true sugars (carbohydrates) according to their constitution (p. 402), and can only be obtained pure in an indirect manner.

Glyceric acid, dioxypropionic acid, $C_3H_6O_4$ or $CH_2 \cdot OH-CH \cdot OH-COOH$, is produced on the oxidation of glycerine by dilute nitric acid. It forms an inactive sirup which is readily soluble yin water and alcohol. It is converted into the lævorotatory modification by Penicillium glaucum, and on being heated it splits into water and pyroracemic acid:

$$C_3H_6O_4 = H_2O + C_3H_4O_3.$$

Tartronic acid, oxymalonic acid, $HOOC-CHOH-COOH$ or $C_3H_4O_5$, is obtained by oxidizing glycerine with potassium permanganate and forms color'ess readily soluble crystals.

Mesoxalic acid, $HOOC-CO-COOH$, is produced in the oxidation of glycerine esters by cold nitric acid and by the action of baryta-water upon alloxan or dibrommalonic acid. It can only be obtained with 1 mol. H_2O; hence it may also be considered as dioxymalonic acid, $HOOC-C(OH)_2-COOH$ (p. 424).

Glycerine ethers may be obtained by the action of potassium alcoholates upon the haloid esters of glycerine. They are colorless liquids having a faint ether-like odor:

$$\text{Ethylin, } C_3H_5 \Big\langle \begin{smallmatrix} OH)_2 \\ (OC_2H_5) \end{smallmatrix}, \qquad \text{Diethylin, } C_3H_5 \Big\langle \begin{smallmatrix} (OH) \\ (OC_2H_5)_2 \end{smallmatrix},$$

$$\text{Triethylin, } C_3H_5(OC_2H_5)_3$$

Glycerine anhydride, glycid alcohol, $\overbrace{O \cdot CH_2-CH-CH_2} \cdot OH$, is produced from monochlorhydrin (see below) by the action of $Ba(OH)_2$. It is a liquid, boiling at 162°, whose hydrochloric acid ester, $\overbrace{O \cdot CH_2-CH-CH_2}Cl$, epichlorhydrin, as well as dichlorhydrin (see below), has techincal uses.

Glycerine **hydrochloric acid ester,** chlorhydrins, are produced by the action of HCl upon glycerine. Dichlorhydrin, $CH_2Cl-CHOH-CH_2Cl$, is a solvent for resins, colors, cellulose nitrates, etc.

Glycerine trinitrate, $C_3H_5(NO_3)_3$, incorrectly called nitroglycerine, is obtained by dropping glycerine into a cooled mixture of sulphuric and nitric acids. It is a colorless, odorless, poisonous, viscous liquid which is difficultly soluble in water but readily soluble in alcohol, ether, chloroform, and crystallizes at $-20°$ C. It is used in medicine under the name glonoinum. On suddenly heating, as well as by a blow or shock, it explodes violently (Nobel's explosive oil), and when mixed with infusorial earth (p. 195) it forms a pasty mass called dynamite which does not explode by shock alone or by ignition, but is ignited by mercury fulminate; hence it serves as an important explosive. Mixtures with cellulose nitrates are used for the same purposes as dynamite.

1 gramme of glycerine trinitrate yields on explosion 1300 cc. gas measured at 0° and 760 mm. pressure. As a rise in temperature takes place on explosion, the gases expand to about 10,500 cc. (p. 207).

Glycero-phosphoric acid, $C_3H_5(OH)_2(H_2PO_4)$, occurs in the urine and is the basis on which the lecithins are formed, and certain of its salts are used in medicine

Lecithins occur as such or as a constituent of the lecithalbumins (see Proteids) and protagons (see Glucosides) in all animal fluids, and especially in the nerve and brain substance, thymus glands, egg-yolk (7–10 per cent.), and to a less extent in semen, pus, blood, milk, yeast, corn, peas, wheat, etc.

It is a wax-like neutral solid, readily soluble in alcohol and ether, readily decomposable, and combines with acids and alkalies.

On boiling with acids or baryta-water they decompose into choline, $(HO \cdot C_2H_4)N(CH_3)_3(OH)$, glycero-phosphoric acid, palmitic and stearic acids, or oleic acid, $C_{18}H_{34}O_2$, and therefore the following formula can be given as their structure:

$$(C_NH_{2N-1}O_2)_2 = C_3H_5 - HPO_4 - C_2H_4 \cdot N(CH_3)_3 \cdot OH.$$

Glyceryl palmitate, palmitin, $C_3H_5(C_{16}H_{31}O_2)_3$, crystallizes in shining leaves which melt at 63° C.

Glyceryl stearate, stearin, $C_3H_5(C_{18}H_{35}O_2)_3$, also crystallizes in shining leaves and melts at 67° C.

Glyceryl oleate, olein, $C_3H_5(C_{18}H_{33}O_2)_3$, forms an oily liquid (p. 440) which solidifies at $-6°$ C.

These three esters may be obtained by heating glycerine with the respective acids and mixed together form the chief constituent of all animal and vegetable fats, which are divided into three classes, namely, tallows (solid), butter and lard (semi-solid), and oils (liquids). Oils contain chiefly olein, while the tallows contain stearin chiefly.

The Fats in General. The fats are obtained by pressure or by extraction with ether, carbon disulphide, etc. Butter is prepared by violently beating milk, when the fat globules conglomerate. Artificial butter or margarine is obtained by trying out animal fat and allowing it to cool, and then exposing it to pressure, which frees the fat of stearin and yields an oil (oleomargarine-oil). This oil, which is a mixture of olein and palmitin is strongly shaken with warm milk in a method similar to the manufacture of butter, when the artificial butter separates. This is colored yellow and treated with butyric acid esters.

The fats when pure are colorless, odorless, and tasteless neutral bodies insoluble in water, not readily soluble in alcohol, but readily soluble in ether and carbon disulphide. They are lighter than water and produce a transparent stain when applied to paper which does not disappear on warming (differing from the ethereal oils). They can be heated to 300° C. without suffering any change, but at higher temperatures they decompose, yielding various products having an unpleasant odor, chief amongst which is acrolein.

Many fats when exposed to the air are gradually converted into a hard transparent mass. These are called drying oils; for instance, hemp-seed oil, croton-oil, linseed-oil, poppy-oil, nut-oil, which consist chiefly of the glycerides of linoleic acid, $C_{18}H_{32}O_2$ (p. 445), as well as castor-oil. These unsaturated glycerides become solid on absorbing oxygen.

The other fats do not change when pure, but as they generally consist of mixtures, especially with proteids, they gradually suffer decomposition becoming disagreeable in taste and smell and having an acid reaction, i.e., they become rancid. This depends upon the setting free of fatty acids and the formation of aldehydes by the action of the oxygen of the air and light, which cause the odor and taste. The fats are split into glycerine and fatty acids or fatty acid salts by superheated steam, strong acius, or bases (see Soaps), as well as by the steatolytic enzyme of the pancreatic juice.

Solid Fats. Lard contains about 60 per cent. olein.

Mutton tallow contains about 75 per cent. palmitin and stearin and 25 per cent. olein. When mixed with 2 per cent. salicylic acid it forms salicylic tallow.

Cacao-butter, obtained from the seeds of the Theobroma Cacao by pressure, consists of olein, palmitin, stearin, and theobromic acid glycerine (p. 377).

Nutmeg-butter, obtained from the nutmeg by pressure, consists of myristicin, olein, and ethereal oils.

Cocoanut-oil, the fat of the seed-kernel of the Cocos nucifera, consists of laurin, myristin, and palmitin.

Liquid Fats. Olive-oil is obtained from the olives by pressure. Olive-oil treated with 6 per cent. oleic acid is used in medicine under the name lipanin.

Almond-oil is obtained by pressing the sweet or bitter almond.

Poppy-oil is produced by pressing the poppy-seed.

Linseed-oil, from the flaxseed.

Laurel-oil, obtained from the fresh ripe laurel by pressure, consists of laurin.

Rape-seed oil, from the seeds of the rape.

Castor-oil, from the seeds of the Ricinus communis, contains chiefly glycerids of ricinoleic acid.

Croton-oil, from the seeds of the Croton tiglium, also contains the poisonous glyceride of the little-known crotonolic acid, but not of crotonic acid (p. 440).

Cod-liver oil, from the fresh livers of the codfish by gently heating.

Sesame oil, from the seeds of the Sesamum orientale, is added to artificial butter in Germany in order to be able to detect the same (see Furol). Its bromine and iodine addition products are used in medicine as bromipin and iodipin.

Soaps. On heating glycerides with alkali hydroxides. and other strong bases they are decomposed in a manner similar to the esters of the mono- and dihydric alcohols, i.e., they decompose into salts of the acids contained in the glycerides and into glycerine (p. 432). The decomposition of the esters, i.e., the fats, by means of caustic alkalies is called *saponification* (p. 357, 2). Soaps are the alkali salts of palmitic, stearic, and oleic acids. Sodium soaps are hard, while the potassium soaps are soft.

The soaps are soluble in water and alcohol, but insoluble in common salt solutions ("salting out" of soaps is the separation of the soaps from the sticky substances formed in the saponification). The soaps have a solvent action on bodies otherwise insoluble in water, such as hydrocarbons, resins, phenols, fats, etc. By the action of large quantities of water the soaps are in part decomposed into free alkali and acid salts of the fatty acids. The action of soaps in washing depends upon this property, as the alkali set free and the acid salt of the fatty acids which form the froth readily remove the fat in the form of an emulsion, and the froth envelops the dirt and removes it.

The other salts of the above-mentioned fatty acids are mostly soluble in alcohol but insoluble in water, hence water rich in calcium salts forms a precipitate with soap solutions forming insoluble calcium salts of the fatty acids which is the reason why hard waters are not suitable for washing purposes.

Official soap is pure sodium soap; Green soap is pure potassium soap, whose alcoholic solution is called spiritus saponatus.

Soap liniment, opodeldoc, is a gelatinous mixture of alcohol, sodium soap, camphor, ammonia, and ethereal oils; liquid opodeldoc is a solution of potassium soaps and camphor in alcohol. Ammonia liniment is a mixture of fatty oils with ammonia, and in the presence of camphor-oil it is called camphorated ammonia liniment.

Plasters. The lead salts of the fatty-acids are called lead plasters. These are obtained directly by boiling fats with lead oxide and water (p. 432).

On mixing lead plaster with different substances we obtain mercurial plaster, soap plaster, adhesive plaster, etc. The tough mixtures of resins, wax, and oils with active bodies which are used externally are also called plasters. These differ from the salves only in their consistency.

Salves are the semi-plastic mixtures of fats or oils with wax, resin, etc., which have either solid bodies as powders or solutions mixed therewith.

Emulsions. If water contains bodies, such as plant mucilage, albumin, gums, etc., in solution, they give a mucilaginous character and greater viscosity to the liquid If oils are thoroughly mixed with such liquids, they remain suspended as very small globules and the liquid retains a milky appearance. Such mixtures are called emulsions.

4. Monohydric Compounds of Trivalent Alcohol Radicals.

Trihydric alcohol radicals may also appear monovalent, but not free, and form unsaturated compounds which have the property, like all other unsaturated compounds, of readily taking up 2 atoms of H, Br, etc., and being converted into saturated compounds. The radical C_2H_3 or $CH_2{=}CH^-$ is called vinyl, C_3H_5 or $CH_2{=}CH^-CH_2^-$ is called allyl, etc.

a. Alcohols, etc.

Vinyl alcohol, $C_2H_3{-}OH$ or $CH_2{=}CH \cdot OH$, occurs in commercial ether and is very unstable.

Vinyl sulphide, $C_2H_3{-}S{-}C_2H_3$, found in the ethereal oil of garlic, is a colorless liquid having a garlic-like odor and boils at $101°$.

Neurine, trimethylvinylammonium hydroxide, $(CH_3)_3N(C_2H_3)OH$, has been treated of on page 380.

Allyl iodide, C_3H_5I or $CH_2=CH-CH_2-I$, is obtained on the distillation of phosphorus iodide with glycerine: $C_3H_5(OH)_3 + PI_3 = C_3H_5I + H_3PO_3 + 2I$. It is a colorless liquid having a mustard-like odor, and from which the various allyl esters can be obtained by heating with the silver salts.

Allyl sulphide, $(C_3H_5)_2S$, as well as allyl disulphides, for example, allyl propyl disulphide, $C_3H_5-S-S-C_3H_7$, forms the chief constituent of the ethereal oil of the onion, garlic, asafœtida, etc.

Allyl Isosulphocyanic Ester, C_3H_5-NCS, Allyl Mustard-oil, Mustard-oil.

Preparation. 1. Ordinarily by allowing powdered black mustard-seeds (Sinapsis nigra) to stand in contact with water, when the contained glucoside sinigrin (also called potassium myronate) is split by the action of the ferment myrosin contained therein into allyl isosulphocyanic ester, potassium hydrosulphate, and glucose. The allyl isosulphocyanic ester can be separated by distillation:

$$C_{10}H_{16}NS_2O_9K + HOH = C_3H_5-NCS + KHSO_4 + C_6H_{12}O_6.$$

2. By the action of carbon disulphide upon allylamine

$$CS_2 + NH_2-C_3H_5 = C_3H_5-NCS + H_2S.$$

3. By heating allyl iodide with potassium sulphocyanide (see below).

Properties. Colorless liquid boiling at 148-152°, having an extremely pungent odor and producing blisters when applied to the skin. It is slightly soluble in water but readily soluble in alcohol (Spiritus sinapis).

Allyl sulphocyan, C_3H_5-SCN, is obtained from allyl iodide by the action of potassium sulphocyanide in the cold:

$$NCSK + C_3H_5I = NCS-C_3H_5 + KI.$$

It is a liquid having a leek-like odor and boiling at 161°, and undergoing a molecular rearrangement into allyl mustard-oil on warming:

$$N \equiv C-S(C_3H_5) = S=C=N-(C_3H_5).$$
Allyl sulphocyan.   Allyl isosulphocyanic ester.

Allyl Alcohol, C_3H_5-OH or $CH_2=CH-CH_2 \cdot OH$. *Preparation.* Obtained on gradually heating glycerine with oxalic acid to 260°, when the oxalic acid is split into CO_2 and formic acid (p. 352), and this latter body forming with the glycerine at 190° its monoformic acid ester, which on distillation decomposes into allyl alcohol, carbon dioxide, and water:

$$H \cdot COO^-C_3H_5(OH)_2 = C_3H_5^-OH + H_2O + CO_2.$$

2. From the allyl esters obtained from allyl iodide by heating with caustic alkali.

3. By the action of nascent hydrogen upon allyl aldehyde.

Properties. Colorless liquid boiling at 97°, having a pungent odor and which on oxidation with silver oxide yields allyl aldehyde and then the corresponding acrylic acid (p. 440).

Allyl Aldehyde, C_3H_4O or $CH_2{=}CH^-CHO$, Acrolein. *Preparation.* By the moderate oxidation of allyl alcohol and by strongly heating glycerine or fats:

$$C_3H_5(OH)_3 = C_3H_4O + 2H_2O.$$

The decomposition of glycerine takes place completely if it is heated with dehydrating substances, such as phosphoric anhydride or potassium bisulphate.

Properties. Colorless liquid boiling at 52° and having an unpleasant pungent odor which causes irritation of the mucous membranes. It is not readily soluble in water. The odor of burnt fats as well as of the smouldering tallow candle is due to acrolein. On keeping, acrolein undergoes polymerization and is converted into amorphous white disacryl or metacrolein. On oxidation it yields acrylic acid, $C_3H_4O_2$, the first member of the following series of acids:

Citronellol, $C_{10}H_{19}^-OH$, an homologous alcohol of allyl alcohol and its aldehyde, and

Citronellal, $C_{10}H_{18}O$, are found in many etheral oils (p. 445).

b. Oleic acid series.

General formula $C_NH_{2N-2}O_2$.

A series of monobasic, unsaturated acids are derived from allyl alcohol and its little-known homologues. These acids are obtained by the oxidation of the corresponding alcohol and aldehyde or by treating the monohalogen derivatives of the fatty acids with alcoholic caustic alkali:

$$\underset{\text{Chlorpropionic acid.}}{C_3H_5ClO_2} + KOH = \underset{\text{Acrylic acid.}}{C_3H_4O_2} + KCl + H_2O.$$

They are very similar to the fatty acids, but differ from these especially by the power they have of taking up hydrogen or halogens by addition and thus being transformed into fatty acids or their substitution products:

$$CH_2\text{-}CH\text{-}COOH + H_2 = CH_3\text{-}CH_2\text{-}COOH;$$
Acrylic acid.   Propionic acid.

$$CH_2\text{-}CH\text{-}COOH + Br_2 = CH_2Br\text{-}CHBr\text{-}COOH.$$

Fats which contain glycerides of the oleic acid series and other unsaturated acids on standing with an alcoholic solution of iodine and mercuric chloride form addition products. From the amount of combined iodine, which can be readily estimated, we can determine in a fat the proportion of unsaturated glycerides to the saturated glycerides, which do not combine with the iodine (Hübl's iodine equivalent). We thus obtain a knowledge of the nature of the fat and whether it is adulterated with other fats.

From crotonic acid upward we have two stereoisomers (p. 308c) besides the structural isomers.

Acrylic acid, $C_3H_4O_2$, is produced on warming a watery solution of acrolein with silver oxide. The silver separates as a mirror (p. 350), and at the same time silver acrylate is produced, from which the acid can be set free by H_2S. It is a pungent acid liquid boiling at 140°.

Crotonic acids, $C_4H_6O_2$, form three fluids and one solid body. Two of them are stereoismers and occur in crude wood alcohol, the third in ethereal camomile-oil. The acids were incorrectly considered as constituents of croton-oil (p. 436).

Angelic acid, $C_5H_8O_2$, occurs free with valerianic acid in the angelica root and as butyl and amyl ester in Roman camomile-oil.

Tiglic acid, $C_5H_8O_2$, occurs as a glyceride in croton-oil and Roman camomile-oil, and is a stereoisomer of angelic acid.

Hypogæic acid, $C_{16}H_{30}O_2$, as glyceride in the oil of the earth-nut (p. 377) and in whale-oil. It forms colorless crystals.

Oleic acid, $C_{18}H_{34}O_2$, elæic acid, occurs as glyceride in most fats and forms the chief constituent of the non-drying oils. It is obtained in the manufacture of stearine candles, where the free fatty acids (p. 432), which form a semi-solid mass, are pressed between warm plates, whereby the liquid oleic acid is pressed out, while the remaining solid palmitic and stearic acids are moulded into candles. It is a colorless oily liquid which does not redden litmus and crystallizes at 4°, and which oxidizes in the air, turning yellow and rancid. By nitrous acid it is converted into its crystalline stereoisomer.

Elaidic acid, $C_{18}H_{34}O_2$, which melts at 45° (detection of non-drying oils, see Linoleic Acid).

Lead oleate, $Pb(C_{18}H_{33}O_2)_2$, is soluble in ether (separation of oleic acid from stearic and palmitic acids).

Sodium oleate, eunatrol, is a fine white powder which is used in medicine.

Rapinic acid, $C_{18}H_{34}O_3$, occurs as a glyceride in rape-seed oil.

Erucic acid, $C_{22}H_{42}O_2$, exists as glyceride in the fatty oils of the varieties of Eruca, Brassica, and Sinapis, and forms colorless crystals which are transformed into

Brassidic acid, $C_{22}H_{42}O_2$, when treated with nitrous acid. It is the stereoisomer of erucic acid.

Acids closely related to the Oleic Acid Series.

Ricinoleic acid, $C_{18}H_{34}O_5$, oxyoleic acid, is a thick oily liquid which occurs as a glyceride in castor-oil. Nitrous acid converts it into its stereoisomer,

Ricinelaidic acid, $C_{18}H_{34}O_3$, which forms the Turkey-red oil of the dyer which consists of ricinoleic sulphuric acid $(C_{18}H_{33}O_2)^-HSO_4$.

COMPOUNDS OF TETRAVALENT ALCOHOL RADICALS.

1. Tetravalent Alcohol Radicals.

General formula C_N $_{2N-2}$.

Acetylene	C_2H_2	Gas	.	Valerylene	C_5H_8	Liquid
Allylene	C_3H_4	Gas		Hexoylene	C_6H_{10}	Liquid
Crotonylene	C_4H_6	Liquid		etc.		

These radicals are also called ethin, propin, butin, pentin, etc. They are, like the divalent radicals, known in the free state. They are obtained by heating the alkylene halogen compounds $C_NH_{2N}X_2$ with alcoholic caustic alkali:

$$C_2H_4Br_2 + 2KOH = C_2H_2 + 2KBr + 2HO_2.$$

They combine directly with the halogens or with nascent hydrogen, forming saturated compounds.

In the isomers of this series, which, like acetylene, contain the ^-CH group, this hydrogen can be replaced by metals. Mineral acids develop from these metallic compounds the pure hydrocarbon C_NH_{2N-2} (for method of preparation see Acetylene).

Acetylene, ethin, C_2H_2 or $CH{=}CH$.

Formation. 1. It is the only hydrocarbon with the exception of methane and ethane which can be prepared by the direct union of its elements, as by passing hydrogen through a vessel containing two carbon poles between which the electric arc is playing.

2. It is also formed in the incomplete combustion (by passing the vapors through red-hot tubes) of many carbon compounds (such as alcohol, ether, methane, ethylene), and hence is also found in illuminating-gas. From this latter it can be· obtained in large quantities by allowing the Bunsen flame to retreat in the burner.

3. It is also produced in the electrolysis of alkali salts of fumaric and maleïc acid (see p. 421),

$$KOOC-C_2H_2-COOK = HC \equiv CH + 2CO_2 + 2K,$$

Potassium fumarate.   Acetylene.

as well as from bromoform or iodoform on heating with powdered silver:

$$2CHI_3 + 6Ag = 6AgI + C_2H_2.$$

Preparation. From calcium carbide (p. 223) by decomposition with water: $CaC_2 + 2HOH = Ca(OH)_2 + C_2H_2.$

Properties. Penetrating, colorless, poisonous gas soluble in equal volumes of water and very readily soluble in acetone ($\frac{1}{25}$ part). It burns with a strongly illuminating and smoky flame and is liquefied at $0°$ and 26 atmospheres pressure. Decomposes into its elements with explosion when ignited by mercury fulminate. If acetylene is slowly passed through a faint red-hot tube, it is converted into benzene, C_6H_6, the most important compound of the isocarbocyclic group:

$$3C_2H_2 = C_6H_6.$$

The use of burning acetylene, with its intensely illuminating flame, in illumination has the advantage that it can be very readily produced from calcium carbide at the locality where it is to be used. On the other hand it has the disadvantage that the explosion limit of a mixture of air with acetylene lies very much farther apart than with illuminating-gas.

Air with 3 to 65 per cent. acetylene explodes in contact with a flame.

Liquid acetylene, as well as the gas under a pressure of 2 atmospheres, is decomposable with violent explosion. Acetylene obtained from calcium carbide contains H_2S, PH_3, and often P_2H_4, which makes it spontaneously inflammable (p. 166). Infusorial earth impregnated with chromic acid-sulphuric acid (heratol), cuprous chloride-hydrochloric acid (frankolin), or a mixture of chloride of lime with lead chromate (akagin), or porous pieces of chloride of lime (puratylen) serve to purify acetylene used for illuminating purposes.

In order to prepare chemically pure acetylene the gases containing the acetylene are passed into a solution of silver nitrate (see Metallic Derivatives).

Metallic derivatives of acetylene. Acetylene precipitates red explosive cuproacetylene, $C_2Cu_2 + H_2O$, from an ammoniacal cuprous chloride solution, and also white explosive acetylene silver, $C_2Ag_2 + H_2O$, from an ammoniacal silver nitrate solution. Acids evolve pure acetylene from these compounds: $C_2Cu_2 + 2HCl = C_2H_2 + 2CuCl$. If sodium is heated with acetylene gas, sodium acetylene, C_2HNa and C_2Na_2, are obtained as a black powder, which with water is converted into acetylene and NaOH:

$$C_2HNa + H_2O = C_2H_2 + NaOH \text{ (see Carbides).}$$

2. Tetrahydric Alcohols.

General formula $C_N H_{2N-2}(OH)_4$.

These have two asymmetric C atoms and hence, like their derivatives, two different stereoisomeric modifications are possible.

Erythrite, phycite, $HOH_2C-CH(OH)-CH(OH)-CH_2OH$ or $C_4H_6(OH)_4$, is the only tetrahydric alcohol occurring in nature, and is found free in the Protococcus vulgaris, and as an ester erythrin (which see) in several algæ and lichens. It can be obtained from these by caustic alkali. It forms large, optically inactive crystals readily soluble in water, but with difficulty in alcohol, and has a sweet taste like all the polyhydric alcohols. On warming with HI it is reduced to secondary butyl iodide: $C_4H_6(OH)_4 + 7HI = C_4H_9I + 4H_2O + 6I$, and on oxidation it yields like all polyhydric alcohols mixed compounds (P. 337). Nitro-sulphuric acid converts it into the explosive erythrin nitrate, $C_4H_6(NO_3)_4$.

3. Derivatives of Tetrahydric Alcohols.

Erythrose, $C_4H_8O_4$ or $HOH_2C-CH(OH)-CH(OH)-CHO$, is obtained on the careful oxidation of erythrite, and as an aldose is a carbohydrate (p. 447), but it is unfermentable and forms a colorless sweet sirup.

Erythritic acid, $HOH_2C-(HO)HC-CH(OH)-COOH$ or $C_4H_8O_5$, trioxybutyric acid, is obtained in the moderate oxidation of erythrite and lævulose, and forms colorless deliquescent crystals.

Mesotartaric acid, $HOOC-(HO)HC-CH(OH)-COOH$ or $C_4H_8O_6$ (p. 428), is the next product on the oxidation of erythrite.

Dioxytartaric acid, $HOOC-CO-CO-COOH$, theoretically the final oxidation product of erythrite, has thus far been obtained only indirectly from tartaric acid and always as $C_4H_2O_6 + 2H_2O$; hence it has received the name dioxytartaric acid, $C_4H_6O_8$ (p. 428).

COMPOUNDS OF PENTAVALENT ALCOHOL RADICALS.

1. Pentavalent Alcohol Radicals.

General formula $C_N H_{2N-3}$.

In common with all the radicals with uneven valence they are unknown in the free state.

2. Pentahydric Alcohols.

General formula $C_N H_{2N-3}(OH)_5$.

These have two asymmetric C atoms, and therefore they exist, like their derivatives in two stereoisomeric modifications (p. 306). They form crystals having a sweet taste, and are readily soluble in water. They are obtained from their aldehydes or ketones by the action of nascent hydrogen.

Arabite, $C_5H_{12}O_5$ or $CH_2OH-(CHOH)_3-CH_2OH$, is inactive.

Adonite, $C_5H_{12}O_5$, the stereoisomer of arabite, occurs in the Adonis vernalis, is inactive, and is obtained from ribose (p. 444).

Xylite, $C_5H_{12}O_5$, stereoisomer of arabite, is inactive.

Rhamnite, $C_6H_{14}O_6$ or $C_5H_6(OH)_5(CH_3)$, is dextrorotatory.

3. Derivatives of Pentahydric Alcohols.

The aldoses and ketoses, $C_5H_{10}O_5$ or $CHO(CHOH)_3CH_2OH$, and $CH_2OH-CO-(CHOH)_2CH_2OH$, derived from the alcohols $C_5H_{12}O_5$, are carbohydrates (p. 447), but unfermentable, and are called *pentoses*. They form colorless, very sweet crystals which on heating with dilute mineral acids yield furol, $C_5H_{10}O_5 = 3HOH + C_5H_4O_2$, which is used in their quantitative estimation and detection from other carbohydrates. They are known, like their alcohols, in numerous stereoisomeric modifications. They occur in the plants as so-called *pentosanes*, $C_5H_8O_4$, which on warming with dilute acids are converted into pentoses by taking up water.

The following pentoses are aldoses:

Arabinose, $C_5H_{10}O_5$, gum-sugar, is obtained from its pentosane araban on boiling cherry gum, gum arabic, or sugar-beets with dilute acids. It is dextrorotatory. Inactive arabinose (urine pentose) occurs sometimes in the urine.

Xylose, $C_5H_{10}O_5$, wood sugar, is produced from most woods, leaves, bark, etc., by the action of boiling dilute acids upon the pentosane xylan (wood gum). Xylose is dextrorotatory.

Ribose, $C_5H_{10}O_5$, obtained synthetically, is optically inactive.

Lyxose, $C_5H_{10}O_5$, is lævorotatory and is obtained synthetically.

Rhamnose, $C_5H_9(CH_3)O_5$, isodulcite, methyl pentose, is obtained from certain glucosides. It is dextrorotatory. Other methyl pentoses are called *fucose* and *quinovose*.

Arabonic acid, xylonic acid, ribonic acid, and lyxonic acid, $C_5H_{10}O_6$ or $HOH_2C-(CH \cdot OH)_3COOH$, are the stereoisomeric acids corresponding to the above aldehydes. They form colorless crystals.

Trioxyglutaric acids, $C_5H_8O_7$ or $HOOC(CH \cdot OH)_3COOH$, is produced on the further oxidation of the above aldehydes. Four stereoisomers are known.

Saccharonic acids, $C_6H_{12}O_6$ or $HO \cdot H_2C-CH_2-(CHOH)_3-COOH$. Of these pentavalent monobasic acids three are known. They are produced by the action of $Ca(OH)_2$ upon the carbohydrates galactose, glucose, lævulose (p. 449). They are unstable and quickly change into their internal anhydrides (lactones), the saccharons, $C_6H_{10}O_5$. (This must not be confounded with the sweet substance called saccharin.)

4. Monohydric Compounds of Pentavalent Alcohol Radicals.

As unsaturated monohydric alcohols, etc., can be derived from vinyl, C_2H_3, allyl, C_3H_5, and their homologues (p. 437), so also we may have the same from propargyl, C_3H_3, and its homologues.

a. Alcohols, etc.

Propargyl alcohol, C_3H_3-OH or $HC \equiv C-CH_2OH$, is synthetically obtained as a colorless liquid boiling at 114°.

Linalool, $C_{10}H_{17}-OH$ is an optically active tertiary alcohol; with dilute acids it is converted into its isomer,

Geraniol, $C_{10}H_{17}OH$, rhodinol, which is an optically active primary alcohol and the chief constituent of rose-oil. On oxidation it is converted into its aldehyde,

Geranial, $C_{10}H_{16}O$, citral, the odoriferous substance of oil of lemons.

These compounds are colorless liquids and occur singly or together, often also mixed with citronellal and citronellol (p. 439), in many plants or their ethereal oils; thus in oil of balm-mint, citronella, geranium, rose, lavender, lemon, linaloe, etc. As they are isomeric with certain terpenes, and, like the olefines, have a double bondage of the C atoms, they are also called *olefinic terpenes.* Linalool and geraniol are readily transformed into terpinhydrate, $C_{10}H_{20}O_2 + H_2O$, which can readily be obtained from the pinenes, which explains the simultaneous presence of these two fatty bodies with terpenes in many plants.

b. Propiolic Acid Series.

The acids of this series have the general formula $C_NH_{2N-4}O_2$, and may have either a treble or two double bonds in the molecule: $HC{\equiv}C-COOH$, propiolic acid, $C_2H_4{=}CH-CH{-}CH-COOH$, sorbinic acid. They are produced from the sodium compounds of acetylene by the action of CO_2 and are very similar to oleic acids, and may be converted into them by nascent hydrogen, and then into fatty acids.

Propiolic acid, $C_3H_2O_2$, propargylic acid, and

Tetrolic acid, $C_4H_4O_2$, are obtained synthetically.

Sorbinic acid, $C_6H_8O_2$, occurs in the unripe fruit of the mountain-ash.

Geranic acid, $C_{10}H_{16}O_2$, obtained by oxidizing geranial.

Linoleic acid, $C_{18}H_{32}O_2$, occurs as glycerid as chief constituent of drying oils, and is a yellowish oil which is not changed by nitrous acid. (Hence if a non-drying oil does not become solid with nitrous acid it must be a mixture of some drying oil.)

COMPOUNDS OF HEXAVALENT ALCOHOL RADICALS.

1. Hexavalent Alcohol Radicals.

General formula C_NH_{2N-4}.

Valylene, C_5H_6, obtained from $C_4H_8Br_2$ by alcoholic caustic alkali (p. 348, 4), as well as by distilling Cannel coal. It is a liquid boiling at 50° and having a leek-like odor.

2. Hexahydric Alcohols.

General formula $C_NH_{2N-4}(OH)_6$.

There bodies have 4 asymmetric C atoms, and hence they occur in numerous stereoisomeric modifications (p. 306). These alcohols and their derivatives are generally obtained synthetically. In the following only those alcohols occurring in nature will be discussed.

Mannite, $C_6H_{14}O_6$ or $CH_2OH-(CH-OH)_4-CH_2OH$, is widely distributed in plants, especially in the larch and manna-ash (whose dried juice is called manna), in celery, sugar-cane, oyster-plant, quitch-grass, olives, etc., as well as in normal dog urine. It is obtained from manna by boiling with alcohol and evaporating the solution to crystallization. It is also produced in the mucilaginous fermentation of sugars and by the action of nascent hydrogen upon the aldehyde alcohols, mannose and glucose. It forms white, sweet, dextrorotatory needles which are readily soluble in water and alcohol and which on oxidation yield mannose, $C_6H_{12}O_6$, then mannonic acid, $C_6H_{12}O_7$, and finally mannosaccharic acid, $C_6H_{10}O_8$. Nitro-sulphuric acid converts it into the explosive mannite nitrate, $C_6H_8(NO_3)_6$.

Dulcite, melampyrite, $C_6H_{14}O_6$, stereoisomer of mannite occurs, in the Madagascar manna, in varieties of Melampyrum, Scrophularia, Evonymus, Rhinantus. . It forms colorless prisms which are less soluble in water than mannite and nearly insoluble in alcohol. It may be artificially prepared by the action of nascent hydrogen upon its aldehyde alcohol, galactose, $C_6H_{12}O_6$, as well as from milk-sugar. On oxidation it yields mucic acid, $C_6H_{10}O_8$, then racemic acid, $C_4H_6O_6$. Dulcite is the only polyhydric alcohol which reduces alkaline solutions of copper oxide. Because of its configuration, which is similar to mesotartaric acid (p 307), it is inactive and cannot be split into active modifications.

Sorbite, $C_6H_{14}O_6$, stereoisomer of mannite, occurs in the mountain-ash and the fruit of many Rosaceæ. It is produced by the action of nascent hydrogen upon its ketone alcohols sorbinose and lævulose, or upon its aldehyde alcohol glucose. It forms small dextrorotatory crystals with $\frac{1}{2}$ mol. H_2O, and on oxidation it yields glucose, $C_6H_{12}O_6$, then gluconic acid, $C_6H_{12}O_7$, and finally saccharic acid, $C_6H_{10}O_8$.

3. Derivatives of Hexahydric Alcohols.

Mannose, glucose, galactose, $HOH_2C-(CH \cdot OH)_4-CHO$ or $C_6H_{12}O_6$, are the three stereoisomeric aldehyde alcohols of the above-mentioned hexahydric alcohols (see Carbohydrates).

Lævulose and sorbinose, $CH_2OH-(CH \cdot OH)_3-CO-CH_2OH$ or $C_6H_{12}O_6$, are the two stereoisomeric ketone-alcohols of the above-mentioned hexahydric alcohols (see Carbohydrates).

These aldoses and ketoses, $C_6H_{12}O_6$, and their anhydride condensation products form the most important compounds of the carbohydrate group (p. 451)

Mannonic, gluconic, galactonic acids, $C_6H_{12}O_7$ or $HOH_2C-(CH \cdot OH)_4-COOH$. These three stereoisomeric hexon acids are obtained on the oxidation of mannose, glucose, galactose with chlorine- or bromine-water. Their lactones (p. 404, a), yield the above-mentioned carbohydrates on reduction.

Glycuronic acid, $C_6H_{10}O_7$ or $OHC-(CHOH_4)-COOH$, stands between mannonic and saccharic acids. It is found in the urine after partaking of various substances, such as camphor, chloral, naphthalene, turpentine, combined with these bodies, and can be obtained from these combinations by treatment with acids, or can be obtained by the reduction of dextrosaccharic acid (see below). Glycuronic acid is a sirup which is readily soluble in water and alcohol and dextrorotatory, while the conjugated glycuronic acids are lævorotatory. It occurs in the artist pigment Indian yellow, in the form of magnesium euxanthinate, which splits by HCl into euxanthic acid, $C_{19}H_{18}O_{11}$, which in turn decomposes at 125° into euxanthon, $C_{13}H_8O_4$, and glycuronic acid, $C_6H_{10}O_7$.

Saccharic acid, $C_6H_{10}O_8$ or $HOOC-(CH \cdot OH)_4-COOH$, stereoisomeric mannosaccharic acid, is produced in the oxidation of cane-sugar, glucose, starch, or mannite, by means of nitric acid. It forms deliquescent gummy masses which yield tartaric acid on further oxidation. The dextro- or lævosaccharic acid is obtained according to the material we start from, and these unite, forming the inactive modification.

Mucic acid, $C_6H_{10}O_8$, stereoisomer of saccharic acid, is obtained on the oxidation of dulcite, galactose, plant mucilages, and certain varieties of gums by nitric acid. It forms a white crystalline powder nearly insoluble in water, which on further oxidation yields racemic acid, $C_4H_6O_6$. It is optically inactive and does not split into active modifications.

COMPOUNDS OF HEPTAVALENT AND HIGHER ALCOHOL RADICALS.

1. Alcohol Radicals.

A few of these with even valence are known; e.g., diacetylene, C_4H_4, also dipropargyl, and dimethyl acetylene, C_6H_6, both isomers of benzene.

2. Alcohols and their Derivatives.

Perseït, $C_7H_{16}O_7$ or $C_7H_9(OH)_7$, is contained in the fruit of the Laurus Persea. It forms colorless crystals which melt at 188°.

Volemite, $C_7H_{16}O_7$, found in the Lactarius volemus, forms colorless dextrorotatory needles which melt at 150°.

Volemose, $C_7H_{14}O_7$, is the aldehyde of volemite.

Glucoheptite, $C_7H_{16}O_7$ or $C_7H_9(OH)_7$, **glucooctite,** $C_8H_{18}O_8$ or $C_8H_{10}(OH)_8$ **glucononite,** $C_9H_{20}O_9$ or $C_9H_{11}(OH)_9$. These alcohols form colorless crystals and are produced by the action of nascent hydrogen upon their aldehyde alcohols

Glucoheptose, $C_7H_{14}O_v$, **glucooctose,** $C_8H_{16}O_x$, **glucononose,** $C_9H_{18}O_9$, are synthetically prepared from the hexoses $C_6H_{12}O_6$ (p. 451, 4) and yield

Glucoheptonic acid, $C_7H_{14}O_0$, **glucooctonic acid,** $C_8H_{16}O_9$, and **glucononic acid,** $C_9H_{18}O_{10}$, on oxidation. They may also be prepared from the glucoses, $C_6H_{12}O_6$, by the addition of hydrocyanic acid, as mentioned on p. 451, 4.

CARBOHYDRATES.

The name carbohdyrate depends upon the fact that all these compounds contain hydrogen and oxygen in the same proportion as they exist in water, but this proportion also exists in many other compounds, hence it is not characteristic of the carbohydrates.

In the broad sense the term carbohydrates is applied to all aldehyde alcohols (*aldoses*) and ketone alcohols (*ketoses*) of the polyhydric alcohols which contain a HO group attached to the C atoms neighboring the aldehyde or ketone groups, as well as their anhydride-like condensation products.

These compounds may be considered in connection with their corresponding polyhydric alcohols, and this plan has been adopted in connection with the carbohydrates containing less than six atoms of carbon. According to the number of C atoms the carbohydrates are now also called *bioses* ($C_2H_4O_2$, p. 402), *trioses* ($C_3H_6O_3$, p. 433), *tetroses*

$(C_4H_8O_4$, p. 443), *pentoses* $(C_5H_{10}O_5$, p. 444), *hexoses* $(C_6H_{12}O_6)$, *heptoses* $(C_7H_{14}O_7)$, *octoses* $(C_8H_{16}O_8)$, *nonoses* $(C_9H_{18}O_9)$.

This nomenclature is confusing because the varieties of sugars with 6 C atoms are often called monoses, and correspondingly those with 12 C atoms bioses and those with 18 C atoms trioses.

Carbohydrates in the narrow sense, or *saccharides* (from *saccharum,* sugar), are those carbohydrates with 6 or x6 carbon atoms occurring in nature. These are divided into the following groups:

> *Monosaccharides,* $C_6H_{12}O_6$.
> *Disaccharides,* $C_{12}H_{22}O_{11}$.
> *Trisaccharides,* $C_{18}H_{32}O_{16}$.
> *Polysaccharides,* $(C_6H_{10}O_5)x$.

All aldoses and pentoses with 5, 6, etc., C atoms have the same structure but different configuration (p. 303) and are therefore stereoisomers of each other. The stereoisomers occur dextro- and lævorotatory, inactive but cleavable (racemic, p. 39), and inactive but not cleavable. Correspondingly the presence of 3-4 or more asymmetric C atoms makes the number of possible stereoisomers very large (p. 307). Only a few of these are found in nature, while most of them have been prepared synthetically.

Occurrence. Up to the present time only pentoses and saccharides have been found in nature. They are especially very widely distributed in the plant kingdom and certain of them form important constituents of all plants. Some are found in the animal kingdom, partly under normal conditions and partly under pathological conditions.

Properties. The mono-, di-, and trisaccharides have a sweet taste and are crystallizable while the polysaccharides are not sweet, amorphous or are organized compounds. Those occurring naturally are soluble in water and are optically active while those carbohydrates prepared from optically inactive compounds are also optically inactive but can be decomposed into optically active modifications (pp. 39 and 306).

The carbohydrates are confusingly designated not only according to the rotation, for instance, *l-* or *d-* (p. 330), but also all carbohydrates obtained from *l-* or *d-*compounds even when they have another rotation. Thus the lævorotatory lævulose obtained from *d-*dextrose is designated *d-*lævulose, etc.

They are indifferent, i.e., they are neither acids nor bases. On heating they decompose into bodies of simpler constitution and leave carbon as a residue. On oxidation they are transformed into hexon acids (p. 446)—respectively saccharic acid, mannosaccharic acid, or mucic acid (p. 446)—and on stronger oxidation (fusion with caustic alkalies, etc.), they all yield oxalic acid. On boiling with dilute acids the di- tri- and polysaccharides take up H_2O (see Hydrolysis, p. 87) and are converted into monosaccharides, and this transformation may also be brought about by different enzymes (which see). With nitric acid they form, according to the temperature, etc., either nitrates or are oxidized to simpler compounds (p. 322). On heating with concentrated mineral acids they yield levulinic acid and humus substances.

Certain carbohydrates with six or one with nine C atoms readily suffer a deep cleavage by means of organized ferments which we call *fermentation* (p. 325). The chief products produced in this cleavage are alcohol, lactic acid or butyric acid, according to the variety of ferment. Each ferment can only split a compound of a certain configuration (p. 303); thus yeast can only ferment dextrorotatory monosaccharides and lævorotatory lævolose, while the disaccharides are first converted into the fermentable monosaccharides by the enzymes of the yeast. Many carbohydrates dissolve metallic oxides and form saccharates corresponding to the alcoholates (p. 343). The hydrogen of the OH groups may also be replaced by organic acids, alcohol radicals, etc. The latter compounds are widely distributed in the plants and are called *glucosides* (which see). The glucosides are split into their components by the action of acids, alkalies, and ferments, at the same time taking up water. Similar compounds to $C_6H_{11}O_6(CH_3)$ may also be obtained synthetically.

Preparation. Only the monosaccharides and those carbohydrates closely related thereto, but not found in nature, containing 2–9 atoms of carbon, as well as the disaccharides maltose and isomaltose, have been obtained synthetically.

The discovery of E. Fischer, that phenylhydrazin forms insoluble compounds called *osazones* with the above-mentioned carbohydrates, has made it possible to precipitate and identify the artificially obtained varieties of sugars prepared according to the methods to be described. The separation of the sugars from the accompanying by-products in the past was very difficult, as the impure sugars are crystallizable with difficulty.

The carbohydrates in question unite, on account of their containing aldehyde or ketone groups, with 1 molecule of phenylhydrazin: with the elimination of H_2O, and form the readily soluble hydrazones (p. 351, 11).

$$C_6H_{12}O_6 + H_2N-NH(C_6H_5) = C_6H_{12}O_5 = N-NH(C_6H_5) + H_2O.$$
d-Glucose.　Phenlhydrazin.　*d*-Glucosohydrazone.

The hydrazones or the carbohydrate itself when in acetic acid solution and warmed with an excess of phenylhydrazin unites with a second molecule of phenylhydrazin and yields yellow, insoluble, crystalline compounds called osazones or dihydrazones; thus, *d*-glucosohydrazone yields *d*-glucosazone:

$$CH_2-OH-(CH\cdot OH)_4-CH=N-NH(C_6H_5) + H_2N-NH(C_6H_5) =$$
$$H_2O + H_2 + CH_2\cdot OH-(CHOH)_3-C(=N-NH\cdot C_6H_5)-CH=N-NH(C_6H_5).$$

The melting-point of the osazones serves in the characterization of the different sugars. The sugar cannot be obtained directly from the osazones, but only indirectly; thus:

a. The osazones are decomposed by concentrated HCl, taking up $2H_2O$, into phenylhydrazin and osones:

$$C_6H_{10}O_4(=N-NH-C_6H_5)_2 + 2H_2O = C_6H_{10}O_6 + 2H_2N-NHC_6H_5.$$
d-Glucosazone.　　　　*d*-Glucosone.　Phenylhydrazin.

The osones contain 2 atoms H less than the sugar from which the osazones are obtained and yield the corresponding ketose by reduction:

$$C_6H_{10}O_6 \quad + \quad H_2 \quad = \quad C_6H_{12}O_6.$$
d-Glucosone.　　　　Lævulose (lævorotatory).

The sugars obtained are either active or inactive. These latter kinds can be decomposed by various methods into the two oppositely active modifications of which they are constructed.

The preceding example shows the conversion of the aldose (glucose) into the ketose (lævulose).

b. The osazones yield glucosamines on direct reduction (the glucosamines or amido sugars are sugar where one OH group is replaced by an NH_2 group):

$$C_6H_{10}O_4(=N-N-C_6H_5)_2 + 2H_2 + H_2O =$$
$$C_6H_{11}O_5(NH_2) + H_2N-C_6-H_5 + H_2N-NH(C_6H_5).$$

The glucosamines (see Glycoproteids) exchange, on treatment with HNO_2, the NH_2 group for the OH group (P. 378), producing the free sugar: $C_6H_{11}O_5(NH_2) + HNO_2 = C_6H_{12}O_6 + N_2 + H_2O.$

1. Mixtures of monosaccharides called methylenitan can be obtained by the polymerization of paraldehyde (P. 350) and formose by the polymerization of formaldehyde (p. 350). On the condensation of glycerine aldehyde with dioxyacetone (p. 433), $2C_3H_6O_3 = C_6H_{12}O_6$, or from acroleindibromide and baryta-water, $2C_3H_4Br_2O + 2Ba(OH)_2 = C_6H_{12}O_6 + 2BaBr_2$, we obtain *a*-acrose.

2. Monosaccharides and the related carbohydrates with 2 to 9 C atoms can be obtained by the careful oxidation of the respective alcohols and their separation as osazones or glucosamines, from which they can be set free in the manner given (p. 450) and finally transformed into their optically active modification (p. 39).

3. Monosaccharides can be prepared by hydrolysis of the di-, tri-, and polysaccharides by boiling with dilute acids or by ferments (p. 324).

4. In order to obtain a carbohydrate rich in C from one poorer in C we proceed as follows: The carbohydrates, on account of their containing the aldehyde and ketone groups, unite with hydrocyanic acid (p. 351, 9), and the nitriles produced are readily transformed into acids (p. 346, 2):

$$CH_2OH-(CH\cdot OH)_4CHO + HCN = CH_2\cdot OH(CH\cdot OH)_4-CH\cdot OH-CN;$$
$$\text{Dextrose.} \qquad\qquad \text{Dextroso-cyanhydrin.}$$

$$CH_2\cdot OH(CH\cdot OH)_4-CH\cdot OH-CN + 2HOH$$
$$= CH_2OH-(CH\cdot OH)_4-CH\cdot OH\cdot COOH + NH_3.$$
$$\text{Glucoheptonic acid, } C_7H_{14}O_8.$$

The acids thus obtained yields, on splitting off H_2O, good crystalline lactones (p. 404, a): $OH_2C-(CH\cdot OH)_4-CH\cdot OH)CO;$ and these by reducing agents yield the corresponding aldose:

$$HO\cdot H_2C(CHOH)_4-CH\cdot OH-CHO.$$

By this method heptoses, $C_7H_{14}O_7$, octoses, $C_8H_{16}O_8$, and nonoses, $C_9H_{18}O_9$, have been obtained; thus, from arabinose, $C_5H_{10}O_5$, we obtain arabinose cyanhydrin, $C_5H_{10}O_5(HCN)$, from which the arabinose carbonic acid, $C_6H_{12}O_7$ or $C_5H_{11}O_5(COOH)$, is derived and from this by internal anhydride formation, arabinose carbonic acid lactone, $C_6H_{10}O_5(CO)$ or $C_6H_{10}O_6$, which on reduction yields the sugar d-mannose, $C_6H_{12}O_6$.

ꞏ 1. Monosaccharides, $C_6H_{12}O_6$.

Glucose, Lævulose, Galactose, Sorbinose, Mannose.

The members of this group are also called *glucoses, monoses, hexoses;* they contain 5 HO groups and occur in part in nature and some of them are obtained synthetically. The synthetically prepared carbohydrates containing 2–9 carbon atoms also belong to this group on account of their behavior. They all contain 2 atoms H less than the corresponding alcohols, and as mentioned on p. 445, they can be converted into alcohols by nascent hydrogen and hence they are the aldehydes or ketones of these alcohols. They all reduce alkaline solutions of copper and silver, and like all aldehydes and ketones, unite with an excess of phenylhydrazin, forming yellow, crystalline osazones (p. 450), which are insoluble in water. When warmed with alkali hydroxides they turn yellow, then brown, and finally become resinous.

Dextrose, grape-sugar, *d*-glucose, glycose, also called diabetic sugar, starch-sugar.

Occurrence. In many sweet fruits and in honey mixed with lævulose, and in smaller amounts in many organs of the animal body, also in certain pathological urines (to 10 per cent.).

Preparation. By the action of dilute acids or unorganized ferments upon cane-sugar (accompanied by lævulose), also upon starch, cellulose, and many glucosides. It is prepared on a commercial scale by boiling starch with dilute sulphuric acid under pressure and recrystallizing the product.

Properties. Glucose crystallizes in warty, colorless masses with one molecule H_2O. From its solution in methyl alcohol it crystallizes in fine, anhydrous prisms which melt at 146°. It is about half as sweet as cane-sugar and dissolves in cold sulphuric acid without blackening, also in an equal weight of water. The fresh solution has twice the dextrorotatory power of an older solution (so-called multi- or birotation). It reduces metallic silver, as a mirror, from an ammoniacal silver solution and red cuprous oxide, from an alkaline cupric salt solution (Fehling's solution, p. 236), slowly in the cold and immediately on heating.

d-Glucose is converted into d-sorbite by nascent hydrogen and on oxidation it yields acids having the same amount of carbon, gluconic acid, $C_6H_{12}O_7$, and saccharic acid, $C_6H_{10}O_8$. Glucose therefore contains an aldehyde group and is an aldose having the structure $CH_2OH(CH\cdot OH)_4$–CHO.

Lævulose, fructose, diabetin, d-lævulose (so-called on account of its preparation from d-glucosazone, although it is lævorotatory (p. 448).

Occurrence. It is found with grape-sugar in most sweet fruits and in honey.

Preparation. Accompanied with d-dextrose from cane-sugar by the action of unorganized ferments or by boiling with dilute inorganic acids. From inulin by boiling with dilute mineral acids or mixed with d-mannose from d-mannite by oxidation.

Properties. It differs from dextrose only by its melting-point (95°), in the property it has of lævorotation (hence the name lævulose) and by being less crystallizable; but it occurs ordinarily as a colorless sweet sirup which is not very soluble in water but readily soluble in alcohol.

With nascent hydrogen, lævulose yields d-mannite, as it is first converted into d-mannose; on oxidation it yields glycollic and racemic acids, p o   having less C; hence it is a ketose of the structure $CH_2OH(CH\cdot OH)_3$–CO–CH_2OH.

d-**Galactose,** cerebrose, a stereoisomer of d-glucose, occurs in the brain and is produced, accompanied with *d*-glucose, from milk-sugar as well as from dextrorotatory varieties of gums on warming with dilute acids. It is more dextrorotatory than *d*-glucose and is insoluble in alcohol. With nascent hydrogen it yields inactive dulcite, and on oxidation galactonic acid, $C_6H_{12}O_7$, and then mucic acid, $C_6H_{10}O_8$, are produced.

l-Sorbinose, sorbin, sorbose, is a ketose stereoisomeric with lævulose, and occurs in the juice of the mountain-ash when it has stood for a long time. It forms colorless crystals whose solutions reduce alkaline copper solutions, but is not fermentable with yeast. It is not changed on boiling with acids and yields trioxyglutaric acid, $C_5H_8O_7$ (p. 425), and with nascent hydrogen it is converted into *d*-sorbite, $C_6H_{14}O_6$ (p. 446).

d-Mannose, seminose, a stereoisomer of *d*-glucose, is obtained with *d*-lævulose on the careful oxidation of mannite as well as by boiling the carbohydrate *seminin,* occurring in the earth-nut, with dilute acids. It yields mannonic acid, $C_6H_{12}O_7$ (p. 446), on oxidation and mannite (p. 445) on reduction with nascent hydrogen.

2. Disaccharides, $C_{12}H_{22}O_{11}$.

Saccharose, Lactose, Maltose, Mycose, Melebiose, Isomaltose.

The bodies of this group are also called *saccharoses* or *bioses* (p. 448) and contain 8 hydroxyl groups. They are anhydrides of two generally different monosaccharides (hence the name disaccharides) and decompose on heating with dilute acids into monosaccharides, at the same time taking up one molecule H_2O (by hydrolysis):

$$C_{12}H_{22}O_{11} + H_2O = C_6H_{12}O_6 + C_6H_{12}O_6.$$

Saccharose, cane-sugar, beet-root sugar, is the anhydride of *d*-glucose and *l*-lævulose. *Occurrence.* In the juice of many plants, especially in the sugar-cane (to 18 per cent.) and the sugar-beet (to 20 per cent.), from which it is chiefly obtained by evaporation. It is principally found in the stem or the roots of the plants, while glucose and lævulose occur to the greatest extent in the fruits.

Preparation. The juice from the sugar-cane or the sugar-beet is heated to boiling with milk of lime (calcium hydroxide), whereby the plant acids are neutralized (p. 454) and the proteids are coagulated and separate as a scum. At the same time a part of the calcium oxide forms with the sugar a soluble calcium saccharate, $C_{12}H_{22}O_{11} + CaO$. The juice is now treated with carbon dioxide, when a large part of the lime is precipitated as calcium carbonate and at the same time a considerable amount of contamination is precipitated out. After the separation of the precipitate the hot juice is filtered through bone-black, which removes the coloring matters and some contained lime and a part of the salts. The filtrate is evaporated to a sirupy consistency in vacuum-pans, when on cooling the cane-sugar crystallizes out. The sirupy, brown mother-liquor, *molasses,* contains still about 50 per cent. of sugar, which is prevented from crystallizing by the contained salts and organic substances (about 30 per cent.).

Cane-sugar molasses has a pure sweet taste and is often used instead of sugar, as well as in the preparation of rum (P. 355).

Beet-root molasses is used for feeding cattle, or alcohol is prepared therefrom by fermentation, or the sugar it contains is abstracted by the following method: It is boiled with an excess of strontium hydrate, which precipitates the sugar as strontium saccharate, $C_{12}H_{22}O_{11} + SrO$, which quickly settles and which can be transformed into crystallizable sugar and strontium carbonate by means of carbon dioxide. Another method is to remove the salts, preventing the crystallization of the sugar by means of dialysis (so-called diffusion method). The residue left after fermentation or after the abstraction of the sugar is incinerated in order to obtain the potash (p. 207).

Properties. White crystalline masses or white crystalline powder. Cane-sugar crystallizes, on slow evaporation, in large monoclinic prisms (rock-candy), has a sweeter and purer taste than grape-sugar, and is not very soluble in alcohol but readily soluble in water, forming a colorless, sweet, dextrorotatory sirup. On heating to 160° it melts and solidifies on cooling, forming an amorphous vitreous mass (barley-sugar) which after a certain time becomes crystalline and then opaque. On heating to 190–200° it is converted into *caramel*, $C_{12}H_{18}O_9$, an amorphous, not sweet, unfermentable, brown mass which is readily soluble in alcohol and used as *sugar color* in the coloring of liquors, etc. On further heating it decomposes with the generation of inflammable vapors and leaves porous shining carbon. On boiling with dilute acids (even with organic acids, hence the neutralization of these in the preparation of sugar, p. 453) it decomposes into a mixture of dextrose and lævulose, so-called *invert-sugar*, which is lævorotatory, because the lævulose turns the plane of polarized light stronger to the left than an equal amount of dextrose does to the right:

$$C_{12}H_{22}O_{11} + H_2O = C_6H_{12}O_6 + C_6H_{12}O_6.$$

When heated with alkalies cane-sugar does not turn brown, differing from glucose and milk-sugar.

Concentrated sulphuric acid carbonizes it even in the cold with the generation of SO_2. On warming with HNO_3 saccharic acid is produced: $C_{12}H_{22}O_{11} + 6O = 2C_6H_{10}O_8 + H_2O$; and on boiling with HNO_3 oxalic acid is obtained: $C_{12}H_{22}O_{11} + 18O = 6C_2H_2O_4 + 5H_2O$.

Alkaline solutions of silver or copper are only reduced after inversion (after boiling for a long time). Cane-sugar is not directly fermentable, but if yeast is added to a solution of cane-sugar it is transformed into fermentable invert-sugar by the *invertase* existing in the yeast. The aqueous solution of cane-sugar dissolves many metallic oxides to a great extent (see preparation); these solutions having a bitter taste and strong alkaline reaction.

Lactose, milk-sugar, lactobiose, is the anhydride of *d*-glucose and *d*-galactose.

Occurrence. In milk, amniotic fluid, in certain pathological secretions, in the urine of sucking animals.

Preparation. Milk which has been freed from casein and fat (the whey) is evaporated to crystallization and the milk-sugar thus obtained purified by recrystallization.

Properties. It forms crystalline masses with 1 mol. H_2O, or a crystalline powder which dissolves in 7 parts cold water and 1 part boiling water, producing a faintly sweet, dextrorotatory, not sirupy solution which shows birotation (p. 452). Lactose is nearly insoluble in dilute alcohol (used in detecting admixture with cane-sugar).

On boiling with dilute acids lactose is converted into a mixture of *d*-galactose and *d*-glucose; by nascent hydrogen it is transformed into mannite and dulcite. It does not ferment with pure yeast, although it readily undergoes lactic-acid fermentation by the action of certain fungi, especially in milk. It undergoes alcoholic fermentation by the kephir and tyrocola fungus (p. 325). On warming with nitric acid it is oxidized to mucic acid, and on boiling with HNO_3 oxalic acid is obtained. It is not decomposed by cold concentrated sulphuric acid. It reduces an ammoniacal silver solution even in the cold, but an alkaline copper solution is only reduced on heating (differing from glucose). On boiling for a long time with dilute H_2SO_4 levulinic acid is produced (p. 371). On heating to 180° it is converted into lactocaramel (p. 454).

Maltose, malt-sugar, maltobiose, is the anhydride of *d*-glucose. It is found in the contents of the small intestine, and is the sugar formed besides dextrins by the action of the enzyme diastase (malt) upon starch:

$$3C_6H_{10}O_5 + H_2O = C_{12}H_{22}O_{11} + C_6H_{10}O_5.$$
$$\text{Maltose.} \qquad \text{Dextrin.}$$

It can be synthetically prepared from *d*-glucose. It crystallizes with 1 mol. H_2O as hard white masses which consist of needles. It has a greater dextrorotatory action and is less soluble in alcohol than dextrose, and is split into 2 mol. *d*-glucose by boiling with dilute H_2SO_4, as well as by the action of the enzyme *maltase*. Maltose reduces alkaline copper solutions even in the cold, but to a less degree than dextrose. It readily ferments with yeast, as the enzyme maltase of the yeast first splits it into fermentable *d*-glucose.

Melibiose, the anhydride of *d*-glucose and *d*-galactose, is obtained, besides *d*-lævulose, in the inversion of melitriose and is stereoisomeric with milk-sugar.

Isomaltose is prepared synthetically from *d*-glucose, and is produced with maltose by the action of diastase upon starch. It has a weaker reducing power than maltose, but is not fermentable.

Mycose, trehalose, is found in certain fungi, in ergot, in the Manna trehala.

3. Trisaccharides, $C_{18}H_{32}O_{16}$.

Melitriose, Gentianose, Stachyose, Melezitose.

The members of this group are to be considered as formed by the union of equal molecules of the sugars, $C_6H_{12}O_6$ and $C_{12}H_{22}O_{11}$, with the elimination of one melocule of water. On boiling with dilute acids they take up water and decompose into different molecules.

Melitriose, gossypose, melitose, raffinose, the anhydride of melibiose and d-lævulose, $C_{18}H_{32}O_{16}+5H_2O$, occurs in the Eucalyptus manna, in the sugar-beet to a slight extent, and in the cottonseed. As it is more soluble than cane-sugar, it is found in cane-sugar molasses. The presence of melitriose in cane-sugar causes it to have a greater rotation, because melitriose has a greater dextrorotatory power than cane-sugar. It does not reduce copper solutions, but does ferment with yeast.

Gentianose, $C_{10}H_{32}O_{16}$, occurs in the roots of the Gentiana lutea.

Stachyose, $C_{18}H_{32}O_{16}+3H_2O$, in the Stachys tubifera.

Melezitose, $C_{18}H_{32}O_{16}+2H_2O$, is found in the juice of the larch-tree.

4. Polysaccharides, $(C_6H_{10}O_5)$.

Cellulose, Starch, Lignin, Inulin, Dextrin, Glycogen, Lichenin, Gums, Plant-mucilages, Pectine Bodies.

The members of this group may be considered as complicated anhydrides of one glucose, as they yield only one variety of sugar on heating with dilute acids. Their molecular weight is at all events $(C_6H_{10}O_5)x$, hence they are called polysaccharides. They differ from the other two groups by not being crystalline, but are either amorphous or organized and then insoluble in water.

Cellulose, Lignose. *Occurrence.* Forms with lignin (p. 458) and the pentosanes (p. 444) the crude fibre, the chief constituent of the cell-walls of all plants, and has an organized structure (p. 4). Purified cotton and filter-paper are nearly pure cellulose. Paper is more or less pure cellulose which has been freed from lignin, etc., by heating with concentrated caustic alkali or alkali sulphide solution or calcium bisulphite solution under pressure. The calcium bisulphite solution obtained is called lignosulphite and is used in medicine.

Preparation. Plant-fibres (cotton wool or filter-paper) is treated consecutively with dilute caustic alkali, dilute sulphuric acid, water, alcohol, and ether, which removes all impurities and leaves pure cellulose.

Properties. White amorphous powder, only soluble without change in an ammoniacal solution of copper oxide (p. 236). It is precipitated from this lævorotatory solution by acids. Concentrated sulphuric acid transforms cellulose after short action, without dissolving it, into a substance which is colored blue by iodine (detection of cellulose). After a longer action it dissolves in concentrated H_2SO_4 without blackening, and from this solution colloidal cellulose may be precipitated by water; this precipitate turns blue with iodine and shows by this reaction, as well as by the formation of closely related bodies, many similarities with starch and hence has been called *amyloid* (not to be confounded with the amyloid substance, a proteid). If sulphuric acid is allowed to act a still longer time, dextrin is produced, and if this is diluted with water and boiled, the dextrin is transformed into dextrose.

If unglazed paper is dipped for a short while into dilute sulphuric acid and then washed with water, the surface of the paper is transformed into amyloid and forms parchment-paper which is used extensively because of its similarity to parchment. On boiling with nitric acid or on fusing with caustic alkalies, cellulose is oxidized into oxalic acid. On putrefaction CO_2 and CH_4 are produced, these products occurring in the intestine from the cellulose of the food (p. 346). Animal cellulose in the covering of the tunicates is very similar to cellulose.

Hydrocelluloses are the compounds produced by the action of sulphuric acid or hydrochloric acid of certain concentration upon the celluloses (hydration products), and occur in plants as so-called hemicelluloses.

Oxycelluloses are oxidation products of the celluloses which contain the carboxyl, aldehyde, and ketone groups and which also occur in the p ants.

Cellulose nitrates, incorrectly called nitro-celluloses, are formed when cold nitric acid acts upon cellulose, such as cotton. The properties of these esters depend upon the length of action and the strength of the nitric acid used. In appearance they do not differ from the cotton or cellulose.

With caustic alkali or calcium sulphide or ferrous chloride solution they yield cellulose:

$$C_6H_8O_3(NO_3)_2 + 6FeCl_2 + 6HCl = C_6H_8O_3(OH)_2 + 6FeCl_3 + 2NO + 2H_2O.$$

Guncotton, pyroxylin, cellulose trinitrate, $C_6H_7NO_2(NO_3)_3$, is formed by the action of concentrated nitric acid ($HNO_3 + H_2SO_4$) and burns without explosion, but explodes violently when confined in an enclosed space by shock or by ignition with mercury fulminate. It is insoluble in a mixture of alcohol and ether.

Smokeless powder consists of guncotton which has been converted into an amorphous transparent mass by moistening with acetone, and explodes more slowly when granular than the original guncotton.

One gram guncotton yields 860 c.c. explosion-gases (p. 207), which expand to 7800 c.c. the moment they are set free by the heat generated.

Collodium cotton, colloxylin, cellulose nitrate, $C_6H_9O_4(NO_3)$, and *cellulose dinitrate,* $C_6H_8O_3(NO_3)_2$, are formed by the action of less concentrated nitric acid, are not explosive, and are soluble in alcohol-ether mixture, forming collodium.

Collodium leaves the colloxylin, on evaporation, as a transparent film (celloidin) Collodium cantharidatum contains the ethereal extract of the Spanish-fly. Collodium elasticum contains some castor-oil and turpentine. Zapon varnish is a solution of collodium cotton in acetone or amyl acetate. Celluloid, the substitute for hard rubber, is collodium cotton impregnated with camphor which is pressed and rolled.

Dualin, lithofracteur, Brain's powder, is a mixture of glycerine trinitrate with sawdust, etc., which has previously been treated with nitric acid or sulphuric acid. Explosive gelatine is guncotton impregnated with glycerine trinitrate. Artificial silk, which is very similar to natural silk, is prepared by pressing collodion into water and treating the very fine fibres of collodion wool thus obtained with calcium sulphide, which reduces them to silky cellulose fibres.

Lignin, xylogen, incrusting substance, occurs with cellulose as the chief constituent of wood, and is similar to it. It dissolves readily in $HNO_3 + KClO_3$, which serves in separating it from cellulose. Bodies containing lignin turn yellow in the air and light and also with aniline sulphate, and beautifully red with a solution of phloroglucin in concentrated HCl (detection of lignin in paper).

Starches, Amylum. *Occurrence.* They are found in nearly all plants, although not always, as microscopic granules of an organized structure whose size and configuration differ in the various species of plants, and this fact is used to differentiate between the various starches.

The starch granule consists chiefly of starch, which is called starch granulose and starch cellulose, also called farinose, which forms the structure of the grains and remains unchanged by the action of water, etc.

Preparation. On a large scale it is chiefly prepared from potatoes, wheat, or rice, which are macerated with water to a paste and then kneaded on sieves with water. The starch is hereby washed out and passes through the meshes of the sieve, while the cell membrane, the gluten, etc., remain behind. The starch is now allowed to settle and dried.

Properties. White odorless powder which is soluble in an aqueous solution of chloral hydrate but insoluble in cold water, alcohol, and ether. It attracts water from the air, and when treated with boiling water it is converted into a slimy mass which forms a paste on cooling and a hard mass on drying. If this paste is boiled for a long time with considerable water, the starch dissolves and alcohol precipitates from this solution a white amorphous powder which is soluble in water (*soluble starch, amidulin*). The solution is dextrorotatory, and on heating to 160–200° the starch is transformed into dextrin (starch gum). On boiling with dilute acids starch is converted into dextrin and finally into dextrose, at the same time taking up water. By the action of malt diastase it is first converted into soluble starch and then into dextrin, and finally into maltose and isomaltose.

Concentrated sulphuric acid dissolves starch with the formation of starch-sulphuric acid, which forms salts with bases. Concentrated nitric acid dissolves starch, and on diluting with water xyloidin, $C_{12}H_{19}O_9(NO_3)$, precipitates out; this xyloidin is explosive like the cellulose nitrates. On heating with nitric acid oxalic acid is produced. The deep blue coloration obtained by an aqueous solution of iodine (p. 143) with starch is characteristic of dissolved starch, as well as the starch in the grains. This coloration disappears on heating and reappears on cooling, and is used in the detection of starch.

The most important forms of starch are wheat starch, potato starch, arrowroot starch, sago and tapioca or cassava starch.

Inulin, found in the roots of Inula helenium, is soluble in hot water and turns yellow with iodine.

Glycogen, liver-starch, animal starch, occurs in the liver, in all developing animal cells, in many fungi, and in certain higher plants. It exists to a greater extent in horse-muscles, in the foetus, and in mollusks. On the death of the animal it is quickly transformed into dextrose. It is an amorphous colorless powder which turns reddish brown with iodine and is soluble in hot water, giving a dextrorotatory power to the solution.

Lichenin, moss starch, occurs in Iceland moss, is soluble in hot water, and turns blue with iodine.

Dextrins, Starch Gum. *Occurrence and Formation.* Dextrins is the name given to a series of intermediary products produced in the transformation of the starches into sugar, and of these only *amylo-*, *erythro-*, *malto-*, and three *achroodextrins* have been closely studied up to the present time. Dextrins are formed by gently roasting starch, as well as by the short action of malt diastase or saliva upon starch (occurrence in the crust of bread and in beer). It is prepared on a large cale (as a substitute for gum as an adhesive body) by moistening starch with 2 per cent. nitric acid and drying in the air and then heating to 110°.

Properties. Yellow amorphous masses readily soluble in water and dextrorotatory, but insoluble in alcohol. Most dextrins do not reduce alkaline copper solutions even on boiling, and are only fermentable by certain kinds of yeast. They are readily transformed into dextrose by dilute acids and converted into maltose by diastase. On oxidation they yield oxalic acid. Amylodextrin turns violet with iodine, erythrodextrin red, while the others do not change in color.

Gums, arabin, arabic acid is the chief constituent of the gums occurring in many plants, and is amorphous and readily soluble in water, but insoluble in alcohol. The aqueous solutions do not reduce alkaline copper solutions, but are precipitated by basic lead acetate. By dilute acids the lævorotatory gums are converted into arabinose, while the dextrorotatory ones yield galactose (p. 453). Nitric acid oxidizes it into mucic acid or oxalic acid, while iodine does not produce any color. Gum arabic consists of the calcium and potassium compounds of arabin. Gum mucilage is a sirupy solution of gum arabic in water.

Animal gum, occurring in the mucin of different organs, in chondrin, in the brain-tissue, etc., is similar to the plant-gums.

Plant-mucilage, Bassorin, shows the general properties of the gums, from which it differs by forming a mucilaginous solution with water which cannot be filtered. It readily dissolves in alkalies. It forms the chief constituent of gum tragacanth, of Bassora gum, of cherry and plum gum. Certain seeds, tubers, and roots, such as linseed, quince-kernels, salep, etc., are rich in vegetable mucilages.

Pectine Bodies. In certain fruits, roots, and barks we find non-nitrogenous amorphous bodies which are precipitated from their aqueous solution by alcohol as gelatinous precipitates and called pectine bodies, vegetable gelatine, or pectine. On account of the presence of these bodies many fruits solidify to a jelly after boiling (fruit jellies). The pectine bodies are closely related to the vegetable mucilages, but have less characteristic properties and are readily changed. They are derived from a body, *pectose*, which is insoluble in water and which forms with cellulose the chief mass of many fruits. Pectose is transformed by dilute acids or alkalies, or by an enzyme *pectase* which occurs in the ripe fruit, into *pectic acids*, which by hydrolysis are split into acids and pentoses or hexoses and which are related to the oxycelluloses.

II. ISOCARBOCYCLIC COMPOUNDS.

CONSTITUTION.

Isocarbocyclic compounds (P. 326) are those compounds whose molecule contains a ring-formed group of C atoms and are all derived from benzene, C_6H_6, whereby its H atoms are partly or entirely replaced by monovalent atoms (P. 465) or by groups of atoms (side chains), thus:

C_6H_5Cl	$C_6H_3(OH)_3$	$C_6H_2(CH_3)_4$	C_6Cl_6
Monochlorbenzene.	Pyrogallol.	Durene.	Hexochlorbenzene.

All isocarbocyclic compounds therefore contain a group of six C atoms united together, 18 of the 24 valences having been satisfied, while 6 valences are free to unite with other atoms, etc. Although in these compounds each of the 6 C atoms only has one free valence, still they behave somewhat like saturated compounds and cannot be converted into these without destruction of the molecule. For example, the saturated hydrocarbon corresponding to benzene, C_6H_6, must have the formula C_6H_{14} (P. 297), while benzene only slowly takes up H atoms, etc., by addition and forms bodies up to C_6H_{12}, so that of the 24 valences of its 6 C atoms, 12 are always mutually satisfied in the molecule, but still in a different way from the isomeric unsaturated aliphatic compounds, the olefines (P. 394).

This peculiar behavior of this carbon group of existing with only 6 free valences, as well as the chemical behavior of all the compounds belonging thereto (see Substitution), and above all the isomeric possibilities of these compounds (see Isomerism, p. 466), can be best explained by admitting that the 6 C atoms are alternately united to each other by one or two bonds, and that the last C atom is united with the first, so that the 6 C atoms form a closed ring-like chain, a so-called benzene ring (Kékulé's benzol theory), thus:

461

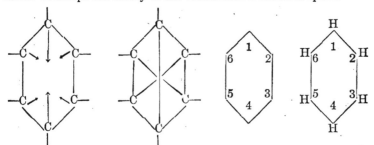

The constitution of the benzene ring is not the same in all benzene derivatives, but dependent upon the nature and position of the atoms or atomic groups introduced, so that recently different benzene formulæ have been suggested. Of these Clauss's diagonal and Baeyer's central formula correspond best with the facts; still these formulæ require some modifica·tion if we accept the theory of the tetrahedral C atoms in·space.

Central Formula.   Diagonal Formula.   Hexagon Scheme.   Benzene, C_6H_6.

In writing the structural formulæ of benzene derivatives we often do not express the mutual bonds between the C atoms, and only use a simple hexagon in which each angle represents a carbon atom with its one free valence.

In benzol one H atom is united to each of the six carbon atoms. These six H atoms may be replaced by atoms or atomic groups (substitution, p. 464), thereby producing benzene derivatives. Nevertheless in all these compounds the carbon ring can only be ruptured with the greatest difficulty by chemical action, i.e., the cyclic compounds are very stable and can only in a few instances be transformed into aliphatic compounds containing the same number of C atoms. The benzenes are only completely destroyed by very energetic oxidation with the formation of carbon dioxide, formic and acetic acids.

Besides the isocarbocyclic compounds with one C ring in the molecule we have those with several C rings chained and condensed together (p. 298).

Monovalent atoms, etc., may also, as mentioned on p. 461, unite directly with benzene and its derivatives, although not more than six are possible. The benzene ring remains closed in these compounds, and the double bonds of the C atoms are all of them or in part changed into single bonds:

Benzene, C_6H_6. Dihydrobenzene, C_6H_8. Hexahydrobenzene, C_6H_{12}.

$$
\begin{array}{ccc}
\underset{\displaystyle|}{\overset{\displaystyle H}{C}} & \underset{\displaystyle|}{\overset{\displaystyle H}{C}} & \underset{\displaystyle\|}{\overset{\displaystyle H_2}{C}} \\
H\text{-}C \diagup\diagdown C\text{-}H & H\text{-}C\diagup\diagdown C\text{-}H & H_2\text{-}C\diagup\diagdown C\text{=}H_2 \\
H\text{-}C\diagdown\diagup C\text{-}H & H_2\text{=}C\diagdown\diagup C\text{-}H & H_2\text{=}C\diagdown\diagup C\text{-}H_2 \\
\underset{\displaystyle H}{C} & \underset{\displaystyle H_2}{C} & \underset{\displaystyle H_2}{C}
\end{array}
$$

The benzene ring in C_6H_6 is also called tertiary, in C_6H_{12} secondary or reduced, in C_6H_8 and C_6H_{10} partially reduced.

These hydrogen addition products, also called *hydrocarbocyclic* compounds, such as hexahydrotoluene, $C_6H_{11}(CH_3)$, hexahydroxylene, $C_6H_{10}(CH_3)_2$, form with the naphthenes (see below) Caucasian petroleum. Most of them may be transformed into the corresponding isocyclic compound by oxidation. These addition products with their respective derivatives differ markedly from their mother-substances, as they behave like the aliphatic compounds and hence belong to the alicyclic compounds (p. 327); still in order to make the subject clear they will be treated in connection with their mother substances.

For example, di- and tetrahydrobenzene, C_6H_8 and C_6H_{10}, as well as their derivatives which still contain a few C atoms with double bonds, behave like the corresponding unsaturated aliphatic compounds. Hexahydrobenzene, C_6H_{12}, which contains the C atoms all with single bonds, behaves, with its derivatives, like a saturated aliphatic compound (see below). The most important compounds of this class are those of the terpene group, which will be considered later.

Closely related to the isocarbocyclic compounds and forming a connection with the aliphatic compounds we have those alicyclic compounds which form the simplest constituted compounds with C rings, namely, the *polymethylenes* or *naphthenes*, which consist of three or more methylene, groups, $^-CH_2^-$, or, as they are isomerides of the olefines, they **are also** called *cyclo-olefines;* e.g.,

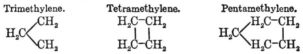

Trimethylene.   Tetramethylene.   Pentamethylene.

Hexamethylene, C_6H_{12}, is identical with the above-mentioned hexahydro-benzene, $C_6H_6(H)_6$ (p. 463).

They form the chief constituents of Caucasian petroleum (p. 341), and also occur in coal-tar and shale-oil, and in the resin-oil obtained in the distillation of colophonium. They behave like saturated aliphatic compounds, and differ from the isomeric olefines by their inability of forming addition products and their stability towards $KMnO_4$.

The acids of these hydrocarbons also occur in petroleum and are called *petrolic acids;* e.g., $C_6H_{10}(CH_3)(COOH)$. These acids are isomerides of the oleic acids, but cannot, like these, be converted into fatty acids.

SUBSTITUTION.

The most essential difference between the aliphatic and cyclic compounds is shown by substitution.

In the aliphatic hydrocarbons the hydrogen can only be directly replaced by other elements with difficulty. The halogens only have the power of expelling the hydrogen and of taking its place; hence these compounds are used in order to obtain new derivatives. In the cyclic hydrocarbons and their derivatives, on the contrary, the hydrogen of the benzene ring can not only be directly replaced by the halogens with ease, but also by the action of nitric acid or sulphuric acid, whereby the nitro group, $^-NO_2$, or sulphonic acid group, $^-SO_3H$, replaces the hydrogen, e.g.,

$$C_6H_6 + HO^-SO_2^-OH = C_6H_5^-SO_2OH + H_2O.$$
Benzene.   Sulphuric acid.   Benzenesulphonic acid.

$$C_6H_6 + 2{HO \atop HO}{>}SO_2 = C_6H_4{<}{SO_2^-OH \atop SO_2^-OH} + H_2O.$$
Benzenedisulphonic acid.

$$C_6H_6 + HO^-NO_2 = C_6H_5^-NO_2 + H_2O.$$
Nitric acid.   Nitrobenzene.

$$C_6H_5OH + HO^-SO_2^-OH = C_6H_4(OH)(SO_3H) + H_2O.$$
Phenol.   Sulphuric acid.   Phenolsulphonic acid.

$$C_6H_5OH + HO^-NO_2 = C_6H_4(OH)^-(NO_2) + H_2O.$$
Phenol.   Nitric acid.   Nitrophenol.

These sulphonic acid and nitro compounds are isomeric with the sulphurous acid and nitrous acid compounds. Still in the sulphonic acids and nitro bodies the sulphur or the nitrogen is directly united to the carbon atom, while in the sulphites and nitrites the binding takes place through the oxygen atoms: $C_2H_5{}^-O{}^-SO^-OH$, ethyl sulphite; $C_2H_5{}^-ONO$, ethyl nitrite (differentiation by nascent hydrogen, see p. 319).

That portion of the benzene molecule remaining after substitution is called the benzene nucleus.

With the aliphatic bodies the sulphuric acid and nitric acid act only upon the alcohols or unsaturated hydrocarbons and form esters with them (pages 357 and 395); e.g.,

$$C_2H_5OH + HO{-}SO_2{-}OH = C_2H_5{-}O{-}SO_2{-}OH + H_2O.$$
Ethyl alcohol. Ethylsulphuric acid.

The sulphonic acid and nitro compounds of the aliphatic compounds can only be obtained in an indirect manner; thus, by the action of the alkyl iodides upon silver sulphite or nitrite:

$$AgNO_2 + C_2H_5I = C_2H_5{-}NO_2 + AgI;$$
$$Ag_2SO_3 + 2C_2H_5I = C_2H_5{-}SO_2{-}O{-}C_2H_5 + 2AgI.$$
Ethyl sulphonic acid ethyl ester.

On heating the last ester with water, alcohol is split off and an aliphatic sulphonic acid is obtained.

The union of the halogen atoms in the benzene ring is much firmer than in the open C chain of the aliphatic compounds, so that they generally cannot be replaced by other groups by double decomposition.

A polyvalent element never replaces several hydrogen atoms in one benzene molecule, therefore compounds like $C_6H_4{=}O$ or $C_6H_3{\equiv}N$ are unknown. On the contrary, a polyvalent radical can replace several H atoms in a benzene molecule (see Terpenes and Alkaloids).

The amido bodies are obtained from the nitro bodies by reduction (by nascent hydrogen):

$$C_6H_5NO_2 + 6H = C_6H_5NH_2 + 2H_2O.$$

In this reduction *azo* compounds (*azote* = nitrogen) appear as intermediate products. They contain the divalent group $^-N{=}N^-$, consisting of two trivalent N atoms, which is united by both valences with two cyclic radicals.

Closely related to these we have the *diazo* compounds which contain the group $^-N{\equiv}N$, consisting of one pentavalent and one trivalent N atom, which is united with only one valence with one C atom of one cyclic radical; e.g.,

$C_6H_5{-}N{=}N{-}C_6H_5$ $(NO_3)(C_6H_5){=}N{\equiv}N$
Azobenzene.   Diazobenzene nitrate.

If the hydrogen atoms of the benzene ring of benzene and its derivatives are replaced by hydroxyl groups, we obtain the phenols which are comparable with the alcohols (p. 474):

$$C_6H_5(OH), \qquad C_6H_4(OH)_2, \qquad C_6H_3(OH)_3.$$

The phenols contain, like the tertiary alcohols, the group $\equiv C^-OH$ (p. 334) and do not yield aldehydes, ketones, and acids on oxidation, as the C atom of this group is only monovalent:

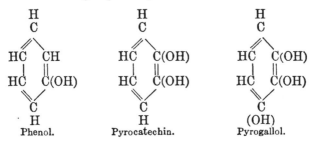

Phenol.   Pyrocatechin.   Pyrogallol.

On replacing the H atoms in benzene consecutively by alkyls we obtain homologues of the benzene hydrocarbons which are richer in carbon (p. 472):

Methyl benzene, $C_6H_5(CH_3)$;
Dimethyl benzene, $C_6H_4(CH_3)_2$;
Trimethyl benzene, $C_6H_3(CH_3)_3$, etc.

In these compounds as well as their derivatives (see Isomerism) the benzene residue retains the properties of the benzene and can be readily replaced by halogens, $^-NO_2$, $^-SO_3H$, etc. On the other hand, the aliphatic side groups behave in a manner similar to the aliphatic hydrocarbons. While, for example, the halogen atoms contained in the benzene nucleus are very firmly united, those in the side chain behave similar to the aliphatic derivatives, and we can therefore readily replace them by monovalent groups. All the homologues of benzene, on the contrary, can be readily oxidized into benzene carboxyl acids in contradistinction to the paraffins (p. 473).

ISOMERISM.

On substitution either in the benzene ring or in the side chains we obtain two series of isomeric compounds. For example, from methyl benzene or toluene the following series of isomers are derived:

Monochlortoluene, $C_6H_4Cl-CH_3$; Benzyl chloride, $C_6H_5-CH_2Cl$;
Cresol, $C_6H_4(OH)-CH_3$; Benzyl alcohol, $C_6H_5-CH_2-(OH)$.

If the hydrogen atoms of the aliphatic side chain are replaced by hydroxyl, we obtain true alcohols of the benzene series, which on oxidation yield aldehydes, ketones, and acids:

$C_6H_5-CH_3$, $C_6H_5-CH_2OH$, C_6H_5-CHO, C_6H_5-COOH.
Methyl benzene. Benzyl alcohol. Benzaldehyde. Benzoic acid.

Substitution in the nucleus is indicated as *endo*-substitution, and in the side chain as *exo*-; thus, endochlortoluene, $C_6H_4Cl-CH_3$, and exochlortoluene, $C_6H_5-CH_2Cl$; ω- denotes substitution at the last C atom of the side chain ($C_6H_5-CH_2-CH_2Cl$, ω-chlorethyl benzene), while α-, β-, etc., denote substitution at the succeeding C atoms; thus, $C_6H_5-CH_2-CHCl-CH_3$ is α-chlorpropyl benzene).

From this it follows that, by substitution in the benzene ring or in the side-chain, we may have a great variety of benzene derivatives, and the number becomes still greater on account of the isomerism possible on account of the structure of benzene.

If any hydrogen atom in the benzene molecule is replaced by another atom or an atomic group, each compound thus obtained can only exist in one modification. Hence we have only one chlorbenzene, one nitrobenzene, one methyl benzene, etc. The 6 H atoms of benzene have the same value, depending upon their mutual position.

If, on the contrary, two hydrogen atoms of the benzene are replaced by two similar or different monovalent atoms or atomic groups, then it follows that three modifications of such a compound are possible, and in fact all three theoretically possible isomers of most compounds are known and are designated as follows:

Ortho and Meta and Para compounds.

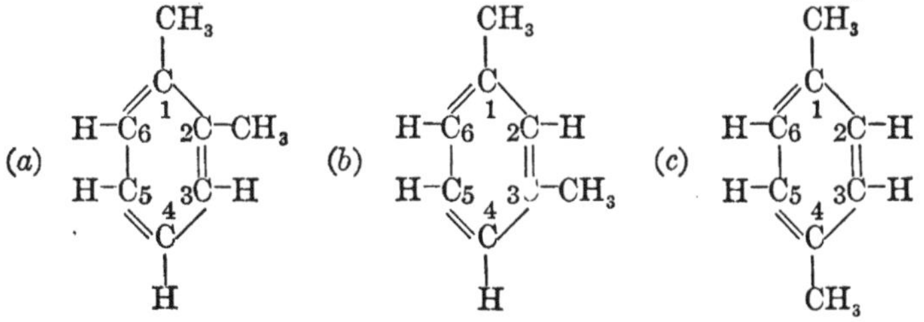

If, for example, we admit that two CH_3 groups are substituted for 2 H atoms, then these methyls take the following positions if we number the C atoms from 1 to 6 as above: (*a*) 1:2, (*b*) 1:3, (*c*) 1:4.

Other positions are not possible, as position 1:6 is the same as *a,* and 1:5 the same as *b.*

In the *ortho compounds* the neighboring hydrogen atoms are substituted (position 1:2 or 1:6).

In the *meta compounds* a hydrogen atom exists between the substituted hydrogen atoms (position 1:3 or 1:5).

In the *para compounds* two H atoms exist between the substituted H atoms; that is, two opposite hydrogen atoms are substituted (position 1:4 or 2:5 or 3:6).

We designate these isomerides by prefixing *o-, m-, p-,* or $1\cdot2$, $1\cdot3$, $1\cdot4$ to the formula; thus, $p\text{-}C_6H_4Cl_2$ or $1\cdot4$ $C_6H_4Cl_2$ is paradichlorbenzene.

These isomers are also called *position* or *nucleus isomers* (p. 302). Those isomers which are obtained by substitution once in the benzene ring and another time in the side-chain are called *mixed isomers;* thus, $C_6H_4Br\text{-}CH_3$ and $C_6H_5CH_2Br$. Isomers produced by substitution in the side-chain are called *side-isomers;* e.g., $C_6H_5\text{-}CH_2\text{-}CH_2Cl$ and $C_6H_5\text{-}CHCl\text{-}CH_3$.

By various chemical operations, for example, oxidation, further substitution, condensation, etc., it is possible to determine the relative position of the substituted groups in the benzene ring. It follows that mesitylene, $C_6H_3(CH)_3$, has a systematic structure, i.e., the methyl groups have the position $1:3:5$, because mesitylene is formed by condensation from 3 molecules of acetone. If we heat acetone, $CH_3\text{-}CO\text{-}CH_3$, with sulphuric acid, O and H_2 are removed as water, and three residues $=C(CH_3)\text{-}CH=$ unite, forming mesitylene just similar to the condensation of 3 mol. acetylene, forming benzene:

$$
\begin{array}{c}
\underset{3}{\overset{\text{CH}_3\text{H}}{\underset{=\text{C——C}=}{\Big|\quad\Big|}}}
\end{array}
\;=\;
\begin{array}{c}
\text{CH}_3\;\;\text{H}\;\;\text{CH}_3\;\;\text{H}\;\;\text{CH}_3\;\;\text{H}\\
\Big|\quad\;\;\Big|\quad\;\Big|\quad\;\;\Big|\quad\;\Big|\quad\;\;\Big|\\
\text{C——C}=\text{C——C}=\text{C——C}
\end{array}
$$

The ortho compounds are characterized by readily forming from **one** molecule so-called internal anhydrides on splitting off of water:

$$C_6H_4<^{NH_2}_{CH_2COOH} \quad \text{yields} \quad C_6H_4<^{NH}_{CH_2}\geqslant CO + H_2O.$$

o-amidophenyl acetic acid.   Oxindol.

$$C_6H_4<^{CO\text{-}COOH}_{NH_2} \quad \text{yields} \quad C_6H_4<^{CO}_{N}\geqslant COH + H_2O.$$

Isatinic acid.   Isatin.

Many derivatives of orthophenylendiamine, for example, the acetyl, propionyl, benzoyl, etc., derivatives, yield benzenediazoles (which see):

$$C_6H_4<^{NH(CH_3\cdot CO)}_{NH_2} \quad \text{yields} \quad C_6H_4<^{N}_{NH}\geqslant C\cdot CH_3 + H_2O.$$

Acetyl-o-phenylendiamine.   Ethenylphenylenamidine.

From the first two processes it follows that the splitting off of H_2O from o-amido acids takes place where only one H atom of the NH_2 group goes out with the OH group, which is designated as *lactame formation*, or both H atoms of the NH_2 group are taken away, which is called *lactime formation.*

If three or four hydrogen atoms of benzene are replaced by equivalent atoms or radicals, then necessarily three isomers are possible, while if five or six hydrogen atoms of benzene are substituted, only one modification is possible; thus we have only one pentachlorbenzene, C_6HCl_5, and one hexachlorbenzene, C_6Cl_6, etc.

The isomers of the tri- and tetra-substitution products are designated by v (*vicinus, neighboring*), s (*symmetrical*), a (*asymmetrical*). The following figures show the relative positions of substitution:

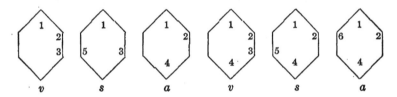

v s a v s a

If the 3, 4, etc., substituted atoms or radicals are different, then the possible isomers are still greater; thus, six isomers of the formula $C_6H_3(X)_2(Y)$ and ten isomers of the formula $C_6H_3(X)(Y)(Z)$ are possible.

As with the corresponding bi-derivatives (p. 467), the C atoms of the benzene nucleus are numbered from 1 to 6, and the group containing the atom having the smallest atomic weight united directly with the nucleus, occupies the position 1. The remaining substituted groups are further indicated in their order, so that the increased atomic weight of the atoms directly united with the nucleus is shown. If two equal atoms are united with the nucleus, then the other atoms of the group are considered according to their atomic weights.

With several side-chains containing carbon the one which causes the least increase in molecular weight takes first place; for example,

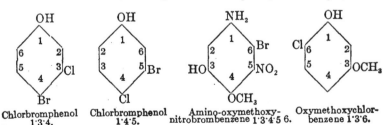

Chlorbromphenol 1·3·4.   Chlorbromphenol 1·4·5.   Amino-oxymethoxy-nitrobrombenzene 1·3·4·5 6.   Oxymethoxychlorbenzene 1·3·6.

With the alicyclic compounds (P. 327) the number of isomers is often still greater as soon as the mutual position of the double bonds is taken into consideration.

In order to designate the position of the double bonds, $\varDelta$ with the number of the C atoms where the double bonds exist is placed before the name of the substance; thus, dihydrobenzene, p. 463, $\varDelta \cdot 2 \cdot 6$ (arrangement of the C atoms, pp. 462 and 467).

RELATIONSHIP BETWEEN ISOCARBOCYCLIC AND ALIPHATIC COMPOUNDS.

The formation of isocarbocyclic compounds from aliphatic ones is possible only in a relatively small number of instances, and is designated as nucleus synthesis.

Many aliphatic hydrocarbons and alcohols when passed through red-hot tubes in the form of vapor yield isocarbocyclic compounds; thus, methane, ethyl alcohol and acetylene yield benzene, C_6H_6. On distilling allylene with dilute sulphuric acid we obtain mesitylene: $3C_3H_4 = C_6H_3(CH_3)_3$, and the same from acetone: $3C_3H_6O = C_6H_3(CH_3)_3 + 3H_2O$. In the same manner many homologues of acetylene, also many ketones, ketone aldehydes, ketonic acids, ketonic acid esters, and aldehyde acids yield isocarbocyclic compounds.

In sunlight propiolic acid yields trimesic acid: $3C_3H_2O_2 = C_6H_3(COOH)_3$; bromacetylene yields tribrombenzene under the same conditions:

$$3C_2HBr = C_6H_3Br_3.$$

On the oxidation of wood charcoal or graphite we obtain mellitic acid $C_6(COOH)_6$, and by the action of CO upon K with heat hexaoxybenzene potassium, $C_6(OK)_6$, is formed (p. 189). Triethylphloroglucin is obtained by the action of $AlCl_3$ upon butyryl chloride: $3CH_3-CH_2-CH_2-COCl = C_6H_3(O \cdot C_2H_5)_3 + 3HCl$; and by the action of potassium bisulphate upon geranial we obtain cymene: $C_{10}H_{16}O = C_6H_4(CH_3)(C_3H_7) + H_2O$. The formation of aliphatic compounds from isocarbocyclic compounds is known only in a few instances.

Most of the isocarbocyclic compounds are very resistant towards the action of high temperatures. When benzene is passed through red-hot tubes it yields acetylene in part, and as this latter also forms benzene it is possible that both reactions must proceed until an equilibrium is established.

Nevertheless it is possible to rupture the benzene ring (especially the phenols, quinones, and their derivatives) by chemical means. In these cases alicyclic intermediary products are produced, and these are isolated with difficulty.

On strong oxidation compounds with 1 and 2 C atoms are produced, for example CO_2, formic acid, oxalic acid, while on milder oxidation (dilute $KMnO_4$ solution) phenol, $C_6H_5(OH)$, is converted into mesotartaric acid, $C_4H_6O_6$, and oxalic acid, while pyrocatechin, $C_6H_4(OH)_2$, and dioxybenzoic acid, $C_6H_3(OH)_2(COOH)$, are oxidized by HNO_2 into dioxytartaric acid, $C_4H_6O_8$ ($= C_4H_2O_6 + 2H_2O$).

On treatment with chloric acid a simultaneous oxidation and chlorination takes place. Thus from benzene we obtain trichloracetylacrylic acid, $CCl_3-CO-CH=CH-COOH$ (p. 440), from phenol, $C_6H_5(OH)$, or salicylic acid, $C_6H_4(OH)(COOH)$, trichlorpyroracemic acid, $CCl_3-CO-COOH$ (p. 428), and chlorpicrin, $CCl_3(NO_2)$, from picric acid, $C_6H_2(OH)(NO_2)_3$.

Chlorine often acts in a similar manner. Di- and trioxybenzenes are converted by chlorine into chlorinated aliphatic acids or ketones with 4 and 5 C atoms; thus, resorcin, $C_6H_4(OH)_2$, is converted into pentachlorglutaric acid, $HOOC-CCl_2-CHCl-CCl_2-COOH$ (p. 425).

By reduction (nascent H) from the phenolcarboxyl acids, for example from salicylic acid, $C_6H_4(OH)(COOH)$, and its derivatives, we obtain pimelic acid, $HOOC-(CH_2)_5-COOH$, and its derivatives. Dihydroresorcin, $C_6H_4(OH)_2H_2$, yields acetylbutyric acid, $CH_3-CO-CH_2-CH_2-CH_2-COOH$, when heated with $Ba(OH)_2$. Benzene yields hexane, C_6H_{14}, when heated to 280° for a long time with HI.

The formation of alicyclic compounds (p. 327) with 6 C atoms in the ring may, corresponding to their intermediary position, take place either from the aliphatic or from the isocarbocyclic compounds.

To these belong the addition products of benzene and their derivatives, of which, for example, hexahydrobenzene, C_6H_{12}, can be produced by the addition of H to benzene, as well as by the action of Na upon dibrompropane: $2CH_2Br-CH_2-CH_2Br+2Na = 2NaBr+C_6H_{12}$. The polymethylenes (p. 463) also belong to the alicyclic compounds, also certain aliphatic compounds with ring-shaped closed atoms closely related to the heterocyclic compounds (which see).

NOMENCLATURE.

The nomenclature of the isocarbocyclic compounds corresponds in general with that of the aliphatic compounds. The differences as well as the designation of certain groups not mentioned in connection with the aliphatic compounds have been given in the preceding pages. In regard to the significance of the letters α-, β-, ω-, etc., see p. 467; of o-, m-, p-, see p. 468; of s-, a-, v-, see p. 469; of $\varDelta$, see p. 470. In regard to the meaning of $1\cdot2$, $1\cdot3$, $1\cdot4$, see p. 468, and of exo- and endo-, see p. 467.

The radical $-C_6H_5$ is called *phenyl*, $C_6H_5-CH_2$, *benzyl*, etc. The monovalent isocarbocyclic hydrocarbon radicals are called *alphyles*.

CLASSIFICATION.

In the following pages those compounds with one benzene ring will be considered in special groups according to the number of aliphatic hydrocarbons introduced; at the same time those compounds standing in close genetic relationship to them will also be considered. After this the compounds containing several benzene rings will follow, each divided into groups according to the mutual binding of the benzene rings. Then follow the alicyclic terpene groups, and finally those compounds belonging to the group of glucosides, bitter principles, and coloring matters which, on account of their not sufficiently known constitution, cannot be placed in the other groups.

THE MOST IMPORTANT ISOCARBOCYCLIC COMPOUNDS IN GENERAL.

1. Hydrocarbons.

General formula $C_N H_{2N-6}$.

Benzene,	C_6H_6.	—	—
Toluene,	$C_6H_5(CH_3)$.	—	—
Xylenes,	$C_6H_4(CH_3)_2$.	Ethyl benzene,	$C_6H_5(C_2H_5)$.
Mesitylene,			
Pseudocumene,	$\Big\} C_6H_3(CH_3)_3$.	Propyl benzenes,	$C_6H_5(C_3H_7)$.
Hemimellitene,			
Durene,		Butyl benzenes,	$C_6H_5(C_4H_9)$.
Isodurene,	$\Big\} C_6H_2(CH_3)_4$.	Cymene,	$C_6H_4 < ^{CH_3}_{C_3H_7}$.
Prehnitene,			
Pentamethyl benzene,	$C_6H(CH_3)_5$.	Amyl benzenes,	$C_6H_5(C_5H_{11})$.
Hexamethyl benzene,	$C_6(CH_3)_6$.	Triethyl benzenes,	$C_6H_3(C_2H_5)_3$.

In these homologues of benzene the isomers differ chiefly in their boiling-points, as well as in their behavior on oxidation (p. 473).

The unsaturated benzene hydrocarbons are derived from the polyvalent hydrocarbons of the olefine and acetylene series, etc., in which hydrogen is replaced by the phenyl radical, C_6H_5, or derivatives of the same; e.g., $C_6H_5-CH=CH_2$, phenyl ethylene; C_6H_5-C CH, henyl acetylene. These are readily transformed into saturated compounds by addition.

Occurrence. Many occur in coal-tar. Benzene, toluene, and their hydrogen addition products (p. 463) occur with naphthenes (p. 464); also in Caucasian petroleum.

Formation. 1. By the action of sodium upon a mixture of brombenzenes and alkyl bromides (p. 340, 1):

$$C_6H_5Br + CH_3Br + 2Na = C_6H_5(CH_3) + 2NaBr;$$
$$C_6H_4Br-CH_3 + CH_3Br + 2Na = C_6H_4(CH_3)_2 + 2NaBr.$$

2. By the action of alkyl chlorides upon isocyclic hydrocarbons in the presence of anhydrous aluminium chloride (Friedel-Craft's synthesis, p. 320):

$$C_6H_6 + CH_3Cl = C_6H_5-CH_3 + HCl.$$

3. From diazo compounds (which see) by heating with alcohols.

4. On the dry distillation of the corresponding isocyclic acid with caustic lime (p. 340, 2):

$$\underset{\text{Benzoic acid.}}{C_6H_5COOH} + CaO = \underset{\text{Benzene.}}{C_6H_6} + CaCO_3.$$

5. On the dry distillation of various non-volatile carbon compounds, such as wood, resin, bituminous shale, and especially coal. They are also produced from volatile fatty bodies (such as methane, alcohol, ether, petroleum) when their vapors are passed through red-hot tubes.

Preparation. They are chiefly obtained on a large scale from coal-tar (p. 323), which contains over 40 cyclic compounds and also a few aliphatic compounds. They are separated therefrom by fractional distillation.

Coal-tar can be separated by fractional distillation into the following four parts:

a. The light oil, having a specific gravity of 0.8–0.9, contains the products boiling below 170°, and consists chiefly of benzene, indene, styrene, toluene, xylene, trimethylbenzene pyridine, and thiophene.

b. The middle or carbolic oil, sp. gr. 0.9–0.98, boiling between 170° and 230°, consists chiefly of phenol, cresols, aniline, naphthalene, and quinoline bases.

c. The heavy oil, which sinks in water and boils between 230° and 270°, contains cresols, xylenoles, pyridine and quinoline bases, and about 50 per cent. naphthalene.

d. The anthracene oil, boiling above 270°, contains acenaphthene, anthracene, phenanthrene, pyrene, chrysene, etc., and is used, under the name carbolineum, as a preservative agent for wood.

Properties. Durene, penta- and hexamethyl benzene form colorless crystals, while the others are colorless liquids which are volatile without decomposition, and insoluble in water but soluble in alcohol and ether. They have a peculiar odor and burn with a smoky flame. With sulphuric acid they yield sulphonic acids (p. 464) and nitro bodies with nitric acid (p. 464), and with nascent hydrogen they are transformed into hydrocyclic compounds. As the homologues of benzene contain alkyls, they have the properties of the aliphatic compounds as well as of the cyclic compounds, and may also yield the corresponding derivatives (p. 467). Although the isomers of the benzene hydrocarbons are very similar in behavior, still they are readily differentiated by oxidation, as in this operation the benzene remains unchanged, while each aliphatic side-chain is transformed into a carboxyl group irrespective of the number of C atoms existing therein.

Thus methyl benzene, $C_6H_5-CH_3$, ethyl benzene, $C_6H_5-C_2H_5$, amyl benzol, $C_6H_5-C_5H_{11}$, phenyl ethylene, $C_6H_5-C_2H_3$, etc., yield the same monocarbonic acid, C_6H_5-COOH (benzoic acid).

Dimethyl benzene, $C_6H_4(CH_3)_2$, diethyl benzene, $C_6H_4(C_2H_5)_2$, methyl propyl benzene, $C_6H_4(CH_3)(C_3H_7)$, etc., yield the corresponding dicarbonic acid, $C_6H_4(COOH)_2$ (phthalic acid, p. 476), etc.

As the halogens act upon the benzene nucleus as well as upon the side-chain, we obtain, according to conditions, isomers which have entirely different properties (p. 467). While the halogen in the side-chain, as in all aliphatic compounds, is readily replaceable, those in the benzene nucleus are very firmly combined and cannot be removed by either alcoholic or aqueous caustic potash, nor by silver salts or ammonia, but only by metallic sodium or nascent hydrogen.

Substitution in the benzene nucleus takes place when halogens act thereon in the cold with the exclusion of direct sunlight or in the presence of iodine. These products have an aromatic odor and their vapors do not irritate the eyes or nose. According to the length of action of the halogens, the hydrogen is in part or entirely substituted: C_6H_5Br, $C_6H_4Cl_2$, $C_6H_3I_3$, $C_6H_2I_4$.

Substitution in the side-chain occurs by the action of halogens in the warmth or in cold with direct sunlight (without the addition of iodine). These products have a pungent odor, and the vapors cause great irritation of the eyes and nose.

With benzene itself, independent upon temperature, the formation of addition products (P. 463) always takes place in direct sunlight with the formation of $C_6H_6Cl_2$, $C_6H_6Cl_4$, and finally $C_6H_6Cl_6$. In diffused sunlight or in the presence of iodine substitution products are produced. The iodine derivatives are only obtained in the presence of oxidizing bodies, especially HgO and HIO_3, which destroy the HI produced (p. 322).

In regard to the formation of the benzene halogens from the phenols see page 475.

2. Phenols.

a. *Monohydric Phenols.*

Benzophenol,	C_6H_5OH.
Cresols,	$C_6H_4(CH_3)(OH)$.
Xylenols,	$C_6H_3(CH_3)_2(OH)$.
Mesitol, Pseudocumenol,	$\}C_6H_2(CH_3)_3(OH)$.
Durenols,	$C_6H(CH_3)_4(OH)$.
Thymol, Carvacrol,	$\}C_6H_3{<}^{CH_3}_{C_3H_7}(OH)$.
Pentamethyl phenol,	$\}C_6H(C_3)_5(OH)$.

b. *Dihydric Phenols.*

Hydroquinone, Pyrocatechin, Resorcin,	$\}C_6H_4(OH)_2$.
Orcin, Homopyrocatechin,	$\}C_6H_3(CH_3)(OH)_2$.
Dioxyxylenes,	$C_6H_2(CH_3)_2(OH)_2$.
Mesorcin,	$C_6H(CH_3)_3(OH)_2$.
Thymohydroquinone,	$\}C_6(CH_3)(C_3H_7)(OH)_2$.

c. *Polyhydric Phenols.*

Pyrogallol, Oxyhydroquinone, Phloroglucin,	$\}C_6H_3(OH)_3$.

Methylpyrogallol,	$C_6H_2(CH_3)(OH)_3$.
Tetraoxybenzenes,	$C_6H_2(OH)_4$.
Hexaoxybenzene,	$C_6(OH)_6$.

Properties. The benzene compounds with the hydroxyl substituted in the benzene ring are not called alcohols, but phenols, as they differ essentially from the alcohols. (True alcohols of the benzene series, see pages 467 and 475.) The phenols cannot be oxidized to the corresponding aldehydes and acids (p. 466). They are colorless liquids or solids, generally characterized by their odor, soluble in alcohol and ether, some readily soluble in water and others with difficulty, and can be distilled without decomposition. They also have more acid-like properties and combine readily with metallic oxides, forming salt-like compounds. Thus phenol, C_6H_5OH, dissolves in caustic soda, forming sodium phenylate, C_6H_5-ONa; lead oxide dissolves in phenol. forming lead phenylate $(C_6H_5O)_2Pb$: mercuric oxide dissolves in phenol forming mercuric phenylate, $(C_6H_5O)_2Hg$.

If Cl, Br, or NO_2 is introduced into the benzene nucleus besides hydroxyl, we obtain compounds which behave nearly like true acids. The hydrogen of the hydroxyl can be readily replaced by alcoholic or acid residues, producing mixed ethers or esters:

$$C_6H_5-O(CH_3), \quad (C_6H_5O)_2CO, \quad or \quad (C_6H_5)_2CO_3.$$

Methylphenol.   Phenyl carbonic acid ester.

On heating with zinc powder the phenols are converted into aromatic hydrocarbons: $C_6H_5OH + Zn = C_6H_6 + ZnO$.

On treating the phenols with nitric or sulphuric acid we do not obtain the corresponding ester as with the alcohols: $C_2H_5OH + HNO_3 = C_2H_5NO_3 + H_2O$, but obtain instead nitro compounds or sulphonic acids of the phenols (p. 464):

$$C_6H_5OH + HNO_3 = C_6H_4 < {NO_2 \atop OH} + H_2O;$$

$$C_6H_5OH + H_2SO_4 = C_6H_4 < {SO_3H \atop OH} + H_2O.$$

The phenols exchange their hydroxyl on treatment with phosphorus chloride, bromide, and iodide for chlorine, bromine, and iodine:

$$3C_6H_5OH + PCl_3 = 3C_6H_5Cl + H_3PO_3;$$

while on the other hand the direct action of the halogens causes a substitution of the hydrogen of the benzene ring; e.g.,

$$C_6H_5OH + Cl_2 = HCl + C_6H_4Cl-OH.$$

The halogen acids do not act upon the phenols. With sulphuric acid containing nitrous acid the phenols give an intense coloration (Liebermann's reaction).

With ferric chloride they yield a blue, green, or violet coloration as long as the H of the OH group is not substituted.

Preparation. 1. From diazo compounds, by heating with water.

2. By fusing the sulphonic acids with caustic alkali:

$$C_6H_5-SO_3K + KOH = C_6H_5-OH + K_2SO_3.$$

3. On the dry distillation of the oxyacids of the benzene series with caustic lime:

$$C_6H_4(OH)(COOH) + CaO = C_6H_5-OH + CaCO_3.$$

4. The halogen derivatives of the benzenes are not attacked by alkali hydroxides (p. 465); but if nitro groups are present at the same time, then they are transformed into nitrophenols:

$$C_6H_4Cl-NO_2 + KOH = C_6H_4(OH)-NO_2 + KCl.$$

3. Aromatic Alcohols.

a. Alcohols.

Benzyl alcohol,	$C_6H_5(CH_2OH)$	or	$C_7H_8O.$
Tolyl alcohol,	$C_6H_4(CH_3)(CH_2OH)$	or	$C_8H_{10}O.$
Phenyl ethyl alcohols,	$C_6H_5(CH_2-CH_2OH).$		
Phenyl propyl alcohol,	$C_6H_5(C_2H-CH_2OH)$	or	$C_9H_{12}O.$
Cumyl alcohol,	$C_6H_4(C_3H_7)(CH_2OH)$	or	$C_{10}H_{14}O.$
Xylylene alcohols,	$C_6H_4(CH_2OH)_2$	or	$C_8H_{10}O_2.$

b. Alcohol Phenols.

Saligenin,	$C_6H_4(OH)(CH_2OH)$	or	$C_7H_8O_2.$
Protocatechuic alcohol,	$C_6H_3(OH)_2(CH_2OH)$	or	$C_7H_8O_3.$
Gallic alcohol,	$C_6H_2(OH)_3(CH_2OH)$	or	$C_7H_8O_4.$

c. *Unsaturated Alcohols.*

Cinnamyl alcohol, $C_6H_5(CH=CH-CH_2OH)$ or $C_9H_{10}O$.

Preparation and Properties. The aromatic alcohols (p. 467) correspond in properties and preparation to the aliphatic alcohols (see also Benzyl Alcohol and Benzaldehyde). As they are on the other hand benzene derivatives, they may also be made to undergo the transformations which are common to benzene. By the introduction of $-OH$ groups into the benzene nucleus, we obtain the alcohol phenols.

4. Aromatic Acids.

a. *Monobasic Acids.*

Benzoic acid,	$C_6H_5(COOH)$	or $C_7H_6O_2$.
Phenylacetic acid,	$C_6H_5(CH_2.COOH)$	or $C_8H_8O_2$.
Toluic acid,	$C_6H_4(CH_3)(COOH)$	or $C_8H_8O_2$.
Mesitylenic acid, } o-, p-Xylic acids, }	$C_6H_3(CH_3)_2(COOH)$	or $C_9H_{10}O_2$.
Ethylbenzoic acids,	$C_6H_4(C_2H_5)(COOH)$	or $C_9H_{10}O_2$.
Phenylpropionic acids,	$C_6H_5(C_2H_4.COOH)$	or $C_9H_{10}O_2$.
Cumic acid,	$C_6H_4(C_3H_7)(COOH)$	or $C_{10}H_{12}O_2$.

b. *Dibasic Acids.*

Phthalic acids,	$C_6H_4(COOH)_2$	or $C_8H_6O_4$.
Uvitic acid, } Xylidic acid, }	$C_6H_3(CH_3)(COOH)_2$	or $C_9H_8O_4$
Cumidic acid,	$C_6H_2(CH_3)_2(COOH)_2$	or $C_{10}H_{10}O_4$.

c. *Tri- and Multibasic Acids.*

Trimesic acid, } Trimellitic acid, } Hemimellitic acid, }	$C_6H_3(COOH)_3$ or $C_9H_6O_6$.	
Pyromellitic acid, } Prehnitic acid, } Mellophanic acid, }	$C_6H_2(COOH)_4$ or $C_{10}H_6O_8$.	
Benzenepentacarbonic acid,	$C_6H(COOH)_5$	or $C_{11}H_6O_{10}$.
Mellitic acid,	$C_6(COOH)_6$	or $C_{12}H_6O_{12}$.

d. *Monobasic Phenol Acids.*

Oxybenzoic acids, (salicylic acid,)	$C_6H_4(OH)COOH$	or $C_7H_6O_3$.
Oxytoluic acids,	$C_6H_3(CH_3)(OH)(COOH)$	or $C_8H_8O_3$.
Dioxytoluic acids,	$C_6H_2(CH_3)(OH)_2(COOH)$	or $C_8H_8O_4$.
Oxymesitylenic acid,	$C_6H_2(CH_3)_2(OH)(COOH)$	or $C_9H_{10}O_3$.
Oxyphenylpropionic acid, (hydrocumaric and melilotic acid,)	$C_6H_4(OH)(C_2H_4COOH)$	or $C_9H_{10}O_3$.
Dioxybenzoic acids, (pyrocatechuic acid,)	$C_6H_3(OH)_2(COOH)$	or $C_7H_6O_4$.
Orsellic acid,	$C_6H_2(CH_3)(OH)_2(COOH)$	or $C_8H_8O_4$.
Gallic acid, } Pyrogallic carbonic acid, } Phloroglucin carbonic acid, }	$C_6H_2(OH)_3(COOH)$	or $C_7H_6O_5$.

e. Bibasic Phenol Acids.

Oxyphthalic acids, $C_6H_3(OH)(COOH)_2$ or $C_8H_6O_5$.
Dioxyphthalic acids, $C_6H_2(OH)_2(COOH)_2$ or $C_8H_6O_6$.

f. Alcohol and Ketone Acids.

Mandelic acids, $C_6H_5(CHOH–COOH)$ or $C_8H_8O_3$.

Tropic acid, $C_6H_5–CH <^{CH_2OH}_{COOH}$ or $C_9H_{10}O_3$.

Phenyllactic acid, $C_6H_5(C_2H_3\cdot OH\cdot COOH)$ or $C_9H_{10}O_3$.
Benzoylacetic acid, $C_6H_5–CO(CH_2–COOH)$ or $C_9H_8O_3$.

g. Unsaturated Acids.

Cinnamic acid, $C_6H_5(CH=CH–COOH)$ or $C_9H_8O_2$.

Atropic acid, $C_6H_5C <^{CH_2}_{COOH}$ or $C_9H_8O_2$.

Phenylpropiolic acid, $C_6H_5(C\equiv C–COOH)$ or $C_9H_6O_2$.
Oxycinnamic acid, $C_6H_4(OH)(CH=CH–COOH)$ or $C_9H_8O_3$.
Dioxycinnamic acid, $C_6H_3(OH)_2(CH=CH–COOH)$ or $C_9H_8O_4$.
  (caffeic and umbellic acids,)
Trioxycinnamic acid, $C_6H_2(OH)_3(CH=CH–COOH)$ or $C_9H_8O_5$.

Properties. They nearly all form crystals which are soluble with difficulty in water, and hence are precipitated from the solution of their salts by mineral acids. Those with simple constitution can be sublimed and distilled, while the others generally split off CO_2 on heating. This cleavage takes place with all of them on heating with alkali hydroxides. In other respects they correspond to the aliphatic acids, but as they are, on the other hand, benzene derivatives, therefore they may be made to undergo the transformations which are common to benzene. By the introduction of HO groups into the benzene nucleus of these acids we produce phenol acids, also called oxyacids, and when HO groups are introduced in the side-chain we obtain alcohol acids which correspond in most properties with the phenol acids.

Preparation. 1. By the oxidation of the corresponding aldehyde and primary alcohol.

2. By the oxidation of the homologues of benzene (p. 473). If the homologues also contain the $^-OH^-$, NO_2^-, NH_2 groups, etc., acids with these groups are obtained.

3. On the saponification of the corresponding nitrile; e.g., $C_6H_5CN + 2H_2O = C_6H_5–COOH + NH_3$.

4. By the action of carbon dioxide derivatives of benzene upon certain benzene derivatives (see Benzoic Acid and Salicylic Acid).

5. The unsaturated acids are obtained according to Perkin's reaction by the action of cyclic aldehydes upon fatty acids (see Cinnamic Acid), also from the monohalogen derivatives of the saturated acids by means of alcoholic caustic alkali (p. 323).

Occurrence. They occur in many resins and balsams and as NH_2 derivatives in the animal organism.

COMPOUNDS WITH SIX CARBON ATOMS UNITED TOGETHER.

1. Benzene Compounds.

Benzene, C_6H_6, benzol, is produced from most organic bodies by very high temperatures and hence is contained in coal-tar, as well as small amounts in illuminating-gas, and can be prepared from acetylene (p. 442). It is produced also on the distillation of all benzene carboxylic acids which contain only COOH side-groups, with calcium oxide. Ordinarily it is obtained from that part of coal-tar boiling between 80–85° by fractional distillation or by cooling. It is therefore also called tar or coal benzine, in contradistinction to petroleum benzine (p. 341). It is prepared pure by distilling benzoic acid with caustic lime.

$$C_6H_5{}^-COOH + CaO = C_6H_6 + CaCO_3.$$

It is a colorless refractive liquid boiling at 80° and crystallizing at 0°. It is readily inflammable and burns with a luminous flame. It is insoluble in water, soluble in alcohol and ether, and it dissolves resins, fats, sulphur, iodine, phosphorus.

Halogen derivatives of benzene (p. 473).

Nitrobenzene, $C_6H_5(NO_2)$, is obtained by dropping benzene into fuming nitric acid. If water is added to this mixture, the nitrobenzene precipitates as a yellow, poisonous, heavy liquid. It has an odor similar to bitter almonds and hence is used in perfumery as oil of Mirbane, or artificial oil of bitter almonds. In most cyclic nitro compounds, as well as in the aliphatic compounds, the NO_2 group is not replaceable by other groups. By reducing agents the nitro group is converted in acid solution into the amido group:

$$C_6H_5{}^-NO_2 + 6H = C_6H_5{}^-NH_2 + 2H_2O.$$

In neutral solution hydroxylamins are obtained, and in alkali solution we produce azoxy-, azo-, and hydrazo-compounds (see Azo Compounds).

Benzenesulphonic acid, $C_6H_5(SO_3H)$, is obtained by heating benzene with concentrated H_2SO_4 and forms deliquescent crystals.

Benzenedisulphonic acid, $C_6H_4(SO_3H)_2$, is produced on warming benzene with sulphuric acid anhydride, and occurs in two isomeric modifications. The third one can only be obtained indirectly.

2. Oxybenzene Compounds.

a. *Phenol,* $C_6H_5{}^-OH$, *and its Derivatives.*

Benzophenol, phenol, carbolic acid, phenylhydroxide, $C_6H_5(OH)$, occurs to a slight extent in urine, in castoreum, among the putrefactive products of the proteids, and forms the chief constituent of the coal-tar distilling between 170–230°. This product, which contains 30 per cent. phenols (p. 473), is shaken with caustic alkali, and the alkaline solution separated from the impurities floating on top and treated with hydrochloric acid. The phenols, which separate and rise to the surface, are removed and again distilled, and the portion distilling between 178–182° consists of pure phenol. It forms crystalline masses, having a characteristic odor, which melt at 40–42° and which often turn red, due probably to impurities. It has a sharp burning taste and causes blisters on the skin, and dissolves in 15 parts water and with great readiness in alcohol and ether. It is poisonous and prevents putrefaction, hence is used as a disinfecting agent.

Aqueous solutions of phenol turn violet with ferric salts, and bromine-water causes from very dilute solution a precipitate of yellowish-white tribromphenol, $C_6H_2B_3(OH)$, besides $C_6H_2Br_3{}^-OBr$.

Crude carbolic acid contains chiefly cresols (p. 491).
Liquid carbolic acid contains about 10 per cent. water.
Carbolic water, is a 2 per cent. watery solution of phenol.
Bismuth tribromphenolate, $(C_6H_2Br_3O)_2=BiOH+Bi_2O_3$, is used in medicine as *xeroform.*
Trinitrophenol, picric acid, $C_6H_2(NO_2)_3{}^-(OH)$. Concentrated nitric acid converts phenol into mono-, di-, or trinitrophenol, depending upon the length of action. Trinitrophenol is also obtained as an end-product by the action of concentrated nitric acid upon many other aromatic bodies, such as indigo, aniline, resins, silk, wool, leather. It forms yellow leaves which are poisonous, without odor, intensely bitter in taste, soluble in water, and burning without explosion when ignited. It is very sensitive to any shock, and explodes when ignited with mercury fulminate. Pure picric acid or the potassium and ammonium compounds (picrate powder), or mixed with carbon and saltpeter or guncotton, etc., are used as explosives (*melinite, lydite, ecrasite.* etc.). Picric acid colors silk and wool beautifully yellow and is used in dyeing and in microscopical technic. Vegetable fibers (cotton) are not colored. It behaves just like an acid and forms metallic derivatives which form well-defined crystals and readily explode: $C_6H_2(NO_2)_3(ONa)$ and alkyl derivatives such as $C_6H_2(NO_2)_3(OC_2H_5)$.

Amidophenols, $C_6H_4(NH_2)(OH)$, are produced in the reduction of the corresponding nitrophenols. The *p*-compound is used as *rodinal*, as a developer in photography.

Diamido phenol, $C_6H_3(NH_2)_2(OH)$, is also used for the same purpose.

Phenol sulphonic acids, $C_6H_4(OH)(SO_3 \cdot H)$, obtained by dissolving phenol in concentrated sulphuric acid, forms colorless crystals.

Zinc phenolsulphonate, $(C_6H_4 \cdot OH \cdot SO_3)_2Zn + 7H_2O$, forms colorless, crystals.

Sozoiodolic acid, $C_6H_2I_2(OH)(SO_3 \cdot H) + 3H_2O$, forms prisms which are readily soluble in water and alcohol, which like the corresponding sodium salt $(C_6H_2I_2(OH) \cdot SO_3 \cdot Na)$ is used as an antiseptic.

Phenylsulphuric acid, $C_6H_5 \cdot O \cdot SO_2 \cdot OH$, is an isomer of phenolsulphonic acid and is unstable. $C_6H_5 \cdot O \cdot SO_2 \cdot OK$ is found in the urine of herbivora and also in human urine after partaking phenol.

Phenolmethyl ether, *anisol*, $C_6H_5 \cdot O - CH_3$, is obtained from anisic acid by splitting off CO_2 or by the action of methyl iodide upon potassium phenolate. It is a liquid boiling at 152°.

Phenolethyl ether, *phenetol*, $C_6H_5O \cdot C_2H_5$, is prepared in an analogous manner. It is a liquid boiling at 172°.

p-Phenetol carbamide, *dulcin*, $H_2N - CO - NH(C_6H_4O \cdot C_2H_5)$, *sucrol*, is a substitute for sugar, as it is 250 times sweeter than cane-sugar; forms white needles which are not very soluble.

Phenyl ether, $C_6H_5 - O - C_6H_5$, is produced in the dry distillation of copper benzoate, forms long, colorless needles.

Thiophenol, $C_6H_5 - SH$, is produced in the reduction of benzenesulphonic acid, $C_6H_5 - SO_2H$, and of benzenesulphochloride, $C_6H_5 - SO_2Cl$. It is a disagreeably smelling liquid, boiling at 173°.

b. *Dihydric Phenols*, $C_6H_4(OH)_2$, *and their Derivatives.*

The three possible dioxybenzenes, namely, pyrocatechin, resorcin, and hydroquinone, are produced from the corresponding phenolsulphonic acids by fusing them with alkali hydroxides and are used in photography as developers (p. 241).

Pyrocatechin, *o*-dioxybenzene, $C_6H_4(OH)_2$, occurs among the products of the dry distillation of catechu extracts, many gum resins, and wood. It is also found in the urine and feces of herbivora, where it is produced from protocatechuic acid (p. 494), which exists extensively in the plant kingdom. It forms colorless prisms which melt at 104° and which are soluble in water, alcohol, ether, with strong reducing action. The solution turns dark green with ferric chloride and with alkalies it undergoes oxidation, turning green, then brown and black.

Guaiacol, $C_6H_4(OCH_3)(OH)$, pyrocatechin monomethyl ether, is produced in the dry distillation of guaiac and wood and forms colorless crystals melting at 32°, and which are not very soluble in water, but readily soluble in alcohol. It forms creosote (p. 493) when mixed with the homologous creosols.

The benzoic acid ester, *benzosol*, the carbonic acid ester as *duotal*, the salicylic acid ester, *guaiacolsalol*, the valerianic acid ester, *geosot*, also potassium guaiacolsulphonate, *thiocol*, are used in medicine.

Veratrol, $C_6H_4(O \cdot CH_3)_2$, occurs in the seeds of the sabadilla.

Resorcin, *m*-dioxybenzol, $C_6H_4(OH)_2$, is obtained on fusing asafœtida, ammoniacum, galbanum, and other gum-resins with caustic alkali. It is produced in large quantities by fusing potassium benzenedisulphonate with KOH:

$$C_6H_4(SO_3K)_2 + 2KOH = C_6H_4(OH)_2 + 2K_2SO_3.$$

It forms colorless sweet crystals which have a reducing action and melt at 110° and which are soluble in 1 part water, and in alcohol, ether, and glycerine. The watery solution gives a deep-violet coloration with ferric chloride and a brown with alkalies.

On fusion with phthalic anhydride it dissolves in caustic alkali with a green fluorescence (formation of fluorescein). On fusion with $NaNO_2$ it yields resorcin blue or *lacmoid*, $C_{12}H_9O_3N$, which acts like the litmus pigment (p. 493).

Trinitroresorcin, styphnic acid, $C_6H(NO_2)_3(OH)_2$, is obtained by the action of nitric acid upon resorcin as well as upon galbanum and other resins and forms yellow crystals.

Hydroquinone, *p*-dioxybenzol, $C_6H_4(OH)_2$, occurs in the glucoside arbutin and is obtained in the dry distillation of quinic acid and by the action of sulphur dioxide upon quinone, $C_6H_4O_2$:

$$C_6H_4O_2 + SO_2 + H_2O = C_6H_6O_2 + H_2SO_4.$$

It forms colorless crystals which have a strong reducing power and which are soluble in hot water, alcohol, ether, and melt at 169°. The watery solution turn temporary blue with ferric chloride and become dark quickly in the air, especially in the presence of alkali, due to the absorption of oxygen. It is converted into quinone, $C_6H_4O_2$, by all oxidizing agents (by boiling with ferric chloride). This quinone is characterized by its odor. Monochlor- and mono-bromhydroquinone, *adurol*, are used as photographic developers (p. 241).

Quinone, $C_6H_4O_2$, is produced on the oxidation of aniline (ordinary preparation) as well as many *p*-diderivatives of benzene, thus of hydroquinone, of phenolsulphuric acid, etc. It forms yellow poisonous prisms having a characteristic odor and which color the skin brown. It is a double ketone, $OC <^{CH=HC}_{CH=HC}> CO$, and is the type of a large group of compounds called *quinones*. The quinones are strong oxidizing agents and combine with the hydroquinones, forming *quinhydrones*, which are strong colors:
$C_6H_4O_2 + C_6H_6O_2 = C_{12}H_{10}O_4.$

From nitrous acid and phenols we obtain nitrosophenols or quinon-oxims (p. 331), $C_6H_4O=(N-OH)$.

The quinonchlorimides, $C_6H_4O(NCl)$, and quinondichlorimides, $C_6H_4(NCl)_2$, are also derived from the quinones by replacing $=O$ by $=NCl$.

The aniline residue, $-NH \cdot C_6H_5$, can replace hydrogen in many quinones, producing anilidoquinones; e.g., $C_6H_2(NH \cdot C_6H_5)_2O_2$, or the aniline residue, $=N \cdot C_6H_5$, can replace oxygen, forming quinonaniles; e g., $C_6H_4(N \cdot C_6H_5)_2$.

Tetrachlorquinone, chloranil, $C_6 \cdot Cl_4O_2$, is obtained from quinone and many other benzene compounds by the action of chlorine and forms yellow crystalline scales which are soluble in water.

c. Trihydric Phenols, $C_6H_3(OH)_3$, and their Derivatives.

All three possible trioxybenzenes are known, namely, pyrogallol (HO position $1:2:3=v$), phloroglucin (HO position $1:3:5=s$), oxyhydroquinone (HO position $1:2:4$ a, p. 469).

Pyrogallol, pyrogallic acid, $C_6H_3(OH)_3$, is produced on heating gallic acid: $C_6H_2(OH)_3COOH = C_6H_3(OH)_3 + CO_2$. Its dimethyl ether occurs in beech-wood tar. It forms colorless crystals having a bitter taste, melting at 131°, and soluble in 1.7 parts water. Its alkaline solution absorbs oxygen energetically, turning brown, and is used in determining oxygen in gaseous mixtures. Pyrogallol has a strong reducing action being oxidized into acetic and oxalic acids (used as a developer in photography). It colors the skin and hair brown and the aqueous solution turns dark blue with ferrous salts and red with ferric salts.

Phloroglucin, $C_6H_3(OH)_3$, occurs in a few plants and is obtained from different resins by fusing them with KOH. It forms sweet, colorless crystals with 2 mol. H_2O. Its watery solution becomes deep violet with ferric chloride and is converted into *phloroglucite*, $C_6H_3(OH)_3H_6$, by nascent hydrogen.

Filicic acid, $C_{14}H_{16}O_5$, the active constituent of the male-fern used as a tape-worm remedy, and

Cotoin, $C_6H_2(C_6H_5 \cdot CO)(O \cdot CH_3)(OH)_2$, the constituent of the coto-bark, are phloroglucin derivatives.

Oxyhydroquinone, $C_6H_3(OH)_3$. is produced on fusing hydroquinone with caustic potash. The aqueous solution of the colorless crystals turn bluish-green with ferric chloride.

d. Tetra and Polyhydric Phenols and their Derivatives.

Tetraoxybenzenes, $C_6H_2(OH)_4$. The s-compound is known free and the others only as methyl ethers. One hexahydro-derivative is called *betite*, $C_6H_2(OH)_4H_6$, and is found in beet-root molasses.

Pentaoxybenzene, $C_6H(OH)_5$, is not known, but its hexahydro-deriva-tive (P. 463).

Quercite, $C_6H_{12}O_5$ or $C_6H(OH)_5(H)_6$. is the sweetish substance found in acorns.

Hexaoxybenzene, $C_6(OH)_6$, forms grayish-white crystals which turn violet when exposed to the air. Potassium hexaoxybenzene, $C_6(OK)_6$, is the explosive potassium carbon monoxide produced on passing CO over heated potassium as well as in the manufacture of potassium.

Inosite, phasæomannite, $C_6H_{12}O_6$ or $C(OH)_6(H)_6$, hexahydrohexoxybenzene, occurs in the liver, spleen, kidneys, lungs, brain, and heart muscle and muscular tissue and in various plants, grape-juice, and wine, and especially in unripe beans, from which it can be obtained by extraction with water and precipitation with alcohol. It forms with 2 mol. water colorless, sweet, crystals, and exist as a dextrorotatory, a lævorotatory, and an inactive modification.

Methyl inosite, pinite, carthartomannite, $C_7H_{14}O_6$ or $C_6(CH_3)(OH)_6(H_5)$, is obtained from the resin from Pinus lambertiana and from the senna leaves. It forms colorless, sweet crystals.

3. Amido Benzene Compounds.

The amines or amido derivatives of benzene may be considered as produced by replacing H of benzene by amido groups,

$C_6H_5(NH_2)$	$C_6H_4(NH_2)_2$	$C_6H_3(NH_2)_3$
Amidobenzene	Diamidobenzene	Triamidobenzene
(a monamine).	(a diamine).	(a triamine).

or by replacing the hydrogen of ammonia by phenyl, C_6H_5:

$(C_6H_5)NH_2$	$(C_6H_5)_2NH$	$(C_6H_5)_3N$
Phenylamine	Diphenylamine	Triphenylamine
(primary amine).	(secondary amine).	(tertiary amine).

Corresponding amido derivatives are obtainable from the benzene derivatives.

Properties. The monamines are colorless liquids or crystalline solids which are not readily soluble in water, and which are volatile with steam. The polyamines are nearly all crystalline, more soluble in water than the monamines, and not volatile with steam. They are quite similar in behavior to the aliphatic amines and form analogous compounds. Quaternary amines, $(C_6H_5)N(CH_3)_3-OH$, are known. *o*-Phenylendiamine and its homologues yield benzodiazole (p. 468), with acids and the elimination of water, and benzodiazine with aldehydes; thus,

$$C_6H_4\underset{NH_2}{\overset{NH_2}{\Big\langle}} + \overset{COH}{\underset{COH}{|}} = 2H_2O + C_6H_4\underset{N=CH}{\overset{N=CH}{\Big\langle}}.$$
$$\text{Quinoxaline.}$$

They unite in the same way with aldehyde acids, ketonic acids, diketones, when these contain neighboring CO groups. The compounds of *m*- and *p*-diamines are unstable. With nitrous acid the primary amines yield diazo compounds (p. 465), the secondary amines yield nitrosamines (p. 330), $(C_6H_5)(CH_3)=N-NO$, the tertiary amines, nitroso compounds, which contain the NO groups in the benzene nucleus, $C_6H_4(NO)-N=(CH_3)_2$.

Preparation. The primary monamines, diamines, and triamines are obtained from the corresponding nitro compound by reduction, especially in acid solution (see Azo Compounds). The secondary and tertiary monamines, which contain alkyls, are obtained by the action of alkyl iodides upon primary cyclic amines: $C_6H_5-NH_2+CH_3I=C_6H_5-NH-CH_3+HI$.

Secondary and tertiary monamines are obtained by heating the primary cyclic amines with their salts:

$$C_6H_5-NH_2 + C_6H_5-NH_2(HCl) = (C_6H_5)_2NH + NH_4Cl.$$

Aniline, amidobenzene, phenylamine, $C_6H_5-NH_2$, is produced by the dry distillation of many organic bodies such as coal (occurrence in coal-tar), of bones, indigo (from Indigofera anil, hence the name). It is chiefly prepared by the reduction of nitrobenzene in acid solution (see p. 483) by warming with iron filings and hydrochloric acid. The reduction may also be brought about by zinc or tin and hydrochloric acid or with ammonium sulphide when the H_2S acts as the reducing substance:

$$C_6H_5-NO_2 + 3H_2S = C_6H_5-NH_2 + 2H_2O + 3S.$$

Aniline is a colorless, poisonous liquid having a peculiar odor and boiling at 184°. It turns brown in the air and is but slightly soluble in water but readily soluble in alcohol and ether. It can be easily detected even in very dilute solution by the temporary deep-violet color it gives with chloride of lime solution as well as the blue coloration with potassium dichromate and sulphuric acid. Aniline has a neutral reaction, but combines directly, like all amine bases, with acids, forming well-defined colorless salts; i.e., $(C_6H_5 \cdot NH_2)HNO_3$, $(C_6H_5 \cdot NH_2)_2H_2SO_4$. The aniline is set free from these salts by alkalies.

The aniline dyes are not derived from aniline, but have been called so because aniline is used in the preparation of certain of them.

If aniline is heated with fuming sulphuric acid, we obtain *p*-amido benzene sulphonic acid, or *sulphanilic acid:* $C_6H_4(SO_3H)(NH_2)$.

If the hydrogen of the NH_2 group in aniline is replaced by alcohol radicals we obtain the *anilines,* whose preparation and behavior correspond to the amines:

$$C_5H_5-N<^H_H$$
Aniline.

$$C_6H_5-N<^H_{CH_3}$$
Methylaniline.

$$C_6H_5-N<^{CH_3}_{CH_3}.$$
Dimethylaniline.

$$C_6H_5-N\lessgtr^{(CH_3)_3}_{OH}$$
Trimethylphenyl ammonium hydroxide.

$$C_6H_5N<^H_{C_6H_5}$$
Diphenylamine.

$$C_6H_5N<^{C_6H_5}_{C_6H_5}.$$
Triphenylamine.

If the hydrogen of the NH_2 group of aniline is replaced by acid radicals we obtain the *anilides* (see below), which are compounds similar to the amides and which are prepared in the same manner.

If the hydrogen of the benzene ring of aniline is replaced by $^-OCH_3$ the *anisidines* or methoxyanilines are obtained, while if $^-OC_2H_5$ is replaced we obtain *phenetidines* or ethoxyanilines (anisol and phenetol, see p. 480).

Acetanilide, *antifibrine,* $C_6H_5{}^-NH(CH_3{}^\cdot CO)$. Just as ammonium acetate decompose on heating into acetamide and water so aniline acetate decomposes into acetanilide and water on heating:

$$C_6H_5{}^-NH_2(CH_3{}^\cdot COOH) = C_6H_5NH(CH_3CO) + H_2O.$$
Aniline acetate.   Acetanilide.

It forms leaves which are not readily soluble in water but easily soluble in chloroform, alcohol, and ether, melting at 113° and boiling at 295°.

As a primary amine it gives the isonitrile reaction (p. 391). If it is boiled with HCl for a short time and then treated with a phenol solution and then a chloride of lime solution added a violet-blue coloration is obtained which on saturation with NH_3 becomes blue (indophenol reaction).

*p-*Bromacetanilide, *antisepsin,* $C_6H_4Br^-NH(CH_3{}^\cdot CO)$,
Methylacetanilide, *exalgin,* $C_6H_5{}^-N(CH_3)(CH_3{}^\cdot CO)$,
Formanilide, *phenylformamide,* $C_6H_5{}^-NH(H^\cdot CO)$,
Gallic acid anilide, *gallanol,* $C_6H_5{}^-NH-CO-C_6H_2(OH)_3$,
Metarsenic acid anilide, *atoxyl,* $C_6H_5{}^-NH-AsO_2$, are all used in medicine.

Carbanilide, diphenyl urea, $(C_6H_5)NH-CO-NH(C_6H_5)$, is produced by replacing 2 H atoms by $^=CO$ in 2 mol. aniline.

*p-*Acetylphenetidine, *phenacetine,* acetphenetidine, $C_6H_4(OC_2H_5)^-NH^-(CH_3CO)$, forms a white crystalline powder without odor or taste, nearly insoluble in water and melting at 135°. It is obtained by boiling *p-*amidophenetol, $C_6H_4(O^\cdot C_2H_5)^-NH_2$, with acetic acid, when water is split off. On shaking with HNO_3 it turns yellow.

*p-*Amidoacetylphenetidine, $C_6H_4(O^\cdot C_2H_5)NH(OC^\cdot CH_2{}^\cdot NH_2)$, *phenocoll* and its salicylate, *salocoll* are used in medicine

*p-*Lactylphenetidine, $C_6H_4(O^\cdot C_2H_5)^-NH(CH_3{}^\cdot CHOH^\cdot CO)$, *lactophenine,* forms colorless crystals.

*m-*Phenylendiamine, $C_6H_4(NH_2)_2$, is used in the detection of nitrous acid (p. 157), and forms colorless crystals.

Diphenylamine, $H_5C_6{}^-NH^-C_6H_5$, forms white scales, which are nearly insoluble in water but readily soluble in alcohol and ether. The solution of diphenylamine in sulphuric acid becomes deep blue with oxidizing agents (p. 167). The following dyes may be considered as derivatives of diphenylamine, and may also be considered as derivatives of quinonimide (i.e., from quinone, $C_6H_4O_2$, in which both O atoms are replaced by the imido group NH).

Indophenol, $O^-H_4C_6-N-C_6H_4-OH$, is red in alcoholic solution and blue in alkalies. In regard to its formation from acetanilide, see above.

Indoaniline, $O-H_4C_6-N-C_6H_4-N(CH_3)_2$, phenol blue,

Indamine, $HN-H_4C_6-N-C_6H_4-NH_2$, phenylene blue and its derivatives are unstable colors which appear as intermediary products in the preparation of saffranin and methylene blue.

p-**Diethoxylethenyldiphenylamidine,** *holocain,* $CH_3-C<{N-C_6H_4-O-C_2H_5 \atop NH-C_6H_4-O-C_2H_5}$, forms colorless crystals which are used as a local anæsthetic instead of cocain.

Acoin, $(H_3C\cdot\cdot C_6H_4)HN-C=N(C_6H_4\ OC_2H_5)-NH(C_6H_4\cdot O\cdot CH_3)$, phenetyldianisylguanidine, is also used as a substitute for cocain.

4. Hydrazine Compounds.

The cyclic hydrazines are very similar to the aliphatic hydrazines (p. 331), but are less basic. They reduce, like these, alkaline copper solutions, are readily oxidizable, but more resistant towards reducing agents. They combine directly with acids, forming salts, and also with aldehydes and ketones, with the elimination of water (P. 351. 11); hence they form with the sugars crystalline compounds, the hydrazones and osazones (p. 450).

Mixed hydrazines are also known; i.e., $(C_6H_5)(C_2H_5)N-NH_2$. Symmetrical hydrazines, i.e., $(C_6H_5)HN-NH(CH_3)$, are called *hydrazo* compounds. Hydrazides correspond to the amides. They are obtained by the reduction of the diazo compounds (p. 509) or from the neutral solutions of nitro compounds (p. 478).

Phenylhydrazine, $C_6H_5-HN-NH_2$, serves as a reagent in the detection of aldehydes, ketones, sugars, and melts at 18°.

Benzoyl hydrazide, $(C_6H_5\cdot CO)HN-NH_2$, serves in the preparation of hydrazoic acid, $HN<{N \atop N}\gg$ Nitrous acid converts it into benzoylhydrazoate (benzazide), $(C_6H_5\cdot CO)HN-NH_2+HNO_2=(C_6H_5\ CO)N_3+H_2O$, which on boiling with caustic alkali yields

$$C_6H_5\cdot CO-N{<}{N \atop N}\ \|\ +2NaOH=C_6H_5-COONa+NaN{<}{N \atop N}\ \|\ +H_2O.$$

Benzoylhydrazoate.　　　　Sodium benzoate.　Sodium hydrazoate.

From this last substance the very explosive hydrazoic acid, N_3H (p. 151), is set free by treatment with sulphuric acid.

COMPOUNDS WITH SEVEN CARBON ATOMS UNITED TOGETHER.

1. Toluene Compounds.

Toluene, methyl benzene, toluol, $C_6H_5-CH_3$, is produced besides benzene, etc., in the dry distillation of certain resins, especially Tolu balsam, also of wood and coal. It is obtained from coal-tar (p. 473) by fractional distillation as a liquid similar to benzene and boiling at 110°.

If in toluene the hydrogen of the benzene ring is substituted we obtain toluene compounds of purely benzene character, while if the hydrogen of the aliphatic side-chain be substituted we obtain compounds of a more aliphatic character. By substitution in the benzene ring and in the side-chain we obtain compounds of a mixed character. Derivatives of toluene containing the radical $C_6H_5^-CH_2^-$ are called *benzyl*, the radical $C_6H_5^-CO$, *benzoyl*, the radical $^-C_6H_4^-CH_3$, *toluyl* or *tolyl* compounds. In regard to the further nomenclature of the toluene derivatives see p. 471.

Chlortoluenes, $C_6H_4Cl(CH_3)$; all three isomers are known as well as the brom- and iodotoluenes (Preparation, p. 474).

Nitrotoluenes, $C_6H_4(NO_2)CH_3$; all three isomers are known. By the action of concentrated nitric acid upon toluene we obtain the crystalline para- and the liquid orthonitrotoluene.

Toluidines, amidotoluenes, $C_6H_4(NH_2)(CH_3)$; the three isomers are formed in the reduction of the three nitrotoluenes. Ortho- and meta-toluidine are fluids and paratoluidine a solid.

Benzyl chloride, $C_6H_5^-CH_2Cl$, as well as the bromide and iodide, form colorless liquids with a pungent odor (Preparation, p. 474).

Benzal chloride, $C_6H_5^-CHCl_2$. and **benzotrichloride,** $C_6H_5^-CCl_3$, are formed on the further action of chlorine upon boiling toluene.

Benzylamine, $C_6H_5^-CH_2NH_2$, is produced by heating benzyl chloride with ammonia and consists of a colorless liquid with a strong alkaline reaction.

Benzylcyanide, $C_6H_5^-CH_2CN$, forms the ethereal oil of the nasturtium and garden cresses (Lepidium sativum) accompanied by

Benzylisosulphocyanide, benzyl mustard-oil, $C_6H_5^-CH_2^-NCS$.

Benzylalcohol, $C_6H_5^-CH_2OH$, occurs as the cinnamic and benzoic acid benzyl ester in storax, balsam of Peru, and Tolu and is prepared from benzaldehyde by the action of alkali hydroxides or nascent hydrogen or by heating benzylacetate with caustic alkalies (p. 344, 3). It is a colorless thick liquid, boiling at 207°:

$$C_6H_5^-CH_2Cl + C_2H_3O_2K = C_6H_5^-CH_2(C_2H_3O_2) + KCl;$$
Benzylchlorid.   Pot. acetate.   Benzylacetate.

$$C_6H_5^-CH_2(C_2H_3O_2) + KOH = C_6H_5^-CH_2OH + C_2H_3O_2K.$$
Benzylacetate.   Benzyl alcohol.

Benzylaldehyde, oil of bitter almond, $C_6H_5^-CHO$, is analogous to the aliphatic aldehydes and is obtained by distilling a salt of benzoic acid with a salt of formic acid (p. 351, 12), also by heating benzal-chloride with water: $C_6H_5^-CHCl_2 + HOH = C_6H_5^-CHO + 2HCl$.

Ordinarily it is prepared from bitter almonds by the decomposition of the glycoside amygdalin contained therein (p. 384). Thus obtained it is generally mixed with hydrocyanic acid as $C_6H_5^-CHO(CN)$ (see p. 351, 9) and forms the oil of bitter almonds of the druggist. Benzaldehyde is a colorless liquid, boiling at 180°, having a characteristic odor, and giving all the reactions of the aldehydes, yielding benzoic acid on oxidation and benzyl alcohol by reduction (sodium amalgam).

The aromatic aldehydes are prepared in an analogous manner to the aliphatic aldehydes or by the action of $CO + HCl$ in the presence of $AlCl_3$ upon the cyclic hydrocarbons (Gattermann's method).

Cyclic aldehydes do not, like the aliphatic aldehydes, yield aldehyde resins when treated with alkali hydroxides, but yield acids and primary alcohols; i.e.,

$$2C_6H_5-CHO + KOH = C_6H_5-COOK + C_6H_5-CH_2 \cdot OH.$$

On warming aromatic aldehydes with an alcoholic solution of potassium cyanide, bodies having double the molecular weight of the aldehyde are produced, these bodies being derivatives of *dibenzyl*, $H_2C-H_5C_6-C_6H_5-CH_2$; thus benzyl aldehyde, $2C_6H_5-CHO = C_6H_5-CO-CH(OH)-C_6H_5$, *benzoïn*. Anisaldehyde similarly yields *anisoïn*, cuminaldehyde, *cuminoïn*, etc.

On oxidation these bodies give diketones; thus benzoïn yields *benzil*, $C_6H_5-CO-CO-C_6H_5$, cuminoïn, yields *cuminil*, anisoïn, *anisil*, etc. These on heating with caustic alkalies form alcohol acids; thus, $(C_6H_5)_2-C(OH)-COOH$, *benzilic* acid, etc.

Benzoic acid, phenyl formic acid, C_6H_5-COOH, occurs in many resins, especially in the gum benzoïn, also in balsam of Peru and Tolu, and as a component of hippuric acid in fresh herbivorous urine and free in decomposed urine and as a constituent of the glucoside populin in the silver poplar.

Benzoic acid is produced on the oxidation of all hydrocarbons, alcohols, aldehydes, ketones, ketonic acids, etc., which are derived from benzene by replacing one H atom by a monovalent side-chain. Thus $C_6H_5-CH_3$, $C_6H_5-C_2H_5$, $C_6H_5-C_3H_7$, etc., all yield C_6H_5-COOH on oxidation (p. 473). It may be obtained synthetically from monobrombenzene, sodium, and carbon dioxide,

$$C_6H_5Br + 2Na + CO_2 = C_6H_5-COONa + NaBr,$$

and the sodium benzoate decomposed by HCl.

It is also obtained from benzaldehyde (which see) by the action of alkali hydroxides. It is ordinarily prepared by heating gum benzoin when the benzoic acid sublimes or by boiling hippuric acid (p. 490) with hydrochloric acid whereby the hippuric acid decomposes into glycocoll and benzoic acid (Benzoic acid from urine) or by heating benzotrichloride with water:

$$C_6H_5-CCl_3 + 2HOH = C_6H_5-COOH + 3HCl.$$

Benzoic acid forms large, shining crystals which have a faint aromatic odor and which are readily sublimable. These crystals melt at 121° and are soluble with difficulty in cold water, but more readily in hot water and are easily soluble in alcohol and ether.

The salts of benzoic acid or *benzoates* are mostly readily soluble in water. Ferric chloride precipitates neutral benzoate solutions forming reddish-yellow ferric benzoate, $(C_6H_5-COO)_3Fe$, which is soluble in fatty oils.

Paramidobenzoic acid, $C_6H_4(NH_2)(COOH)$. Its ethyl ester is called *anæsthesin* and is used as a local anæsthetic.

Benzamide, $C_6H_5-CO-NH_2$, corresponds to acetamide and by splitting off H_2O is converted into

Benzonitrile, C_6H_5-CN, a liquid having an odor similar to oil of bitter almond. The cyclic nitriles behave like the aliphatic nitriles (p. 330).

Orthoamidobenzoic acid, $C_6H_4(NH_2)(COOH)$, *anthranilic acid*, is produced on the oxidation of indigo and is prepared from phthalimide (p. 497), and serves in the synthesis of indigo. Its methyl ester is a constituent of the ethereal oil of orange flowers, pomegranate, and jasmin flowers, and is used in the preparation of perfumes.

Orthosulphaminbenzoic acid, $C_6H_4 < {}^{COOH}_{SO_2-NH_2}$, when heated yields

Orthosulphaminbenzoic acid anhydride, orthobenzoic acid sulphimide, *saccharin, sycose,* $C_6H_4 < {}^{CO}_{SO_2} > NH$, forms white crystals which are difficultly soluble in water but readily soluble in alcohol and ether, and which are 500 times sweeter than cane-sugar, and. hence used as a sweetening agent.

The corresponding *m-* and *p-*compounds do not taste sweet. The sodium compound is the "readily soluble saccharin" and contains water of crystallization and is called *crytallose, saccharose.*

Benzyl benzoate, $C_6H_5-COO(C_6H_5\cdot CH_2)$, is used in medicine under the name *peruscabin*, and its solution in oil is called peruol.

Benzoylperoxide, $C_6H_5\cdot CO\cdot O\cdot O\cdot OC\cdot C_6H_5$, and

Benzoperacid, $C_6H_5\cdot CO\cdot OOH$, are to be considered as H_2O_2, in which the H atoms have been replaced by benzoyl. The first melts at 110° and the other at 42°.

Benzoyl chloride, C_6H_5-COCl, is obtained from PCl_3 and benzoic acid as a fuming, colorless liquid which is used in introducing the benzoyl group into other compounds:

$$C_6H_5-OH + C_6H_5-COCl = C_6H_5-OOC-C_6H_5 + HCl;$$
$$C_6H_5-NH_2 + C_6H_5COCl = C_6H_5-NH-CO-C_6H_5 + HCl.$$

This takes place readily in the presence of caustic alkalies (Schotten-Baumann reaction) and serves in detecting $-NH_2$, $-OH$, and $=NH$ in aromatic and aliphatic compounds.

Ornithuric acid, $C_{19}H_{20}N_2O_4$, is benzoylornithin, $(C_4H_7)(NH\cdot C_6H_5\cdot CO)_2-$ COOH, occurs in bird urine, especially on feeding with benzoic acid and decomposes on heating with mineral acids into ornithin (p. 375) and benzoic acid.

Hippuric acid, $C_9H_9N_3O$, is benzoyl glycocoll:

$$(COOH \cdot CH_2)^- NH_2; \qquad (COOH \cdot CH_2)NH^- (C_6H_5{}^- CO^-).$$
Glycocoll.   Hippuric acid.

It is abundantly found in urine of herbivora and in small amounts in the urine of carnivora. In the passage of benzoic acid, toluene, cinnamic acid and other aromatic bodies, which on oxidation yield benzoic acid, through the animal organism they are transformed into hippuric acid and are eliminated as such by the urine. On exclusive vegetable diet the quantity of hippuric acid in the urine of carnivora and herbivora is the same. Hippuric acid is artificially obtained by heating:

$$(C_6H_5 \cdot CO)NH_2 + CH_2Cl^- COOH = (C_6H_5 \cdot CO)^- NH^- (CH_2 \cdot COOH) + HCl.$$
Benzamide.   Monchloracetic acid.

It is prepared on a large scale from fresh urine of herbivora by evaporating it to a sirup and after cooling (p. 488) treating with hydrochloric acid. The hippuric acid which separates out is purified by recrystallization. Hippuric acid forms colorless columns which are not soluble in cold water but slightly soluble in hot water, and very soluble in alcohol but insoluble in petroleum ether (differing from benzoic acid). On boiling with acids or alkalies it decomposes into glycocoll and benzoic acid:

$$\begin{array}{l} CH_2{}^- NH(C_6H_5 \cdot CO) \\ | \\ COOH \end{array} + H_2O = \begin{array}{l} CH_2 \cdot (NH_2) \\ | \\ COOH \end{array} + \begin{array}{l} C_6H_5 \\ | \\ COOH \end{array}$$

The same decomposition can be brought about by micrococcus ureæ in the alkaline fermentation of urine, hence only benzoic acid is found in putrid horse urine.

Of the salts of hippuric acid the yellow ferric hippurate is characterized by its insolubility.

Many cyclic acids combine in the animal organism with glycocoll, splitting off H_2O, and these compounds are eliminated by the urine; thus, salicylic acid as *salicyluric acid*, $(HOOC \cdot CH_2)- NH(C_6H_4 \cdot OH.CO)$; tolulic acid as *toluric acid*, $(HOOC \cdot CH_2)- NH(C_6H_4 \cdot CH_3 \cdot CO)$, cymene as *cuminuric acid*, $(HOOC \cdot CH_2)- NH(C_6H_4 \cdot C_3H_7 \cdot CO)$, etc. *Phenaceturic acid*, $(HOOC \cdot CH_2)- NH(C_6H_5 \cdot CH_2CO)$, occurs as a regular constituent of the urine of herbivora. These compounds must not be mistaken for the urea derivatives (p. 414), which have a similar termination.

2. Oxytoluene Compounds.

a. *Monohydric Phenols,* $C_6H_4(OH)(CH_3)$, *and their Derivatives.*

, All three possible oxytoluenes, the cresols, are known. The three alcohol phenols, $C_6H_4(OH)(CH_2OH)$, related thereto are called saligenin, *m-* and *p*-oxybenzyl alcohol, and yield on oxidation three acids, $C_6H_4(OH)(COOH)$, salicylic acid, *p*· and *m*·oxybenzoic acids.

Cresols, methyl phenols, $C_6H_4(OH)(CH_3)$. All three isomers occur with phenol in the heavy coal-tar oil and wood-tar oil and form as alkali cresol-sulphates the chief portion of the phenols of the urine of carnivora and herbivora. The cresols form colorless prisms having an odor similar to phenol and are difficultly soluble in water.

A fluid mixture of the three cresols (pure *tricresol, enterol*) contain‹ ing 10–15 per cent. hydrocarbons is called crude carbolic acid and forms the portion of the coal-tar which distils over between 187° and 210°.

The addition of soaps increases the solubility of the cresols, hence such mixtures are used as more or less poisonous disinfectant agents under the names *lysol* (containing 50 per cent. cresols), *bacillol, desinfectol, creolin, cresolin, lysitol, phenolin, sapocarbol, saprol.* In *solveol* the solubility of the cresols in water is increased by sodium cresotate and in *solutol* by sodium cresol. Cresylic acid is *m*-cresol

Dinitro-orthocresol potassium, $C_6H_2(NO_2)_2(CH_3)(OK)$, as *antinonnin* is used to kill caterpillars and worms.

Europhen, $(HO)(CH_3)(C_4H_9)C_6H_2-C_6H_2(C_4H_9)(CH_3)(O·I)$, diisobutyl orthocresol monoiodide, is a yellow amorphous powder insoluble in water.

Methylamidocresol, $C_6H_3(NH·CH_3)(CH_3)(OH)$, *amidol*, serves as a developer; its salts are also called *metol*.

Saligenin, *o*-oxybenzyl alcohol, $C_6H_4(OH)(CH_2·OH)$ is obtained by the decomposition of the glucoside salicin (which see) occurring in the bark of the willow-tree by the action of dilute acids. It forms shining leaves which are soluble in hot water, alcohol, and ether and its aqueous solution turning deep blue with ferric chloride.

m-Amido-o-oxybenzyl alcohol, *edinol*, is used as a photographic developer.

Salicyl aldehyde, $C_6H_4(OH)(CHO)$, occurs in the volatile oil of the varieties of Spiræa and is produced by the oxidation of saligenin. It forms a liquid which is not readily soluble in water and this solution turns deep violet with Fe_2Cl_6.

All aromatic oxyaldehydes may also be prepared by the action of chloroform and caustic alkali upon the phenols (Reimer's synthesis):

$$C_6H_5-OH+CHCl_3+3KOH=C_6H_4(OH)(CHO)+3KCl+2H_2O.$$

Salicylic acid, $C_6H_4(OH)^-COOH$, *o*-oxybenzoic acid, a phenol acid, occurs free, besides salicyl aldehyde, in various varieties of Spiræa, and as its methyl ester in oil of wintergreen (p. 349). It is obtained by the oxidation of saligenin or salicyl aldehyde and on a large scale by allowing sodium to act upon phenol and saturating the sodium phenolate obtained with CO_2 in the cold and under pressure and then heating the sodium phenol carbonate, $C_6H_5^-O^-COONa$, to 130° when it is converted into sodium salicylate (Kolbe's synthesis of cyclic oxyacids). The sodium salicylate thus obtained is decomposed by an inorganic acid.

Salicylic acids form colorless needles or a crystalline powder which is not readily soluble in cold water (in 500 parts) but readily soluble in alcohol and ether. The solutions of the acid as well as of its salts, the salicylates, turn violet with ferric chloride. It is used as an antifermentive agent, and is non-poisonous, and as it is a phenol it forms three series of esters, thus:

$$C_6H_4<^{OH}_{COOCH_3} \qquad C_6H_4<^{OCH_3}_{COOH} \qquad C_6H_4<^{OCH_3}_{COOCH_3}.$$

Salicylic acid   Methyl   Salicylic acid
methyl ester.   salicylic acid.   dimethyl ester.

Sodium salicylate, $C_6H_4(OH)COONa$, forms white soluble scales having a sweet-salty taste.

Lithium salicylate, $C_6H_4(OH)COOLi+H_2O$, is a white crystalline powder soluble in water.

Mercuric salicylate, $C_6H_3(OH)COOHg$, has a different structure from the other salicylates in that one valence of the mercury atom is combined with a carbon. It forms a white odorless, tasteless powder, insoluble in water but soluble in alkali chlorides, hydroxides, and carbonates.

Basic bismuth salicylate, $C_6H_4(OH)·COO^-Bi(OH)_2$, forms a white crystalline odorless and tasteless powder which is insoluble in water and alcohol.

Methyl salicylate, $C_6H_4(OH)^-COO(CH_3)$, forms the oil of wintergreen, and is used in medicine.

Phenyl salicylate, *salol*, $C_6H_4(OH)·COO-C_6H_5$, is prepared by heating sodium salicylate with sodium phenolate in the presence of $POCl_3$ or $COCl_2$. It forms a white crystalline powder which is nearly insoluble in alcohol, ether, and chloroform, melting at 42°, and having a faint aromatic odor and taste.

p-Amidophenylacetyl salicylate, *salophene*, $C_6H_4(OH)·COO-C_6H_4-NH(CH_3-CO)$, form leaves which are insoluble in water.

Acetyl salicylic acid, $C_6H_4(O \cdot OC\ CH_3)-COOH$, *aspirin*, forms crystals readily soluble in water.

Dithiosalicylic acid, $HOOC(HO)H_3C_6-S-S-C_6H_3(OH)-COOH$. The basic bismuth salt, *thioform*, is used in medicine.

Methylamidooxybenzoate, $C_6H_3(NH_2)(OH)(COOCH_3)$. The *p*-amido-*m*-oxy-compound is called *orthoform*, the *m*-amido-*p*-oxy-compound, *new orthoform*, the *p*-amido-*o*-oxy-compound combined with dimethylglycocoll, *nervanin*; all are used in medicine.

Anisyl alcohol, *o*-methyloxybenzyl alcohol, $C_6H_4(O \cdot CH_3)(CH_2OH)$, is produced from its aldehyde by reduction (by caustic alkali, see Benzaldehyde) and forms crystals melting at 45°

Anisic aldehyde, $C_6H_4(O \cdot CH_3)(CHO)$, is prepared by the oxidation of anethol (which see) and is the odoriferous principle of the hawthorn and forms a liquid which boils at 248°.

Anisic acid, $C_nH_4(O \cdot CH_3)(COOH)$, produced by the oxidation of anisic aldehyde or anethol, forms crystals which melt at 185°.

b. *Dihydric Phenols*, $C_6H_3(OH)_2(CH_3)$, *and their Derivatives.*

All six of the possible dioxytoluenes are known, but only orcin and homopyrocatechin will be treated of. The alcohol phenols or dioxybenzyl alcohols, $C_6H_3(OH)_2(CH_2OH)$, related to these are not known free, nevertheless six possible dioxybenzyl aldehydes and dioxybenzoic acids have been prepared synthetically.

Orcin, $C_6H_3(OH)_2(CH_3)$, exists free to a slight extent in the lichens of the varieties Rocella, Evernia, and Lecanora, and can be obtained from the ester, erythrin (p. 496), contained in these lichens. Orcin forms colorless sweet crystals whose watery solution turns bluish violet with ferric chloride and temporary deep violet with a solution of chloride of lime.

If an ammoniacal orcin solution is allowed to stand in the air it absorbs oxygen and nitrogen and a red crystalline pigment, *orcein*, $C_{28}H_{24}N_2O_7$, the chief constituent of the *orseille pigments* (called when pure persio, cud-bear, red indigo) is produced. These also may be obtained directly by fermenting the above-mentioned lichens with ammonia.

If an ammoniacal orcin solution is allowed to stand with alkali carbonate, *litmus pigment*, which becomes red with acids and blue again with alkalies is obtained. This consists chiefly of *azolitmin*, $C_7H_7NO_4$.

Homopyrocatechin, $C_6H_3(OH)_2(CH_3)$, obtained from creosol by heating with HI, forms hygroscopic crystals.

Creosol, $C_6H_3(CH_3)(OCH_3)(OH)$, methyl homopyrocatechin, is a liquid boiling at 220°, which is very similar to guaiacol (p. 480), and forms, when mixed with this,

Creosote, which is obtained from beech-wood tar by fractional distillation between 200° and 220°. It forms a yellowish liquid having a penetrating smoky odor. Its alcoholic solution turns deep blue with a small quantity of ferric chloride solution and dark green with a larger quantity. *Creosotal* is the name given to a mixture of guaiacol and creosol carbonates.

Protocatechuic alcohol, $C_6H_3(OH)_2(CH_2OH)$, is unknown.

Protocatechuic aldehyde, $C_6H_3(OH)_2-CHO$, obtained from vanillin (p. 494) or from pyrocatechin by means of the chloroform reaction (p. 492). It forms colorless crystals which are soluble in water, and whose aqueous solution turns deep green with ferric chloride.

Vanillin, methyl protocatechuic aldehyde, $CH_3(O \cdot CH_3)(OH)(CHO)$, the odoriferous constituent of the vanilla bean. It is obtained by extracting the vanilla bean with ether. It is also found in the Siam benzoin and in certain other resins, in the sugar-beet, asparagus, etc., to a slight extent.

It is artificially prepared from coniferyl alcohol, contained in the glucoside coniferin (which see), by oxidation:

$$C_6H_3{\Big\langle}{\overset{\displaystyle OCH_3}{\underset{\displaystyle C_3H_4OH}{OH}}} + 6O = C_6H_3{\Big\langle}{\overset{\displaystyle OCH_3}{\underset{\displaystyle CHO}{OH}}} + 2CO_2 + 2HOH.$$

$$\underset{\text{Coniferyl alcohol.}}{} \qquad\qquad \underset{\text{Vanillin.}}{}$$

It is also produced from guaiacol and chloroform (p. 492):

$$C_6H_4{<}{\overset{\displaystyle OCH_3}{OH}} + CHCl_3 + 3KOH = C_6H_3(OH){<}{\overset{\displaystyle OCH_3}{CHO}} + 3KCl + 2H_2O.$$

It is prepared on a commercial scale by the oxidation of isoeugenol, $C_6H_3(O \cdot CH_3)(OH)(CH : CH \cdot CH_3)$ (see Eugenol, p. 499).

Vanillin crystallizes in long characteristically smelling needles, which are soluble in water, alcohol, and ether, and which, when heated with HCl, decompose into methyl chloride and protocatechuic aldehyde:

$$C_6H_3(OH){<}{\overset{\displaystyle OCH_3}{CHO}} + HCl = C_6H_3{<}{\overset{\displaystyle (OH)_2}{CHO}} + CH_3Cl.$$

Piperonal, $C_6H_3(CHO){<}{\overset{\displaystyle O}{O}}{>}CH_2$, methylenprotocatechuic aldehyde, forms the perfume *heliotropin,* used to be prepared by the oxidation of piperic acid (p. 501), but now from safrol (p. 500).

Protocatechuic acid, pyrocatechin carbonic acid, $C_6H_3(OH)_2–COOH$, one of the six known oxysalicylic or dioxybenzoic acids, is obtained by fusing gum catechu, kino, benzoin, asafœtida, myrrh, etc., with caustic alkalies, also by the oxidation of its aldehydes. It forms colorless needles which are soluble in water, alcohol, and ether, and whose solution turns green with ferric chloride.

Hydroquinone carbonic acid, gentisic acid, $C_6H_3(OH)_2–COOH$, is also one of the six oxysalicylic acids (p. 493).

Veratric acid, $C_6H_3(OCH_3)_2–COOH$, pyrocatechuic acid dimethyl ester, occurs in the seed of the sabilla and forms colorless crystals.

c. *Trihydric Phenols,* $C_6H_2(OH)_3CH_3$, *and their Derivatives.*

The six possible trioxytoluenes and their respective alcohols, aldehydes, and acids are not all known.

Methylpyrogallol, $C_6H_2(OH)_3(CH_3)$, is obtained synthetically. Its dimethyl ether occurs in beech-wood tar.

Trioxybenzyl alcohol, $C_6H_2(OH)_3(CH_2 \cdot OH)$, gallic alcohol, and **Trioxybenzaldehyde,** $C_6H_2(OH)_3(CHO)$, are also obtained synthetically.

Gallic Acid, $C_7H_6O_5 + H_2O$, one of the three known trioxybenzoic acids, $C_6H_2(OH)_3(COOH)$, occurs free with tannin in the oak-gall, pods of the Divi-Divi, in tea and the rind of the pomegranate root and many plants. It is also found combined, and indeed generally as a glucoside, in certain tannic acids. It is obtained from tannin by boiling with dilute acids (see below) or by fusing bromprotocatechuic acid with caustic alkali:

$$C_6H_2Br \lessgtr {(OH)_2 \atop COOH} + KOH = C_6H_2(OH) \lessgtr {(OH)_2 \atop COOH} + KBr.$$

Gallic acid consists of white fine needles which are soluble in hot water, alcohol, and ether, and has a strong reducing action. Its alkaline solution absorbs oxygen, and turns brown.

Ferric chloride produced in a watery solution of gallic acid a dark-blu ⋅ precipitate, gelatine solution gives no precipitate, while potassium cyanide gives a red coloration (differing from tannin). On heating gallic acid it decomposes into CO_2 and pyrogallol (p. 482), $C_6H_3(OH)_3$, and on oxidation it yields ellagic acid, $C_{14}H_6O_8$ (p. 496), which is used in medicine under the name *gallogen*.

Basic bismuth gallate-iodide, *airoform*, *airogen*, *airol*, $C_6H_2(OH)_3^-$ $COO^-Bi(OH)I$, and

Basic bismuth gallate, *dermatol*, $C_6H_2(OH)_3COO^-Bi(OH)_2$, are yellow amorphous powders without odor and taste and insoluble in water, alcohol, and ether.

Digallic acid, ordinary tannic acid, tannin, $C_{14}H_{10}O_9 + 2H_2O$, an anhydride of gallic acid (structure below) is found in oak-galls, in the sumach, and in red wine. It is obtained from gallic acid by dehydrating agents and can be recomposed into gallic acid by boiling with dilute acids and alkalies:

$$C_6H_2(OH)_3^-COO^-C_6H_2(OH)_2^-COOH + HOH = 2C_6H_2(OH)_3^-COOH.$$

It is ordinarily obtained by extracting oak-galls with ether containing alcohol and evaporating the same.

Tannic acid consists as a colorless, amorphous, neutral, dextrorotatory powder, which is readily soluble in water, alcohol, glycerine, but not in ether. It has a strong reducing action and when dissolved in the presence of ferric salts, tannic acid gives a deep-blue coloration. When heated it decomposes, yielding pyrogallol. For further properties, see the tannic acids (below).

Tannigen, diacetyl tannin, $C_{14}H_8(CH_3 \cdot CO)_2O_9$, and

Aluminium tannate-tartrate, $Al_2(C_4H_4O_6)_2(C_{14}H_9O_9)_2$, are brown powders soluble in water.

Mercurous tannate forms colorless, tasteless, insoluble scales which have a deep green color.

Lead tannate, obtained from basic lead acetate and tannic acid, forms when mixed with lard an ointment used in pharmacy.

Tannic acids, tanning bodies, tannins, are the names given to the bodies widely disseminated in the vegetable kingdom, which are soluble in

water, have an astringent taste, give black or green precipitates with
Iron salts (inks), precipitate proteids and gelatine solutions, and form
insoluble compounds with animal hides, which prevent these from
putrefying (leather manufacture). The tanning bodies do not form
one group, but they are divided into *tannogens* (cyclic oxyacids),
tannoids (anhydrides, oxidation, and condensation products of the tannogens), and *glucotannoids* (compounds of the tannoids with the sugars).
According to their origin we differentiate between *ellagic, coffee, catechu,
tormentilla, kino, quina, moringa, oak* (= tea), *quinova, rheum, filix, ratan-
hia* tannic acids which with dilute acids form colored, resin-like compounds which are called *phlobaphenes.*

Quinic acid, hexahydrotetraoxybenzoic acid, $C_6H(OH)_4(COOH)H_6$,
occurs widely distributed in the plant kingdom, especially in the cinchona bark, sugar-beets, coffee-beans, and forms active prisms. Its
lithium salt is called *urosin,* its piperidin salt *sidonal,* its urea salt *urol,*
its anhydride *neusidonal,* which are all used in medicine.

COMPOUNDS WITH EIGHT CARBON ATOMS UNITED TOGETHER.

1. Dimethyl Benzene Compounds.

Dimethyl benzenes, xylenes, xylols, $C_6H_4(CH_3)_2$. From coal-tar by
fractional distillation at 140° we obtain a colorless liquid which consists
of a mixture of the three isomeric xylenes, which cannot be separated
from each other by fractional distillation, and hence they must be prepared synthetically (p. 472). On oxidation they yield, according to the
energy of oxidation, either toluic acids, $C_6H_4(CH_3)-COOH$, or phthalic
acids, $C_6H_4(COOH)_2$. In regard to the preparation of the halogen derivatives see page 474.

Oxyxylenes, xylenols, $C_6H_3(OH)<{}^{CH_3}_{CH_3}$, have all the properties of the
phenols and occur in beech-wood tar. All six isomers are known and
four of them can be obtained by fusing the isomeric xylene sulphonic acids
with caustic alkali.

Oxytoluic acids, cresotic acids, $C_6H_3(OH)<{}^{CH_3}_{COOH}$. All ten possible isomers have been prepared.

Oxyphthalic acids, $C_6H_3(OH)<{}^{COOH}_{COOH}$. The six possible isomers have
also been prepared.

Dioxyxylenes, $C_6H_2(OH)_2<{}^{CH_3}_{CH_3}$; five are known. They are called *m-*
and *p-*xylorcin (the latter also *β*-orcin), *o-, m-,* and *p-*xylohydroquinone
(the last also hydrophlorone).

Dioxytoluic acids, $C_6H_2(OH)_2<{}^{COOH}_{CH_3}$. The most interesting of the
known isomers is orcin carbonic acid or *orsellic acid,* which exists
in certain lichens as diorsellic acid erythrite ester (Erythrin, p. 493),
$${}^{C_6H_2(OH)_2(CH_3)COO}_{C_6H_2(OH)_2(CH_3)COO}>C_4H_6(OH)_2,$$ and which decomposes on boiling

with Ba(OH)$_2$ into erythrite and orsellic acid. The latter splits on heating into orcin, C$_6$H$_3$(OH)$_2$CH$_3$ and CO$_2$.

Numerous other acids are found in lichens, so-called *lichen acids*, and are derived from orcin; still the mother substance of these acids has not been investigated. To this group belong *cetraric acid*, C$_{26}$H$_{20}$O$_{12}$; *protocetraric acid*, C$_{30}$H$_{22}$O$_{15}$, contained in the Iceland moss, and yielding fumaric and cetraric acids on cleavage; *evernic acid*, C$_{17}$H$_{16}$O$_7$; *lecanoric acid* (orsellic acid), C$_{16}$H$_{14}$O$_7$; *lichsteric acid*, C$_{18}$H$_{30}$O$_5$; *usninic acid*, C$_{18}$H$_{16}$O$_7$; *parellic acid*, C$_{21}$H$_{16}$O$_2$, etc.

Dioxyphthalic acids, C$_6$H$_2 \cdot$(OH)$_2 <^{\text{COOH}}_{\text{COOH}}$; six are known.

Meconinic acid, C$_6$H$_2$(O$\cdot$CH$_3$)$_2 <^{\text{CH}_2\text{OH}}_{\text{COOH}}$, and

Opianic acid, C$_6$H$_2$(O$\cdot$CH$_3$)$_2 <^{\text{CHO}}_{\text{COOH}}$, and

Hemipinic acid, C$_6$H$_2$(O$\cdot$CH$_3$)$_2 <^{\text{COOH}}_{\text{COOH}}$,

are produced on the oxidation of the plant-base narcotine (which see).

Tolyl alcohols, methylbenzyl alcohols, C$_6$H$_4 <^{\text{CH}_3}_{\text{CH}_2\cdot\text{OH}}$;

Toluic aldehydes, methylbenzyl aldehydes, C$_6$H$_4 <^{\text{CH}_3}_{\text{CHO}}$;

Toluic acids, methylbenzoic acids, C$_6$H$_4 <^{\text{CH}_3}_{\text{COOH}}$;

Toluylene alcohols, phthalyl alcohols, C$_6$H$_4 <^{\text{CH}_2\text{OH}}_{\text{CH}_2\text{OH}}$;

Oxymethylbenzoic acids, C$_6$H$_4 <^{\text{CH}_2\text{OH}}_{\text{COOH}}$;

are known in all three isomeric forms, namely, *o*-, *m*-, and *p*-compounds.

Phthalic acids, benzendicarbonic acids, C$_6$H$_4 <^{\text{COOH}}_{\text{COOH}}$, are produced by the oxidation of the corresponding three xylenes, as well as all isocyclic compounds which contain two alkyl radicals (p. 473).

Ordinary or *o*-phthalic acid is prepared by the oxidation of naphthalene. From this it follows that the carboxyls occupy the positions 1 : 2 (see Naphthalene).

Phthalic acid forms colorless crystals which melt at 213°, and are soluble in hot water, alcohol, ether, and which decompose into phthalic anhydride, C$_6$H$_4 <^{\text{CO}}_{\text{CO}}>$O, and water on heating

Iso- or *m*-phthalic acid forms needles which melt at 300° and sublime without decomposition, and are difficultly soluble in hot water.

Tere- or *p*-phthalic acid is obtained by the oxidation of cymene, turpentine, etc. It forms an amorphous powder, nearly insoluble in water, alcohol, or ether, and sublimes without decomposing and without melting.

Phthalimide, C$_6$H$_4 <^{\text{CO}}_{\text{CO}}>$NH, is produced by passing NH$_3$ over phthalic anhydride, and yields anthranilic acid on oxidation (p. 489).

2. Ethyl Benzene Compounds.

Ethyl benzene, $C_6H_5(C_2H_5)$ or $C_6H_5-CH_2-CH_3$, is a colorless liquid boiling at 134°, and is only obtained synthetically.

Phenyl ethyl alcohol, $C_6H_5(C_2H_4\cdot OH)$, forms the chief constituent of the odoriferous principle of the rose and hence also of the ethereal oil of roses.

Methyl phenylketone, $C_6H_5-CO-CH_3$, acetophenone, a colorless liquid having a peculiar odor.

All aromatic ketones may be obtained by warming an aromatic hydrocarbon with an acid chloride in the presence of $AlCl_3$ (Friedel-Craft's ketone synthesis):

$$C_6H_6 + CH_3-COCl = C_6H_5-CO-CH_3 + HCl.$$

If a benzoate is distilled with another organic salt, then mixed ketones are obtained:

$$C_6H_5-COONa + CH_3-COONa = Na_2CO_3 + C_6H_5-CO-CH_3$$

Ethenyl benzene, styrene, cinnamol, $C_6H_5-CH=CH_2$, is produced by the distillation of cinnamic acid with lime, when CO_2 is evolved. It occurs in the storax balsam and is a colorless liquid which boils at 146° and which undergoes polymerization readily.

Phenylacetic acid, alphatoluic acid, $C_6H_5-CH_2-COOH$, the only isomeric toluic acids of ethyl benzene known, occurs in the urine and in the putrefactive products of proteids.

Oxyphenylacetic acid, $C_6H_4(OH)-CH_2-COOH$, occurs in urine and is produced in the putrefaction of tyrosin.

Dioxyphenylacetic acid, $C_6H_3(OH)_2-CH_2-COOH$, homogentisic acid (p. 494), occurs sometimes in the urine, which causes it to turn dark on standing.

Oxyphenylamido acetic acid, $C_6H_4(OH)-CH(NH_2)-COOH$, *glycin,* is used as a developer in photography.

Mandelic acid, phenylglycollic acid, $C_6H_5-CHOH-COOH$, is obtained from benzaldehyde and $HCN + HCl$ (p. 405) or from amygdalin (see below). The lævo and the inactive modifications are known.

Mandelic acid nitrile diglucose, $C_6H_5-CHO(C_{12}H_{21}O_{10})-CN$, amygdalin, the glucoside of the bitter almond, etc. (p. 384), forms colorless crystals which on warming with dilute acids or on standing with water by the action of the ferment emulsin contained in the plants, decomposes into glucose, hydrocyanic acid, and benzaldehyde (oil of bitter almonds): $C_{20}H_{27}NO_{11} + 2H_2O + 2C_6H_{12}O_6 + HCN + C_6H_5-CHO$. With zymase it decomposes into glucose and mandelic acid nitrile glucose, $C_6H_5-CHO(C_6H_{11}O_5)-CN$, which is only further split by emulsin like amygdalin. If amygdalin is boiled with bases a development of NH_3 takes place and amygdalic acid, $C_{20}H_{24}O_{13}$, is produced, which on heating with acids decomposes into glucose and mandelic acid (see above).

COMPOUNDS WITH NINE CARBON ATOMS UNITED TOGETHER.

1. Trimethyl Benzene Compounds.

The three trimethyl benzenes, $C_6H_3(CH_3)_3$, mesitylene (position 1, 3, 5), pseudocumene (position 1, 2, 4), and hemellithene (position 1, 2, 3), occur in coal-tar.

Mesitylene, $C_6H_3(CH_3)_3$, is obtained when acetone is heated with sulphuric acid; hence it has the structure (1, 3, 5) (p. 470). It is a liquid having a peculiar odor, boiling at 163°, and which yields the following three acids on successive oxidation:

Mesitylenic acid, $C_6H_3(CH_3)_2COOH$, isomer of the xylic acids.
Uvitic acid, $C_6H_3(CH_3)(COOH)_2$, isomer of methylphthalic acid.
Trimesic acid, $C_6H_3(COOH)_3$, isomer of trimellitic acid.

Pseudocumene, $C_6H_3(CH_3)_3$, a liquid boiling at 169° and which first yields the following two isomeric acids:

Xylic acids, $C_6H_3(CH_3)_2COOH$, and then on further oxidation
Methylphthalic acid, $C_6H_3(CH_3)(COOH)_2$, and its isomer,
Xylidic acid, $C_6H_3(CH_3)(COOH)_2$, which is readily oxidized to
Trimellitic acid, $C_6H_3(COOH)_3$.

Hemellithene, $C_6H_3(CH_3)_3$, a liquid boiling at 175° and which yields on oxidation

Hemellitic acid, $C_6H_3(CH_3)_2COOH$, and finally
Hemillitic acid, $C_6H_3(COOH)_3$, an isomer of the benzene tricarboxy acids trimesic and trimellitic acids.

2. Allyl Benzene Compounds.

The radical C_3H_5 may act trivalent as glyceryl and monovalent as allyl (p. 437). As the derivatives of the allyls of the fatty series appear as unsaturated compounds and pass readily over into the saturated propyl compounds, so also the aromatic derivatives of the allyls are converted into the saturated propylbenzene derivatives by the action of nascent hydrogen.

Allyl benzene, $C_6H_5-C_3H_5$ or $C_6H_5-CH=CH-CH_3$, is obtained by the action of bromallyl upon brombenzene in the presence of sodium (p. 472). It is a pleasant-smelling liquid.

Anethol, aniscamphor, $C_6H_4(O \cdot CH_3)(C_3H_5)$, the chief constituent of oil of anise, fennel, estragon, crystallizes on cooling these oils as colorless scales which melt at 21°, and boil at 223°.

Eugenol, $C_6H_3(OH)(O \cdot CH_3)(C_3H_5)$, contained in the ethereal oil of cloves and pimenta (besides $C_{10}H_{16}$ see terpenes); forms an aromatic-smelling liquid which boils at 252°, and which forms the isomer isoeugenol with alcoholic caustic potash.

Coniferylic alcohol, $C_6H_3(OH)(O \cdot CH_3)(C_3H_4 \cdot OH)$, occurring in the glucoside coniferin, melts at 85°, and yields vanillin on oxidation (p. 494).

Asarone, $C_6H_2(OCH_3)_3C_3H_5$, contained in the ethereal oil of Asarum europæum, Matiko, and Calamus, forms shining crystals which melt at 61°.

Apiol, $C_6H(O\cdot CH_3)_2(O\cdot CH_2\cdot O)(C_3H_5)$, contained in the ethereal oil of the parsley, forms shining crystals which melt at 32°.

Safrol, $C_6H_3(O\cdot CH_2\cdot O)(C_3H_5)$, allylpyrocatechin methylene ether, is the chief constituent of the ethereal oil of sassafras and of the camphor-oil obtained in the preparation of camphor.

Cubebin, $C_6H_3(O\cdot CH_2\cdot O)(C_3H_4\cdot OH)$, occurs in cubebs.

Cinnamyl alcohol, cinnamic alcohol, phenylallylalcohol, $C_6H_5{}^-C_3H_4(OH)$ or $C_6H_5{}^-CH{}^=CH{}^-CH_2OH$, occurs as cinnamic acid ester in storax and is separated therefrom by distillation with caustic alkali. It forms needle-shaped crystals which melt at 33°, have an odor similar to the hyacinth, and are not very soluble in water. On oxidation it yields cinnamic aldehyde and cinnamic acid.

Cinnamic aldehyde, $C_6H_5{}^-CH{}^=CH{}^-CHO$, the chief constituent of the oil of cinnamon and cassia, forms a colorless liquid boiling at 246°, is insoluble in water, and has an odor similar to cinnamon. It is oxidized to cinnamic acid even in the air.

Cinnamic acid, β-phenylacrylic acid, $C_6H_5{}^-C_3H_3O_2$ or $C_6H_5{}^-CH{}^=CH{}^-COOH$, occurs in certain gum benzoins, in storax, Peru and Tolu balsam. It is prepared by boiling storax with caustic alkali and precipitating the alkali cinnamate by HCl or by boiling benzaldehyde with dry sodium acetate and acetic anhydride, which abstracts water (Perkin's reaction, which may be made to undergo numerous modifications): $C_6H_5{}^-CHO + CH_3{}^-COONa = C_6H_5{}^-CH{}^=CH{}^-COONa + H_2O$; or from benzal chloride and sodium acetate:

$$C_6H_5{}^-CHCl_2 + CH_3{}^-COONa = C_6H_5{}^-CH{}^=CH{}^-COOH + NaCl + HCl.$$

Cinnamic acid forms colorless and odorless crystals which melt at 133° and which are difficultly soluble in cold water. Ferric chloride precipitates yellow ferric cinnamate from its solution. On oxidation it yields benzaldehyde and benzoic acid (p. 473).

Sodium cinnamate, $C_6H_5{}^-CH{}^=CH{}^-COONa$, is used in medicine as *hetol.* Besides the ordinary cinnamic acid three modifications of the same structure are known which are readily convertible into ordinary cinnamic acid, while this latter cannot be transformed into the other modifications. These are called *allocinnamic* and *artificial isocinnamic acid,* which is obtained synthetically, and *natural isocinnamic acid,* which is found with ordinary and allocinnamic acid in the cleavage acids of the alkaloids associated with cocaine.

Besides these we have four dicinnamic acids having the same structure, namely, the diphenyltetramethylendicarboxyl acids or *tuxillic acids*, $(C_9H_8O_2)_2$, which are also cleavage acids of the alkaloids accompanying cocaine.

Atropic acid, α-phenylacrylic acid, $C_9H_8O_2$ or $CH_2=C(C_6H_5)-COOH$, the isomer of cinnamic acid, is produced on the cleavage of atropine or apo-atropine and yields two diatropic acids, $(C_9H_8O_2)_2$, isomers of the truxillic acids (see above), on heating with water.

Cinnamein, $C_6H_5-CH=CH-COO(C_6H_5-CH_2)$, benzyl cinnamate, a constituent of storax, balsam of Peru and Tolu, forms colorless needles.

Styracin, $C_6H_5-CH=CH-COO(C_6H_5-C_3H_4)$, cinnamyl-cinnamate, a constituent of storax and Peru balsam, forms colorless crystals.

o-Oxycinnamic acid, coumaric acid, $C_6H_4(OH)-CH=CH-COOH$ or $C_9H_8O_3$, occurs in Melilotus officinalis, forms colorless needles, and is obtained from coumarin (see below).

Methoxyloxycinnamic acids, $C_6H_3(O\cdot CH_3)(OH)-CH=CH-COOH$. *Ferulic acid,* contained in asafœtida, and the *isoferulic acid* and *hesperitic acid,* cleavage products of the glucoside hesperidin, belong to this group.

o-Oxycinnamic anhydride, coumarin, $C_6H_4\diagup\begin{smallmatrix}O-\\C_2H_2\end{smallmatrix}\diagdown CO$ or $C_9H_6O_2$, in sweet-scented woodruff, in the Tonka bean, in clover, sweet-scented grass, as well as in other plants. It forms colorless prisms having the odor of the sweet-scented woodruff, and is produced by heating salicyl aldehyde with acetic anhydride:

$$C_6H_4(OH)CHO + (CH_3-CO)_2O = C_6H_4\diagup\begin{smallmatrix}O-\\C_2H\end{smallmatrix}\diagdown CO + CH_3-COOH + H_2O.$$

On boiling with caustic alkalies coumarin takes up H_2O and is converted into coumaric acid.

Umbelliferone, $C_9H_6O_3$, is an oxycoumarin, occurs in the spring flax, and is formed in the dry distillation of many umbelliferous resins; for example, of asafœtida, galbanum, etc.

Daphnetin and **Æsculetin,** $C_9H_6O_4$, are dioxycoumarins, split off from the glucosides daphnin and æsculin.

Dioxycinnamic acids, $C_9H_8O_4$ or $C_6H_4(OH)_2-CH=CH-COOH$. *Caffeic acid* is obtained as yellow leaves from the glucoside coffee-tannin occurring in coffee. The isomer *umbellic acid* is produced from umbelliferone (see above) by abstracting water.

Closely related to these we have *piperic acid,* $C_{12}H_{10}O_4$ or $C_6H_3(O\cdot CH_3\cdot O)-CH=CH-CH=CH-COOH$, a cleavage product of piperin (which see).

Dimethyltrioxycinnamic acid, sinapic acid, $C_{11}H_{12}O_5$ or $C_6(CH_3)_2(OH)_3-CH=CH-COOH$, is a cleavage product of sinapin (which see).

3. Propyl Benzene Compounds.

The allyl derivatives are readily converted into the propyl benzene derivatives by means of nascent hydrogen.

Cumene, cumol, isopropyl benzene, $C_6H_5-C_3H_7$, occurs in the ethereal oil of the Roman caraway; it is a colorless liquid boiling at 151°.

Phenylpropyl alcohol, $C_6H_5-C_2H_4-CH_2OH$, is produced by the action of nascent hydrogen (sodium amalgam) upon cinnamyl alcohol.

β-Phenylpropionic acid, $C_6H_5-C_2H_4-COOH$, hydrocinnamic acid, is obtained by the action of nascent hydrogen upon cinnamic acid. It is a putrefactive product of the proteids.

Tropic acid, $C_6H_5-C_2H_3(OH)-COOH$, α-phenyl-β-oxypropionic acid. This inactive acid, which can be converted into the two active modifications, is obtained on the cleavage of atropine and hyoscyanine.

β-Phenyl-α-amidopropionic acid, $C_6H_5-C_2H_3(NH_2)-COOH$, phenylalanin, is a cleavage product of the proteids.

Mellotic acid, $C_6H_4(OH)-C_2H_4-COOH$, o hydrocoumaric acid, one of the six existing oxyphenylpropionic acids, occurs with coumarin in clover. It forms needles which melt at 38°.

p-Hydrocoumaric acid, phloretic acid, $C_6H_4(OH)-C_2H_4-COOH$, occurs in urine and as a putrefactive product of tyrosin.

p-Oxyhydrocoumaric acid, $C_6H_4(OH)-C_2H_3(OH)-COOH$, is found in the urine in acute atrophy of the liver and in phosphorus poisoning.

Tyrosin, $C_6H_4(OH)-C_2H_3(NH)_2-COOH$ or $C_9H_{11}NO_3$, amidohydrocoumaric acid, p-oxyphenylamidopropionic acid, occurs in cochineal, in dahlia tubers, and in the animal organism under pathological conditions, nearly always accompanied with leucin. It is a cleavage product of the horn substances and the proteids (not of gelatine) by the action of panceatic juice, by boiling with acids or fusing with alkali hydroxides, as well as in their putrefaction, and hence is also found in old cheese (τυρός). It can also be prepared synthetically. It crystallizes in fine, characteristically grouped needles which are lævorotatory, soluble with difficulty in water and insoluble in alcohol and ether, and melt at 235°. Dextrorotatory tyrosin occurs in the sugar-beet sprouts.

Orthonitrophenylpropiolic acid, $C_6H_4(NO_2)-C\equiv C-COOH$, is produced from the dibromide of o-nitrocinnamic acid by alcoholic caustic alkali:

$$C_6H_4(NO_2)-CHBr-CHBr-COOH + 2NaOH$$
$$= C_6H_4(NO_2)-C\equiv C-COOH + 2NaBr + 2H_2O.$$

It crystallizes in colorless needles, and yields indigo blue on warming with alkaline reducing agents.

COMPOUNDS WITH TEN OR MORE CARBON ATOMS COMBINED TOGETHER.

Tetramethyl benzenes, $C_6H_2(CH_3)_4$. These are called durols, isodurols, prehnitol, and are prepared synthetically.

Benzene tetracarbonic acids, $C_6H_2(COOH)_4$, all three of the theoretically possible ones are known and are called pyromellitic acid, prehnitic acid, and mellophanic acid.

p-Methylisopropyl benzene, cymene, cymol, $C_{10}H_{14}$ or $C_6H_4(C_3H_7)(CH_3)$ (structure, p. 519), occurs in the ethereal oils of the Roman caraway and thyme, and is produced by heating the terpenes, $C_{10}H_{16}$, with iodine, and the camphors, $C_{10}H_{16}O$, with P_2O_5 (p. 520). It is a liquid boiling at 175°, which on oxidation yields *p*-tolulic acid, $C_6H_4(CH_3)(COOH)$ (p. 497), and then terephthalic acid, $C_6H_4(COOH)_2$ (pp. 497–473).

Cumyl alcohol, $C_6H_4(C_3H_7)(CH_2OH)$, and

Cumic aldehyde, cuminol, $C_6H_4(C_3H_7)(CHO)$, occur in Roman caraway-oil and cicuta-oil, and are colorless liquids.

Cumic acid, $C_6H_4(C_3H_7)(COOH)$, forms plates which melt at 160° and is produced by the oxidation of the above compounds.

Thymol, $C_6H_3(OH)(CH_3)(C_3H_7)$, thyme camphor, thyminic acid, is a *p*-methylisopropyl phenol, and occurs with cymene and thymene in oil of thyme and forms crystals which melt at 51° and having an odor similar to thyme. When fused the crystals boil at 230° and are volatile with steam. Thymol is insoluble in water.

Dithymoldi-iodide, $(C_3H_7)(CH_3)(IO)H_2C_6-C_6H_2(OI)(CH_3)(C_3H_7)$, aristol, is produced by treating an alkaline solution of thymol with an iodine solution. It is a reddish, odorless, insoluble powder.

Carvacrol, $C_{10}H_{14}O$, isomer of thymol, occurs in the ethereal oils of the varieties of Origanum and Satureja. It forms colorless crystals melting at 0°. It is produced by heating camphor with iodine, as well as by heating the isomeric carvoe (see terpene group) with phosphoric acid.

Cantharidin, $C_{10}H_{12}O_4$ or $C_7H_9(\cdot CH_2 \cdot COOH)(\cdot O \cdot CO \cdot)$, a lactone acid (p. 404), is the blister-producing substance of the Spanish fly (cantharides); forms crystals which melt at 218°, and is readily converted into the isomer cantharic acid and on heating with alkali hydroxides yields the salts of cantharinic acid, $C_{10}H_{14}O_5$, which is unstable when free.

Terpenes, $C_{10}H_{16}$. These compounds, which are very widely distributed in nature, will be discussed at the end of the isocarbocyclic compounds, as they form a very large group (analogous to the carbohydrates). The ketone and alcohol derivatives, the camphors, and a series of compounds derived therefrom will also be discussed at that time.

m-Tolyloxybutyric acid, $CH_3-C_6H_4-CH \cdot OH-C_2H_4-COOH$, forms a lactone (p. 404) which forms the active principle of the Indian hemp called *cannabinol*, $C_{21}H_{26}O_2$.

Pentamethyl benzene, $C_6H(CH_3)_5$, and

Hexamethyl benzene, $C_6(CH_3)_6$, form colorless crystals. (Preparation, p. 473.)

Mellitic acid, $C_6(COOH)_6$, occurs in mellite or honey-stone, $C_6(COO)_6Al_2 + 18H_2O$, as yellow crystals in lignite deposits. Mellitic acid crystallizes in white needles which are soluble in water and alcohol. On heating they melt and at higher temperatures they decompose into pyromellitic acid, $C_6H_2(COOH)_4$, and $2CO_2$. With nascent hydrogen it yields hydromellitic acid, $C_6H_6(COOH)_6$ (p. 463).

Trinitrobutyl toluene, $C_6H(NO_2)_3(CH_3)(C_4H_9)$, forms white needles, which occur in commerce as artificial musk.

Ionone, $C_{13}H_{20}O$, obtained from citral by means of acetone, and

Irone, $C_{13}H_{20}O$, prepared from iridin (see Glucosides), give the odor to the violet-root and to the orris-root. They form colorless liquids having the odor of violets, and are both tetrahydrotrimethylbutylene **ketone** benzenes, $C_6H_2(CH_3)_3(CH=CH-CO-CH_3)(H_4)$.

COMPOUNDS WITH SEVERAL BENZENE RINGS.

The benzene molecules have the property of uniting together directly or by means of C atoms or other atoms. The hydrocarbons thus produced, like benzene, are the starting-point of a large series of derivatives which may be divided into the following four groups.

1. Compounds containing Benzene Rings Directly United.

1. **Diphenyl,** $C_{12}H_{10}$ or $H_5C_6-C_6H_5$, is obtained by the action of sodium upon brombenzene. It crystallizes in colorless crystals and forms, like benzene, the starting-point in the preparation of numerous derivatives. By the action of the halogens, HNO_3 or H_2SO_4, upon diphenyl we obtain mono- and disubstitution products, such as $C_{12}H_9Br$, $C_{12}H_9-SO_3H$, $C_{12}H_8(SO_3H)_2$, $C_{12}H_9NO_2$, etc. By reduction of the nitrodiphenyl we obtain amidodiphenyl, $C_{12}H_9(NH_2)$, and diamidodiphenyl, $C_{12}H_8(NH_2)_2$.

Hexaoxydiphenyl, $(HO)_3H_2C_6-C_6H_2(OH)_3$, is the mother-substance of cœrulignone, $C_{16}H_{16}O_6$, which is obtained in the purification of crude wood-vinegar. It forms blue needles.

p-**Diamidodiphenyl,** benzidine, $H_2N-H_4C_6-C_6H_4NH_2$, is produced from the isomeric hydrazobenzene by molecular rearrangement and forms the mother-substance of the benzidine dyes (see Azo Compounds). It is also used in the quantitative estimation of sulphuric acid as insoluble benzidine sulphate,

$$(C_6H_4-NH_2)_2H_2SO_4.$$

Phenyltolyl, $C_6H_5-C_6H_4(CH_3)$, is produced by the action of sodium upon a mixture of brombenzene and bromtoluene:

$$C_6H_5Br+C_6H_4Br(CH_3)+2Na=C_6H_5-C_6H_4(CH_3)+2NaBr.$$

It forms a colorless liquid which on oxidation yields
Diphenyl carbonic acid, $C_6H_5-C_6H_4-COOH$.
Ditolyl, $(CH_3)C_6H_4-C_6H_4(CH_3)$, obtained by the action of sodium upon bromtoluene, yields on oxidation
Diphenyl dicarbonic acid, $(COOH)C_6H_4-C_6H_4(COOH)$.

Diphenylenimide, carbazole, dibenzopynol, $C_{12}H_9N$ or $<^{C_6H_4}_{C_6H_4}>NH$,
occurs in coal-tar and is obtained from *o*-diamidodiphenyl, $C_{12}H_8(NH_2)_2=$ $C_{12}H_9N+NH_3$. It yields carbazol yellow.

Diphenylenoxide, dibenzofurane, $C_{12}H_8O$ or $<^{C_6H_4}_{C_6H_4}>O$, is derived from dioxydiphenyl, $C_{12}H_8(OH)_2$, and forms colorless crystals.

Diphenylensulphide, dibenzothiophene, $C_{12}H_8S$ or $<^{C_6H_4}_{C_6H_4}>S$, is derived from dithiodiphenyl, $C_{12}H_8(SH)_2$, and forms colorless crystals.

2. Compounds with Benzene Rings united by One Carbon Atom.

Diphenyl methane, $C_{13}H_{12}$ or C_6H_5-CH_2-C_6H_5, is produced by warming benzylchloride and benzene with aluminium chloride (p. 320):

$$C_6H_5CH_2Cl + C_6H_6 = C_6H_5-CH_2-C_6H_5 + HCl.$$

It forms colorless needles which melt at 26° and have an odor similar to oranges. Its derivatives are obtained from benzylchloride and toluene, xylene, and other hydrocarbons, also with phenols.

$$C_6H_5-CH_2Cl + C_6H_5(CH_3) = C_6H_5-CH_2-C_6H_4(CH_3) + HCl;$$
Benzyl chloride.   Benzyl toluene.

$$C_6H_5-CH_2Cl + C_6H_5(OH) = C_6H_5-CH_2-C_6H_4(OH) + HCl.$$
Benzyl chloride.   Benzyl phenol.

On oxidation these hydrocarbons are converted into ketones, where the CH_2 or $CH-CH_3$ groups are transformed into CO; thus:

$$\begin{matrix}C_6H_5\\C_6H_5\end{matrix}>CH_2 \quad \text{or} \quad \begin{matrix}C_6H_5\\C_6H_5\end{matrix}>CH\text{-}CH_3 \quad \text{form} \quad \begin{matrix}C_6H_5\\C_6H_5\end{matrix}>CO.$$
Diphenylmethane.   Diphenylethane.   Diphenylketone.

If the benzene nucleus still contains alkyls they are oxidized to carboxyl groups; thus:

$$C_6H_5-CH_2-C_6H_4(CH_3) \quad \text{forms} \quad C_6H_5-CO-C_6H_4(COOH).$$
Benzyl toluene.   Benzoyl benzoic acid.

Diphenylcarbinol, $C_6H_5-CH(OH)-C_6H_5$, benzhydrol, forms crystals which melt at 68° and on oxidation yields

Diphenylketone, $C_6H_5-CO-C_6H_5$, benzophenone (Preparation, p. 498), forms crystals which melt at 27°. It forms the mother-substance of the yellow dye auramine. $C_{17}H_{24}N_3HCl$, which has found medical use and of Michler's ketone, $(CH_3)_2N-H_4C_6-CO-C_6H_4-N(CH_2)_2$, which is used in the preparation of dyes belonging to the fuchsin series (p. 507). The yellow or brown plant bodies, *maclurin*, $C_{13}H_{10}O_6$, of fustic, *genistein*, $C_{14}H_{10}O_5$, of dyers' broom, and *catechin*, $C_{15}H_{10}O_6$, of catechu, are complicated benzophenone derivatives.

Diphenylenmethane, fluorene, $<\begin{matrix}C_6H_4\\C_6H_4\end{matrix}>CH_2$, occurs in coal-tar and forms fluorescent leaves. (Structure, see p. 513.)

Tetramethyldiamidodiphenylmethane, $(CH_3)_2N-C_6H_4-CH_2-C_6H_4-N(CH_3)_2$, is used in the differentiation of ozone from other gases (p. 112).

Triphenylmethane, $C_{19}H_{16}$ or $(C_6H_5)_3 \equiv CH$, is produced by heating benzal chloride and benzene in the presence of zinc dust or aluminium chloride (p. 320),

$$C_6H_5-CHCl_2 + 2C_6H_6 = C_6H_5-CH(C_6H_5)_2 + 2HCl,$$

or from chloroform and benzene under the same conditions:

$$3C_6H_6 + CHCl_3 = (C_6H_5)_3 \equiv CH + 3HCl.$$

It forms colorless leaves melting at 93° and boiling at 360°. It is insoluble in water and on oxidation yields

Triphenyl Carbinol, $(C_6H_5)_3 \equiv C^-OH$, which forms colorless prisms melting at 159°.

Triphenylmethane and triphenylcarbinol, as well as their homologues, form beautifully colored derivatives by the introduction of ^-OH, $^-NH_2$, ^-COOH, which are generally called the aniline colors. This misnomer (p. 484) arises from the fact that the anilines are used in the preparation of certain of these dyes.

Aniline colors are also found in nature. The mollusk Aplysia depilans secretes aniline red. The red and blue color obtained on allowing food to stand is due to the formation of aniline colors by means of bacteria (blood bread, blue milk, etc.).

The dyes of this group, like nearly all organic pigments (which see), may be converted into colorless *leuco* compounds by reducing agents, these compounds yielding pigments again on oxidation. The leuco bases contain one O atom less or generally two H atoms more than the corresponding pigment base. *Leucaniline,* $C_{20}H_{21}N_3$, and *para-leucaniline,* $C_{19}H_{19}N_3$, the leuco bases of rosaniline and pararosaniline (see below) are precipitated from their salts by ammonia and form colorless crystals.

The triphenylmethane colors may be classified into the following groups:

a. Triamido Derivatives or Rosaniline Group.

Rosaniline, triamidodiphenyltolylcarbinol, $C_{20}H_{20}N_3(OH)$ or $(H_2N \cdot H_4C_6)_2 = C(OH)^-C_6H_3(CH_3)(NH_2)$, and

Pararosaniline, triamidotriphenylcarbinol, $C_{19}H_{18}N_3(OH)$, or $(H_2N^-H_4C_6)_2 = C(OH)^-(C_6H_4NH_2)$, the free bases of the rosaniline dyes, are produced when the solution of their salts is treated with caustic alkali. They form white needles which combine with acids, forming red-colored salts with the elimination of water. Even with CO_2 of the air pararosaniline turns red. They are trivalent bases stronger than NH_3.

Rosaniline salts, for example, $C_{20}H_{20}N_3Cl$, and

Pararosaniline salts, for example, $C_{19}H_{18}N_3(C_2H_3O_2)$, form green, metallic-looking crystals which are mostly soluble in water and alcohol and occur in the trade as *fuchsin* and *aniline red*. Their solutions are crimson red and dye animal fibers directly, while the vegetable fibers are only dyed after the use of a mordant (p. 250).

As the hydrogen of the amido groups in the rosaniline and para rosaniline salts are replaceable by alkyls or phenyls it is possible to obtain various colored compounds.

The rosaniline salts are obtained by the oxidation of a mixture of analine and *o*- and *p*-toluidine (so called aniline oil) with arsenic acid or other oxidizing agents with the aid of heat until the mass becomes metallic in appearance. If only paratoluidine is used, then the para-rosaniline salt is obtained:

$$C_6H_5-NH_2 + 2C_6H_4(CH_3)NH_2 + 3O = C_{20}\cdot H_{21}N_3O + 2H_2O.$$
$$\text{Aniline.} \qquad \text{Toluidine.} \qquad \text{Rosaniline.}$$

The arsenic acid is reduced to arsenious acid, which then forms rosaniline arsenite. This is removed from the fused mass by means of water and the solution treated with salt, when rosaniline hydrochloride crystallizes out. As this product contains arsenic, nitrobenzene has recently been suggested as the oxidizing medium, it taking part at the same time in the formation of rosaniline:

$$2C_6H_4(CH_3)(NH_2) + C_6H_5-NO_2 = C_{20}H_{21}N_3O + H_2O.$$
$$\text{Toluidine.} \qquad \text{Nitrobenzene.} \qquad \text{Rosaniline.}$$

Aniline blue, Lyons blue, is $C_{20}H_{17}(C_6H_5)_3N_3Cl$. It is similar to gentian blue, Parisian blue, Poirier's blue, spirit blue, and is produced by introducing phenol groups into rosaniline and pararosaniline. Its compounds soluble in water are called alkali blue, water blue, light blue, etc.

Methyl violet, $C_{19}H_{12}(CH_3)_6N_3(Cl)$, is used in medicine. Methylated rosanilines are called Hoffmann's violet, crystal violet, dahlia violet, while methylated pararosanilines are called Parisian violet or gentian violet.

Methyl green, light green, is $C_{19}H_{12}(CH_3)_6N_3(Cl) + CH_3Cl$.

b. Diamido Derivatives, or the Malachite-green Group.

Malachite green, oil-of-bitter-almond green, is derived from $C_6H_5-C(OH)=[C_6H_4-N(CH_3)_2]$, which base forms with zinc chloride or oxalic acid green crystals which are soluble in water. The homologues of this are called brilliant green, Victoria green, and Helvetia green.

c. Trioxy Derivatives, or the Rosolic Acid Group.

Aurine, $C_{19}H_{14}O_3$ or $(C_6H_4\cdot OH)_2=\overline{C-C_6H_4\cdot}O$, the anhydride of trioxy-triphenylcarbinol, $(C_6H_4\cdot OH)_2=C(OH)-(C_6H_4\cdot OH)$, is soluble in alcohol and acids, forming a yellowish-red solution, and in alkalies, producing a fuchsin-red solution. It is prepared from pararosaniline in the same way as rosolic acid is obtained from rosaniline.

Pittacal, eupitton, hexamethoxyaurine, $C_{12}H_6(OCH_3)_6O_2$, occurs in beech-wood tar.

Rosolic acid, $C_{20}H_{16}O_3$ or $(C_6H_4 \cdot OH)_2 \text{-} \overset{\rule{1cm}{0.4pt}}{C} \text{-} C_6H_3(CH_3)(O)$, dissolves in alcohol and acids, forming a yellow solution, and a red solution in alkalies. It is prepared by the action of nitrous acid upon rosaniline, and decomposing the diazo compound produced by means of water (p. 512).

Corallines are the red (päonines) and yellow dyes produced from a mixture of aurine and rosolic acid.

d. *Carbonic Acid Derivatives, or the Phthaleïn Group.*

On taking up 2 H atoms these compounds are converted into their leuco derivatives, the phthalins; thus, phenolphthaleïn into phenolphthalin, fluoresceïn into fluorescin, galleïn into gallin, etc.

Phenolphthaleïn, $C_{20}H_{14}O_4$ or $(C_6H_4 \cdot OH)_2 \text{-} \overset{\rule{1cm}{0.4pt}}{C} \text{-} C_6H_4 \text{-} COO$, obtained from phthalic acid anhydride and phenol, is soluble in alcohol, forming a yellow solution, and also in alkalies, producing a beautiful red solution. Acids decolorize the solutions.

Tetraiodophenolphthaleïn, $C_{20}H_{10}I_4O_4$, nosophen, is a yellow powder, almost odorless, used as a substitute for iodoform.

Fluoresceïn, $C_{20}H_{12}O_5 + H_2O$ or $O < {C_6H_3(OH) \atop C_6H_3(OH)} > \overset{\rule{1cm}{0.4pt}}{C} \text{-} C_6H_4COO$, resorcinphthaleïn, is obtained from phthalic acid anhydride and resorcin; it forms yellowish-red crystals which are soluble with a yellowish red color in alcohol, and red with alkalies having a beautiful green fluorescence.

Potassium tetrabromfluoresceïn, eosin, $C_{20}H_6K_2Br_4O_5$, colors silk yellowish red with a fluorescence.

Tetraiodofluoresceïn, erythrosin, iodeosin, dianthin, pyrosin, $C_{20}H_8I_4O_4$, are used as indicators (p. 87), as the dilute colorless solution in ether turns rose-colored with traces of alkali. Other eosin colors are called phloxin, primrose, rose-bengal, safrosin.

Galleïn, $C_{20}H_{10}O_7$, pyrogallolphthaleïn, is soluble in alkalies with a blue color.

Cöruleïn, $C_{20}H_8O_6$, anthracene green, is an olive-green soluble dye obtained from galleïn.

Rhodamine, $C_{28}H_{33}N_2O_3$, a beautiful red, fluorescent, soluble pigment.

3. Compounds with Benzene Rings United by Several C Atoms.

Dibenzyl, $C_6H_5 \text{-} CH_2 \text{-} CH_2 \text{-} C_6H_5$, sym-diphenylethane, is produced by the action of sodium upon benzyl chloride and forms large prisms which melt at 52° and which when heated to 500° in a sealed tube yield stilbene (see below) and toluene.

Hydrobenzoïn, $C_6H_5 \text{-} CH(OH) \text{-} CH(OH) \text{-} C_6H_5$, on oxidation yields

Benzoïn, $C_6H_5 \text{-} CO \text{-} CH(OH) \text{-} C_6H_5$, so-called oil-of-bitter-almond camphor (p. 488), and on further oxidation yields

Benzil, $C_6H_5 \text{-} CO \text{-} CO \text{-} C_6H_5$, diphenylketone (see p. 488).

Stilbene, $C_6H_5 \text{-} CH = CH \text{-} C_6H_5$, diphenylethylene, toluylene, is produced by the action of sodium upon benzalaldehyde.

Tolane, $H_5C-C\equiv C-C_6H_5$, diphenylacetylene, is obtained from stilbene bromide by boiling with alcoholic caustic alkali.

Diphenyldiacetylene, $C_6H_5-C \doteq C-C-C-C_6H_5$, and

Hydrocinnamoïn, $C_6H_5-CH=CH-CH\cdot OH-CH\cdot OH-CH=CH-C_6H_6$, are examples of the combination of two benzene rings by means of more than two C atoms.

Triphenylethylene, $(C_6H_5)CH=C(C_6H_5)_2$, and

Tetraphenylethylene, $(C_6H_5)_2C=C(C_6H_5)_2$, are prepared synthetically and form colorless crystals.

4. Compounds with Benzene Rings United by Nitrogen Atoms.

Azoxy-, Azo-, Hydrazo-Compounds. The aliphatic nitro-compounds are directly transformed into amines by reduction, while with the cyclic nitro-compounds the reduction can be so conducted that a series of intermediary compounds may be obtained. Thus on reduction in acid solution (by Fe or $Zn + HCl$ or H_2SO_4) we obtain amines; e.g., $C_6H_5NO_2 + 6H = 2H_2O + C_6H_5-NH_2$ (phenylamine). On reduction in neutral solution (by magnesium or aluminium amalgam, p. 243) hydroxylamine compounds are obtained: $C_6H_5NO_2 + 4H = H_2O + C_6H_5-NH\cdot OH$ (phenylhydroxylamine), from which the nitroso-compounds are obtained on oxidation : $C_6H_5-NH\cdot OH + O = H_2O + C_6H_5-NO$ (nitrosobenzene). On reduction of the cyclic nitro-compounds in alkaline solution, depending upon the reducing agent (ammonium sulphide, sodium amalgam, sodium methylate, zinc-dust and caustic alkali), we obtain, before the formation of amines, a series of intermediary products in which two C atoms of two alphyls are united together by means of two nitrogen atoms and are called azoxy-, azo-, and hydrazo-compounds; e.g.,

$$
\begin{array}{cccccc}
C_6H_5-NO_2 & C_6H_5-N & C_6H_5-N & C_6H_6-NH & C_6H_5-NH_2 \\
 & \quad\big|\!\!\searrow\!O & \quad\| & \quad| & \\
C_6H_5-NO_2 & C_6H_5-N & C_6H_5-N & C_6H_5-NH & C_6H_5-NH_2 \\
\text{Nitrobenzene.} & \text{Azoxybenzene.} & \text{Azobenzene.} & \text{Hydrazobenzene.} & \text{Amidobenzene.}
\end{array}
$$

The azoxy- and azo-bodies form yellow or red crystals, while the hydrazo-bodies are colorless. All three groups are insoluble or slightly soluble in water and readily soluble in alcohol, and are of an indifferent nature, i.e., neither acids nor bases. Only the azo-compounds can be distilled without decomposition.

The azo-compounds, the most important of these compounds, contain the divalent group $^-N^-N^-$, formed from two trivalent nitrogen atoms. This $^-N^=N^-$ group is combined by each of its bonds with one C atom of a cyclic radical. Mixed azo-compounds containing a cyclic and an aliphatic radical are known: $CH_3^-N^=N^-C_6H_5$, azophenylmethyl.

Compounds are also known which contain the azo-group $^-N^=N^-$ two, three, or four times, and are called *dis-*, *tris-*, *tetrazo-*compounds, e.g., $C_6H_5^-N^=N^-C_6H_4^-N^=N^-C_6H_4 \cdot OH$, disazobenzenephenol. All azo-compounds on oxidation yield azoxy-compounds, and by reduction we obtain hydrazo- and then amido-compounds.

Besides by reduction of the nitro-compounds, they are also obtained by the action of weak reducing agents (sodium amalgam) upon azoxy-compounds, or by the oxidation of the amidoazo-compounds, which may be obtained from the diazo compounds (see below).

Azo Pigments. Many of the so-called coal-tar or aniline colors are azo-compounds, for by introducing OH^- or NH_2 groups into azo-compounds we obtain bodies having the properties of dyes (see p. 530), producing oxyazo-compounds: $C_6H_5^-N^=N^-C_6H_4(OH)$, $C_6H_5^-N^=N^-$ $C_6H_3(OH)_2$, etc., and amidoazo-compounds: $C_6H_5^-N^=N^-C_6H_4(NH_2)$, $C_6H_5^-N^=N^-C_6H_3(NH_2)_2$, etc. By the introduction of alkyl and phenyl groups into these compounds they change from yellow or red color to violet or blue. On the introduction of NH_2 or OH groups we produce from the indifferent azo-compounds derivatives which are basic or acid in character and have the property of combining with the fiber to be dyed. Many azo pigments are also used as indicators in analytical chemistry (p. 87). The azo pigments are commercially prepared by the action of cyclic amines or phenols upon diazo-compounds (p. 512).

The hydrochlorides of amidoazo-compounds are orange to brown in color and are called aniline yellow, chrysoidin, phenylene brown (Bismarck brown, vesuvin), Manchester brown, etc. The alkali and ammonium salts of the oxyazo- and amidoazo-compounds are orange, red, yellow, or brown, and are called true yellow, helianthin, methyl orange, resorcin yellow, ethyl orange, tropæolins, etc.

The azo-compounds of naphthalene, C_1H_8, form the pigments orange II, ponceau, true red, brilliant black, naphthalene blue.

Dis-, tris-, tetrazo pigments (see above) are called azoblack, Biebricher scarlet, crocein scarlet, etc. Many of the compounds, especially those obtained from naphthalene with benzidine, diparamidodiphenyl, $(H_2N)H_4C_6-$ $C_6H_4(NH_2)$, so-called benzidine pigments, such as benzazurin, benzopurpurin, Congo red, chrysamine yellow, also diamin black, red, and brown, brilliant yellow (curcumin w), etc., differ from nearly all other dyes in that they dye cotton without mordants (substantive dye) and hence have replaced the other dyes more and more.

Diazo=Compounds. According to constitution these compounds do not belong here, but on account of their relationship to the azo-compounds they will be discussed at this place. The diazo-compounds contain the divalent diazo group, $>$N N, formed from one penta-valent and one trivalent N atom, which is combined with only one valence to the C atom of one cyclic radical and the other valence is combined with a monovalent group (but not with the C atom of the group even if it contains C atoms) or a halogen; e.g.,

$$\underset{\text{O}_4\text{SH}}{\overset{\text{H}_5\text{C}_6}{>}}\text{N} \overset{\text{III}}{\underset{}{-}}\text{N} \qquad \underset{\text{Cl}}{\overset{\text{H}_5\text{C}_6}{>}}\overset{\text{V}}{\text{N}} \equiv \overset{\text{III}}{\text{N}} \qquad \underset{(\text{H}_5\text{C}_6)\text{HN}}{\overset{\text{H}_5\text{C}_6}{>}}\overset{}{\text{N}} \equiv \overset{\text{III}}{\text{N}},$$

Diazobenzene sulphate.   Diazobenzene chloride.   Diazoamidobenzene.

While the primary aliphatic amines, by the action of nitrous acid under all conditions, replace their NH_2 groups directly for HO groups (p. 378) and form alcohols, so the primary cyclic amines (or their salts) when strongly cooled (otherwise phenols are produced) yield with nitrous acid in the presence of an acid (with $NaNO_2+$ acid) the corresponding diazo salts; e.g., $C_6H_5NH_2(HCl) + HNO_2 = (C_6H_5)(Cl)^-N \equiv N + 2H_2O$, which on heating with water yield the HO derivatives, the phenols; e.g., $(C_6H_5)(Cl)^-N \equiv N + H_2O = C_6H_5^-OH + 2N + HCl$. If nitrous acid is allowed to act in the absence of other acids (by passing $NO+NO_2$, p. 156) upon cyclic amines, diazoamido-compounds are produced, which form only unstable salts with acids, but which readily rearrange themselves into strongly basic amidoazo-compounds. The diazo salts may, like the ammonium salts of $(HO)HN \equiv H_3$, be derived from their basic, very unstable diazohy-droxides, e.g., from $(HO)(C_6H_5)N \equiv N$, and may therefore also be called *diazonium salts.*

All the diazo-compounds are crystalline, generally colorless solids, soluble in alcohol and insoluble in water. They explode by shock or on being heated, while their aqueous solution can be handled with impunity. They are extensively used in the preparation of azo pigments (p. 512), as they do not have to be removed from their solution for this purpose.

They are of the greatest importance, as their nitrogen can be readily replaced by $^-$H, $^-$Br, $^-$I, $^-$Cl, $^-$OH, $^-$CN, $^-$NO$_2$, $^-$SO$_3$H, $^-$SCN, $^-$OCN (Gries's reaction).

The nitrogen of the diazo-compounds can be replaced in the following manner·

1. On heating with water they yield phenols (P. 475).
2. On heating with alcohols, cyclic hydrocarbons are produced:

$$(NO_3)(C_6H_5)^-N\equiv N + C_2H_6O = C_6H_6 + N_2 + C_2H_4O + HNO_3.$$

3. By warming with CuBr, CuCl, CuCN, the diazo group is replaced by Br, Cl, CN (Sandmeyer's reaction):

$$2(NO_3)(C_6H_5)^-N\equiv N + 2CuBr = 2C_6H_5Br + 2CuNO_3 + 2N.$$

4. With reducing agents (P. 322) they yield hydrazins:

$$(Cl)(C_6H_5)^-N\equiv N + 4H = (C_6H_5)HN-NH_2 + HCl.$$

5. On heating with phenols or cyclic amines the azo-compounds (p. 510) are produced:

$$(Cl)(C_6H_5)^-N\equiv N + C_6H_5-NH_2 = C_6H_5-N=N-C_6H_4-NH_2 + HCl;$$
$$(Cl)(C_6H_5)^-N\equiv N + C_6H_5OH = C_6H_5-N=N-C_6H_4-OH + HCl.$$

6. With primary or secondary cyclic amines diazoamido compounds are produced:

$$(Cl)(C_6H_5)^-N\equiv N + C_6H_5-NH_2 = HCl + (C_6H_5)(C_6H_5\cdot NH)^-N\equiv N.$$

The nitrosamins (p. 330) as well as the isodiazo-compounds are isomeric with the diazo-compounds. The isodiazo-compounds have an acid character and do not yield azo-compounds with phenols. By acids they are retransformed into diazo-compounds. They have, like the azo-compounds, the $-N=N-$ group and exist in two stereoisomeric forms, as syn and anti compounds (p. 309). The first are obtained by the action of caustic alkali upon diazo-compounds and are very unstable, generally explosive, and are quickly converted into the stable, non-explosive anti-isodiazo-compounds.

Certain isodiazo-compounds are also known among the aliphatic compounds; thus, isodiazoacetic acid ester, $(N=N)=CH-COO(CH)_3$ (p. 369, 4). This ester is converted by caustic alkali into salts, $(N=N)=CH-COONa$, which with acids decomposes into hydrazin and oxalic acid:

$$3(N=N)=CH-COONa + 6H_2O + 3HCl = 3N_2H_4 + 3C_2H_2O_4 + 3NaCl$$

(Preparation of the Hydrazins, p. 151).

COMPOUNDS WITH CONDENSED BENZENE RINGS.

Condensed compounds (P. 298) and their derivatives agree completely in their chemical behavior with the corresponding benzene compounds. The following compounds contain two benzene rings with two mutual C atoms:

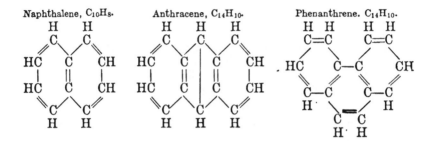

Fluoranthrene, $C_{15}H_{10}$, pyrene, $C_{16}H_{10}$, chrysene, $C_{18}H_{12}$, naphthacene, $C_{18}H_{12}$, picene, $C_{22}H_{14}$, all have a similar structure and are found in that part of coal-tar boiling above 360°. They are all colorless crystals and their derivatives yield even a larger number of isomers than the benzenes.

Condensed compounds of benzene rings with rings poorer in C (p. 463) which have two mutual C atoms are also known, namely:

Indene, C_9H_8. Hydrindene, C_9H_{10}. Fluorene, $C_{13}H_{10}$.

Besides these we also have condensed compounds of the isocarbocyclic with the heterocyclic compounds (see the latter). Condensed compounds which consist of two ring systems which have 3 or 4 mutual atoms are also known (see Terpenes and Alkaloids).

1. Naphthalene Compounds.

These compounds contain two benzene rings with two common C atoms. Naphthalene may yield two series of monosubstitution derivatives (α and β) according as the substituting body is introduced next to one of the two hydrogen-free C atoms or removed therefrom, thus:

$$
\begin{array}{ccc}
\text{a-Compounds.} & \text{β-Compounds.} & \text{o-Phthalic acid (see below).}
\end{array}
$$

If in naphthalene two hydrogen atoms are substituted by the same body, then ten isomers are possible, as follows, making use of the figures given above: 1:2; 1:3; 1:4; 1:5; 1:6; 1:7; 1:8; 2:3; 2:6; 2:7. The positions 2:4 and 1:3; 2:5 and 1:6; 2:8 and 1:7 are identical. On the substitution by different bodies the number of isomers is still greater.

Naphthalene, $C_{10}H_8$. *Preparation.* On the dry distillation of many carbon compounds, especially if their vapors are passed through red-hot tubes (hence it is a chief constituent of coal-tar). On cooling that portion of coal-tar boiling between 210° and 240° impure naphthalene separates, which is purified by sublimation.

It may be prepared synthetically in many ways; thus (analogous to anthracene) by passing the vapors of phenylbutylene over red-hot lead oxide: $C_6H_5-C_4H_7=C_{10}H_8+4H$.

Properties. Shining colorless plates which melt at 80° and boil at 218°, but which slowly volatilize at 15° as well as with steam. It has a peculiar odor and a burning taste, and is insoluble in water but soluble in alcohol, ether, chloroform, and fatty oils. It is oxidized by dilute nitric acid into orthophthalic acid (p. 497), from which it follows that this acid takes the ortho position (see above). This body, and hence naphthalene indirectly, is the starting-point in the commercial manufacture of artificial indigo (p. 548).

With concentrated nitric acid we obtain mono-, di-, or trinitro naphthalene, depending upon the extent of action. These nitro derivatives are readily converted by reduction into the corresponding amido naphthalenes (naphthylamines). By concentrated sulphuric acid we obtain the naphthalene sulphonic acids, $C_{10}H_7-SO_3H$, which on fusion with alkali hydroxides yield the naphthols, $C_{10}H_7(OH)$. Chlorine first produces addition products rupturing the double bonds of the C atoms, and then it forms substitution products.

Naphthols, $C_{10}H_7(OH)$. On heating naphthalene with sulphuric acid we obtain two isomeric naphthalene sulphonic acids, $C_{10}H_7(SO_3H)$, which when fused with alkali hydroxides yield the corresponding naphthols. β-naphthol forms colorless plates which are not readily soluble in water but easily soluble in alcohol. It melts at 122°, and its solutions turn green with ferric chloride and give a violet fluorescence with NH_3 and a white cloudiness with chlorine-water. α-Naphthol is much more poisonous and melts at 95°.

β-Naphthol **ethylether,** nerolin, $C_{10}H_7(O\cdot C_2H_5)$, has an odor similar to orange-flowers and is used as a perfume.

β-Naphthyl benzoate, benzonaphthol, $C_6H_5-COO(C_{10}H_7)$, is used as an intestinal disinfectant.

Amidonaphtholmonosulphonic acid, $C_{10}H_5(OH)(NH_2)(SO_3\cdot H)$. Its sodium salt is used as a photographic developer and called eikonogen.

Calcium β-naphtholdisulphonate, $C_{10}H_5(OH)(SO_3)_2Ca$, asaprol, abrastol, is a non-poisonous preservative agent.

Dinitronaphthol, $C_{10}H_5(NO_2)_2(OH)$. Its sodium salt is called naphthalene yellow or Martius's yellow. Its sulphonic acid is called naphthol yellow.

Naphthoquinone, $C_{10}H_6O_2$, occurs in the volatile form, having an odor similar to quinone, when it is designated α, and an odorless, non-volatile form, called β, both of which form yellow plates.

Dioxynaphthoquinone, $C_{10}H_4(OH)_2O_2$, naphthazarine, alizarine black, is an important dyestuff.

Acenaphthene, $C_{10}H_6=C_2H_4$, occurs in coal-tar, forming colorless prisms which melt at 95°.

Santonin, $C_{15}H_{18}O_4$, the active principle of the so-called worm-seed, is a hexahydrodimethylen naphthalene derivative. It forms colorless bitter crystals which become yellow in the light, melt at 170°, and are insoluble in water but soluble in alcohol and chloroform. When dissolved in alkalies it forms the salts of santonic acid, $C_{15}H_{20}O_4$, which on boiling with baryta-water is converted into the isomer santoic acid.

Helenin, $C_{15}H_{20}O_2$, alantol-lactone, the bitter stuff of the elecampane-root (Inula helenium), is a naphthalene derivative.

Naphthalene dyes. Victoria blue is a dye having a constitution similar to the rosanilines, and naphthol blue like the indoaniline. The naphthols and naphthylamines are used in the preparation of red, reddish-brown, and brown azo dyes (p. 510).

2. Anthracene Compounds.

These compounds contain three benzene rings, each with 2 C atoms in common; e.g.,

Anthracene, $C_{14}H_{10}$, or $H_4C{<}^{CH}_{CH}{>}C_6H_4$ (p. 513).

Preparation. It is produced from many carbon compounds by heating them to a high degree, and hence it occurs in that portion of coal-tar having a high boiling-point, the so-called anthracene oil (p. 473), from which it is separated from the isomeric phenanthrenes by distillation with potassium carbonate, and from phenanthrene by treatment with carbon disulphide, which only dissolves phenanthrene. It is also produced in the distillation of alizarin or purpurin with zinc-dust.

It may be prepared synthetically by several methods; e.g., by passing *o*-benzyltoluene over heated lead oxide:

$$C_6H_4\!\!\begin{array}{c}\diagup CH_3 \\ \diagdown CH_2{-}C_6H_5\end{array} + O_2 = 2H_2O + C_6H_4\!\!\begin{array}{c}\diagup CH \diagdown \\ | \\ \diagdown CH \diagup\end{array}\!\!C_6H_4,$$

or by heating benzene with tetrabromethane and $AlCl_3$ (p. 320):

$$C_6H_6 + \begin{array}{c} BrCHBr \\ | \\ BrCHBr \end{array} + C_6H_6 = C_6H_4\!\!\begin{array}{c}\diagup CH \diagdown \\ | \\ \diagdown CH \diagup\end{array}\!\!C_6H_4 + 4HBr.$$

Properties. Scaly crystals having a bluish fluorescence, melting at 213° and boiling at 360°. They are insoluble in water and soluble with difficulty in alcohol or ether, but readily soluble in benzene. Anthracene is not nitrated by nitric acid, but is oxidized to anthraquinone.

The number of possible isomeric anthracene derivatives is very great. Three mono substitution products are possible and fifteen disubstitution derivatives with the same groups are possible, etc.

Anthraquinone, diphenylendiketone, $C_{14}H_8O_2$ or $H_4C_6\!\!<\!\!\begin{array}{c}CO\\CO\end{array}\!\!>\!\!C_6H_4$, is produced in the oxidation of anthracene and forms yellow needles which melt at 277° and yield anthraquinone sulphonic acid, $C_{14}H_7O_2(SO_3H)$, and anthraquinone disulphonic acid, $C_{14}H_6O_2(SO_3H)_2$, with fuming sulphuric acid.

Alizarin, dioxyanthraquinone, $C_{14}H_8O_4$ or $C_6H_4\!\!<\!\!\begin{array}{c}CO\\CO\end{array}\!\!>\!\!C_6H_2(OH)_2$.

Occurrence. Alizarin is a coloring matter which like indigo-blue does not exist already formed in plants, but is obtained first from a glucoside, *ruberythric acid*, which is contained in the root of the madder and which is split into alizarin and glucose by fermentation of the powdered root or by treatment of the same with dilute acids or alkalies:

$$\underset{\text{Ruberythric acid.}}{C_{26}H_{28}O_{14}} + 2H_2O = \underset{\text{Alizarin.}}{C_{14}H_8O_4} + \underset{\text{Glucose.}}{2C_6H_{12}O_6}.$$

Preparation. In the past it used to be prepared from powdered madder-root (see above), but at the present time alizarin is chiefly obtained by fusing anthraquinone sulphonic acid with caustic alkalies:

$$C_{14}H_7O_2(SO_3H) + 2KOH = KHSO_3 + HOH + C_{14}H_7(OK)O_2;$$
$$C_{14}H_7(OK)O_2 + KOH = H_2 + C_{14}H_6(OK)_2O_2;$$

which last compound is decomposed by HCl into alizarin.

Properties. Technical alizarin is a yellowish-brown paste, and when pure it forms red needles which are not readily soluble in water but readily soluble in alcohol and ether, producing a yellow solution. It behaves like a substituted diphenol, as an acid (p. 474), and dissolves in alkalies with a purple-red color. Aluminium and stannic salts form a red precipitate, ferric salts a dark-violet with solutions of alizarin (madder-lakes). In calico-printing the figures are printed on the material with the above-mentioned salts and the goods are then dipped into water in which alizarin is suspended, when the colored compound of alizarin with the metal deposits at the place mordanted. If cotton is mordanted with alum and oil then alizarin produces the beautiful turkey-red color.

Aloïn, of the aloes, which differs according to the variety of aloes, also called barbaloin, $C_{16}H_{16}O_7 + 3H_2O$, capaloin, $C_{16}H_{16}O_7$, nataloin, $C_{18}H_{18}O_7$, all of which are derivatives of anthraquinone.

Methyldioxyanthraquinones, $C_{14}H_5(CH_3)(OH)_2O_2$ or $C_{15}H_{10}O_4$. One of these is the chrysophanic acid which is found in the senna leaves, in the root of the rhubarb, and in certain lichens. It forms golden-yellow needles which are soluble in alkalies with a purple-red color. Chrysarobin, $C_{30}H_{26}O_7$, the active constituent of the drug called Goa or Arroroba powder. It is a yellowish crystalline powder, which is not readily soluble in water and ammonia, but forms a yellow solution with caustic alkalies. On shaking the alkaline solution with air it turns red and contains then chrysophanic acid: $C_{30}H_{26}O_7 + 4O = 2C_{15}H_{10}O_4 + 3H_2O$.

Trioxyanthraquinones, $C_{14}H_5(OH)_3O_2$. One of these, called purpurin, occurs with alizarin in the roots of the madder and may be formed from alizarin by oxidation. It dyes wool in a similar manner to alizarin.

Methyltrioxyanthraquinones, $C_{14}H_4(OH)_3(CH_3)O_2$. One of these, emodin, occurs in the rhubarb roots, aloes, bark of the black alder, and in senna leaves. It forms orange-yellow needles.

Methyltetraoxyanthraquinones, $C_{14}H_3(CH_3)(OH)_4O_2$. One of these, rheïn, occurs sometimes with emodin.

Anthracene dyes. Many artificial dyes are derived from alizarin by introducing $-OH$ and NH_2 groups, producing alizarin carmin, alizarin orange, alizarin blue, alizarin bordeaux, alizarin cyanin, alizarin green, flavopurpurin, anthrapurpurin, anthracene blue, anthracene brown.

3. Phenanthrene Compounds.

These, like the anthracene compounds, contain three benzene nuclei with two common C atoms.

Phenanthrene, $C_{14}H_{10}$ or $\begin{array}{c} C_6H_4-CH \\ \| \\ C_6H_4-CH \end{array}$ (structure, p. 513), is obtained from many carbon compounds at a white heat, and hence occurs in coal-tar with anthracene (separation, p. 515) and also by heating benzofurane (p. 550) with benzene:

$$C_6H_4\underset{O-}{\overset{CH}{\diagdown}}CH + C_6H_6 = \begin{array}{c} C_6H_4-CH \\ \| \\ C_6H_4-CH \end{array} + H_2O.$$

It forms colorless crystals which melt at 89° and boil at 340°, and on oxidation yields phenanthraquinone, $\begin{array}{c} C_6H_4-CO \\ | \\ C_6H_4-CO \end{array}$, and then diphenic acid, $\begin{array}{c} C_6H_4-COOH \\ C_6H_4-COOH \end{array}$ (o-biphenyldicarboxyl acid).

Retene, $C_{18}H_{18}$, methylisopropylphenanthrene, occurs in the tar from coniferæ and certain earth resins, and

Perhydroretene, $C_{18}H_{32}$, fichtelite, contained in the turf from fossil pines; both form colorless crystals.

4. Indene and Fluorene Compounds.

These contain benzene and pentamethylene rings (p. 464) with common C atoms (structure, p. 513).

Indene, C_9H_8, occurs in coal-tar and is a liquid, boiling at 180° and having an odor similar to naphthalene and yielding o-phthalic acid (p. 514) on oxidation. Its name is derived from the fact that on substituting an $=CH_2$ group of indene by $=NH$ we obtain indol (P. 547).

Hydrindene, C_9H_{10}, accompanies pseudocumene (P. 490) and is produced by the action of nascent hydrogen upon indene. It boils at 177°.

Fluorene, $C_{13}H_{10}$, has already been discussed on p. 505.

Carminicacid, $C_{24}H_{22}O_{14}$, carmine red, contains probably two hydrindene derivatives in the molecule. It occurs in the flowers of the Monarda didyma, in cochineal, in the kermes berries, and forms red masses, which are soluble in water and alcohol, with a yellowish-red color and which turn crimson red in the presence of alkalies. The carmine of commerce is a clay lake (p. 250) of carminic acid.

Compounds of the Terpene Group.

This group includes the terpenes and the varieties of camphors. Terpenes are the isomeric hydrocarbons having the formula $C_{10}H_{16}$, or their polymers which are either dihydrocymenes (p. 519) or are camphenes. The latter have a benzene ring in which two C atoms are united bridge-like with another C atom.

The terpenes contain only one or two double bonds between the C atoms in the benzene ring, so that only two or four univalent atoms or groups of atoms can be attached thereto, when all the C atoms of the benzene ring will have single bonds.

The terpenes are also hydrocarbocyclic and hence also alicyclic compounds (p. 327), the camphors are terpenes or hydroterpenes with a ketone or alcohol group, which generally contains the C atoms in the benzene ring with single bonds. Camphors with several ketone or alcohol groups are not known in nature but may be prepared synthetically.

The terpenes and camphors, which are derived from the dihydrocymenes are called terpanes, as they may also be derived from hexahydrocymene or terpane, $C_{10}H_{20}$. The terpenes and camphors which are derived from the camphenes are called *camphanes*, as they may also be derived from dihydrocamphene or camphane, $C_{10}H_{18}$.

We also know of aliphatic compounds, isomeric with the terpenes and camphors, which like these occur in the ethereal oils (see below) and which are called *olefinic terpenes*. These can be readily transformed into cyclic terpenes (p. 445).

As the variation in structure cannot be shown with most of these bodies having the same empirical formula, it is very difficult to give a structural formula for them. The following formulæ show the relationship that certain of these compounds bear to each other:

Cymene, $C_{10}H_{14}$.   Limonene, $C_{10}H_{16}$.   Menthol, $C_{10}H_{20}O$.

Camphene, $C_{10}H_{16}$.   Camphor, $C_{10}H_{16}O$.   Camphoric acid. $C_{10}H_{16}O_4$.

Contrary to the other isocarbocyclic compounds the atoms are in véry unstable equilibrium, so that the various compounds can very readily be transformed into each other and the ketones may also, by displacement of the atoms, be converted into acids. That also cymene can be split off from compounds of the second series depends upon the fact that by the chemical action the C atoms on the dotted lines are separated producing an isopropyl group in the *p*-position to the methyl group.

The terpenes boil at 160°–190° and their boiling-points are so close together that they cannot be separated by fractional distillation, while this can be done on the contrary by their ability of forming well-defined crystalline compounds with HCl, HBr, Br_2, or N_2O_3. These compounds may be differentiated by their different melting-points. They are distinguished as *terpenes*, $C_{10}H_{16}$, *sesquiterpenes*, $C_{15}H_{24}$, *diterpenes*, $C_{20}H_{32}$, and *polyterpenes*, $(C_{10}H_{16})x$, and form (with the exception of the camphenes and the polyterpenes, which are solids) colorless liquids having characteristic odors. They have different boiling-points, various densities, great difference in their odor, and are most of them optically active.

Nearly every terpene has a known dextro- and lævorotatory modification, from which an optically inactive form (P. 39) may be obtained by mixing equal parts of each. By repeated distillation or shaking with small amounts of concentrated sulphuric acid inactive or polymeric terpenes are produced. On heating with iodine many yield cymene, $C_{10}H_{16} + 2I = C_{10}H_{14} + 2HI$, and on oxidation with dilute nitric acid they yield *p*-toluic acid, $C_6H_4(CH_3)(COOH)$, and terephthalic acid, $C_6H_4(COOH)_2$, which are also obtained on the oxidation of cymene (p. 503). The terpenes inflame by the energetic action of the halogens as well as by nitric acid. They also readily absorb oxygen from the air and are converted into acids, as well as solids, which have great similarity to the natural resins. By addition of HCl the terpenes, $C_{10}H_{16}$, produce the hydrochlorides, $C_{10}H_{17}Cl$, which readily replace the Cl for ^-OH and are transformed into the corresponding camphors.

The camphors (terpene alcohols and terpene ketones) are optically active and generally form crystals, and seldom liquids, having a characteristic odor. The alcohol camphors can have their HO groups readily replaced by Cl, and the chlorine derivatives are easily transformed into terpenes by the action of alcoholic caustic alkali, splitting off HCl. The HO groups are also replaceable by NH_2, producing amine bases. The ketone camphors are converted into alcohol camphors by nascent hydrogen, and these latter yield ketone camphors on oxidation with potassium bichromate and sulphuric acid..

The terpenes occur generally mixed together in many plants and form the odoriferous principles of the same, namely, the ethereal oils. The camphors also occur in many plants, especially dissolved in the ethereal oils of the given plant. Many terpenes and camphors can be obtained only synthetically.

Ethereal oils is the name given to a large number of organic compounds occurring in the vegetable kingdom. On account of their volatility they can be obtained from the respective plants by distillation with steam. They have a specific odor and a burning taste, and are most of them liquid at ordinary temperatures and nearly insoluble in water. They have no relationship to the actual oils (fats), and, with the exception of the ethereal oils produced from the terpenes, form no connecting groups. They are extensively used in perfumery, medicine, and in the preparation of cordials.

Nearly all ethereal oils rotate the polarized ray of light and produce a temporary transparent spot on paper and cotton material. The name ethereal oils is derived from this last property and from their often oily consistency, as well as their volatility. They are readily soluble in alcohol, ether, chloroform, and fatty oils, and burn with a smoky flame. They absorb oxygen from the air and thicken, being converted into resins, or are transformed into acids.

The solids obtained from the ethereal oils by cooling were formerly called *stearoptenes*, and the liquid remaining called *elaoptenes*.

On rubbing 1 part of an ethereal oil with 50 parts powdered sugar, we obtain what is called oil-sugar.

Oxygen-free ethereal oils consist nearly entirely of terpenes; e.g., oil of turpentine, oil of savine, oil of bergamot, oil of lavender, oil of rosemary, oil of orange-flowers, oil of calamus, oil of juniper, etc.

The ethereal oils containing oxygen are mixed in the crude state with small or large quantities of terpenes. They consist of alcohols, aldehydes, phenols, compound ethers, camphors, etc., which are split off in their preparation by the steam from the esters or ethers of peculiar alcohols (the *oleols*) contained in the various plants. To this group belongs the oil of bitter almond, which contains benzaldehyde, oil of cinnamon, which contains cinnamic aldehyde, oil of wintergreen, which contains methyl salicylate. Oil of anise contains anethol, oil of fennel contains fenchone, oil of cajuput contains cajuputol, oil of nutmeg contains myristicol, $C_{10}H_{16}O$, oil of caraway contains carvol, $C_{10}H_{14}O$, oil of cloves contains eugenol, oil of peppermint contains menthol, oil of thyme contains thymol, oil of rose contains geraniol, oil of lemons contains besides terpene also linalool, citral, and citronellal, oil of sandal-wood contains besides santalene also santalol, $C_{15}H_{26}O$. See also p. 445.

The ethereal oils containing sulphur are compounds of alcohol radicals with $^{-}$NCS or with sulphur; thus, mustard oil contains isosulphocyan-allyl, oil of garlic and oil of onions contain allyl sulphide, etc.

1. Terpenes.

The camphanes, to which pinene, camphene, fenchene, belong, can take up by addition 1 mol. HCl, HBr, or Br_2 (p. 519); the terpanes, such as linomene, terpinolene, sylvestrene, can take up 2 mols. HCl, HBr, Br_2; while the terpanes phellandrene and terpines take up only N_2O_3.

Pinene boils at 156°. *l*-Iinene, terebentene, occurs in the German, French, and Venetian turpentine oil, Canada balsam, incense, and many ethereal oils. *d*-Pinene, or australene, occurs in English and American turpentine oil and also in many ethereal oils. *i*-Iinene is produced synthetically.

Pinene hydrochloride, $C_{10}H_{16}.HCl$, is produced by passing HCl gas into pinene, forming crystals having an odor similar to camphor, and is called "artificial camphor."

Turpentine oil is obtained from the turpentine, which exudes from various coniferæ, by distillation with water. It is a colorless liquid having a peculiar odor boiling at 160° and which is insoluble in water but soluble in alcohol, ether, and fatty oils. It dissolves sulphur, phosphorus, resins, and caoutchouc. On distilling with lime-water we obtain rectified oil of turpentine, which is free from acids and resins. When exposed to the air turpentine oil absorbs oxygen and becomes resinous (used in the preparation of resin varnishes and oil-paints).

Terebene, a mixture of inactive terpenes, is produced by the action of sulphuric acid upon turpentine oil and subsequently distilling with steam. It smells like thyme.

i-, *d-*, *l*-Camphenes are produced by the action of alkalies upon *i-*, *d-*, *l*-pinene hydrochloride and are colorless crystals melting at about 50°, having an odor similar to camphor.

i-Fenchene, obtained from fenchone (p. 524), boils at about 160°.

Limonene (structure, p. 519) boils at 175°. *d*-Limonene, citrene, carvene, cajeputene, hesperidene, smells like lemons and occurs in oil of dill, caraway, orange-flowers, bergamot, and lemons.

l-Limonene occurs with *l*-pinene in pine-needle oil.

i-Limonene, dipentene, cinene, boils at 180°, and is found with sylvestrene in Swedish and Russian turpentine oil, in camphor oils, and oil of Cinæ, etc. It is produced on heating the active limonenes to 300° or by splitting off water from terpine hydrate.

i-Terpinols are formed by boiling terpine hydrate, terpineol, cineol, with dilute sulphuric acid. Boil at 1£5°.

d-Sylvestrene, the chief constituent of Swedish oil of turpentine, boils at 176°.

i-Sylvestrene, carvestrene, is prepared synthetically.

Phellandrene boils at 170°. *d*-Phellandrene occurs in ethereal oil of fennel, water-fennel, elemi, oil of eucalyptus; *l*-phellandrene also occurs in the two last-mentioned oils.

i-Terpinene is produced by boiling most terpenes with dilute sulphuric acid. Boils at 1£0°, and smells like lemons.

2. Sesqui-, Di-, Polyterpenes.

Cedrene, cardinene, clovene, caryophyllene, santalene, are sesquiterpenes, $C_{15}H_{24}$, and occur in the ethereal oils of cubebs, savine, patchouli, and santal. They boil between 250°–280°.

Colophene, retinol, are both diterpenes, $C_{20}H_{32}$, are found in balsam of copaiba, are produced by the distillation of rosin, and boil above 300°.

Caoutchouc (India-rubber), the dried juice of the tropical Euphorbiaceæ, Apocynæ, etc., and

Gutta percha, the dried juice of the tropical Sapotaceæ, both bodies polyterpenes, $(C_{10}H_{16})_N$, are nearly insoluble in alcohol, but soluble in CS_2, $CHCl_3$, C_6H_6, and turpentine oil. At ordinary temperatures they are tough and elastic, and in the cold they become hard.

The elasticity of caoutchouc is increased by introducing sulphur (vulcanization). Vulcanized rubber contains 2–4 per cent. sulphur. If the quantity of sulphur is increased, then the rubber becomes hard, horny, and is used as ebonite, or vulcanite, for the making of combs, probes, electrical machines, etc.

Rolled gutta percha is called gutta-percha paper. Purified gutta percha is white and generally occurs in rolls.

3. Camphors.

Borneol, fenchone, thujone have a camphene structure, while the others have a cymene structure.

a. Camphors $C_{10}H_{14}O$.

Carvone, carvol, the ketone of limonene (p. 522), the isomer of carvacrol (p. 503), boils at 225°.

d-Carvone is found in the ethereal oil of dill and caraway.

l-Carvone occurs in the ethereal oil of curomoji.

b. Camphors $C_{10}H_8O$.

d-Camphor, Japan camphor, laurinene camphor, is a ketone (p. 519); it occurs in the tree Laurus camphora, and is obtained therefrom by sublimation. It forms white crystalline masses having a characteristic odor and burning taste. It is volatile at ordinary temperatures, melts at 175°, and boils at 204°. Camphor is insoluble in water, readily soluble in alcohol (spirits of camphor), in ether, acetic acid, ethereal and fatty oils. A solution in olive-oil forms the camphorated oil.

l-Camphor, in ethereal oils of Matricaria Parthenium, is produced on oxidizing l-borneol.

On warming a solution of camphor with sodium we obtain sodium camphor and sodium borneol: $2C_{10}H_{16}O + 2Na = C_{10}H_{15}NaO + C_{10}H_{17}NaO$, which are decomposed by water into caustic soda, camphor, and d-borneol, $C_{10}H_{18}O$ (p. 524).

By dehydrating agents (P_2O_5, $ZnCl_2$, P_2S_5) camphor is converted into cymene: $C_{10}H_{16}O = C_{10}H_{14} + H_2O$.

Heated with iodine camphor is transformed into the cymene phenol, carvacrol (p. 503), $C_{10}H_{14}O$: $C_{10}H_{16}O + 2I = C_{10}H_{14}O + 2HI$.

Oxycamphor, $C_{10}H_{16}O_2$, forms colorless crystals which are soluble in 50 parts water and which smell like pepper.

Camphoric acid, $C_{10}H_{16}O_4$ (structure, p. 519), is produced besides camphanic acid, $C_{10}H_{16}O_5$, and camphoronic acid, $C_9H_{14}O_6$, on boiling camphor with HNO_3, and forms colorless and odorless crystals which melt at 186° and are not readily soluble in cold water, but easily soluble in alcohol and ether.

Alantole, alant camphor, in the roots of the Inula Helenium.

Fenchone, in the ethereal oil of fennel and oil of thuja.

Absinthole, wormwood camphor, in the ethereal oil of wormwood.

Myristicol, in the ethereal oil of the nutmeg.

Thujone, tanacetone, in the ethereal oil of thuja.

Sabinole in the ethereal oil of savine.

c. *Camphors* $C_{10}H_{18}O$.

d-Borneol, Borneo camphor, occurs in the Dryobalanops camphora tree, in the ethereal oil of rosemary and spike. It differs from the Japan camphor ·by its crystalline form, by being harder, melting at 208°, and boiling at 212°. It is the alcohol corresponding to the Japan camphor and is produced therefrom by the action of nascent hydrogen or of sodium (see above), and is converted into Japan camphor by mild oxidation. l-Borneol occurs as Ngai camphor obtained from the Blumea balsamifera as well as with i-borneol in certain ethereal oils.

Cineol, eucalyptol, cajeputol, occurs in the ethereal oils of eucalyptus, cajeput, worm-seed, and is a colorless, inactive liquid.

Menthon is produced by oxidizing menthol (see below).

Terpineol is contained in various ethereal oils, melts at 37°, and smells like lilac flowers.

d. *Camphors* $C_{10}H_{20}O$, $C_{10}H_{20}O_2$.

Menthol, menthol camphor, $C_{10}H_{20}O$ (structure, p. 519), is obtained from peppermint-oil by strongly cooling, when it separates out. It forms colorless crystals having a melting-point of 43°. It yields the ketone menthon, $C_{10}H_{18}O$, on oxidation, and menthen, $C_{10}H_{18}$, on splitting off of H_2O. Menthol valerianate, called validol, is used in medicine. The chlormethyl menthyl ether, $C_{10}H_{19}(\cdot CH_2Cl)O$, is called forman and decomposes with water into menthol, formaldehyde, and hydrochloric acid.

Terpine hydrate, $C_{10}H_{20}O_2 + H_2O$, is produced by allowing pinene, terpineol, or linalool to stand in contact with dilute mineral acids, as well as from terpine, by the action of water, and from oil of turpentine by the action of alcohol and nitric acid. It forms colorless and odorless crystals which are soluble with difficulty in water, ether, chloroform, but readily soluble in alcohol, and melting at 116°, when it loses water (formation of terpine), and the melting-point falls to 102°.

Terpine, $C_{10}H_{20}O_2$, is produced by continuously warming terpine hydrate.

4. Resins

is the name given to a number of amorphous, brittle, yellowish to brown bodies of unknown constitution which contain only C, H, O, which are closely related to the terpenes and which occur with these in the plants and perhaps formed from the terpenes by oxidation in the air. They consist of an indefinite amount of ethereal oils, etc., and a mixture of resins which are difficult of separation. Chemically they act like acids, they dissolve in alkalies, forming so-called resin soaps, which form a lather like soaps and from which the resins are precipitated by acids. They cannot be distilled without decomposition and they burn with a luminous flame.

On fusion with alkalies most of the resins yield protocatechuic acid, phloroglucin, paraoxybenzoic acid, pyrocatechin, resorcin, volatile fatty acids. With hot HNO_3 they give picric acid, phthalic acid, and finally oxalic acid. On distillation with reducing agents, toluene, xylene, methylanthracene, etc., are obtained.

The chief constituents of the resins are the resin esters (*resines*), resin acids (*resinolic acids*, e.g., abietic acid, $C_{19}H_{28}O_2$, and pimaric acid, $C_{20}H_{30}O_2$), and indifferent aromatic compounds, *resenes*, of which little is known. The resin esters contain peculiar alcohols, the resin alcohols (p. 526). These latter are divided into the colorless *resinols* and the colored *resinotannol*, which give the reactions for the tannins. Only a few resins contain resines, resinolic acids, and resenes at the same time.

Hard resins are amorphous, generally hard and brittle, containing little or no ethereal oil, insoluble in water, soluble in alcohol, and most of them soluble in ether and ethereal oils. The solution of many resins in alcohol or turpentine is used as lacquer or varnish. The most common hard resins are common rosin or colophony, gum benzoin (contains benzoic acid), gum dammar, jalap resin, podophyllin, alces (also contain aloin, p. 517), gum lac (purified shellac), amber (also contains ethereal oils and succinic acid), gum guaiac (contains also guaiacic acid and guaiac resin acid), and gum mastic.

Soft resins (balsams) are soft or semi-fluid and are mixtures of ethereal oils and resin. They are insoluble in water, soluble in alcohol and ether. As the ethereal oil absorbs oxygen the soft resins gradually become hard in the air. The most important are balsam of copaiba, turpentine, pine resin, and elemi resin.

The true balsams (balsam of Peru, p. 501, balsam of Tolu, p. 501, storax, p. 501) contain resin and ethereal oil only in small amounts and consist chiefly of aromatic acids, alcohols, and esters.

Gum resins are amorphous mixtures of gum, plant mucilage, resin, and ethereal oils. They are only partly soluble in water (the gum), and in alcohol (the resin). To this group belong gum ammoniacum, gum galbanum, gum asafœtida, gum gutta, gum euphorbium, gum olibanum, and gum myrrha.

Fossil resins. The most important are amber, which consists chiefly of a resinol ester of succinic acid, and asphalt, which is produced by the gradual oxidation of petroleum.

5. Cholesterins or Cholesterols

are the monohydric alcohols having the formula $C_{27}H_{46}O$, which are very widely distributed in the animal and plant kingdoms, and which were first discovered in the bile ($\chi o\lambda\acute{\eta}$, bile; $\varsigma\tau\epsilon\rho\epsilon\acute{o}\varsigma$, solid). They are dextro- or lævorotatory and seem to be closely related to the terpenes or the camphors. They form odorless and tasteless crystals having various melting-points, are insoluble in water, dilute acids, and even in concentrated caustic alkalies; they are very readily soluble in boiling alcohol, ether, and fatty oils.   If concentrated H_2SO_4 is added to their solutions in chloroform, the chloroform becomes purplish-red, while the sulphuric acid has a greenish fluorescence.   If the red chloroform solution is evaporated it becomes blue, then green, and finally yellow (Salkowski's reaction).   The solution of the cholesterins in acetic anhydride turns violet and then deep green when treated with concentrated sulphuric acid (Liebermann's reaction).

Animal cholesterins, ordinary cholesterin, $C_{27}H_{46}O$, is lævorotatory, and occurs to a slight extent in all fats, in blood, and nearly all animal fluids, in feces, but seldom in the urine, and sometimes in the gall-bladder as round masses (gall-stones).   It is found abundantly in the brain, egg yolk, and the nerve substance.   It forms compounds with fatty acids which correspond to the fats and which occur in animal cutaneous formations (hair, hoofs, skin, etc.), and to a greater extent in wool fat.   *Isocholesterin* is also found in wool fat, *koprosterin*, or *sterconin*, in normal human feces, etc.

Fatty-acid esters of cholesterin are prepared from wool fat, and when pure the substances are used as a basis for salves, as they do not become rancid and because they combine with considerable water.   A mixture with 25 per cent. water is called *lanolin*.   Thilanin is wool-fat with chemically combined sulphur.

Plant cholesterins, phytosterins, occur to a less extent in all plant fats, *paracholesterin* in tan-bark, *caulosterin* in etiolated lupin sprouts, *onocerin*, in the Radix ononidis, etc.; *betasterin* in the sugar-beet and *sitosterin* in the wheat and rye sprouts.   Besides these bodies we find a series of alcohols having a high molecular weight which must also be classed with the cholesterins because of the correspondence of their color reaction, although they differ in constitution.   These bodies are *cyanchol, quebrachol, cupreol, lactucerin, amyrin, lupeol,* the *resin alcohols*, etc.

COMPOUNDS OF THE GLUCOSIDE GROUP.

Glucosides, sometimes also incorrectly called saccharides (p. 448), are the widely distributed class of compounds occurring chiefly in the plant kingdom and seldom in the animal world.   On boiling with dilute acids or alkalies, sometimes even on boiling with water alone, also by certain organized and unorganized ferments, they decompose into varieties of sugar and one or more other, generally cyclic, compounds.   They are to be considered as ether-like compounds of the sugars.

According to the sugar split off we differentiate between *glucosides, galactosides,* and *rhamnosides*, etc.

They are colorless and odorless and not volatile without decomposition.   They are optically active, generally crystallizable, having a bitter taste, soluble in water or alcohol, and with the exception of achillein, amygdalin, glycyrrhyzin, indican, sinigrin, solanin, the cerebrosides and glycoproteids are free from nitrogen. With the exception of achillein, · solanin, glycoproteids, which are hetero-carbocyclic compounds, they are all isocarbocyclic compounds whose exact constitution has not been thoroughly investigated. Those of known constitution have been treated of in connection with their mother-substance.

Absinthin, $C_{30}H_{40}O_8$, found in the wormwood; decomposes into glucose and a resinous body.

Æsculin, $(C_{15}H_{16}O_2)_2 + 3H_2O$, in the horse-chestnut; decomposes into glucose and æsculetin, $C_9H_6O_4$ (p. 501).

Apiin, $C_{27}H_{30}O_{15}$, occurring in the parsley, splits into glucose and apigenin, $C_{15}H_{10}O_5$.

Arbutin, $C_{12}H_{16}O_7$, with methyl arbutin, in the leaves of the Uvæ ursi; yields glucose and hydroquinone or methyl hydroquinone on cleavage.

Cerebrosides, $C_{70}H_{38}N_2O_{22}$, and homologues of this formula, *cerasin*, *cerebrin*, and *encephalin*, split into galactose and bodies which have been little studied, but which split into fatty acids and ammonia. The cerebrosides are found combined with lecithin as bodies called *protagons* in the brain and nerve substance.

Quinovin, $C_{30}H_{48}O_8$, found in cinchona bark; splits into a methyl pentose, quinovose, $C_6H_{12}O_5$, and quinovic acid, $C_{24}H_{38}O_4$.

Cathartic acid, $C_{30}H_{37}NO_{15}$, the active principle of the senna leaves (Folia sennæ) and of the buckthorn (Cortex frangulæ); decomposes into glucose and cathartogenic acid, $C_{24}H_{26}NO_{10}$.

Cnicin, $C_{42}H_{56}O_{15}$, in the leaves Centaurea benedicta (blessed thistle); decomposes into hexose, fatty acids, and phenol and aldehyde compounds.

Colocynthin, $C_{56}H_{84}O_{23}$, the active body in the fruit of Citrullus colocynthis; splits into glucose and colocynthein, $C_{44}H_{64}O_{13}$.

Coniferin, $C_{16}H_{22}O_8 + 2H_2O$, in the asparagus ana, in the cambium of several of the coniferæ; separates as colorless crystals on the evaporation of the same. By emulsin it decomposes into glucose and coniferyl alcohol:

$$C_{16}H_{22}O_8 + H_2O = C_6H_{12}O_6 + C_{10}H_{12}O_3 \text{ (p. 494)}.$$

Convolvulin, $C_{54}H_{96}O_{27}$, the active principle of jalap resin (Resina jalapæ); decomposes into glucose, rhodiose (a methyl pentose), methyl-ethyl acetic acid, $C_5H_{10}O_2$, oxylauric acid, $C_{12}H_{24}O_3$, decylenic acid, $C_{10}H_{18}O_2$, and convolvulinolic acid, $C_{15}H_{30}O_3$.

Digitalis glucosides. The dried leaves of Digitalis purpurea contain the glucosides *digitonin*, $C_{28}H_{47}O_{14}$ (cleavable into *digitogenin*, $C_{16}H_{25}O_3$, glucose and galactose), *digitalin*, $C_{35}H_{56}O_{14}$ (splits into *digitalose*, $C_7H_{14}O_5$, (ethyl pentose), *digitaligenin*, $C_{22}H_{30}O_3$, and glucose, $C_6H_{12}O_6$), and *digitoxin*, $C_{34}H_{54}O_{11}$ (which may be split into *digitoxigenin*, $C_{22}H_{32}O_4$, and *digitoxose*, $C_6H_{12}O_4$). Commercial digitalin consists of a mixture of the above three glucosides. If it is dissolved in concentrated sulphuric acid and then treated with a drop of bromine water a violet-red coloration is obtained. If the digitalin is dissolved in acetic acid and a trace of $FeCl_3$ added and then an equal volume of H_2SO_4, without mixing, an intense red zone is obtained.

Ericolin, $C_{26}H_{34}O_3$, with arbutin in the leaves of Uvæ ursi, decomposes into 2 mols. ericinol, $C_{10}H_{16}O$, and glucose.

Frangulin, $C_{21}H_{20}O_9$, the yellow body of the frangula bark; splits into rhamnose and emodin (p. 517).

Gentiopicrin, $C_{20}H_{30}O_{12}$, of the gentian root; decomposes into glucose and gentiogenin, $C_{14}H_{16}O_5$.

Glucotannoids, see Tannins (p. 495).

Glycyrrhizic acid, $C_{44}H_{63}NO_{18}$, is found as ammonium salt in the liquorice-root (Radix glycyrrhizæ or Liquiritiæ) and splits into glycyrrhetin, $C_{32}H_{47}NO_4$, and parasaccharic acid, $C_6H_{10}O_8$.

Helleborein, $C_{37}H_{56}O_{18}$, in the roots of the black hellebore (Helleborus niger, etc.); splits into glucose and helleboretin, $C_{19}H_{30}O_5$, and acetic acid.

Hespiridin, $C_{50}H_{60}O_{27}$, in the orange, lemon, pomegranate, and decomposes into glucose, rhamnose, and hesperitin, $C_{16}H_{14}O_6$ (isoferulic acid, phloroglucin ester, p. 50₁).

Iidin, $C_{24}H_{26}O_{13}$, in the root of the blue flag (Rhizoma iridis); decomposes into glucose and irigenin, $C_{18}H_{16}O_8$ (a polyoxy ketone).

Jalapin, $C_{34}H_{56}O_{16}$, in jalap resin (Resina jalapæ); splits into glucose and jalapinolic acid, $C_{16}H_{30}O_3$ (oxyhexdecylic acid).

Menyanthin, $C_{33}H_{50}O_{14}$, in the buck-bean; splits into 2 mols. glucose and menyanthol, $C_{21}H_{34}O_6$.

Ononin, $C_{25}H_{16}O_{11}$, occurs with the little-known glucosides pseudo-oninin and onon in the Radix ononidis, and splits into glucose and formonetin, $C_{19}H_{14}O_5$, which splits further into ononetin, $C_{18}H_{16}O_5$, and formic acid.

Phlorhizin, $C_{22}H_{24}O_{10} + 2H_2O$, in the cherry-tree, apple-tree, pear-tree, and plum-tree, and splits into glucose and phloretin, $C_{15}H_{14}O_5$, which is the phloroglucin ester of *p*-oxyhydrocoumaric acid and which splits further into phloroglucin (p. 482) and phloretic acid (p. 502).

Populin, benzoylsalicin, $C_{13}H_{17}(C_6H_5-CO)O_7 + 2H_2O$, in the silver poplar; decomposes into salicin (see below) and benzoic acid.

Protagons, see Cerebrosides (P. 527).

Pseudostrophantin, $C_{40}H_{60}O_{16} + H_2O$, occurs in the seeds of the Strophantus hispidus, and splits into pseudostrophantidin, $C_{28}H_{40}O_6$, and a sugar, $C_{12}H_{22}O_{11}$.

Quercitrin, $C_{21}H_{22}O_{12} + 2H_2O$, in the bark of the black oak; decomposes into quercetin (P. 542) and rhamnose (p. 444).

Ruberythr c acid, $C_{26}H_{28}O_{14}$, in madder-root, forms yellow shining needles which split into alizarin and glucose:

$$C_{26}H_{28}O_{14} + 2H_2O = C_{14}H_8O_4 + 2C_6H_{12}O_6.$$

Salicin, $C_{13}H_{18}O_7$, in the bark of the willow-tree and several varieties of poplars; is split into glucose and saligenin (p. 491): $C_{13}H_{18}O_7 + H_2O = C_6H_{12}O_6 + C_6H_4(OH)\cdot CH_2\cdot OH$. On oxidation it yields *helicin*, $C_{13}H_{16}O_7$ (salicyl aldehyde glucose), which may also be prepared synthetically. Helicin is converted into salicin by nascent hydrogen.

Saponins is the name given to a series of compounds having the formula $C_NH_{2N-8}O_{10}$ (N = 17 to 26), which occur in the ordinary soapwort (saponaria officinalis), in the soap-bark (Cortex quillajæ), in sarsaparilla-root (Radix sarsaparillæ) (in this as *smilacin*, also as *sarsasaponin* and *smilasaponin*), and about 130 other plants. They form white amorphous powders having a harsh taste and whose dust causes sneezing and whose watery solution, even when diluted to 1:1000, foams when shaken.

On boiling with dilute acids they split into glucose and *sapogenin*, $C_{14}H_{22}O_2$. The poisonous saponins are also called *sapotoxines*.

Scillain, scillitoxin, $(C_6H_{10}O_3)x$, in the squill (Bulbus scillæ); decomposes into glucose, butyric acid, and isopropyl alcohol.

Xanthorhamnin, $C_{34}H_{42}O_{20}$, rhamnin, rhamnegin, in the fruits of various varieties of rhamnus, and splits into rhamnetin (P. 542) and rhamninose, which can be split into rhamnose (P. 444) and galactose.

BITTER PRINCIPLES

are numerous bodies which consist only of C, H, and O, have a bitter taste, and are colorless or only faintly yellow in color and generally markedly crystalline. They occur ready formed in the plants and form the active constituent of the same. Their behavior towards chemical agents is very diverse, and only a few combine with acids or bases. Most of them are decomposed by these agents. Nearly all dissolve with difficulty in water, but dissolve easily in alcohol and ether. Their constitution has not been sufficiently studied, but they all seem to belong to the isocarbocylic compounds. Those of known constitution will be discussed with their mother-substances.

Agaricin, agaricic acid, $C_{16}H_{30}O_5 + H_2O$, is obtained from the resin of the touchwood as a white amorphous or crystalline powder which melts at 140°.

Angelicin, hydrocarotin, $C_{18}H_{20}O$, with angelic acid in the roots of Angelica archangelica; also in the carrot.

Arnicin, $C_{20}H_{30}O_4$, in the arnica-flowers.

Artemisin, $C_{15}H_{18}O_4$, with santonin in the worm-seed.

Acorin, acoretin, $C_{30}H_{60}O_6$, in the sweet flag.

Bryonin, $C_{48}H_{80}O_9$, in the roots of the bryony (Bryonia alba).

Euphorbon, $C_{15}H_{24}O$, in euphorbium resin.

Erythrocentaurin, $C_9H_{14}O_5$, in the centaury herb.

Hop bitter, $C_{25}H_{36}O_4$, in the strobiles of Humulus lupulus (lupulin).

Capsaicin, $C_9H_{14}O_2$, is the active principle of the Spanish pepper (Fructus capsici).

Cascarillin, $C_{12}H_{18}O_4$, in the Cortex cascarillæ.

Colombin, $C_{21}H_{22}O_7$, in the Radix colombo.

Kosin, kussin, $C_{22}H_{26}O_7$, in the Flores koso.

Pimpinellin, $C_{14}H_{12}O_5$, in the Radix pimpinellæ.

Picrotoxin, in the Cocculus indicus, consists of a mixture of *picrotoxinin*, $C_{15}H_{16}O_6$, and *picrotin*, $C_{15}H_{18}O_7$.

Podophyllotoxin, $C_{15}H_{14}O_6 + 2H_2O$, in Podophyllin.

Quassin, $C_{32}H_{42}O_{10}$, in quassia-wood (Lignum quassiæ).

Urson, $C_{30}H_{48}O_3 + 2H_2O$, in the leaves of the bearberry (Folia Uvæ ursi).

COMPOUNDS OF THE PIGMENT GROUP.

All organic bodies having color are not pigments, i.e., they do not permanently color animal or vegetable fibers. Only those compounds show coloring characteristics which are bases or acids, because they unite with the animal or vegetable fibers forming salt-like combinations. Only certain of the coloring matters have the power of permanently dyeing the vegetable fibers directly (*substantive dyes*), while another group (*adjective dyes*) in order to permanently color vegetable fibers require a fixing agent, which combines not only with the fiber, but also with the pigment.

The fixing agents are called *mordants* and the aluminium, tin, and ferric salts are especially suited for this purpose. The fibers are dipped in these solutions and then heated (steamed), whereby the salts are decomposed and the metallic oxide or metallic hydroxide precipitated in a finely divided state in the fibers. The coloring matter now forms with this an insoluble compound which forms on the fiber. Proteids also serve as mordants, giving the plant fibers the character of an animal fiber, also fats, fluorine compounds, or tannic acid alone, or with antimony salts. Indifferent pigments are those without acid or basic character, which dye only when they are precipitated from their soluble compounds upon the fibers (indigo) or when they are given salt properties by converting them into sulphonic acids (indigo and many artificially prepared dyes).

The pigment nature of a body is dependent upon the presence of certain atomic groups called the *chromophore group*. Compounds containing chromophore groups are called *chromogens* and become true pigments only after the introduction of groups giving salt-forming (i.e., acid or basic) properties to the compound. Many bodies used as dyes are colorless and only become colored with the mordants, acids, or bases.

If the pigment compound produced on the fibers is resistant to the influence of the air, of soap-water, dilute acids, and alkalies, it is called fast, and if not it is called unfast.

The coloring matters are destroyed by oxidizing agents, e.g., by H_2O_2 of the air (grass-bleach), or by chlorine (chlorine-bleach). By reducing agents (H_2S, SO_2, nascent hydrogen) the colors are bleached; still generally no destruction of the pigment occurs, but colorless compounds with hydrogen are produced which on oxidation (often even by the air) yield the coloring matter again and are called *leuco-compounds* (p. 506).

On shaking a solution of a pigment with animal charcoal this latter removes most of the coloring matters. Most of the solutions of coloring matters show characteristic absorption bands when viewed at through a spectroscope.

The artificial dyes, often called coal-tar colors on account of their obtainment from coal-tar, have, because of their beautiful shades and simplicity of use, nearly entirely superseded the natural dyes. They are nearly all of them carbocyclic, nitrogenous compounds and may be divided into the following groups: 1. Di- and triphenylmethane colors (the last also called aniline dyes, p. 505); 2. Azo dyes (p. 510); 3. Nitro and nitroso dyes (pp. 479, 515); 4. Anthracene dyes (p. 517); 5. Quinoline and acridine dyes (p. 540); 6. Quinonimide dyes (p. 485); 7. Azine, oxazine, and thiazine dyes (pp. 543, 544); 8. Azole dyes (p. 552); 9. Benzopyrone dyes (p. 541); 10. Benzopyrrol dyes (p. 546).

The natural dyes occur either already formed (pigments) or they are obtained from bodies that are colorless.

The vegetable colors with the exception of indigo and chlorophyll are free from nitrogen, are isocarbocyclic compounds, and may be divided into the following groups: 1. Pyrone derivatives; 2. Benzophenone derivatives; 3. Flavone and xanthone derivatives; 4. Hydrindene derivatives; 5. Orcine derivatives; 6. Anthracene derivatives; 7. Colors of unknown constitution.

The animal colors, with the exception of the lipochromes, which are also widely distributed in the vegetable kingdom, contain nitrogen, are heterocarbocyclic compounds, and are all pyrrol derivatives.

The coloring matters of known constitution have been discussed with their mother-substances.

Alkannin alkanet red, $C_{15}H_{14}O_4$, occurs in the alkanet-root.

Anthocyanin is found in the violet and corn-flower.

Anthoxanthin, the pigment of fruits with bright-red color (strawberry, etc.).

Bixin, $C_{28}H_{34}O_5$, is the red pigment of the orlean.

Carthamin, $C_{14}H_{16}O_7$, in the safflower (flowers of the Carthamus tinctorius), is a red powder which is soluble in alkalies with a yellowish-red color.

Curcumin, $C_{21}H_{20}O_6$, the coloring matter of turmeric, forms orange-yellow crystals which are soluble in alkalies with a brown-red color. Paper impregnated with turmeric turns brown with alkalies. Acids reproduce the yellow color; with boric acid the paper turns orange red on drying.

Lipochromes are the yellow or red pigments of fatty tissues, the corpora lutea, the cones of the visual epithelium of birds and reptiles (called *chromophan* and *chloro-*, *xantho-*, and *rodophan*, according to the color), of the yolk of the egg (*luteines*), of corn, many stamens, flowers, etc. Closely related by chemical and spectroscopic behavior we have the

Xanthocarotin, xanthophyll, chrysophyll, which accompany chlorophyll and form the yellow pigment of the leaves and many flowers, also

Carotin, $C_{18}H_{24}O$, the coloring matter of the yellow carrots and many other plants. It is the chief constituent of the coloring matters of saffron (Crocus), which was formerly called polychroit or crocin.

Rottlerin, $C_{33}H_{30}O_9$, kamalin, mallotoxin, the active constituent of the kamala, forms flesh-colored crystals whose alkaline solution is red and which dyes silk.

III. HETEROCARBOCYCLIC COMPOUNDS.

CONSTITUTION.

The heterocarbocyclic compounds are those compounds whose molecule contains a ring-formed closed chain (an atomic ring) which is composed of other atoms besides C atoms in the ring. Besides the heterocyclic compounds with one atomic ring we also know of those with double or condensed atomic rings (p. 298).

All these compounds are in behavior quite similar to the isocarbocyclic compounds, and what has already been given in regard to these compounds applies in general to the heterocarbocyclic compounds.

A complete separation of the heterocarbocyclic compounds from the aliphatic compounds is not rational, as a great many have out of necessity been discussed in connection with their aliphatic mother-substances, as they can be obtained from these and also because they can be transformed into the aliphatic compounds. Of these compounds which have already been discussed with the aliphatic compounds we find the anhydrides of many alcohols, acids, and their derivatives; e g., ethylene oxide, $<^{CH_2}_{CH_2}>O$,

succinic anhydride, $<^{CH_2-CO}_{CH_2-CO}>O$; lactides, lactames, lactones, e.g.,

butyllactone, $O<^{CH_2-CH_2}_{CO-CH_2}>$; imides, e.g., succinimide, $<^{CH_2-CO}_{CH_2-CO}>NH$;

salts of polybasic acids with polyvalent metals; e.g., lead malonáte, $CH_2<^{COO}_{COO}>Pb$; also the di- and polyamines, e.g., ethylendiamine,

$HN<^{CH_2-CH_2}_{CH_2-CH_2}>NH$; the imines, e.g., ethylenimine, $<^{CH_2}_{CH_2}>NH$; the

ureids and purin derivatives, e.g., alloxan, $OC<^{NH-CO}_{NH-CO}>CO$.

In the following pages the heterocarbocyclic compounds having 5 and 6 atoms in the ring and having the same bondage as the atoms of the benzene ring will be discussed. These rings, in contradistinction to the rings with more or less atoms, cannot be ruptured in a simple manner, but form compounds corresponding to the benzene compounds, from which the derivatives can be derived without a change in the atomic rings.

The addition products of the heterocarbocyclic compounds will be treated of in connection with the individual members. These, like the addition products of the isocarbocyclic compounds, because of their chemical behavior (p. 326), belong to the aliphatic compounds.

Of the heterocarbocyclic compounds those containing ⁻N⁻ atoms in the ring, the *azocarbocyclic* compounds, are the most important. No element forms such a large and stable number of heterocarbocyclic compounds as nitrogen. They may also be considered as benzene derivatives in the broad sense, produced by replacing the ⁻CH⁻ groups of the benzene ring by ⁻N⁻ atoms.

TRANSITION BETWEEN HETEROCARBOCYCLIC AND ISOCARBOCYCLIC AND ALIPHATIC COMPOUNDS.

Although only a few isocarbocyclic compounds can be obtained from the aliphatic compounds, nearly all heterocarbocyclic compounds can be prepared from aliphatic or isocarbocyclic compounds, and in fact all those that are of most use.

Alkylpyridines are produced by heating aliphatic aldehyde ammonia alone or with aliphatic aldehydes or ketones, thus:

$$4CH_3\text{-}CHO + NH_3 = 4H_2O + HC\!\!\leqslant^{CH=\!\!=\!\!CH}_{C(C_2H_5)\text{-}C(CH_3)}\!\!\geqslant N, \quad \text{methylethyl-pyridine.}$$

Alphylpyridines are prepared by heating certain isocarbocyclic derivatives with hydroxylamine; e.g.,

$$CH_2\!<^{CH_2\text{-}CO\text{-}C_6H_5}_{CH_2\text{-}CO\text{-}C_6H_5} + NH_2OH = 3H_2O + HC\!\!<^{CH=C(C_6H_5)}_{CH\text{-}C(C_6H_5)}\!\!\geqslant N, \quad \text{diphenyl-pyridine.}$$

Alkylate quinolins in the pyridine or benzene nucleus are produced from aliphatic aldehydes by amidobenzene in the presence of H_2SO_4; e.g.,

$$C_6H_5\text{-}NH_2 + 2CH_3\text{-}CHO = 2H_2O + 2H + C_6H_4\!\!<^{CH=CH}_{N=C(CH_3)}\!\!, \text{methylquinoline.}$$

On passing allyl aniline over heated lead oxide we obtain quinoline:

$$C_6H_5\text{-}NH\text{-}CH_2\text{-}CH=CH_2 = 4H + C_6H_4\!\!<^{CH=CH}_{N=CH}\!\!, \text{quinoline.}$$

Pyrrole, furane, and thiophene derivatives are produced from certain aliphatic diketones; e.g., acetonylacetone, $CH_3\text{-}CO\text{-}CH_2\text{-}CH_2\text{-}CO\text{-}CH_3$ yields dimethyl furane, $<^{CH=C(CH_3)}_{CH=C(CH_3)}>O$, on splitting off of H_2O, dimethyl thiophene, $<^{CH=C(CH_3)}_{CH=C(CH_3)}>S$, on distillation with P_2S_5, and dimethyl pyrrole, $<^{CH=C(CH_3)}_{CH=C(CH_3)}>NH$ on distillation with ammonia. From the aliphatic mucic or isosaccharic acid we can obtain furane by dry distillation, thiophene on distillation with BaS, and pyrrole derivatives by distillation of their ammonium salts.

The β-ketonic acid esters (p. 365) can also be used in the p e   ion of the heterocarbocyclic compounds. On heating with aldehyde ammonia they give alkyldihydropyridine carbonic acid ester, e.g.,

$$2CH_3-CO-CH_2-COO-C_2H_5 \atop +CH_3-CHO+NH_3 = CH(CH_3) {\textstyle <} {C(COO \cdot C_2H_5)=C(CH_3) \atop C(COO \cdot C_2H_5)=C(CH_3)} {\textstyle >} NH+3H_2O,$$

with anilines they give alkyloxyquinoline:

$$CH_3-CO-CH_2-COO-C_2H_5+C_6H_5-NH_2=C_6H_4 {\textstyle <} {C(CH_3)=CH \atop N ==C(OH)} |\ +H_2O+C_2H_6O;$$

with phenylhydrazin they yield phenylalkylpyrazolones, e.g.,

$$CH_3-CO-CH_2-COO-C_2H_5 \atop +H_2N-NH(C_6H_5) = {\textstyle <} {C(CH_3)=N \atop CH_2--CO} {\textstyle >} N \cdot C_6H_5+H_2O+C_2H_6O.$$

Diazindiazole (purin) is obtained from uric acid (p. 415).

Triazine, $CH {\textstyle \lessgtr} {N=CH \atop N-CH} {\textstyle \geqslant} N'$ and its derivatives are produced from HCN and its derivatives (p. 385) by polymerization.

Condensed heterocarbocyclic compounds are obtained from the ortho-substitution products of benzene and naphthalene by splitting off of H_2O (pp. 483 and 486).

The conversion of heterocarbocyclic compounds into isocarbocyclic or aliphatic compounds, with the exception of the addition products and the compounds with more or less than 5 and 6 atoms in the ring, can only be done with difficulty. The azocyclic ring is generally more resistant to rupture and more stable than the benzene ring.

Pyridines on heating with HI are converted into paraffines; e.g., pyridine, C_5H_5N, yields hexane, C_6H_{14}.

The benzene ring of the quinolines is destroyed on oxidation, being converted into pyridine carboxylic acid (p. 537). The pyridine ring in α-alkylquinolines is destroyed by $KMnO_4$, yielding the derivatives of o-amidobenzoic acid:

$$C_6H_4 {\textstyle <} {CH=CH \atop N==C(CH_3)} |\ +5O=CO_2+C_6H_4 {\textstyle <} {COOH \atop NH(CO \cdot CH_3)} \quad \text{(Acetyl-o-amido-benzoic acid.)}$$

Pyrrole, furane, and thiophene derivatives yield fumaric or maleic acid derivatives on oxidation, e.g.,

$$ {\textstyle <} {CH=CH \atop CH=CH} {\textstyle >} O+3O=HOOC-CH=CH-COOH.$$

NOMENCLATURE.

In general the nomenclature of these compounds is the same as the isocarbocyclic compounds. The differences will be mentioned in connection with the individual groups. The six-membered azocyclic compounds are called *azines*, the five-membered *azoles* (p. 550). According as in the ring formation, one, two, etc., other atoms take part besides C atoms, we designate them mono-, di-, etc., hetero-atomic rings, and as to the total number of atoms forming the ring we differentiate between three-, four-, five-, etc., membered rings. The compounds produced by condensation of iso- with heterocarbocyclic rings are designated by adding the prefix benzo-, phen-, dibenzo- or diphen-, naphtho-, etc., to the name of the heterocarbocyclic ring (pp. 539, 541, etc.).

CLASSIFICATION.

According to the number of atoms forming the ring, the following six- and five-membered chief groups are differentiated, and these divided into sub-groups according to the number of the different atoms (hetero atoms) in the ring, so that first the mono- and then the di-, tri-, etc., hetero-atomic compounds will be discussed. The addition products and condensed compounds will be treated in connection with these subgroups.

The compounds of the alkaloid and protein groups which are treated of at the end cannot be placed in any of the foregoing groups, as their constitution has not been sufficiently studied to warrant such arrangement for the present.

SIX-MEMBERED COMPOUNDS WITH ONE OTHER ATOM IN THE CARBON RING.

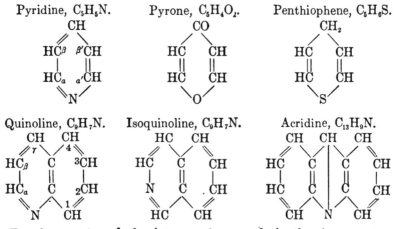

Pyridine, C_5H_5N.   Pyrone, $C_5H_4O_2$.   Penthiophene, C_5H_6S.

Quinoline, C_9H_7N.   Isoquinoline, C_9H_7N.   Acridine, $C_{13}H_9N$.

(For the meaning of the letters and numerals in the rings, see pp. 538 and 540.)

As in these compounds the hydrogen can be replaced by alkyls, it follows that a series of homologous compounds can be derived therefrom, e.g., $C_5H_4(CH_3)N$, $C_5H_3(CH_3)_2N$, etc. From these, as in the case of the corresponding benzene derivatives, on oxidation of the alkyl groups we obtain carboxylic acids; e.g., pyridine mono-, pyridine di-, pyridine tri-, pyridine tetracarboxylic acids and pyridine pentacarboxylic acid. The original hydrocarbon can be obtained from these acids by the splitting off of the carbon dioxide groups.

These compounds, like the benzene derivatives, also form addition products whereby monovalent atoms or radicals attach themselves with the complete or partial rupture of the double bonds (p. 463); thus from pyridine we obtain a hexahydride derivative, $C_5H_5NH_6$, which on oxidation again yields pyridine. This behavior, which is analogous to the benzene nucleus, is explained by its constitution.

Pyridine may be considered as benzene in which a trivalent $^-CH^-$ (methine) group is replaced by one $^-N^-$ atom, while quinoline may be derived from naphthalene, and acridine from anthracene, in a similar manner.

In the same way naphthoquinoline, $C_{13}H_9N$, may be derived from phenanthrene, $C_{14}H_{10}$, by replacing one $^-CH^-$ group by $^-N^-$, and also anthraquinoline, $C_{17}H_{11}N$, from chrysene, $C_{18}H_{12}$. These compounds behave like quinoline bases and may form corresponding derivatives.

The number of isomerides is even greater than with the benzene derivatives, as the position of the replacing groups as compared to the nitrogen must be considered (p. 538).

In all these compounds the $^-N^-$ atom is combined to the carbon by all three valences, hence they may be considered as tertiary amines. As would be expected from this, they are strong bases, which, like ammonia, unite directly with acids, forming salts, and with alkyl iodides, forming analogous compounds with ammonium iodide (p. 379). Hydrochlorauric and hydrochlorplatinic acids give compounds which are difficultly soluble and which are similar to the corresponding ammonium salts.

Penthiophene, C_5H_6S (p. 536), is only known in the form of derivatives which behave like the thiophene derivatives.

Most of the nitrogenous plant bases called alkaloids (p. 552) are hydroderivatives of pyridine, quinoline, and isoquinoline.

1. Pyridine Compounds.

Pyridine, C_5H_5N (structure, p. 536).
Picoline, C_6H_7N (methylpyridine, isomeric with aniline).
Lutidine, C_7H_9N (dimethylpyridine, isomeric with toluidine).
Collidine, $C_8H_{11}N$ (trimethylpyridine, isomeric with xylidine).
Parvoline, $C_9H_{13}N$ (tetramethylpyridine, isomeric with cumidine).

These pyridine bases occur with quinoline bases and their isomers, the anilines, in coal-tar and in bone oil (see below). They are formed on the decomposition of many alkaloids and by heating aliphatic aldehydes with ammonia:

$$4C_2H_4O + NH_3 = C_5H_2(CH_3)_3N + 4H_2O.$$

They form colorless poisonous liquids which are volatile without decomposition and which have a characteristic odor. The lower members are readily soluble in water, while the higher members are not. The homologues of pyridine are readily oxidized into pyridine carboxylic acids (p. 537), while pyridine itself is not attacked by nitric acid, iodine, or oxidizing agents, and by sulphuric acid only on raising the temperature.

The hexahydropyridines, $C_5H_5N(H_6)$, are called *piperidines*, methyl piperidines are called *pipecolines*, the dimethyl piperidines, *lupetidines*, and the trimethyl piperidines, *copelidines*, etc.

Isomerides. Three mono derivatives are possible, which are designated α-, β-, γ-pyridines, according as the substitution takes place near or remote from the N atom. Six biderivatives are possible, which are designated $\alpha\alpha'$-, $\alpha\beta$-, $\alpha\gamma$-, $\alpha\beta'$-, $\beta\gamma$-, $\beta\beta'$-pyridines. Ten biderivatives with two different substitutions are possible (p. 467).

If the attachment takes place at the N atom due to a rupture of the double bond the compound is called *n*-pyridine.

The constitution of pyridine is derived from its formation from quinoline. This last yields a pyridine carboxlyic acid on oxidation, when one benzene nucleus is destroyed, similar to the oxidation of naphthalene (p. 514). Pyridine can be obtained from this pyridine carboxylic acid by splitting off 2 mols. carbon dioxide (p. 537).

Bone oil, animal oil (Dippel's oil), is prepared from the crude animal oil obtained in the dry distillation of bones and other animal substances by rectification. It contains chiefly pyridine bases, also quinoline bases, pyrrole and its homologues (p. 544 , nitriles of the fatty acids and benzene hydrocarbons.

Dipyridyl, $NH_4C_5-C_5H_4N$, corresponding to diphenyl (p. 504); we have several isomers which on oxidation yield nicotinic acids.

Oxypyridines, $C_5H_4(OH)N$, correspond to the phenols and give a yellow or red coloration with ferric chloride or they correspond to the ketones and are then called pyridones, e.g., $HN<^{CH=CH}_{CH=CH}>CO$, r-dioxy-pyridone.

Nicotinic acid, $C_5H_4(COOH)N$, β-pyridine carboxlyic acid, is produced by the oxidation of nicotine (p. 534) and by heating trigonelline (p. 553) with HCl.

Cinchomeronic acid, $C_5H_3(COOH)_2N$, one of the six known pyridine carboxylic acids, is produced by the oxidation of quinine, cinchonine, and cinchonidine (p. 555).

Piperidine, $C_5H_{10}NH$ or $H_2C<^{CH_2-CH_2}_{CH_2-CH_2}>NH$, hexahydropyridine (p. 537), is produced by the action of nascent hydrogen upon pyridine. It is a colorless, alkaline liquid having a pepper-like odor. It may also be prepared from the alkaloid piperine (p. 553) by boiling it with alkalies. Piperidine is the mother-substance of the alkaloids of coca and most of the solanum alkaloids (p. 557).

Propylpiperidine, $C_5H_9(C_3H_7)NH$, is produced by the action of nascent H upon allyl pyridine, and is an inactive substance which can be split into a lævo and a dextro modification. This latter is the substance called coniine (p. 554).

Eucain B, $C_5H_6(CH_3)_3(\cdot O\cdot C_6H_5\cdot CO)-NH\cdot HCl$, the hydrochloride of benzoyltrimethyl oxypiperidine (also called benzoylvinyldiacetonalkamine), is used as a local anæsthetic instead of cocain.

Euphthalmin, $C_5H_6(CH_3)_3(\cdot O\cdot C_6H_5\cdot CHOH\cdot CO)-N(CH_3)HCl$, the hydrochloride of phenyl glycolyl-*n*-methyl-trimethyl oxypiperidine (phenyl glycolyl=mandelic acid radical, p. 498), dilates the pupils and is used in medicine instead of atropine.

2. Quinoline Compounds.

Quinoline,	C_9H_7N (structure, p. 536).
Isoquinoline,	C_9H_7N (structure, p. 536).
Quinaldine,	$C_{10}H_9N$ (α-methyl quinoline).
Lepidine,	$C_{10}H_9N$ (γ-methyl quinoline).
Cryptidine,	$C_{11}H_{11}N$ (dimethylquinoline).

These quinoline or benzopyridine bases occur with the pyridine bases in bone oil (P. 538) and in coal-tar and are prepared from many alkaloids (p. 543) by distillation with caustic alkalies (preparation, p. 540).

They are liquids having peculiar odors and soluble with difficulty in water. Oxidizing agents, iodine and nitric acid only act upon the benzene nucleus and not upon the pyridine nucleus. By rupture of the double bonds in quinoline up to ten H atoms can be attached, producing the hydroquinolines, e.g., $C_9H_7N(H_{10})$, decahydroquinoline.

Two isomerides of quinoline are possible and are also known, according as the N atom exists beside one of the hydrogen-free C atoms or removed therefrom (P. 538). Seven isomers of monoquinoline derivatives are also possible. The substituted H atoms in the benzene ring are designated by 1, 2, 3, 4, or o-, m-, p-, a- (ana), while in the pyridine ring they are designated by α-, β-, γ- (p. 538).

The constitution of the quinolines has been determined by numerous syntheses and also by its analogous behavior to naphthalene on oxidation (p. 537), when it yields pyridine carboxylic acid, $C_5H_3(COOH)_2N$.

It is produced, for example, by passing allylaniline over heated lead oxide (analogous to the preparation of naphthalene from phenyl butylene, p. 514),

$$C_6H_5-NH-CH_2-CH=CH_2+O_2=C_6H_4{\underset{N=CH}{\overset{CH=CH}{<}}}\Big|+2H_2O;$$

also from orthoamidocinnamic aldehyde by splitting off water,

$$C_6H_4{\underset{NH_2}{\overset{CH=CH-CHO}{<}}}=C_6H_4{\underset{N=CH}{\overset{CH=CH}{<}}}\Big|+H_2O;$$

as well as by heating aniline, glycerin, and sulphuric acid with nitrobenzene, which later acts oxidizingly (Skraup's synthesis):

$$C_6H_4{\underset{NH_2}{\overset{H}{<}}}+HOHC{\underset{CH_2OH}{\overset{CH_2OH}{<}}}+O=C_6H_4{\underset{N=CH}{\overset{CH=CH}{<}}}\Big|+4H_2O.$$

If instead of aniline we use its homologues, we obtain the homologous quinolines, while if the halogen, nitro-, etc., substituted amines are used we obtain the halogen, nitro-, etc., substituted quinolines.

α-Oxyquinoline, $C_9H_6(OH)N$, carbostyril, forms asbestos-like crystals. Iodochloroxyquinoline, called vioform, is used as an odorless substitute for iodoform.

Kynurenic acid, $C_9H_5(OH)(COOH)N$, an oxyquinoline carboxylic acid, occurs in dogs' urine and forms colorless crystals which on fusion yield γ-oxyquinoline or kynurine, $C_9H_6(OH)N$.

Iodoxyquinoline sulphonic acid, loretin, $C_9H_4I(OH)(SO_3\cdot H)N$, forms a yellow, crystalline powder which is insoluble in water.

Berberine, papaverine, narcotine, narceine, hydrastine (see Alkaloids) are complicated isoquinoline derivatives.

Quinine, quinidine, cinchonine, cinchonidine, strychnine, brucine, cephæline, aconitine (see Alkaloids) are quinoline derivatives.

Quinoline dyes. By the introduction of amido, alphyl, and alkyl groups into quinoline we obtain various dyes which are called quinoline red, quinoline yellow, flavaniline, and the cyamines.

3. Acridine Compounds.

Acridine, $C_{13}H_9N$ (structure, p. 536), occurs in coal-tar and forms colorless needles which cause an active itching of the skin and melt at 100°. The solution of its salts have a beautiful greenish-blue fluorescence. Acridine and its homologues are weaker bases than the quinoline and pyridine bases. It forms the mother-substance of certain dyes, such as chrysaniline or phosphin, acridine yellow, etc.

4. Pyrone Compounds.

α-Pyrone, coumalin, $C_5H_4O_2$ or $O<^{CO-CH}_{CH=CH}>CH$, is a colorless, neutral liquid, boiling at 209°, and having an odor similar to caraway seeds. It is produced by heating coumalic acid (see below).

γ-Pyrone, pyrocomane, $O<^{CH=CH}_{CH=CH}>CO$, forms colorless, neutral crystals which melt at 32°. The pyrones and their derivatives on warming with NH_3 are converted into pyridones (p. 539), when the oxygen of the ring is replaced by =NH. The properties of many pyrones is peculiar, especially dimethyl pyrone, $C_5H_2(CH_3)_2O_2$, which although it has a neutral reaction, forms salts with acids directly by addition, similar to ammonia. In these salts for every hydrogen atom of the acid one molecule of the pyrone compound takes its place. On account of their analogy to the ammonium salts (pp. 150 and 210) these salts are called *oxonium salts* and we must admit that the formation of salts is caused by the oxygen in the ring being tetravalent; thus, $^{H}_{Cl}>O<^{CH=CH}_{CH=CH}>CO$, dimethyl pyrone hydrochloride.

According to more recent researches numerous other cyclic and aliphatic compounds containing oxygen have the property of forming salts such as the oxonium salts.

The same property has been known for a long time in certain sulphur compounds, the sulphonium or sulphin hydroxides (p. 352), where the sulphur atom appears tetravalent.

Comanic acid and coumalic acid, both $C_5H_3(COOH)O_2$, are pyrone-carboxylic acids. The first is produced from chelidonic acid (see below) and the other from malic acid by heating with concentrated sulphuric acid.

Chelidonic acid, $C_5H_2(COOH)_2O_2$, pyrone-dicarboxylic acid, occurs in the Chelidonium majus, and yields on heating comanic acid and then γ-pyrone.

Meconic acid, $C_5H(OH)(COOH)_2O_2$, oxypyrone-dicarboxylic acid, occurs in opium, and on boiling with water it splits off CO_2 and is converted into

Comenic acid, oxypyrone-carboxylic acid, $C_5H_2(OH)(COOH)O_2$; this splits off more CO_2 on heating and yields

Pyromeconic acid, β-oxy-r-pyrone, $C_5H_3(OH)O_2$.

α-Benzopyrone, $C_6H_4<^{CH=CH}_{O-C}|$. Coumarin (p. 501) is to be considered as α-benzopyrone.

Xanthone, $C_{13}H_8O_2$, diphenylenkentone oxide, is the mother-substance of *euxanthon*, $C_{13}H_8O_4$, and of *euxanthic acid*, $C_{19}H_{18}O_{11}$, both occurring in Indian yellow (p. 446), and also of the yellow plant pigment *gentisin*, $C_{14}H_{10}O_5$, of the gentian root and *datiscetin*, $C_{15}H_{12}O_6$; also of *rhamnochrysine*, $C_{18}H_{12}O_7$, and *rhamnocitrine*, $C_{13}H_{10}O_5$ (see below).

Flavone, $C_{15}H_{10}O_2$, is the mother-substance of numerous yellow plant pigments, for example, *quercetin*, $C_{15}H_{10}O_7$, from the glucoside quercitrin; *fisetin*, $C_{15}H_{10}O_6$, in the fustic wood; *chrysine*, $C_{15}H_{10}O_4$, in the buds of the various poplars; *luteolin*, $C_{15}H_{10}O_6$, in Reseda luteola and digitalis leaves; *galangin*, $C_{15}H_{10}O_5$, and *campferid*, $C_{16}H_{12}O_6$, in the galanga root; *morin*,

$C_{15}H_{10}O_7$, and *maclurin*, $C_{13}H_{10}O_6$, in fustic; *rhamnetin* (methyl quercetin), $C_{16}H_{12}O_7$, in the glucoside xanthorhamnin; *rhamnazin* (methyl rhamnetin), $C_{17}H_{14}O_7$; and *rhamnolutin*, $C_{15}H_{10}O_6$, in the buckthorn; *apigenin*, $C_{15}H_{10}O_5$, in the glucoside apiin; *scoparine*, $C_{14}H_{10}O_5$, in the broom, and *gossypetine*, $C_{16}H_{12}O_8$, in the cotton flower.

Dibenzopyrone   and   Phenylbenzopyrone
or Xanthone   or Flavone.

The following pigments belong to the complex flavone derivatives but their structure has not been well-studied:

Santalin, $C_{15}H_{14}O_5$, contained in the sandalwood.

Brasilin, $C_{16}H_{14}O_5$, in the Brazil wood; forms yellow crystals which turn carmine-red with traces of caustic alkalies or ammonia being oxidized to *brasilein*, $C_{16}H_{12}O_5$.

Hæmatoxylin, $C_{16}H_{14}O_6$, of the logwood; forms pale-yellow prisms which are soluble with a purple color in ammonia and caustic alkalies (delicate reagent for these), being oxidized to *hæmatein*, $C_{16}H_{12}O_6$.

SIX-MEMBERED COMPOUNDS WITH SEVERAL OTHER ATOMS IN THE CARBON RING.

The most important compounds belonging to this six-membered ring are those containing nitrogen and which are derived from pyridine, quinoline, and acridine, when $-N-$ atoms take the place of $=CH-$ groups. They are called *azines*, and the number of N atoms is indicated by the prefixes di-, tri-, tetrazines; *oxazines* contain one N and one O atom, *thiazines* one N and one S in the ring, etc.

According as the N atoms occupy the o-, m-, or p-position relative to each other we designate them *oiazines*, *miazines*, and *piazines*.

1. Azine Compounds.

Orthodiazine, $C_4H_4N_2$ or $HC{\leqslant}^{CH=N}_{CH-CH}{\geqslant}N$, pyridazine, is a liquid smelling like pyridine and boiling at 208°.

Metadiazine, $C_4H_4N_2$ or $HC{\leqslant}^{N=CH}_{CH-CH}{\geqslant}N$, pyrimidine, forms a solid melting at 22° and having a narcotic odor.

Dioxyhydrometadiazine, $HC{\leqslant}^{NH-CO}_{CH-CO}{>}NH$, uracil, is closely related to the purin derivatives (p. 415). Methyl uracil is thymin (p. 565).

Paradiazine, $C_4H_4N_2$ or $N{\leqslant}^{CH=CH}_{CH-CH}{\geqslant}N$, pyrazine, aldine, forms crystals which melt at 55° and having an onion-like odor.

Piperazine, $C_4H_{10}N_2$ or $HN<^{CH_2-CH_2}_{CH_2-CH_2}>NH$, diethylendiamine, is a hexahydropyrazine. It is obtained from ethylene bromide by the action of ethylendiamine, and forms crystals which are soluble in water and melt at 106° and which dissolve large quantities of uric acid. Spermine crystals (Schreiner's crystals), found in semen, seem to be closely related to piperazine.

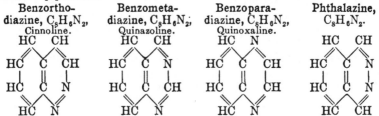

| Benzortho-diazine, $C_8H_6N_2$, Cinnoline. | Benzometa-diazine, $C_8H_6N_2$, Quinazoline. | Benzopara-diazine, $C_8H_6N_2$, Quinoxaline. | Phthalazine, $C_8H_6N_2$. |

The first three of these four isomers are derived from quinoline and the last from isoquinoline, where one ⁼CH⁻ group is replaced by ⁼N⁻. Quinazoline is only known in the form of derivatives, while the others form crystals having strong basic properties.

Phenyldihydroquinazoline, $C_8H_5(C_6H_5)N_2(H_2)$. Its compound with HCl is used in medicine under the name orexin.

Dibenzoparadiazine, phenazine, $C_{12}H_8N_2$, is derived from acridine, where one ⁼CH⁻ group is replaced by ⁼N⁻. It is found in coal-tar and forms yellow needles and by the introduction of amido-, alkyl-, or alphyl groups forms the dyes called indolines (true blue), nigrosines (aniline black), safranines (indoin blue, mauvein, magdala red, indazin blue), toluylene red, toluylene blue, Janus green.

Tribenzoparadiazine, naphthophenazine, $C_{16}H_{10}N_2$, obtained from certain azo compounds, forms colorless crystals and by the introduction of amido-, alphyl-, or alkyl groups forms the dyes called eurhodoles, eurhodines, rosindulines (azocarmin).

s-Triazine, $HC<^{N-CH}_{N-CH}>N$, $C_3H_3N_3$, cyanidine, corresponds to trihydrocyanic acid and is the mother-substance of the cyanuric compounds (p. 382).

a-Triazine, $HC<^{N=CH}_{CH-N}>N$, as well as

v-Triazine, $HC<^{CH-N}_{CH-N}>N$, is only known in the form of derivatives.

In regard to the meaning of s-, a-, v-, see page 469.

Phentriazine, $C_7H_5N_3$, is produced from quinoline, C_9H_7N, by replacing two ⁼CH⁻ groups by two ⁼N⁻ atoms. It forms yellow crystals which melt at 75° and have a narcotic odor.

Tetrazine, $HC<^{N==N}_{CH-N}>N$, as well as its isomers is not known free, but the derivatives are known.

2. Oxazine and Thiazine Compounds.

Oxazine, $O<^{CH=CH}_{CH=CH}>NH$, is not known free.

Tetrahydo-paraoxazine, $O<^{CH_2-CH_2}_{CH_2-CH_2}>NH$, morpholin, is a color-less, basic fluid, boiling at 129° and which has great similarity to piperidine (p. 530). It is the mother-substance of the alkaloids morphine, codeine, and thebaine (see Alkaloids).

Phenoxazine, $C_6H_4<^{NH}_{O}>C_6H_4$, forms colorless leaves and by the introduction of alkylamine groups it forms the oxazine and oxazone dyes (Nile blue and gallocyanin), the blue muscarin pigment.

Phenthiazine, thiodiphenylamine, $C_6H_4<^{NH}_{S}>C_6H_4$, is produced from diphenylamine by heating it with sulphur and forms yellow scales. By the introduction of alkyl or amido groups it forms the thionin dyes, for example, Lauth's violet, methylene blue, methylene green, thiazine red, and thiazine brown, thionin blue and toluidine blue.

FIVE-MEMBERED COMPOUNDS WITH ONE OTHER ATOM IN THE CARBON RING.

These contain the tetrol group, C_4H_4, which with $^-NH^-$, $^-O^-$, or $^-S^-$ form a closed chain:

Pyrrol, C_4H_5N.   Furane, C_4H_4O.   Thiophene, C_4H_4S.

$$\begin{matrix} HC \!-\! CH \\ \| \quad \| \\ HC \quad CH \\ \diagdown \diagup \\ NH \end{matrix} \qquad \begin{matrix} HC \!-\! CH \\ \| \quad \| \\ HC \quad CH \\ \diagdown \diagup \\ O \end{matrix} \qquad \begin{matrix} HC \!-\! CH \\ \| \quad \| \\ HC \quad CH \\ \diagdown \diagup \\ S \end{matrix}$$

The preparation and properties of the derivatives of these compounds are the same as the corresponding benzene derivatives.

1. Pyrrol Compounds.

Pyrrol, C_4H_5N, occurs in coal-tar and in bone-oil, is produced by heating succinimide with zinc dust and water:

$$<^{CH_2-CO}_{CH_2-CO}>NH+4H=<^{CH-CH}_{CH-CH}>NH+2H_2C.$$
Succinimide.   Pyrrol.

It is a colorless basic liquid boiling at 130° and having an odor similar to chloroform and chemical properties similar to pyridine. Its vapors color a pine shaving moistened with HCl purple-red.

If in pyrrol the hydrogen attached to a carbon atom or to a nitrogen atom be replaced by alkyls, we obtain the homologues of pyrrol, the pyrrol bases, which are also found in bone-oil and are faintly basic liquids; e.g., methyl pyrrol, $C_3H_4=N-CH_3$, homopyrrol, $C_4H_3(CH_3)=NH$, ethylpyrrol, $C_4H_4=N-C_2H_5$, dimethylpyrrol, $C_4H_2(CH_3)_2=NH$.

Tetraiodopyrrol, iodol, C_4I_4NH, forms a yellow powder which is used as an antiseptic.

Pyrroline, dihydropyrrol, C_4H_7N or $<^{CH_2-CH_2}_{CH=CH}> NH$, and

Pyrrolidine, C_4H_9N, or $<^{CH_2-CH_2}_{CH_2-CH_2}>NH$, tetrahydropyrrol, are produced by the addition of hydrogen to pyrrol. Pyrrolidine is the mother-substance of most of the coca and sclanum alkaloids.

Pyrrolidine carbonic acid, $C_4H_8(COOH)N$, is a cleavage product of the protein bodies.

Chlorophyll, the pigment of the green parts of plants, consists of C, H, O, N, P, perhaps also iron, and up to the present time is obtained more or less decomposed as soft masses which are soluble with a bluish-green color in alcohol, ether, fatty and ethereal oils. It is decomposed by acids to *phylloxanthin* and then from this into *phyllocyanin*, which on heating with alkalies yields *phylloporphyrin*, $C_{16}H_{18}N_2O$, and this, like the hæmato-porphyrin obtained from hæmoglobin, is reducible to hæmopyrrol (iso-butyl pyrrol).

Blood-pigments. The color of the blood is produced by the chromo-proteids oxyhæmoglobin and hæmoglobin, which are compounds of the colorless proteid globin with the pigments hæmatin or hæmochromogen. Closely related to these are the pigments of the bird's-egg shells, the blue *oocyanin*, the reddish *oorhodin*, the green *oochlorin*, and the yellow *ooxanthin*.

Hæmatin, $C_{34}H_{34}N_4FeO_5$, the cleavage product of oxyhæmoglobin (p. 565), is a bluish-black amorphous powder soluble with a red color in acidified alcohol, and on oxidation yields two bibasic *hæmatinic acids*, $C_8H_{10}O_5$. On treatment with concentrated sulphuric acid or hydrochloric acid it is converted into hæmatoporphyrin (see below). Hæmatin gives a characteristic absorption spectrum (p. 45).

Hæmin, $C_{34}H_{32}N_4FeO_4 \cdot HCl$, hæmatin hydrochloride, Teichmann's blood-crystals, serve in the detection of small amounts of blood. The object containing the blood is extracted with a little cold water and this extract allowed to evaporate on a glass slide and warmed with a trace of salt and a few drops of glacial acetic acid. On cooling, the characteristic bluish-black rhombic crystals of hæmin are seen under the microscope.

Hæmochromogen, $C_{64}H_{70}Fe_2N_{10}O_7$, the cleavage product of hæmoglobin (p. 566), forms dark crystals which are soluble in alkalies yielding a red solution, and then quickly absorb oxygen, forming hæmatin. In acid solution hæmatoporphyrin is produced. It also gives a characteristic absorption spectrum.

Hæmatoporphyrin, $C_{32}H_{36}N_4O_6$, the cleavage product of hæmatin and hæmochromogen (which see) and also often found in the urine, forms a brown amorphous powder soluble with a red color in dilute acids and gives a characteristic absorption spectrum. On oxidation it yields the bibasic *biliverdic acid,* $C_8H_9NO_4$, and on reduction it yields like *phylloporphyrin* (the cleavage product of chlorophyll) and hæmopyrrol (isobutyl pyrrol).

Melanins are the several nitrogenous and often ferruginous pigments, closely related to the blood-pigments, which are found in the skin of the negro, in the choroidea and iris, in the hair, in the urine and blood in certain diseases. They are amorphous black or brown pigments which are insoluble in water, alcohol, ether, chloroform, dilute acids, but are soluble in caustic and carbonated alkalies and give an imperfect absorption spectrum.

Bile-pigments. The color of the bile is due to *bilirubin* and *biliverdin*, derivatives of the blood-pigment. In gall-stones we find other pigments which have been little studied, especially *choletelin*, $C_{16}H_{18}N_2O_9$, *biliprasin*, $C_{16}H_{22}N_2O_6$, *bilifuscin*, $C_{16}H_{20}N_2O_4$, *bilicyanin*, and *bilihumin*. These pigments are weak acids, insoluble in water, produce no absorption spectrum, and give Gmelin's test: Float the liquid to be tested upon nitric acid containing some nitrous acid, when at the point of contact of the two liquids a series of colored rings will be formed from below upward in the following order: a yellowish-red (choletelin), then red, violet-blue (bilicyanin), and on top a green ring (biliverdin).

Bilirubin, $C_{16}H_{18}N_2O_3$, an isomer of hæmatoporphyrin (p. 545), also occurs in gall-stones, in all blood exudates, in icteric urine, and in the contents of the small intestine. It forms dark-red crystals, readily soluble in chloroform, but difficultly soluble in alcohol. The yellowish-red solution is converted into biliverdin in the air. Nascent hydrogen (in e presence of water) converts bilirubin as well as biliverdin into colorless

Hydrobilirubin, $C_{32}H_{40}N_4O_7$, which is also found in the intestine.

Biliverdin, $C_{16}H_{18}N_2O_4$, forms green crystals insoluble in chloroform and soluble with difficulty in alcohol.

Urobilin, $C_{32}H_{40}N_4O_7$, is chemically closely related to bilirubin, and occurs as traces in normal urine and in larger amounts in icterus. It forms an amorphous brown mass soluble in alcohol or chloroform. Sodium amalgam converts it into colorless *urobilinogen*.

2. Benzopyrrol Compounds.

The indol or benzopyrrol compounds contain on one side a benzene nucleus and on the other side a closed chain, such as is also contained in pyrrol, C_4H_5N, consisting of four carbon atoms (of which two belong to the benzene nucleus) and one nitrogen atom (p. 544). We can therefore consider the indol compounds also as being the substitution of $=CH_2$ by $=NH$ in indene, C_9H_8 (p. 513).

By rupturing the pyrrol ring (oxidation, etc.) the indol bodies are converted into ortho derivatives of benzoic acid:

Indol, C_8H_7N.

Indoxyl, C_8H_7NO.

Isatin, $C_8H_5NO_2$.

Indol, benzopyrrol, C_8H_7N, is formed on fusing proteids with caustic alkalies as well as in the putrefaction of the same and hence is found in feces. It is also obtained by the distillation of oxindol with powdered zinc and also by heating orthonitrocinnamic acid with caustic alkali and iron filings:

$$C_6H_4 < {CH=CH-COOH \atop NO_2} = C_6H_4 < {CH \atop NH} > CH + CO_2 + O_2.$$

It forms colorless, faintly basic scales having feces-like odor and which are readily volatile with steam. It can be obtained from feces, accompanied by the skatol and phenols contained therein, by distillation with water. It colors a pine shaving moistened with HCl cherry-red.

Oxindol, C_8H_7NO (see p. 468), obtained by the action of sodium amalgam upon isatin (see below) in acid solution. It forms colorless needles which by reduction are converted into indol, C_8H_7N, and are converted into dioxindol (see below) by oxidation in the air.

Indoxyl, urine indican, C_8H_7NO, isomer of oxindol, occurs as potassium indoxyl sulphate, $C_8H_6N(O-SO_2-OK)$, in small amounts in the urine of carnivora and larger amounts in herbivorous urine. If urine is carefully treated with oxidizing substances the indoxyl sulphuric acid is oxidized to indigo-blue, which can be shaken out by chloroform, giving a blue color to the same. On stronger oxidation it yields isatin.

Preparation. From aniline and chloracetic acid we obtain anilido-acetic acid (phenylglycocoll), which on fusion with alkali hydroxides yields indoxyl:

$$(C_6H_5-NH)CH_2-COOH = H_2O + C_6H_4 < {C(OH) \atop NH-} \geqslant CH.$$

Properties. Yellow crystals soluble in water, alcohol, ether, etc., and melting at 85°. In alkaline solution it is oxidized into indigo-blue, even in the air.

Indoxylic acid, $C_9H_7NO_3$ or $C_8H_6(COOH)NO$. *Preparation.* Brommalonic acid ester and aniline yield anilido-malonic acid ester:

$$CHBr < {COO \cdot C_2H_5 \atop COO \cdot C_2H_5} + C_6H_5 \cdot NH_2 = C_6H_5 \cdot NH \cdot CH < {COO \cdot C_2H_5 \atop COO \cdot C_2H_5} + HBr, \text{ which}$$

at 260° yields alcohol and indoxylic acid ester, $C_6H_5 \cdot NH \cdot CH < {COOC_2H_5 \atop COOC_2H_5} =$

$C_6H_4 < {NH- \atop C(OH)} \geqslant C \cdot COO \cdot C_2H_5 + C_2H_5OH.$ This on oxidation in alkaline solution in the air turns into indigo-blue: $2C_9H_6(C_2H_5)NO_3 + O_2 = C_{16}H_{10}N_2O_2 + 2CO_2 + 2C_2H_5OH.$

See also the technical preparation of indigo (p. 548).

Properties. Colorless prisms which are oxidized into indigo-blue and on heating yield indoxyl and CO_2.

Dioxindol, $C_8H_7NO_2$, is produced by the action of sodium amalgam upon an alkaline solution of isatin. It forms colorless prisms and is readily oxidized in watery solution into isatin.

Isatin, $C_8H_5NO_2$, formed in the oxidation of indigo by nitric acid as yellowish-red prisms soluble in hot water and alcohol. It is artificially prepared from orthonitrophenyl propiolic acid by boiling with alkalies (see also p. 468):

$$C_6H_4 {<}^{C≡C-COOH}_{NO_2} = C_6H_4 {<}^{C(OH)}_{N\underline{\quad}} {\geq} CH + CO_2.$$

Skatol, β-methyl indol, C_9H_9N, occurs as potassium skatoxyl sulphate, $C_9H_8-N-(O{\cdot}SO_2{\cdot}OK)$, in human urine and is formed at the same time as indol in all the methods of formation given for indol. It forms colorless scales which melt at 94°, is much less soluble in water than indol, and does not color a pine chip moistened with HCl. The odor of excrements is chiefly due to skatol and indol.

Indigo-blue, indigo, indigotin, $C_{16}H_{10}N_2O_2$ (structural formula below). *Occurrence.* It does not occur preformed. It is the chief constituent of commercial indigo and may also be obtained in small amounts from urine (see Indoxyl).

Preparation of Natural Indigo. Various varieties of Indigofera of India and America, also Isatis tinctoria (woad), Polygonum and Nerium tinctorium contain the glucoside indican, $C_{14}H_{17}NO_6$, which by fermentation (if the plants are moistened with water and allowed auto-fermentation) or by boiling with dilute acids decomposes into glucose and indoxyl, which is oxidized in the air into indigo: $C_{14}H_{17}NO_6 + H_2O = C_8H_7NO + C_6H_{12}O_6$, the indigo depositing as a blue powder.

Preparation of Pure Indigo. 1. By the careful sublimation of commercial indigo or from indigo white (P. 549).

2. From orthonitrophenyl propiolic acid (p. 502) by reducing agents (grape-sugar, etc.) in alkaline solution:

$$2C_6H_4 {<}^{C≡C-COOH}_{NO_2} \quad + \quad 4H \quad = \quad {\begin{array}{c} C_6H_4-CO-CO-C_6H_4 + 2H_2O \\ | \qquad | \quad | \\ NH--C==C--NH \quad + 2CO_2 \end{array}}$$

o-nitrophenylpropiolic acid, $C_9H_5NO_4$.   Indigo, $C_{16}H_{10}N_2O_2$.

We can therefore consider indigo blue as a double compound of the groups $C_6H_4 {<}^{CO}_{NH} {>} C=$, and in its formation from indol bodies a combination of two indol groups must take place.

3. From *o*-nitrobenzaldehyde and acetone in alkaline solution:
$$2C_6H_4(NO_2)(CHO) + 2C_3H_6O = C_{16}H_{10}N_2O_2 + 2C_2H_4O_2 + 2H_2O.$$

4. Besides these methods indigo blue can be obtained in various ways; i.e., from most indol bodies (p. 547).

Technical Preparation of Indigo. As the above methods for preparing artificial indigo are too expensive, the following method is used in its technical preparation:

Anthranilic acid, $C_6H_4(NH_2)(COOH)$, which can be prepared in quantities sufficiently great and cheaply from naphthalene (p. 514), is converted into phenylglycocoll carboxylic acid by monochloracetic acid, and on heating with alkali hydroxides it yields indoxylic acid:

$$C_6H_4 {<}^{NH{\cdot}CH_2{\cdot}COOH}_{COOH} = C_9H_7NO_3 + H_2O,$$

which in alkaline solution on oxidation by the air precipitates crystalline indigo-blue (process, see above).

Properties. Dark-blue powder with reddish lustre which sublimes at 300°. It is without odor and taste, insoluble in water, alcohol, ether, dilute acids, and alkalies. It is soluble in chloroform, aniline, oil of turpentine, paraffin, phenol, and benzene. On fusion with alkali hydroxides, indoxyl, salicylic acid, anthranilic acid (*o*-amidobenzoic acid), and aniline are obtained according to the temperature. On oxidation it yields isatin, and on reduction we obtain indigo white. It dissolves in very concentrated or better in fuming sulphuric acid, forming

Indigomonosulphonic acid, $C_{16}H_9N_2O_2(SO_3)H$, and

Indigodisulphonic acid, $C_{16}H_8N_2O_2(SO_3 \cdot H)_2$, which are used in the dyeing of wool. Its sodium salt, $C_{16}H_8N_2O_2(SO_3 \cdot Na)_2$, which is soluble in water with a blue color, is called *indigo carmine.*

Indirubin, indigo purpurin, $C_{16}H_{10}N_2O_2$, an isomer of indigo blue, occurs in commercial indigo blue, and is produced besides indigo blue on the decomposition of the indoxyl sulphuric acid of the urine by HCl. It forms brownish-red shining needles.

Indigo white, $C_{16}H_{12}N_2O_2$, is produced from indigo blue by reduction with ferrous sulphate or grape-sugar in alkaline solution. The indigo and these substances are mixed with water and filled completely into a flask, which is closed and allowed to stand (indigo of the dyer). The yellow solution obtained is treated with HCl, the air being excluded, when the indigo white precipitates as a white crystalline powder, which is oxidized to indigo blue in the air (preparation of pure indigo blue from the commercial indigo). Indigo white, in contradistinction to indigo blue, dissolves in alcohol, ether, and alkalies. This last property is made use of in dyeing.

The dyeing is done according to two methods: The material to be dyed is dipped in an aqueous solution of indigo sulphonic acid (Saxonblue dyeing), or the material is dipped in the indigo solution as above described, and then exposed to the air, when the indigo white is oxidized to indigo blue, which deposits in the tissues.

3. Furane Compounds.

Furane, furfurane, tetrol, C_4H_4O, or $< {CH=CH \atop CH=CH} > O$, occurs in the first distillate of pinewood-tar and in the distillation of pyromucic acid, $C_5H_4O_3$, with soda-lime. It is a neutral fluid, insoluble in water, boiling at 32°, and having a peculiar odor. It colors a pine chip moistened with HCl green.

Furanalcohol, $C_4H_3(CH_2OH)O$, found in the products obtained on roasting coffee, is prepared from its aldehyde. It is a colorless liquid boiling at 170°.

Furol, furanaldehyde, furfurol, $C_4H_3(CHO)O$, a decomposition product of certain proteids, occurs in beer and brandy, and is produced in the distillation of bran (furfur), seaweeds, pentoses, and pentosanes (p. 444) with dilute sulphuric acid. It is a colorless aromatic liquid boiling at 162°, and which turns brown in the air. Sesame-oil is colored cherry-red by HCl and an alcoholic solution of furfurol (use of furfurol in the detection of margarine, which in Germany must contain sesame-oil).

Pyromucic acid, furancarboxylic acid, $C_4H_3(COOH)O$, is obtained by the oxidation of furol as well as in the dry distillation of mucic acid, $C_6H_{10}O_8$. It forms colorless needles which melt at 134°.

Furoïn, $C_4H_3O-CO-CH-OH-C_4H_3O$, is produced from furol in a manner similar to benzoïn from benzaldehyde (p. 488). It melts at 135°.

Benzofurane, coumarone, $C_8H_4<^{CH}_{O-}>CH$, bears the same relationship to furane as indol bears to pyrrol. It occurs with its homologues in coaltar and is an indifferent oily liquid.

4. Thiophene Compounds.

Thiophene, C_4H_4S or $<^{CH=CH}_{CH=CH}>S$, occurs with methylthiophene (thiotolene) and dimethylthiophene (thioxene) to a slight extent in the light oils of coal-tar (p. 473) and therefore occurs in crude benzene.

It is produced by heating succinic anhydride with P_2S_3:

$$<^{CH_2-CO}_{CH_2-CO}>O + P_2S_3 = <^{CH=CH}_{CH=CH}>S + P_2O_3 + S_2,$$

or by passing acetylene through boiling sulphur: $2C_2H_2 + S = C_4H_4S$. It is a colorless neutral liquid boiling at 84°, which with isatin and sulphuric acid forms blue indophenin, $C_{12}H_7NOS$ (delicate reaction for thiophene).

Thiophenic acid, $C_4H_3(COOH)S$, very similar to benzoic acid, is, like all thiophene derivatives, known as two isomers (α and β).

Benzothiophene, thionaphthene, $C_8H_4<^{CH}_{S-}>CH$, bears the same relation to thiophene that benzofurane does to furane. It forms colorless crystals.

Thiophtene, $\begin{matrix} CH-\!\!\!-C-\!\!\!-CH \\ \| \quad \| \quad \| \\ CH-S-C-S-CH \end{matrix}$, is related to thiophene in the same manner as naphthalene to benzene, and is produced by heating citric acid with P_2S_3. It forms a thick liquid boiling at 225°.

FIVE-MEMBERED COMPOUNDS CONTAINING SEVERAL OTHER ATOMS IN THE CARBON RING.

The most important compounds of this group are those containing nitrogen, which are derived from pyrrol, furane, thiophene, indol derivatives by replacing =CH− groups by =N− atoms. They are called *azoles*, and according to the number of N atoms in the ring di-, tri-, tetrazoles. *Oxazoles* contain one N and one O atom; *thiazoles*, one N and one S atom in the ring. According as the N atoms occupy the *o*-, *m*-, or *p*-position to each other they are called *oiazoles*, *miazoles*, and *piazoles* (p. 542).

1. Azole Compounds.

Glyoxaline, $C_3H_4N_2$ or $<^{CH=CH}_{N-\!\!=CH}>NH$, *m*-diazole, imidazole, and

Pyrazole, $C_3H_4N_2$ or $<^{CH=CH}_{CH-\!\!=N}>NH$, *o*-diazole, form colorless crystals.

Dihydrypyrazole, pyrazoline, $<^{CH_2-CH_2}_{CH-\!\!-\!\!-N}>NH$, is a basic liquid.

Tetrahydropyrazole, pyrazolidine, $<^{CH_2-CH_2}_{CH_2-NH}>NH$, is only known in the form of derivatives.

Methyldihydroimidazole, ethylenethylenyldiamine, lysidin, $<^{CH_2-CH_2}_{N=C(CH_3)}>NH$, is used as a solvent for uric acid.

Pyrazolon, $<^{CH_2-CO}_{CH=-N}>NH$, pyrazolin ketone, forms colorless crystals and is the mother-substance of

Phenyldimethylpyrazolon, antipyrine,

$$C_{11}H_{12}N_2O \text{ or } <^{CH———CO}_{C(CH_3)-N(CH_3)}>N(C_6H_5),$$

is produced by warming methylphenylhydrazine with acetoacetic ester, when a peculiar rearrangement in the pyrazolon ring takes place:

$$\underset{\substack{\text{Methylphenyl-}\\\text{hydrazine.}}}{C_6H_5 \cdot N_2H_2 \cdot CH_3} + \underset{\substack{\text{Aceto-}\\\text{acetic ester.}}}{C_6H_{10}O_3} = \underset{\text{Antipyrine.}}{C_{11}H_{12}N_2O} + H_2O + \underset{\text{Alcohol.}}{C_2H_6O}.$$

It is a white crystalline powder, melting at 113° and readily soluble in water and alcohol. Its dimethylamido derivative is called pyramidon, its acetyl salicylic acid salt is called acopyrine or acetopyrine.

Salipyrine, $C_{11}H_{12}N_2O(C_6H_4 \cdot OH \cdot COOH)$, forms a white sweetish microcrystalline powder or plates, which are not readily soluble in water, but readily soluble in alcohol. It melts at 92°.

Indazole, $C_7H_6N_2$ or $C_6H_4 <^{CH}_{N}>NH$, benzopyrazole, and its isomer,

Isindazole, $C_7H_6N_2$ or $C_6H_4 <^{CH}_{NH}>N$, is indol, in which one $-CH-$ group is replaced by a $=N-$ atom. They form crystals having weak basic properties and are the mother-substance of many compounds.

Purin, $C_5H_4N_4$ or $HC<^{N=CH-C-NH}_{N——C—N}>CH$, the mother-substance of the uric acid group (p. 416), is a diazindiazole which may be considered as benzopyrrol, C_8H_7N (p. 547), in which three $=CH-$ groups are replaced by three $=N-$ atoms.

Osotriazole, $C_2H_3N_3$ or $<^{CH=N}_{CH=N}>NH$, pyrro (aa') diazole, forms colorless crystals which melt at 22°;

Pyrrodiazole, $C_2H_3N_3$ or $<^{CH=CH}_{N==N}>NH$, pyrro (ab) diazole, melts at 111°;

Triazole, $C_2H_3N_3$ or $<^{N=CH}_{N=CH}>NH$, pyrro (bb') diazole, melts at 121°, are all derived from pyrrol, C_4H_4N, by substituting two $=CH-$ groups by $=N-$ atoms, and hence are called pyrrodiazoles. They with their derivatives are weak bases.

Tetrazole, CH_2N_4 or $<^{CH=N}_{N==N}>NH$, forms crystals which melt at 155°, having an acid character. Its salts explode.

2. Oxazole and Thiazole Compounds.

Oxazole, C_3H_3NO or $<^{CH=CH}_{N=CH}>O.$

Isoxazole, C_3H_3NO or $<^{CH=CH}_{CH=N}>O.$

Azoxazole, $C_2H_2N_2O$ or $<^{CH=N}_{CH=N}>O.$

These three do not exist free, but they are the mother-substance of various compounds and they are derived from furane, C_4H_4O, by replacing one or two $=CH-$ groups by one or two $=N-$ atoms. Oxazole and isoxazole are therefore called furomonoazole and azoxazole is called furodiazole or furazan.

Thiazole, C_3H_3NS or $<^{CH=CH}_{N=CH}>S$, thiomonazole, is thiophene, C_4H_4S, in which one $=CH-$ group is replaced by a $=N-$ atom and forms a basic liquid boiling at 117°.

Benzothiazole, C_7H_5NS or $<^{CH=CH-C-S}_{CH=CH-C-N}>CH$, melts at 134° and forms the thiazole dyes by the introduction of amido, alkyl, and alphyl groups. These dyes are called erica, primulin, thiazole yellow, thioflavin, which dye without mordants.

Benzothiodiazole, $C_6H_4N_2S$ or $<^{CH=CH-C-N}_{CH=CH-C-N}>S$, piazthiol, melts at 44°, seems to be the mother-substance of the sulphur dyes (immedial black, vidal black), which dye without mordants.

ALKALOIDS.

Alkaloids, plant bases, is the name given to nitrogenous carbon compounds occurring in the plants and having pronounced basic (alkali-like) characteristics and which are derivatives of simple or combined heterocarbocyclic or combined iso- and heterocarbocyclic rings and therefore show in their behavior a great similarity with each other. They generally form the physiologically active constituent of the plants and most of them are strong poisons.

All the nitrogenous carbon compounds occurring in the plants and having basic properties, such as betaine, choline, muscarine, caffeine, theobromine, asparagine, etc., used to be considered as alkaloids. The oxygen-free alkaloids coniine, nicotine, piperidine, sparteine have a characteristic odor, are colorless volatile liquids, while the others contain oxygen and are odorless, crystallizable, non-volatile, generally colorless solids. All are soluble in alcohol, benzene, chloroform, amylalcohol, and, with the exception of morphine and narceine, also soluble in ether. All with the exception of nicotine are difficultly soluble in water. The solutions have a bitter taste and an alkaline reaction. They have the character of amine bases and combine, like these, with acids by addition, forming good crystalline salts, which are soluble in water and alcohol and insoluble in ether with the exception of colchicine. The free alkaloid is precipitated from the aqueous solution of its salts by alkali hydroxides and alkali carbonates. They are mostly optically active and indeed lævorotatory. Coniine, quinidine, cinchonine, pilocarpine, are dextrorotatory, and piperine, papaverine, berberine, and atropine are inactive.

They are precipitated from their solution by tannic acid, phospho-molybdic acid, and phospho-tungstic acid (p. 271), potassium-mer-curic iodide, potassium-cadmium iodide, and a great many by iodine. Their hydrochloric acid compounds are precipitated by platinum chloride as a double crystalline salt, similar to ammonium salts (p. 317) and amine bases.

In preparing them the parts of the plants are extracted with dilute hydrochloric acid. The volatile alkaloids can be obtained from this solution by distillation with caustic alkali. In order to separate the non-volatile alkaloids the solution is generally first precipitated by lead acetate in order to remove tannin bodies, pigments, and gluco-sides, and the filtrate freed from lead by H_2S, and the alkaloids pre-cipitated from this filtrate by caustic alkalies. This precipitate is dissolved in alcohol and repeatedly recrystallized.

1. Pyridine and Pyrrol Compounds.

Trigonelline, $C_7H_7NO_2 + H_2O$, nicotinic acid methyl betaine (pp. 369 and 539), occurs in the seeds of Trigonella fœnumgræcum, and the seeds of Strophantus hispudus. It forms neutral prisms.

Hygrine, $C_8H_{15}NO$, $C_4H_7(CH_3)(CO \cdot C_2H_5)$, methylpyrrolidine ethyl ketone, forms a liquid which turns brown in the air and which occurs to a slight extent in coca leaves (p. 558).

Piperine, $C_{17}H_{19}NO_3$, piperylpiperidine, occurs in ordinary and long pepper, decomposes on boiling with alkali hydroxides into piperic acid $C_{12}H_{10}O_4$ (p. 501), and piperidine, $C_5H_{10}NH$ (p. 539).

Pilocarpine, $C_{11}H_{16}N_2O_2$, pilocarpidine, $C_{10}H_{14}N_2O_3$, jaborine, $C_{22}H_{32}N_4O_4$, occur together in the leaves of Pilocarpus pennatifolius. Pilocarpine is soluble with a green color in fuming nitric acid and blackens on rubbing with an equal quantity of mercurous chloride and moistening with alcohol.

Pilocarpine hydrochloride, $C_{11}H_{16}N_2O_2 \cdot HCl$, forms crystals which melt at 194°.

Arecaidine, $C_7H_{11}NO_2$, tetrahydromethyl nicotinic acid (p. 539), areco-line, $C_8H_{13}NO_2$ (methylarecaidine), guvacine, $C_6H_9NO_{21}$ (tetrahydromethyldioxypyridon, p. 539), and arecaine, $C_7H_{11}NO_2$ (methyl guvacine), occur in the areca-nut (Semen arecæ).

Coniine, $C_8H_{17}N$ (*d*-hexahydropropylpyridine, *d*-propyl piperidine, p. 539), conhydrine and pseudo-conhydrine, $C_8H_{17}NO$ (both oxyconiines), as well as

Coniceine, $C_8H_{15}N$ (*i*-tetrahydropropylpiperidine), occur together in the hemlock (Herba conii). Coniine forms a liquid having a narcotic odor, boiling at 168°, dextrorotatory and very poisonous. It turns brown in the air and becomes thick. A trace of coniine gives a bluish-green coloration when heated with metaphosphoric acid.

$$\begin{array}{ccc} CH_2-CH & CH_2 \\ | & | & | \\ CH_2 & N\cdot CH_3 & CO \\ | & | & | \\ CH_2-CH & CH_2 \end{array}$$

Pseudopelletierine, $C_9H_{15}NO$, contains two condensed pyridine rings and forms crystals. It is found with the liquid alkaloids pelletierine and isopelletierine, $C_8H_{15}NO$, and methylpelletierine, $C_9H_{17}NO$, in the pomegranate-root.

Nicotine, $C_{10}H_{14}N_2$, *d*-pyridyl-β-tetrahydro-*n*-methyl pyrrol, is found in the leaves of the tobacco (Nicotiana tabacum) and the seeds of the tobacco. It forms a very poisonous liquid having a narcotic odor and boiling at 247° and soluble in water. It turns brown in the air and when heated with metaphosphoric acid it gives an orange color.

2. Quinoline Compounds.

a. Alkaloids of the Strychnos.

Strychnine, $C_{21}H_{22}N_2O_2$, contains a condensed quinoline-piperidone (= α-ketopiperidine) ring or quinoline-pyrrolidone (= α-ketopyrrolidine) ring and is found with brucine in the Nux vomica, in the St. Ignatius bean, and in the wood of the Strychnos colubrina.

It forms colorless, instenely bitter crystals which with concentrated sulphuric acid and some potassium bichromate give an intense bluish-violet solution which gradually changes to red and then yellow.

Strychnine nitrate, $C_{21}H_{22}N_2O_2\cdot HNO_3$, forms colorless needles having a persistent bitter taste and soluble in 90 parts water.

Brucine, $C_{23}H_{26}N_2O_4 + 4H_2O$, dimethyloxylstrychnine, occurs also in the bark of the Strychnos Nux vomica and forms colorless, lævorotatory crystals, which give a red color with nitric acid (test for nitric acid) which on warming turns yellow. On the addition of $SnCl_2$ the yellow color becomes violet.

Curarine, $C_{19}H_{26}N_2O$, the active constituent of the extracts of various varieties of Strychnos and called curare or urari and which is used as an arrow poison. It forms a brown amorphous powder which turns violet with sulphuric acid and red with nitric acid.

c. Alkaloids of the Cinchona.

The bark of the various species of cinchona contain the four following alkaloids combined with quinic acid (p. 496):

Quinine, $C_{20}H_{24}N_2O_2$;   Quinidine, $C_{20}H_{24}N_2O_2$;

Cinchonine, $C_{19}H_{22}N_2O$;   Cinchonidin, $C_{19}H_{22}N_2O$.

They are derivatives of a quinoline and a complicated piperidine ring (see Quinine).

Quinine and quinidine are readily soluble in ether, cinchonine and cinchonidine are nearly insoluble therein. If to an aqueous solution of a quinine or quinidine salt we add chlorine water and then ammonia we obtain a beautiful emerald-green coloration.

Cinchonine, $C_{19}H_{22}N_2O$ (structure, see Quinine), forms bitter prisms which are difficultly soluble in alcohol and which melt at 255°.

Quinine, $C_{20}H_{24}N_2O_2$, methyloxylcinchonine, exists to about 6 per cent. in the cinchona bark, forms with $3H_2O$, shining bitter needles which are difficultly soluble in alcohol. When anhydrous they melt at 175°. It forms primary and secondary salts, which latter are nearly insoluble in water.

Quinine bisulphate,
$$(C_{20}H_{24}N_2O_2)H_2SO_4 + 7H_2O,$$
forms colorless crystals which dissolve in 10 parts water.

Quinine sulphate, $(C_{20}H_{24}N_2O_2)_2H_2SO_4 + 8H_2O$, is not readily soluble in water but more soluble than the bisulphate by the addition of some sulphuric acid.

Quinine hydrochloride, $C_{20}H_{24}N_2O_2\cdot HCl + 2H_2O$, white crystalline needles, soluble in 34 parts water.

Quinine-iron citrate is readily soluble in water.

Quinine tannate, yellow amorphous powder soluble with difficulty.

Quinine carboxylic acid ethyl ester, $C_{20}H_{23}(COO\cdot C_2H_5)N_2O_2$, *euquinine,* serves as a tasteless substitute for quinine.

Cinchonidine, $C_{19}H_{22}N_2O$, the stereoisomer of *cinchonine,* also obtained from cinchonine by rearrangement of the molecule, forms large prisms which melt at 203° and are readily soluble in alcohol.

Quinidine, $C_{20}H_{24}N_2O_2$, methyloxylcinchonidine, forms prisms with $2\frac{1}{2}$ molecules H_2O and when anhydrous melts at 172°.

3. Isoquinoline Compounds.

Hydrastine, $C_{21}H_{21}NO_6$, occurs in Hydrastis canadensis (Golden seal), forms prisms melting at 132° and on oxidation yields opianic acid, $C_{10}H_{10}O_5$ (p. 497), and

Hydrastinine, $C_{11}H_{13}NO_3$. The salts of this contain the complex $C_{11}H_{11}NO_2$ which is obtained by the removal of water.

Hydrastinine hydrochloride, $C_{11}H_{11}NO_2\cdot HCl$ (see structural formula). This forms yellow needles which melt at 210° and are soluble in water with a blue fluorescence.

Berberine, $C_{20}H_{17}NO_4$, contains a hydrastinine and an isoquinoline ring with a mutual C and N atom. It occurs in the varieties of Berberis, in the columbo root, in the Hydrastis canadensis, in the bark of the Geoffroya jamaicensis, Xanthoxylum clava, and many other plants used as a yellow dye. It forms yellowish-brown, optically inactive needles with 6 molecules H_2O. It melts at 145°.

Papaverine, $C_{20}H_{21}NO_4$, dimethyloxylbenzyldimethyl oxylisoquinoline. About 1 per cent. is found in opium (see below) and forms optically inactive prisms which melt at 147°.

Narcotine, $C_{22}H_{23}NO_7$, methyloxylhydrastine, exists in opium (about 6 per cent.) and forms prisms melting at 176° and which on oxidation yields opianic acid, $C_{10}H_{10}O_5$ (P. 497), and

Cotarnine, $C_{12}H_{15}NO_4$, methyloxylhydrastinine, which forms salts having the complex $C_{12}H_{13}NO_3$; thus, cotarnine hydrochloride, $C_{12}H_{13}NO_3 \cdot HCl$ is used in medicine as *stypticine*.

Narceine, $C_{23}H_{27}NO_8 + 3H_2O$, exists in opium to about 0.2 per cent. and is closely related to narcotine. It is produced from the methyliodide compound of narcotine by heating with bases: $C_{22}H_{23}NO_7 \cdot CH_3I + KOH = C_{23}H_{27}NO_8 + KI$. When anhydrous it forms crystals melting at 145°.

4. Phenanthrene-Morpholine Compounds.

Opium, the dried juice of the unripe poppy, contains over 20 alkaloids combined with meconic acid (p. 541). The most important are papaverine, narcotine, narceine, which have been treated with the isoquinoline derivatives, and morphine, codeine, thebaine, which are phenanthrene-morpholine derivatives (PP. 518, 544).

Morphine, $C_{17}H_{19}NO_3 + H_2O$, exists to about 20 per cent. in opium, forms small prisms which melt at 230° and is the only alkaloid soluble in an excess of alkalies (but not by ammonia). The solution of morphine is lævorotatory and becomes deep blue with ferric chloride. The solution in concentrated sulphuric acid turns blood-red with a trace of nitric acid.

Morphine hydrochloride, $C_{17}H_{19}NO_3 \cdot HCl + 3H_2O$, forms white needles or cubiformed masses soluble in 25 parts water.

Ethyl morphine hydrochloride, $C_{17}H_{18}(C_2H_5)NO_3 \cdot HCl$, *dionin* and

Morphin diacetoacetic ester, $C_{17}H_{17}(CH_3COO)_2NO_3$, *heroin* and

Benzoylmorphin hydrochloride, $C_{17}H_{18}(C_6H_5-CO)NO_3 \cdot HCl + H_2O$, *peronin*, are used in medicine and form colorless crystals.

Apomorphine, $C_{17}H_{17}NO_2$, morphine anhydride, is produced when morphine is heated to 150° for a long time with fuming HCl, when a molecule of water is split off. Amorphous white powder producing emesis and which turns quickly green in the air.

Apomorphine hydrochloride, $C_{17}H_{17}NO_2 \cdot HCl$, is soluble in 40 parts water, immediately reduces silver nitrate solution in the presence of NH_3. Nitric acid dissolves it with a red color.

Codeine, $C_{18}H_{21}NO_3 + H_2O$, methyl morphine, exists in opium (about 0.3 per cent.), forms crystals which when anhydrous melt at 155°. It is soluble in concentrated sulphuric acid and when treated with ferric chloride gives a deep-blue coloration.

Codeine phosphate, $C_{18}H_{21}NO_3 \cdot H_3PO_4 + 2H_2O$, forms white, bitter needles that are soluble in 3.2 parts water.

Thebaine, $C_{19}H_{21}NO_3$, is found in opium (about 0.15 per cent.), forms shining leaves melting at 193° and which are soluble in concentrated sulphuric acid with a deep-red color.

5. Pyrrolidine-Piperidine Compounds.

The Solanaceæ alkaloids, atropine, hyoscyamine, belladonine, scopol-amine, also the coca alkaloids, cocaine and tropacocaine, contain a methylated pyrrol ring (p. 544) in combination with a piperidine ring (p. 539). They may also be considered as a seven-carbon ring (cycloheptane) with a so-called nitrogen bridge. The mother-substance of these alkaloids is tropine, which is evident from the following formulæ:

Tropine, $C_8H_{15}NO$.

$$
\begin{array}{c}
CH \\
H_2C \qquad CH_2 \\
N \cdot CH_3 \; CH \cdot OH \\
H_2C \qquad CH_2 \\
CH
\end{array}
$$

Atropine, $C_{17}H_{21}NO_3$.

$$
\begin{array}{c}
CH \\
H_2C \qquad CH_2 \quad C_6H_5 \\
N \cdot CH_3 \; CH \cdot OOC \cdot C \cdot OH \\
H_2C \qquad CH_2 \quad CH_2OH \\
CH
\end{array}
$$

Ecgonine, $C_9H_{15}NO_3$.

$$
\begin{array}{c}
CH \\
H_2C \qquad CH \cdot COOH \\
N \cdot CH_3 \; CH \cdot OH \\
H_2C \qquad CH_2 \\
CH
\end{array}
$$

Cocaine, $C_{17}H_{21}NO_4$.

$$
\begin{array}{c}
CH \\
H_2C \qquad CH \cdot COO \cdot CH_3 \\
N \cdot CH_3 \; CH \cdot COO \cdot C_6H_5 \\
H_2C \qquad CH_2 \\
CH
\end{array}
$$

Atropine, belladonine, hyoscyamine, scopolamine, cause a dilation of the pupils even in the smallest quantities. When a trace of these bodies is warmed with concentrated sulphuric acid and some water and potassium permanganate added, an odor similar to oil of bitter almonds is developed. If a trace is dissolved in concentrated H_2SO_4 and some sodium nitrate added an orange coloration is obtained which turns reddish violet and then pale pink when treated with an alcoholic caustic alkali solution.

Atropine and hyoscyamine are stereoisomers and decompose on heating with bases into tropine and tropic acid (p. 502):

$$C_{17}H_{23}NO_3 + H_2O = C_8H_{15}NO + C_9H_{10}O_3.$$
$$\text{Tropine.} \qquad \text{Tropic acid.}$$

On warming tropine (tropanol) and tropic acid with dilute hydrochloric acid we only obtain atropine, as any hyoscyamine formed is immediately changed into atropine.

As tropine combines with tropic acid so it also combines with other oxyacids, producing compounds called *tropeines*. With mandelic acid (p. 498) and tropine we obtain *homatropine*, $C_{16}H_{21}NO_4$, which is used in medicine instead of atropine.

Homatropine *hydrobromide,* $C_{16}H_{21}NO_3 \cdot HBr$, forms crystals readily soluble in water.

Atropine, $C_{17}H_{23}NO_3$, *i*-tropic acid-*i*-tropine (structure, see p. 557), occurring especially in the deadly nightshade, thorn-apple, and henbane, crystallizes in colorless prisms which melt at 114°. Natural atropine is inactive, although active atropine has been obtained from dextro- and lævo-tropic acid and *i*-tropine (see p. 557).

Atropine sulphate, $(C_{17}H_{23}NO_3)_2H_2SO_4$, forms colorless crystals which dissolve in 1 part water and melt at 180°.

Hyoscyamine, daturin, $C_{17}H_{23}NO_3$, occurs in henbane (Folia hyoscyami), in the leaves of Duboisia myoporoides with atropine, in the deadly nightshade, and in the thorn-apple. It crystallizes in fine needles which melt at 10.5° and is then converted into atropine and is lævorotatory and is probably *l*-tropic acid tropine.

Belladonine, $C_{17}H_{21}NO_2$, atropanine, atropine anhydride, is contained in the Atropa belladonna, and is stereoisomeric with *apoatropine,* $C_{17}H_{21}NO_2$, which is prepared from atropine and hyoscyamine by splitting off of H_2O and which melts at 60° and is then converted into belladonine.

Scopolamine, $C_{17}H_{21}NO_4$, hyoscine, duboisine, occurs in the scopolia and belladonna root, in henbane and thorn-apple, in the leaves of Duboisia myoporoides. It forms lævorotatory prisms which split on boiling with bases into atropic acid (p. 531) and *scopoline* (oxytropine, *oscine*), $C_{17}H_{21}NO_4 = C_9H_8O_2 + C_8H_{13}NO_2$. It can be converted into *i*-scopolamine (*atroscine*).

Scopolamine hydrobromide, $C_{17}H_{21}NO_4 \cdot HBr$, forms colorless crystals which are readily soluble in water and which melt at 180°.

Cocaine, $C_{17}H_{21}NO_4$, benzoyl ecgonine methyl ester (structure, p. 557), occurs in the South American coca leaves; forms lævorotatory prisms which melt at 98° and produce local anæsthesia. On boiling with water it decomposes into methyl alcohol and benzoyl ecgonine, $C_9H_{14}(C_6H_5 \cdot CO)NO_3$, and on boiling with acids or bases it yields methyl alcohol, benzoic acid, and *l*-ecgonine, $C_9H_{15}NO_3$ tropine carboxylic acid). The reverse may also be brought about from these constituents, and this is of importance, as *l*-ecgonine can be readily obtained in large quantities from the amorphous residues from the cocaine manufacture.

Cocaine hydrochloride, $C_{17}H_{21}NO_4 \cdot HCl$, forms prisms which melt at 183° and are readily soluble in water and alcohol. The mixture of equal parts cocaine hydrochloride and mercurous chloride turn black when moistened with alcohol.

Tropacocaine, $C_{15}H_{19}NO_2$, benzoyl pseudotropine, occurs in the Japanese coca leaves, is a powerful anæsthetic, less poisonous than cocaine, and melts at 49°.

6. Alkaloids of Unknown Constitution.

Aconitine, $C_{34}H_{47}NO_{11}$, occurs besides picroacontine, $C_{32}H_{45}NO_{10}$, in the leaves and the roots of Aconitum napellus (monkshood), and to a less extent also in other varieties of the aconitum genus. It forms colorless crystals having a sharp but not bitter taste. On boiling with water it decomposes partly into acetic acid and picroaconitine and partly into acetic acid, benzoic acid, and *aconine* $C_{25}H_{41}NO_x$. A trace of aconitine warmed with sirupy phosphoric acid yields an intense violet solution. (This reaction is also given by delphinine and digitaline.) Pseudoaconitine, $C_{36}H_{49}NO_{12}$, from the tubers of the Aconitum ferox, japaconitine, $C_{34}H_{49}NO_{11}$, from the Japanese aconitum, are physiologically different from aconitine.

Achilletine, $C_{11}H_{17}NO_4$, is produced from the gluco-alkaloid achillein, $C_{20}H_{38}N_2O_{15}$, which occurs in the Achillea millefolium. Achillein decomposes by the action of acids into glucose, achilletine, ammonia, and aromatic bodies.

Cephæline, $C_{14}H_{20}NO_2$, in the roots of the Cephælis ipecacuanha (Indian physic); forms colorless crystals and is a quinoline derivative.

Emetine, $C_{15}H_{22}NO_2$, in the ipecacuanha root, Radix ipecacuanhæ.

Ergotinine, $C_{35}H_{40}N_4O_6$, in the ergot, Secale cornutum.

Eserine, physostigmine, $C_{15}H_{21}N_3O_2$, in the Calabar bean. Its solutions soon turn red. It dissolves in warm ammonia and turns yellowish red, and then leaves on evaporation a blue- or bluish-green residue which dissolves in alcohol with a blue color and in a small quantity of sulphuric acid with a green color.

Physostigmine salicylate, $(C_{15}H_{21}N_3O_2)C_7H_6O_3$, forms colorless crystals which are not readily soluble in water, while the sulphate, $(C_{15}H_{21}N_3O_2)H_2SO_4$, is readily soluble.

Eseridine, $C_{15}H_{23}N_3O_3$, also occurs in the Calabar bean.

Jervine, $C_{26}H_{37}NO_3$, pseudojervine, $C_{29}H_{43}NO_7$, rubijervine, $C_{26}H_{43}NO_2$ $+H_2O$, form colorless crystals. They are found together in the white hellebore, whose most poisonous alkaloid is

Protoveratrine, $C_{32}H_{51}NO_{11}$, which gives a green, then blue and violet, color with concentrated sulphuric acid.

Colchicine, $C_{22}H_{25}NO_6$, found in the seeds of the Colchicum autumnale, forms an amorphous yellow mass which turns yellow with concentrated sulphuric acid and dissolves in concentrated nitric acid with a violet coloration. On taking up water it splits readily into colchiceine, $C_{21}H_{23}NO_6$, and methyl alcohol.

Sinapine, $C_{16}H_{23}NO_5$, is not known free, but its sulphocyanate is contained in the white mustard-seed. It decomposes by alkalies into neurine (p. 398) and sinapic acid (p. 501).

Solanidine, $C_{40}H_{61}NO_2$, produced from the gluco-alkaloid solanine, $C_{52}H_{93}NO_{18}$, contained in the potato plant, especially in the sprouts, and which splits into glucose and solanidine by the action of acids.

Veratrine, $C_{37}H_{53}NO_{11}$, a white, amorphous mass, occurs in the sabadilla seeds with the following crystalline alkaloids:

Cevadine, $C_{32}H_{49}NO_9$, sabadinine, $C_{27}H_{45}NO_8$, sabadine, $C_{29}H_{51}NO_8$, and the amorphous cevadilline, $C_{34}H_{53}NO_8$. Veratrine of the druggist is a mixture of these alkaloids.

Lycopodine, $C_{32}H_{52}N_2O_3$, occurs in the Lycopodium clavatum (clubmoss).

Lobelline, $C_{18}H_{23}NO_2$, in the Lobelia inflata.

Ricinine, $C_{17}H_{14}N_2O_4$, in the castor seeds.

Yohimbine, $C_{23}H_{32}N_4O_4$, forms with yohimbenine, $C_{35}H_{45}N_3O_6$, the active constituents of the bark of the Tabernæmontana.

PROTEINS OR ALBUMINOUS SUBSTANCES.

These substances, often confusingly called albuminoids, form a large group of compounds found in the plants and animals, which are classified according to their chemical and physiological behavior. The name albuminous bodies comes from the white of the egg (albumin) and the name protein substances from πρῶτος, the first, because of their importance in the construction of living matter. They occur to a slight extent in the plants where the carbohydrates, especially cellulose, exist to the greatest extent, while they are found in large quantities in the animal kingdom. Only the urine, perspiration, and the tears are free from proteids under normal conditions. They are produced chiefly in the plants and suffer, in the animal organism during assimilation, only slight modifications when introduced as plant food. They occur either in solution or as moist, soft solids which are organized in structure or amorphous masses.

The composition of different proteids varies within rather narrow limits and amounts to the following percentages calculated on the ash-free substance:

Carbon..................................	50.6–54.5
Hydrogen................................	6.5– 7.3
Nitrogen................................	15.0–17.6
Oxygen.	21.5–23.5
Sulphur.	0.3– 2.2

The nucleoproteids also contain some phosphorus.

With the exception of the artificially prepared ash-free albumins they always contain inorganic salts.

Because of their unp onounced chemical character and the ready decomposability of these substances no positive empirical molecular formula has been given to them up to this time. No doubt the molecular weight is very high and the formula $C_{72}H_{112}N_{18}O_{22}S$ seems to be approximately correct for the albumin of the egg.

The following groups of compounds must take part in the construction of the proteins. as they are formed in the cleavage of all proteins by continuous boiling with dilute acids and bases as well as by the action of trypsin (see Ferments):

a. Monamido fatty acids (leucin, aspartic acid, glutamric acid, glycocoll, cystein, etc.). Diamido fatty acids (lysin, diamido acetic acid, arginin, histidin, ornithin, cystin).

The fatty acids below caproic acid in the series which are produced by treatment with acids or bases or by energetic oxidation are no doubt produced from the complicated compounds first split off from the proteins, and the same is true for the ammonium carbonate, sulphide, and cyanide formed in the dry distillation of these bodies.

b. Isocarbocyclic compounds (phenol, benzoic acid, tyrosin, phenylglycocoll, phenylpropionic acid, etc.).

c. Heterocarbocyclic compounds (pyridine, indol, skatol, furol, pyrrolidine carbonic acid, etc.).

Superheated steam produces atmidalbumin, which stands between the proteoses and proteins, then atmidproteoses, then peptones, and finally amido acids.

On oxidation with $KMnO_4$ in acid solution only a little NH_3 is produced, but on the contrary large amounts of urea, also oxyprotsulphonic acid, and then peroxyprotsulphonic acid, bodies of unknown constitution but related to the protein substances.

Tryptophan is a chromogen (P. 530) produced on the cleavage of proteins by means of trypsin, etc., which gives a violet pigment on oxidation.

From the decomposition products it seems as if these compounds all contained some common groups having a relatively simple structure and which also exist free in the spermatozoa of fishes and called *protamins;* for example, salmine, $C_{30}H_{57}N_{17}O_6$, of the salmon; the sturine, $C_{36}H_{69}N_{19}O_7$, of the sturgeon. On careful treatment with acids the protamines form *protones* which are similar to the peptones (p. 563) and which by stronger action are converted into *hexon bases,* i.e.:

$$C_{30}H_{57}N_{17}O_6 + 4H_2O = C_6H_9N_3O_2 + 3C_6H_{14}N_4O_2 + C_6H_{14}N_2O_2.$$
Salmin.   Histidin.   Arginin.   Lysin.

The hexon bases can be converted into hexoses (glucose), which explains the splitting off of sugar (glucose, etc.) from different albuminous substances. The complex protein substances are probably produced by the substitution of amido compounds, etc., in the protamins, and the most complicated formed by the combination of these with other atomic groups. For instance, the blood pigments are formed by the combination of a proteid with a pigment and the nucleoproteid by the union of a proteid with the phosphorized nucleic acids.

Properties. In the solid state they are white, flocculent or lumpy masses without odor and taste, and when dry are yellow, transparent, and brittle like horn. Besides the protein crystals which occur in certain seeds we have also been able artificially to obtain ovalbumin and seralbumin in a crystalline state. Only a few are soluble in water, but all dissolve, with partial decomposition, in caustic alkalies, forming alkali albuminates and in concentrated mineral acids, forming acid albuminates (p. 568). The proteids are insoluble in ether, chloroform, carbon disulphide, benzene. Only the proteids of gluten are soluble in alcohol. With the exception of peptone they give insoluble compounds with many aldehydes. The solutions of all proteids rotate the ray of polarized light to the left, with the exception of hæmoglobin, the nucleins, and nucleoproteids, which are dextrorotatory.

Color Reactions. *a.* On heating with conc. nitric acid the proteids or their solution turn yellow and then deep orange on the addition of ammonia (*xanthroproteic reaction*). *b.* On boiling with Miilon's reagent (solution of $Hg(NO_3)_2$ in $HNO_3 + HNO_2$) they become purple-red, or when applied to a solution containing small quantities of proteid it becomes red (*Millon's test*). *c.* If to a solution of proteid in glyoxylic acid (or acetic acid) concentrated sulphuric acid is added, we obtain a violet coloration which shows characteristic absorption bands in the spectroscope (*Adamkiewicz's reaction*). *d.* If a solution of proteid is treated with caustic alkali and a few drops of dilute $CuSO_4$ solution, a bluish-violet coloration is obtained (*biuret test*).

Precipitation Reactions. The proteids are precipitated from their solutions: *a.* By potassium ferrocyanide after acidification with acetic acid. *b.* By nitric acid when added in sufficient amount to the boiling hot solution to make the solution acid. *c.* By tannin, basic lead acetate, and most metallic salts (use of proteids as antidotes in metallic poisoning). *d.* By treating the solution, acidified with acetic acid, with an equal volume of a saturated solution of sodium sulphate and heating to boiling or by saturating the acidified solution with sodium chloride. *e.* Faintly acid solutions, especially with acetic acid, coagulate readily on boiling, especially in the presence of inorganic salts of the alkali metals. The coagulation temperatures are different for the various proteid bodies and can be used in their identification and separation. *f.* Trichloracetic acid, metaphosphoric acid, and many of the so-called alkaloid reagents (see Alkaloids) precipitate the proteids. In these precipitations, with the exception of the neutral salts, the proteins are chemically changed.

Numerous preparations of proteid bodies are used in medicine; thus, halogen compounds as *albacide* and *eigone,* ichthyol compounds as *ichthalbin,* tannin compounds as *tannalbin,* silver compounds as *protargol, largin,* and *argonin,* the iodoform compound as *iodoformogen,* the formaldehyde compound as *formolalbumin,* those of iron as *ferratin, hæmatin, hæmatogen, hæmalbumin, iron somatose, eubiol, dynamogen, roborin, fersan, ferratogen.* Readily digestible products prepared from animal and vegetable proteids are used as foods; thus from the residues from the preparation of meat extracts we obtain *soson* and *tropon,* from skimmed milk, *plasmon, sanatogen, galactogen,* from blood serum, *protoplasmin,* from plant proteids, *eucasein* (ammonium caseate), *nutrose* (sodium caseate), *mutase, aleuronate* (p. 567). As the peptones (p. 568) have a bitter taste and the proteoses (p. 568) are tasteless these latter are used in medicine as foods; thus, *somatose, mietose, Nahrstoff Heyden,* etc. Reduced hæmoglobin is used under the names *hæmogallol* and *hæmol;* the latter also, with iodine and bromine, as *iodohæmol* and *bromhæmol.*

Classification. This is difficult on account of our imperfect knowledge as to the size of the molecule, the constitution of the bodies, etc. The following is the generally accepted classification, but we know of many proteids which belong between these groups and also those which cannot be classified in any of the groups: 1. Simple proteids; 2. Compound proteids; 3. Modified proteids; 4. Albuminoid or proteinoid substances; 5. Enzymes or unorganized ferments (it has not been positively proven that these bodies are protein substances); 6. Poisonous proteids or toxalbumins (more correctly toxoproteins). Under the true, native, or genuine proteins we understand those whose solutions coagulate on heating or by other means and which cannot be then dissolved without further cleavage or without changing their original properties but remain permanently changed (modified).

1. Native or True Proteids.

a. Albumins.

They are soluble in water, acids, alkalies; their aqueous solutions coagulate on warming when they contain neutral salts such as NaCl, $MgSO_4$. They are precipitated even in the cold on saturating their neutral solution with ammonium sulphate, but not on saturating with NaCl, $MgSO_4$, or $ZnSO_4$.

Ovalbumin is precipitated from its solution by ether, coagulates at 56°, is precipitated by excess of acids, and is more difficult of solution than seralbumin.

Seralbumin occurs in blood serum, animal semen, chyle, lymph, and in all serous fluids, and in certain pathological urines. It is not precipitated by ether, coagulates at 40°–90°, depending upon the nature of the solvent, and is precipitated by acids and readily soluble in an excess of the acid.

Myogen, muscle albumin, the chief constituent of the muscle plasma, coagulates at 40° or on the death of the muscle.

Lactalbumin, milk albumin, occurs in milk and colostrum and coagulates at 72°–84°, according to the solvent used.

Plant albumins behave like seralbumin.

Opalisin is found to a great extent only in woman's milk, and is characterized by its opalescent solution.

Legumelin occurs in the cereal grains.

b. *Globulins.*

These are insoluble in water but soluble in the presence of neutral salts. If these solutions are diluted with considerable water or the salts removed by dialysis the globulins precipitate out. Their aqueous solutions coagulate on boiling and are completely precipitated in the cold by saturating with $(NH_4)_2SO_4$, $MgSO_4$, or $ZnSO_4$, and incompletely by NaCl (see Vitellins).

Myosin, muscle globulin, is with myogen the chief constituent of muscle plasma. After the death of the muscle both coagulate and form the rigor mortis. Myosin coagulates at 50°. Myosin-like bodies also occur in many plants.

Serglobulin, blood casein, paraglobulin, fibrinoplastic substance, occurs in blood serum, chyle, lymph, and nearly all fresh transudates; also in albuminous urine besides seralbumin. Its neutral solutions coagulate at 72°-75°.

Colostrum globulin occurs in colostrum.

Fibrinogen, metaglobulin, is contained in all animal fluids which either coagulate spontaneously on standing (fibrin formation, p. 567) or after the addition of a few drops of the fluid expressed from some freshly coagulated blood. It coagulates on warming its neutral solution to 53°-56°.

Fibrin globulin is produced from fibrinogen besides fibrin in its coagulation and also in the digestion of fibrin.

2. Compound Proteids.

These split into simple proteids and other organic compounds.

a. *Nucleoproteids.*

On cleavage these yield proteids and acids of unknown constitution called *nucleic acids* $(C_{29}H_{49}N_9P_3O_{22}$, etc.), which on boiling with dilute acids or alkalies yield phosphoric acid, besides adenine, guanine, hypoxanthine, and xanthine (p. 415), and these are therefore called nuclein bases.

Nucleins, compounds of nucleic acid with proteid, also contain iron and occur in the cell nucleus of animals and plants and are separated from the nucleoalbumins on digestion with pepsin-hydrochloric acid (p. 563). They behave like acids, are only slightly soluble or insoluble in water, dilute mineral acids, and neutral salt solutions, but, on the contrary, are readily soluble in dilute caustic alkalies. Nucleins occurring in fish sperm can be split into nucleic acids and protamins (p. 561).

Nucleoalbumins, compounds of nucleins with proteid, occur in the cell nucleus, often also in the protoplasm and in animal fluids, and leave nucleins on peptic digestion, and decompose, on standing with alkalies or acids, in the cold into albuminates or acid albumins (p. 568). According as to the length of time they are exposed to the alkalies or acids they are split into nucleins, nucleic acids, or their constituents. They behave like acids and dissolve in dilute caustic alkalies and decompose, in contradistinction to the nucleins, on heating their neutral solutions, with the separation of proteid.

To this class also belongs *nucleohiston,* which occurs in all cell nuclei and which readily splits into a proteose-like body, *histon,* and into *leuconuclein.* The nucleic acid of this latter body also occurs in the thymus gland, and hence is called *thymus nucleic acid* or *adenylic acid,* and yields on further cleavage *thymic acid* and then *thymin* (p. 542).

b. Paranucleoproteids.

These on cleavage yield proteid and acids of unknown composition called *para* or *pseudo nucleic acids* ($C_{16}H_{25}N_3P_2O_{12}$, etc.). On boiling the paranucleic acids with alkalies or dilute acids, phosphoric acid, proteid, etc., but no nuclein bases, are produced.

Paranucleins, compounds of paranucleic acid with proteid, also contain iron and are formed from the paranucleoalbumins in their peptic digestion.

Paranucleoalbumins, compounds of paranucleins with proteids. They form the chief constituent of protoplasm and also occur in secretions, etc. They behave like acids and are insoluble in water but dissolve in the presence of traces of caustic alkalies. Their neutral solutions do not coagulate on boiling, are incompletely precipitated by NaCl and completely by $MgSO_4$. On standing with alkalies or acids in the cold the paranucleoalbumins decompose in an analogous manner to nucleoalbumins (which see). To this group belong the mucin of ox bile, *ichthulin* in the carp eggs, and *helicoproteid* in Helix pomata. These latter split into paranuclein and sugar (p. 566, *d*).

The most important paranucleoalbumins are the *caseins.* They are precipitated from their solutions by heating to 130°–150° and also at ordinary temperatures by certain ferments as well as by the careful addition of acids. The casein precipitated by rennin is different from that in solution or that precipitated by acids. Cheese is a putrefactive product of casein.

Milk casein in the milk of all mammals can be precipitated by rennin, acids, and saturation with $MgSO_4$. If a few drops of acid or some rennin is added to milk all the casein separates out immediately and at the same time the fat is carried with it (curdled milk). In the solution we have milk-sugar, albumin, and the salts. This solution is called sweet whey. If milk is allowed to stand for a long time it also coagulates, due to the formation of lactic acid from the milk-sugar. The filtrate from this curd is called acid whey.

The *vitellins* occur as *ovovitellin* in the yolk of the egg, as α and β *crystallin* in the crystalline lens, also as *phytovitellins* (phytoglobulins) in plants, and also as *conglutin, legumin, vicilin, artolin,* etc. *Thyreoglobulin,* the iodized vitellin of the thyroid gland, splits with acids into *thyroiodine* or *iodothyrin,* which does not belong to the proteids.

The *lecithalbumins* are compounds of lecithins (p. 434) with vitellins and other paranucleoalbumins and are found in the mucous membrane of the stomach and in the kidneys, etc.

~ c. Chromoproteids.

They are readily decomposed into proteid and a ferruginous pigment hæmochromogen, which in the presence of oxygen is oxidized to hæmatin (p. 545). They are soluble in water and salt solutions and on heating their solutions we obtain, even below the boiling-point, a brown precipitate of coagulated proteid and hæmatin. Alcohol, alkalies, or acids (even NO_2) produce this cleavage even in the cold.

Oxyhæmoglobin, the pigment of arterial blood, forms with hæmoglobin the chief constituent of the red corpuscles. Both are also found to a slight extent in certain muscles of mammals and in the muscles and in the blood of certain invertebrates. It is obtained from the blood corpuscles and forms microscopic crystals which are soluble in water with a blood-red color.

If a solution of oxyhæmoglobin or arterial blood is placed before the spectroscope we obtain, even when very dilute, a characteristic absorption spectrum consisting of two absorption bands. If a small quantity of a reducing agent (ammonium sulphide or an ammoniacal solution of ferro-tartrate) is added to an oxyhæmoglobin or blood solution the color becomes darker and the two characteristic absorption bands of oxyhæmoglobin disappear and the characteristic one-banded spectrum of hæmoglobin takes their place. The band is broader than the other two and on shaking the reduced solution with air or oxygen the two oxyhæmoglobin bands appear again. Carbon monoxide hæmoglobin gives two absorption bands similar to the oxyhæmoglobin bands, but they are not removed by reducing agents (p. 189).

Hæmoglobin, the pigment of venous blood (other occurrence see above), forms red plates or prisms which when dissolved give one characteristic absorption band (see above). With acids or bases it splits into proteid and hæmochromogen. With H_2S, CO_2, CO, NO, HCN, and C_2H_2 it gives crystalline compounds which are isomorphous with oxyhæmoglobin. If these gases are passed through an oxyhæmoglobin solution or blood they expel the oxygen and combine with the hæmoglobin. If oxygen is again passed through these compounds only the carbon dioxide hæmoglobin is converted into oxyhæmoglobin, the other compounds remaining unchanged. For this reason these gases have a poisonous action, as they make the hæmoglobin incapable of carrying the oxygen necessary for the organism.

Methæmoglobin, isomer of oxyhæmoglobin, is a transformation product of the oxyhæmoglobin, from which it is produced by the action of $KMnO_4$, pyrogallic acid, potassium ferricyanide, etc. It is sometimes found in pathological fluids, as well as in urine and blood after poisoning, and forms brownish-red crystals. The absorption spectrum of the watery or acidified solution is similar to that of hæmatin in acid solution, but is readily differentiated by the fact that the addition of alkali and a reducing substance converts it into the spectrum of hæmochromogen.

d. Glycoproteids.

These split into proteid and carbohydrate.

Mucins are the substances precipitated from their colloidal solutions by acetic acid, which are insoluble in an excess of the acetic acid (in contradistinction to all other proteid substances and the mucoids). On heating with dilute mineral acids they yield proteid and a carbohydrate which reduces alkaline solutions of copper. They occur in many secretions and excretions (human bile, saliva, mucus, synovial fluid, feces, urine, etc.), in the connective tissue as well as in all the organs composed chiefly of cells (glands, etc.). In the lower animals we do not find mucins but mucinogens, which are split into mucins and proteid substances by bases. The various mucins are called *snail mucin, tendon mucin,* and *submaxillary mucin.*

Muciods, mucinoids, are the mucuos bodies not precipitated by acetic acid and occur as *pseudomucin* (*para* or *meta albumin*) in ascitic and ovarial cystic fluids, *collomucoid* in cancerous growths, *ovomucoid* in the egg, *toxomucoid* (p. 572) from cultures of the tubercle bacillus. *Chondromucoid*, contained in the cartilaginous tissues, is a compound of proteid with *chondroitin sulphuric acid,* $C_{18}H_{27}NSO_{17}$, which splits further into *chondroitin* and sulphuric acid. Animal *amyloid* is also a compound of proteid with chondroitin-sulphuric acid and is found in milk and pathologically in various organs. It differs from other proteids in giving a ιed color with iodine solutions and a violet to blue color with iodine solution and su!phuric acid, hence its name.

Hyalogens is the name given to a series of very widely distributed bodies found in the skeleton tissues of lower animals. They have been little studied and decompose by bases into a proteid-like body, and into hyalins containing nitrogen and closely related to the carbohydrates.

The hyalin, *chondroitin,* $C_{18}H_{27}NO_{14}$, decomposes on boiling with dilute acids into acetic acid and *chondrosin,* $C_{12}H_{21}NO_{11}$; this last splits into glycuronic acid and glucosamine (p. 457, *b*); the hyalin *chitin,* $C_{18}H_{30}N_2O_{12}$, splits into acetic acid and *chitosan,* $C_{14}H_{26}N_2O_{11}$; and this further yields acetic acid and glucosamine.

3. Modified Proteids.

These are formed from the native proteids by the action of heat, ferments, or chemical agents and have other properties from their mother-substances and are not reconvertible into them.

Coagulated proteids are produced on heating their neutral or faintly acid solutions or by the action of certain ferments. They are insoluble in water, dilute acids, and alkalies. They form when dry colorless or yellowish hard masses.

Animal fibrin forms in liquids containing fibrinogen when these are removed from the influence of the living normal vessel walls (drawn blood). The fibrinogen on the exit of its solution (of blood) from the animal body is converted into fibrin and fibringlobulin by the *fibrin ferment* which is produced on the destruction of the white blood-corpuscles. Fibrin is insoluble in water and salt solutions, swells up in NaCl solution, in dilute acids and alkalies, without dissolving therein. It forms when moist a white amorphous, elastic mass which on warming to 75° or by alcohol becomes hard and brittle.

Myogen fibrin and myosin fibrin are the solid proteids formed from myogen and myosin, caused by a myosin ferment.

Plant fibrin, gluten, occurs in the plant seeds, especially the cereal grains, and is obtained by kneading the seeds with water, when the plant albumins and starch granules are washed out and the gluten remains as a sticky, tough mass, which when dry forms a yellow powder called *aleuronate* (p. 563). Like an'mal fibrin it does not seem to be preformed and it is probably derived from the plant globulins in the presence of water by a ferment which has not been isolated. Plant fibrin is a mixture of a proteid called *artolin* with a not closely studied body containing phosphorus and calcium. Such mixtures are the proteids *gluten casein, gluten fibrin, gliadin,* and *mucedin* which used to be considered as constituents of gluten. The solubility in 60 per cent. alcohol is characteristic of plant fibrin.

Acid albumins (syntonins) and alkali albuminates (albuminates) are produced by the action of acids or bases upon proteids and form gelatinous compounds which are insoluble in water and soluble in dilute acids and alkalies. Their solutions do not coagulate on boiling but are precipitated on neutralization and by saturating with neutral salts By the action of bases a splitting off of nitrogen as ammonia takes places as well as a removal of sulphur.

Proteoses, hemialbumoses, albumoses, propeptones, are the intermediary products soluble in water, obtained in digestion between the proteids and the peptones. Just as the starches pass through a series of dextrins be ore sugar is produced, so the passage of the proteids into peptones takes place through the formation of proteoses, which in properties deviate more and more from the original proteid. They are sometimes found in the urine in osteomalacia and give the biuret test (p. 562, *d*) even in the cold. Their solutions do not coagulate on boiling. The *primary proteoses* (*proto-, hetero-, dys-proteoses*) are produced first from the proteids and are precipitated by copper sulphate, nitric acid, by acetic acid and potassium ferrocyanide, these precipitates disappearing on warming and reappearing again on cooling. Their neutral solutions are completely precipitated by an equal volume of a saturated ammonium sulphate solution.

Secondary proteoses (*deuteroproteoses*) are formed from the primary proteoses and pass directly into peptones. They are not precipitated by $CuSO_4$, HNO_3, $K_4FeC_6N_6 + C_2H_4O_2$, but on the contrary are precipitated on saturating their solution with powdered ammonium sulphate.

Peptones. All proteids, with the exception of amyloid, metalbumin, nucleins, paranucleins, and keratins, are transformed by pepsin or trypsin (p. 570) into peptones The proteids also yield peptones beside other products by putrefaction and by treatment with strong acids or bases. The formation of peptones is always an intermediary stage to the formation of leucin, tyrosin, and other amido acids. The formation of peptone from other proteids depends upon the taking up of water by the proteid molecule. *Carnic acid*, $C_{10}H_5N_4O_5$, a cleavage product of *phosphocarnic acid*, which is a nucleon, i.e., a nuclein-like substance, is a crystalline peptone free from sulphur, in contradistinction to other peptones.

The peptones differ from all other proteid bodies by the following reactions: *a.* They are diffusible (p. 46) through vegetable and animal membranes. *b.* They are soluble in water in every proportion, the solution does not coagulate on boiling (see Proteoses). *c.* They are not precipitated from their solutions by either acetic acid and potassium ferrocyanide, nor by acids or alkalies, nor by acetic acid and neutral salts, nor on saturation with ammonium sulphate. By this last method or by boiling with ferric acetate in faintly acetic acid solution all proteids may be separated from the peptones. *d.* Peptones are detected in solutions free from other proteids by means of the biuret test (p. 562, *d*), and their precipitation by tannic acid, phosphomolybdic acid, phosphotungstic acid. Mercury peptone solution, iron peptone, iron-manganese peptone, obtained by the action of the respective metallic salts upon peptone solutions, are used in medicine.

4. Albuminoids

are nitrogenous constituents of the animal body which are related to the proteids not only by their elementary composition but also by the correspondence of many react'ons. They occur mostly undissolved as an integrate constituent of tissues and are characterized by their resistance to chemical agents.

Collagen, gelatine-yielding substance, forms the chief constituent of the connective fibres and organic substance of the bony tissue (called *ossein*); also mixed with other substances forms the ground substance of cartilage which was formerly considered as a special body and which was called *chondrin* or *chondringen*. Collagen is insoluble in water salt solutions, dilute acids, alkalies, and on boiling with water it is converted into gelatine. The gelatine-forming tissues combine with tannic acid, alum, or fats and then when dry form a pliable tissue which does not undergo putrefaction. Tissues thus changed are called leather.

Gelatine, glutin, glue, is produced on boiling the collagens with water and when pure is a colorless, transparent amorphous mass which is soluble in hot water: On cooling this solution it solidifies to a gelatinous mass; if the boiling is continued too long, the gelatine loses its property of gelatinizing and is converted into so-called gelatine peptone. Gelatine is soluble in the cold in dilute acetic acid, other acids, and in alkalies, and is not precipitated from its solutions by acids, basic lead acetate, alum, while it is precipitated by tannic acid or alcohol. On boiling with dilute sulphuric acid we obtain glycocoll and leucin. When impure it is called glue and is prepared from animal hides which are free from fat, blood, etc., and softened. The best forms of gelatine are prepared from calves' feet.

Keratin, horn substance, is the chief constituent of whale-bone, of horn tissue, epidermis, nails (hoofs and claws), hair (feathers, quills, tortoise-shell), horns. If these are finely powdered and treated consecutively with water, alcohol, ether, and pepsin hydrochloric acid a body having variable composition remains behind which has been called keratin. Keratin does not putrefy and burns with a characteristic odor, dissolves in caustic alkalies, ammonia, and boiling acetic acid. The sulphur of the keratin is in part loosely combined, hence tissue containing keratin becomes black by lead and silver salts by forming their sulphides (hair-dyeing agents). Horn shavings develop H_2S when moist.

Glutolin, from its behavior, stands between proteids and the albuminoids, occurs in blood serum.

Elastin forms the elastic tissue occurring in higher animals, especially in the connective tissue. It retains the structure of the material used in its preparation. It is yellowish-white and very elastic when moist. It contains its sulphur so loosely combined that it was formerly considered free from sulphur.

Spongin forms the chief mass of sponges.

Conchiolin occurs in the shells of musse's.

Fibroin and sericin are the two chief constituents of raw silk. On boiling with dilute acids, fibroin yields glycocoll and considerable tyrosin, while sericin yields leucin and crystalline serine (p. 407).

Cornein forms the organic substance of the corals.

5. Enzymes.

The unorganized ferments or enzymes (p. 54) all appear to be proteid-like compounds, as they give nearly all the reactions for the proteids; still it is possible that the enzymes attach themselves to the proteids on being precipitated and their separation not being perfect on purification. They are readily soluble in faintly acid or alkaline water as well as in glycerine. They are colorless, powderous bodies which are not prec pi-tated from their solutions on boiling but are precipitated by alcohol. Their solutions lose their activity generally at 60° and some at 100°, but when dry many may be heated above 100° without losing their activity. They are generally obtained by extracting the substances containing them with glycerine and precipitating the extract with alcohol. Their maximum activity lies generally between 35° and 45°.

a. Proteid-digesting or Proteolytic Enzymes.

Ingluvin is the pepsin-like ferment of the hen's gizzard.

Pepsin, the proteolytic ferment of the gastric juice, forms a white powder which is only active in faintly acid solution.

Papain, papayotin, in the juice of the Carica papaya, is most active in faintly alkaline liquids.

Proteolytic enzymes are also found in yeast and other fungi.

Trypsin, pancreatin, occurs in the pancreas and is most active in faintly alkaline solutions.

b. Polysaccharide-splitting Enzymes.

They are also called diastatic, saccharifying, amylolytic ferments on account of their transforming starch, etc., into sugars.

Diastases, amylases, have the power of converting starches into dextrins and maltose (p. 354). They are found rather widely distributed in the higher and lower plan s (plant diastases), also in the animal kingdom (animal diastases), chiefly in the pancreas (*amylopsin*), in the saliva (*ptyalin*), to a slight extent in the liver, bile, blood, chyle, brain, the kidneys, stomach, and intestinal mucosa.

Inulase splits inulin (p. 459) into lævulose and is found in many plants instead of diastase.

Cellulase splits cellulose into hexoses and pentoses, is found in many germinating plants and also in the animal kingdom.

c. Disaccharide-splitting Enzymes.

Invertase, sucrase, invertin, is contained in many plants and extractable from yeast by water. It splits cane-sugar into dextrose and lævulose. A similar enzyme is found in the intestinal contents

Glucase, found in malt and splits maltose (p. 455).

Maltase occurs in certain varieties of yeast and inverts maltose (p. 455).

Lactase, found in the Saccharomyces kefir and Tryocola, splits lactose into dextrose and galactose (p. 455).

Trehalase occurs in the Asperigillus niger, and splits trehalose (p. 455).

d. Monosaccharide-splitting Enzymes.

Zymase, alcoholase, contained in the fluid pressed from previously destroyed yeast-cells, produces alcoholic fermentation (p. 354). Similar enzymes to zymase have also been found in animal cells.

e. Glucoside-splitting Enzymes.

Emulsin, synaptase, of the sweet and bitter almond, splits the glucosides amygdalin and salicin.

Myrosin, of the white and black mustard-seed, splits the glucoside potassium myronate (p. 438).

f. Glyceride-splitting Enzymes.

Lipase, steapsin, steaptase, occurs in the pancreas of all carnivora as well as in certain plants.

g. Coagulating Enzymes.

Rennin, chymosin, coagulates neutral casein solutions, occurs in the gastric juice of the calf and sheep. A similar enzyme is also found in various varieties of Fiscus and other plants.

Fibrin ferment, thrombin, thrombase, converts fibrinogen into fibrin (p. 567) in the presence of neutral salts.

Gluten ferment (?) converts the proteids of flour into gluten (p. 567).

Pictase, which converts the expressed juice of many fruits into a jelly.

h. Oxidizing Enzymes or Oxidases.

These seem to be a mixture of peroxidases which only oxidize in the presence of peroxides and oxygenases, the latter oxidizing only in the presence of oxygen.

Laccase, in the juice of the Japanese lac-tree, and oxidases related to this also occur in many plants.

Tyrosinase, in many fungi, in the dahlia and potato tubers, beets, etc., oxidizes tyrosin but al o many other cyclic compounds.

i. Amide-splitting Enzymes

occur in fungi which cause the fermentation of urine (splitting the urea, p. 410).

j. Reducing Enzymes,

also called *reductases,* cause reduction processes and are found especially in the plants.

Catalase is the name of the enzyme which sets molecular oxygen free from peroxides.

6. Toxalbumins.

In various plants and animals we find bodies which in regard to behavior belong on one side to the food-proteids and on the other side to the enzymes, but they are more or less poisonous. In certain cases their poisonous action is due to toxins which cannot be separated from the proteids by the methods used.

Of those to be mentioned we have *abrin* of the jequirity bean, *crotin* in the Croton Tiglium, r*icin* in the castor-seed, *lupinotoxin* in certain varieties of lupines, *sapotoxin* in soap and senegal roots, *snake poisons*, and the poison of spiders, and certain fishes, etc.

Toxalbumins may also be obtained from the pure cultures of pathogenic bacteria and also from those yielding toxins; thus, from diphtheria, anthrax, typhoid, and tetanus cultures, and *tuberculin* and *tuberculocidin* from the tubercle bacilli, and *anticholerin* from cholera bacilli, *mallein* from glanders bacilli, *peptotoxines* produced in digestion, and *toxomucoid* (p. 567), etc.

INDEX.